Understanding China

The series will provide you with in-depth information on China's social, cultural and economic aspects. It covers a broad variety of topics, from economics and history to law, philosophy, cultural geography and regional politics, and offers a wealth of materials for researchers, doctoral students, and experienced practitioners.

Yuqun Liao

Reconstructing the Image of Medicine in the Qin and Han Dynasties

上海交通大學出版社
Shanghai Jiao Tong University Press

Springer

Yuqun Liao
Institute of Natural Science History
Chinese Academy of Sciences
Beijing, China

Translated by
En Zhou
Shanghai, China

Lin Su
Shanghai, China

Rui Wang
Shanghai, China

Qi Zou
Shanghai, China

ISSN 2196-3134 ISSN 2196-3142 (electronic)
Understanding China
ISBN 978-981-95-6800-0 ISBN 978-981-95-6801-7 (eBook)
https://doi.org/10.1007/978-981-95-6801-7

Jointly published with Shanghai Jiao Tong University Press
The print edition is not for sale in the Chinese Mainland. Customers from the Chinese Mainland please order the print book from: Shanghai Jiao Tong University Press

This work was supported by Chinese Fund for the Humanities and Social Sciences.

This Translation, authorized by Shanghai Jiao Tong University Press, was sponsored by the Chinese Fund for the Humanities and Social Sciences. [Grant Number: 18WZS023]

This Springer imprint is published by the registered company Springer Nature Singapore Pte Ltd.
The registered company address is: 152 Beach Road, #21-01/04 Gateway East, Singapore 189721, Singapore

Acknowledgements

The translation of *Reconstructing the Image of Medicine of the Qin and Han Dynasties* has been a meaningful and rewarding project. We are pleased to present this English edition and hope it will contribute to bringing the study of ancient Chinese medical history to a wider international audience.

We began the translation project in 2019, with sponsorship from the Chinese Fund for the Humanities and Social Sciences (18WZS023). More than a linguistic task, the translation project required sustained engagement with historical sources and classical medical thought. Through careful reading and interpretation, we gained a deeper understanding of the complex theories and practices that shaped medicine in the Qin and Han periods.

This work would not have been possible without the dedication and collaboration of a diverse group of individuals, whose generous contributions we gratefully acknowledge.

We are deeply indebted to Lin Pan, Xuewei Li, Lei Meng, Na Li, Zhenxiong Qian, Xinxin Ye, Xin Huang, Ruonan Zhang, and Yuang Zhou for their translation work. Their tireless efforts and linguistic proficiency ensured the accuracy and readability of the translation. We are grateful to Yue Zhu, Yuerong Zhao, Xin Liu, Qian Yang, Jiali Zeng, Yingqiao Xiong, Yonghui Jiang, Honghong Wang, Fangfang Yu, Wenyi Pu, Biyi Lai, Jingzhou Zhu, Hanxi Lan, and Xuanyou Pan for their meticulous proofreading of the drafts. Their keen attention to detail, passion for the subject matter, and commitment to excellence have greatly enhanced the quality of this work.

Special thanks go to Wei Wang, Dan Li, Ye Li at Shanghai Jiao Tong University Press, whose editorial guidance has shaped the final manuscript into its present form.

We would like to thank Professor Hongjie Guo at Shanghai University of Finance and Economics, whose inspiration, assistance and encouragement have been crucial to this translation project. A special mention goes to our colleagues Professor Chouping Han, Professor Nianqing Ding, Professor Yu Yang, Professor Rongzheng Ren, Professor Yonghong Xu, and Professor Yun Gao, all of whom have an in-depth understanding of translation in the field of traditional Chinese medicine and provided invaluable suggestions to improve this translation.

We extend our sincere thanks to the peer reviewers for their thoughtful, insightful, and constructive suggestions. Any remaining shortcomings are, of course, our own responsibility.

Furthermore, we would like to express our heartfelt gratitude to the entire team at Springer, whose professionalism and dedication to scholarly publishing have ensured the successful publication of this volume.

Finally, we are deeply grateful to Professor Yuqun Liao, the author of the original work, whose pioneering research and significant contributions to the study of ancient Chinese medical history have provided the essential foundation for this translation. We hope that this English edition will further promote international dialogue on the medical thought and cultural heritage of the Qin and Han dynasties.

Contents

Part I
Introduction

During the medical development in the pre-Qin period (before 221 BC) and the Western and Eastern Han dynasties (202 BC-AD 220), it is widely believed that seminal medical works such as the *Huangdi's Internal Classic* (*Huang Di Nei Jing*, 《黄帝内经》), the *Classic of Difficulties* (*Nan Jing*, 《难经》) and the *Shennong's Classic of Materia Medica* (*Shen Nong Ben Cao Jing*, 《神农本草经》) originated in the pre-Qin period. These works laid the foundation for theoretical frameworks in physiology, pathology, and pharmacology, with a focus on "treatment based on pattern identification". They also introduced key therapeutic modalities such as acupuncture, moxibustion, and herbal medicine. Centuries later, during the late Eastern Han Dynasty (AD 25-220), the *Treatise on Cold Damage and Miscellaneous Diseases* (*Shang Han Za Bing Lun*,《伤寒杂病论》) emerged, offering comprehensive treatment principles, methods, formulas, and medications. Notably, Hua Tuo pioneered surgery under anaesthetization using "Mafei Powder" during this era. In addition, Bian Que and Chunyu Yi are frequently referenced in historical accounts. Bian Que is renowned as a physician of the pre-Qin period, while Chunyu Yi is recognized as a medical practitioner who documented the earliest medical cases during the Western Han period.

The four centuries spanning the Western and Eastern Han dynasties mark a crucial epoch in the evolution of Chinese culture, and medicine is no exception to this transformative era. However, if the currently dominant narrative holds true, then within the domain of medicine, these four centuries of the prosperous Han cultural period would seem almost a blank. This gives rise to another question: How did medicine achieve remarkable progress ahead of its time compared to other ancient sciences and technological systems that flourished during the Han dynasties? Is this observation a historical fact or a matter deserving further scholarly investigation? By recognizing the importance of these puzzling questions, we embark on an exploratory journey, contemplating whether the development of medicine during the Western and Eastern Han dynasties represents a gap, a continuation of earlier practices, or an innovation?

To begin with, is it accurate to assert that medicine had already reached an advanced stage during the pre-Qin period, thus implying a lack of notable progress in the subsequent 400 years between the Western and Eastern Han dynasties? If not, then could it be suggested that the core principles of medicine in the Han Dynasty originated from the state of Qin, renowned for its abundance of skilled physicians, as well as from the cultural heritage in the "six states" (state of Qi, Chu, Yan, Han, Zhao, and Wei), despite their political demise?

If these hypotheses fail to hold true, then it follows that medicine in the Han Dynasty underwent a period of significant innovation, mirroring the trajectory of other branches of ancient science and technology. Yet, what precisely constituted this process of innovation?

Through this line of questioning, akin to peeling away layers of a cocoon, the true essence of medicine during the Western and Eastern Han dynasties may gradually come to light, revealing the intricate dynamics and evolution of this essential facet of ancient civilization.

Chapter 1
Almost Nothing: Astonishing Sluggish Development of Medicine in the Han Dynasty

1.1 "General Images" of Medicine in the Qin and Han Dynasties

The pre-Qin period (before 221 BC) and the Han Dynasty (202 BC-AD 220) hold great significance in the establishment of ancient Chinese medical theories and the development of its system. A notable indicator of this is the creation of a series of classic works that were written in those ages and are still in use today. References to the completion dates of these works can be found in medical books, as well as literature and history texts from different dynasties. To simplify matters for readers, Table 1.1 in this paper exclusively focuses on *The History of Chinese Medicine* (*Zhong Guo Yi Xue Shi*, 《中国医学史》) by Chen Bangxian (Fig. 1.1), which incorporates representative viewpoints from scholars across different periods. Table 1.1 clearly illustrates the extensive history, diverse ideas, and considerable time span encompassed by the contents of these classics.

These statements include both general historical opinions and specialized research by medical historians. Each of them will be given one example below. When it comes to the overview of academic development in the pre-Qin period, Lyu Simian, a renowned historian, remarked on this topic,

> There are four medical books passed on to future generations from the pre-Qin period, namely *Plain Questions, Miraculous Pivot, Classic of Difficulties*, and *Shennong's Classic of Materia Medica*.[1]

During the period of the Republic of China (AD 1912-1949), Xie Guan (Fig. 1.2), the president of Shanghai University of Traditional Chinese Medicine (SHUTCM) at that time, was famous for leading over sixty students to compile *Dictionary of Chinese Medicine* (*Zhong Guo Yi Xue Da Ci Dian*, 《中国医学大辞典》) and writing

---

[1] Lyu Simian, *An Overview of Academy in the Pre-Qin Period* (*Xian Qin Xue Shu Gai Lun*, 《先秦学术概论》), Shanghai: Orient Publishing Center, 1985, p. 147.

Y. Liao, *Reconstructing the Image of Medicine in the Qin and Han Dynasties*, Understanding China, https://doi.org/10.1007/978-981-95-6801-7_1

Table 1.1 Statements about the age of medical classics recorded in The History of Chinese Medicine[2]

Name of the classic	Number of references	Time span	Author's preferred opinions
Huangdi's Internal Classic (*Huang Di Nei Jing*,《黄帝内经》)	31	The period of Huang Di—the Warring States period (403–221 BC)—the Qin and the Han Dynasties (221 BC-AD 220) [*Miraculous Pivot* (*Ling Shu*,《灵枢》)] was completed after the Wei and the Jin Dynasties <AD 220–420>.)	"There is no doubt that *Plain Questions* (*Su Wen*,《素问》) was created during the time from the era of Warring States to the Western Han Dynasty (202 BC-AD 8)". "*Miraculous Pivot* appeared later than *Plain Questions*"
Shennong's Classic of Materia Medica (*Shen Nong Ben Cao Jing*,《神农本草经》)	15	The period of Shen Nong—the Shang and the Zhou Dynasties (1600–256 BC)—the Han Dynasty—the Liu Song and the Southern Qi Dynasties (AD 420–502)	"Liang Qichao posited that the appropriate timeframe should span from the decline of the Eastern Han Dynasty (AD 25–220) to the Liu Song and the Southern Qi Dynasties"
Classic of Difficulties (*Nan Jing*,《难经》)	18	The period of Huang Di—the Warring States period—the Six Dynasties (AD 222–589)	"Based on the existing evidence, the notion that Qin Yueren authored the book holds greater credibility. However, Yao Jiheng believed that the book was created during the era of the Six Dynasties. This was significantly divergent from its actual age"

Fig. 1.1 Chen Bangxian. In 1919, Chen Bangxian (AD 1889–1976) authored the first book on the comprehensive history of Chinese medicine. In the middle of the twentieth century, numerous early scholars in Chinese medicine history emerged as his students. Consequently, he is widely recognized as the pioneer and founder of Chinese medicine history research

Treatise on the Origin and Development of Medicine in China (*Zhong Guo Yi Xue Yuan Liu Lun*,《中国医学源流论》). He said,

[2] Chen Bangxian, *The History of Chinese Medicine*, pp. 97–108.

Fig. 1.2 Xie Guan. Xie Guan (AD 1878–1950), also called Li Heng, was born in Wujin, Jiangsu province. He led the compilation of *A Dictionary of Chinese Medicine* and wrote *Treatise on the Origin and Development of Medicine in China*. He was the founder of Shanghai University of TCM

> All Chinese academic achievements are made before the era of Warring states. And since the Qin Dynasty (221-207 BC), there has been no innovation but only extensions of ancient people's established theories in Chinese academic development. It is also true of medicine.[3]

Even if these medical classics were believed to be completed at the end of the Warring States period according to Mr. Xie's opinions, there was limited medical development during the four hundred years of the Han Dynasty. It was until the end of the Eastern Han Dynasty that Zhang Zhongjing's *Treatise on Cold Damage and Miscellaneous Diseases* (*Shang Han Za Bing Lun*, 《伤寒杂病论》) emerged. This comprehensive work extensively introduced principles, methods, formulas, and medicinal herbs. Hua Tuo, a highly skilled doctor who invented "Ma Fei San", an anesthetic used in surgery, was also born in this period. This is the "general image of medicine in the Qin and Han Dynasties" (Fig. 1.3).

After the middle of the twentieth century, a great number of medical books were unearthed from Mawangdui Han Dynasty Tombs in Changsha, Hunan province in 1973. Then another two medical books called *Book on Meridians* (*Mai Shu*, 《脉书》) and *Book of Daoyin* (*Yin Shu*, 《引书》) were found in Zhangjiashan Han Tomb in Jiangling, Hubei province in 1984. These two medical books share a similar geographic location and burial period and exhibit not only content similarities but also several crucial features. Firstly, statements about "meridians" in these books directly align with the classic system of the "twelve meridians" found in today's version of *Huangdi's Internal Classic*.

Therefore, the two books about meridians are considered to be the original version of the chapter "The Conduit Vessels" in *Miraculous Pivot*. Secondly, a large number of medical books unearthed from Mawangdui Han Dynasty Tombs exactly cover four sections of Fangji (medical methods and techniques) which were recorded in *Seven Categories of Writing* (Qi Lve, 《七略》), an imperial bibliography written at the end of the Western Han Dynasty. Thus, in both theory and practice, these precious cultural relics form crucial foundations for reanalyzing the age of classics and the development level and composition of medical knowledge in the early Western Han Dynasty.

[3] Xie Gugan, the chapter "A Study of Confucianism" in the *Treatise on the Origin and Development of Medicine in China*, Cheng Zhai Medical Society, 3rd edition, 1936.

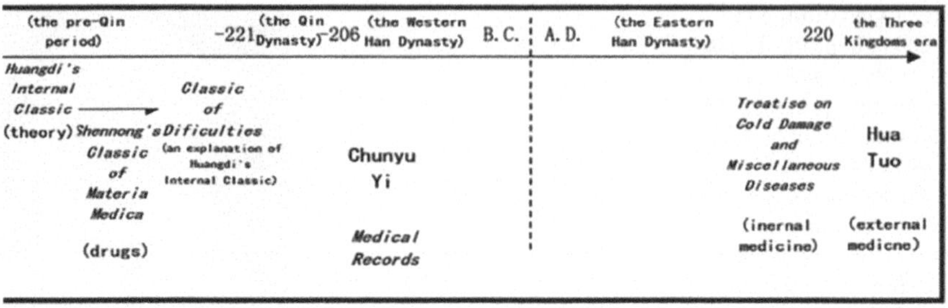

Fig. 1.3 Images of the medical history of the Qin and Han Dynasties-1

In 1993, a wooden sculpture of human figure was unearthed in Mianyang, Sichuan Province. Dating back to the Western Han Dynasty, it featured numerous engraved "lines". While there is no written confirmation of the purpose or the meaning behind the "lines" of the wooden sculpture of human figure, it can be speculated that it was used to display the "meridians of the human body". However, it is challenging to argue that it shares the same system with medical books from Mangwangdui Han Tomb and the chapter "The Conduit Vessels" in *Miraculous Pivot*. In reality, they belong to two completely different systems.

The Records of the Grand Historian (*Shi Ji*, 《史记》) documented 25 *Medical Records* (*Yi An*, 《医案》) written by Chunyu Yi, a doctor who lived in the early Western Han Dynasty. Together with the precious documents and relics from the tombs, in addition to the research value, all these documents provide important clues to analyze the "coexistence" of medical knowledge in different regions during the Western Han Dynasty. In terms of time, constrained by a certain mindset, most people believe that these unearthed documents which are relics of modern times are still old in the dynasty when they were buried. They are considered to be completed in the Spring and Autumn period (770-476 BC) or an even earlier era. Geographically, although the Jianghan Plain is far away from Sichuan Basin, separated by high mountains and lofty hills, the culture and knowledge of both regions are viewed as the same by most people. As these unearthed documents see the light of day, Chunyu Yi should have received more attention. But his *Medical Records* can only be thought of when it comes to listing the "earliest" things, which really underestimated the value of his records. He is only given the honor of "creating the earliest medical records".

Therefore, while the newly painted "images of the medical history of the Qin and Han Dynasties" can include some additional information, they still cannot bridge the gaps in our understanding of the Han Dynasty.

Fig. 1.4 Picture of Shen Nong and a toad on the stone. It belongs to the Eastern Han Dynasty with 105 cm in height and 52 cm in width. The one holding the large bird is intended to represent Shen Nong. Notably, the moon in the image features a rabbit and a toad. Obviously, Shen Nong's identity has changed in this picture

1.2 Artificially Created "Advanced Development"

In fact, any medical history work with hundreds of thousands of words won't describe the medical development process during this long historical period as simply as the above "images". Naturally, it should have specific contents in all aspects. Still talking about *The History of Chinese Medicine* written by Chen Bangxian, which can be regarded as the founding work of Chinese medical history, as an example, let's take a look at its specific description of medicine in the pre-Qin period.

(1) The first chapter, "Medicine in the Primitive Society", mentions Shen Nong (Fig. 1.4) and Huang Di, two pioneers of Chinese medicine, in relation to the "Longshan culture" (the end of the Neolithic Age).
(2) The second chapter delves into the legend of the origin of medicine, featuring Fu Xi, Shen Nong, and Huang Di as central characters. It is normal to use "legends" in historical works, but the author's writing is misleading. Firstly, Tamba Motoyasu (AD 1755–1810), a famous Japanese expert, wrote a book about the history of Chinese medicine called *Yi Sheng* (『醫賸』). In the introduction of this book, the author mentions that in the *Writings of Prince Huainan* (*Huai Nan Zi*, 《淮南子》), "Shen Nong tasting a hundred herbs" aimed to identify which plants were edible instead of finding medicine and this conforms to the original meaning of the name of "Sheng Nong" because "Nong" in Chinese means agriculture. After the Eastern Han Dynasty, Shen Nong evolved into the symbol of the origin of drug knowledge.[4] *Yi Sheng* was written in Chinese and

[4] Tamba Motoyasu (Japan), Chapter "Shen Nong Tasting a Hundred Herbs" in *Yi Sheng*.

was listed in *A Series of Books of Emperor Chinese Medicine* (*Huang Han Yi Xue Cong Shu*, 《皇汉医学丛书》) which was published by Shanghai World Book Company in 1936. This book holds immense significance for individuals within the community of Chinese medical history. Unfortunately, although Tamba Motoyasu had made it clear two hundred years ago, not only Chen Bangxian, but also other Chinese medical historians continued to produce works based on the concept of "Shen Nong grasping medicinal properties after tasting a hundred herbs". The reason why they do this is evident. Secondly, Chen-Bangxian drew upon ancient texts such as *The Generational Origins* (*Shi Ben*, 《世本》) and *Genealogical Records of Emperors and Kings* (*Di Wang Shi Ji*, 《帝王世纪》) to support his assertion that *Huangdi's Internal Classic* consisted of dialogues between Huang Di and Qi Bo. What's more, he chose statements from *Chinese Ancient History* (*Zhong Guo Gu Dai Shi*, 《中国古代史》) completed in modern times as evidence. The book states, "*Plain Questions* and *Miraculous Pivot* both comprised of 81 chapters originate from the conversation between Huang Di and Qi Bo. Shen Nong focuses on medical experience while Huang Di creates medical principles, illustrating the trajectory of medical development". Chen Bangxian further suggests, "This indicates that laboring people have attached great importance to medicine since the end of the Neolithic Age (the period of Yangshao culture <5000–3000 BC>)". On the basis of this, he justified the opinion of historical materialism.

(3) The third chapter is about legendary medical characters, but the description is completely different from the "legendary" style.

Jiu Daiji: Qi Bo's teacher, the earliest doctor, a man who studied inspection of complexion and pulse-taking in the period of Huang Di.

Qi Bo: one of Huang Di's ministers, an influential figure in the field of ancient Chinese medicine.

Lei Gong: one of Huang Di's ministers, a young and promising man, gaining knowledge from Jiu Daiji and Qi Bo.

Yu Fu: a specialist in surgery with excellent operative techniques.

Shao Yu: one of Huang Di's ministers, Yu Fu's younger brother, a doctor whose medical skills were mostly the same as his brother's.

Wu Peng: perhaps a witch doctor proficient in prescribing medicine and providing acupuncture treatment.

Tong Jun: perhaps an herb researcher who wrote a book called *A Record of Picking Herb* (*Cai Yao Lu*, 《采药录》).

Bo Gao: one of Huang Di's ministers who was said to work with Qi Bo to assist Huang Di in explaining meridians and exploring the principles.

Ma Shihuang: a horse doctor who was good at recognizing the physique, qi, living and death of a horse and excels at acupuncture and moxibustion during the period of Huang Di.

Wei Yuqu: one of Huang Di's ministers who played the same role as Bo Gao.

Miao Fu: a man who was proficient at treating disease by incantation and talisman.

Wu Fang: a pediatric doctor who wrote a book called *Book of Fontanelles* (*Lu Xin Jing*, 《颅囟经》).

Wu Xian: a witch doctor.

Legends play a crucial role in historical studies, offering not only utility but also great significance. When utilizing legends, the fundamental principle lies in analyzing the cultural aspects that underlie their creation within a specific time period and region. Descriptions of these 13 characters in *The History of Chinese Medicine* are based on historical materials. Information about Wu Fang comes from *Essential Prescriptions Worth a Thousand Gold for Emergencies* (*Qian Jin Fang*, 《千金方》) in the Tang Dynasty (AD 618–907). The information about Wu Fang is the latest one among all the materials. Records of Tong Jun are based on *Introduction of Materia Medica* (*Ben Cao Xu*, 《本草序》)[5] written by Tao Hongjing in the Southern Dynasties (AD 420–589). Though the source of statements of Wei Yuqu is not explicitly mentioned, it is believed that he introduced the doctrine of the five circuit phases and six atmospheric influences. If these "legends" can be used in a different way, at least the three characters can be viewed as materials to study pediatrics as an independent department, a herbal work and a new explanation of five circuit phases and six atmospheric influences respectively.

(4) Though this book has mentioned that Yi Yin in the Shang Dynasty (1600–1046 BC) was capable of distinguishing medicinal properties and creating decoction, it does not account for what "decoction" is.

(5) During the Spring and Autumn and the Warring States period (770–221 BC), it is evident from the book that the *Classic of Difficulties* was attributed to Bian Que. The author explicitly stated in the book that there are over 22 chapters dedicated to the study of pulse, indicating Bian Que's profound exploration of sphygmology.

(6) At last, the book talks about medical theory which is of vital importance. While references to the "oracle bone script" and excerpts from texts completed during or after the pre-Qin period, such as *Master Lyu's Spring and Autumn Annals* (*Lyu Shi Chun Qiu*,《吕氏春秋》), *The Rites of Zhou* (*Zhou Li*,《周礼》), *Discourses of the States* (*Guo Yu*,《国语》), and *The Book of Rites* (*Li Ji*,《礼记》), have been made, it is important to emphasize that *Plain Questions* and *Miraculous Pivot* serve as the primary sources of evidence supporting the argument that the general concept in *Huangdi's Internal Classic* of yin-yang is not metaphysical. This is particularly evident in *Plain Questions* and *Miraculous Pivot*, where extensive evidence is presented regarding the understanding of diseases, physiological explanations, cause analysis, treatment methods, health preservation, disease prevention, and hygiene during the Spring and Autumn and Warring States period. Over 70 statements in the two books are cited as evidence. When it comes to the knowledge of herbal medicine, the *Classic of Mountains and Seas* (*Shan Hai Jing*, 《山海经》) is mainly cited instead of *Shennong's Classic of*

[5] The name of "Tong Jun" can be seen in *Wupu's Materia Medica* (*Wu Pu Ben Cao*, 《吴普本草》) completed in the period of the Three Kingdoms.

Materia Medica. This conforms to the previous claim of the age of *Shennong's Classic of Materia Medica*.[6]

The subsequent medical history books were essentially shaped after the discovery of ancient medical works in the 1980s. Since then, various types of medical history books emerged, motivated by the booming development of science, culture, academic study, and the publishing industry. However, authors of medical history books face the inevitable problem of explaining why the medical works unearthed from Western Han Dynasty tombs are considered inferior to *Huangdi's Internal Classic*, which originated in the pre-Qin period. In response, some researchers either push the age of these unearthed medical books back to an even earlier era or choose to avoid mentioning them altogether. Additionally, a significant number of scholars have revised the age of *Huangdi's Internal Classic*, proposing a stereotypical formation between the Spring and Autumn period and the Han Dynasty. Such approach is also used to define the age of the *Classic of Difficulties* and *Shennong's Classic of Materia Medica*. The following table clearly shows the age of medical classics recorded in different medical history books written after the unearthing of ancient medical texts in Mawangdui (see Table 1.2).

From the table, it is evident that the prevailing view still supports the notion that *Huangdi's Internal Classic* originated in the pre-Qin period. Others believe that the age of the book spans from the Spring and Autumn and Warring States periods to the Qin and Han Dynasties or that it specifically belongs to the Western Han Dynasty or a later period. The opinions actually suggest that the given time just refers to the one when today's version of the classic was completed and it has a long history and has long existed in the civilizations of the Xia, Shang, and Zhou dynasties.

Thus, it will not significantly correct the "image" of medicine during the Han Dynasty. Instead, it gives researchers who make horizontal comparisons between "medicine" and other scientific and technological fields of the same period the overall impression of the "the advanced development of medicine". Around 1980, after the Chinese scientific history community had spent nearly 30 years constructing the corresponding ancient disciplinary histories according to the categorization of modern natural sciences, it began to compile "general histories" in both vertical and horizontal dimensions. At that time, a team led by Du Shiran compared the development of various fields of ancient Chinese natural sciences and came to such an overall impression: medicine developed ahead of other disciplines during the Spring and Autumn and the Warring States period. The reason is precisely that there were no researchers specializing in the history of medicine in this team, so they could only rely on the prevailing views in the field of medical history. Therefore, in the innovative research titled "*A History of Science and Technology in China*" (*Zhong Guo Ke Xue Ji Shu Shi Gao,* 《中国科学技术史稿》)—which was the first to transcend disciplinary boundaries—the discussion of medical issues during "a very important period in the history of the development of ancient Chinese science and technology" notably omits any mention of the Qin, Western Han, and Eastern Han

[6] Details can be found from page 3 to 54 in *The History of Chinese Medicine* by Chen Bangxian.

Table 1.2 The age of medical classics recorded in different medical history books written after medical documents unearthed in Mangwangdui Han Dynasty Tombs

Name of book	Author and publication year	*Huangdi's Internal Classic*	*Classic of Difficulties*	*Shennong's Classic of Materia Medica*
*The History of Chinese Medicine**	Beijing University of Chinese Medicine, 1978	The Warring States period	Mentioning that Bian Que is the possible writer, but denying it finally	Summarizing the experience since the Qin and the Han Dynasties
The Doctrine of the Schools of Traditional Chinese Medicine (*Zhong Yi Ge Jia Xue Shuo,*《中医各家学说》) *	Ren Yingqiu, 1980	The Warring States period	Written by Qin Yueren	The Eastern Han Dynasty
Volume of Medical History Literature in Chinese Medicine Dictionary (*Zhong Yi Da Ci Dian,*《中医大辞典》)	The editorial committee of *Chinese Medicine Dictionary*, 1981	The Warring States period	Prior to the Eastern Han Dynasty; or during the Qin and Han Dynasties	During the Qin and Han Dynasties; or the Warring States period
A Brief History of Medicine in China (*Zhong Guo Yi Xue Jian Shi,*《中国医学简史》)	Yu Shenchu, 1983	2 BC	A collection of theories and experiences of Bian Que made by people of the Han Dynasty	A summary of materia medica knowledge before the Han Dynasty
Ancient Chinese Medicine (*Zhong Guo Gu Dai Yi Xue,*《中国古代医学》)	Zhao Pushan, 1983	From the Warring States period to the Qin and Han Dynasties	Pre-Qin period or the Eastern Han Dynasty	Later than Huangdi's Internal Classics
History of Chinese Orthopedic Techniques (*Zhong Guo Gu Ke Ji Shu Shi,*《中国骨科技术史》)	Wei Yizong, 1983	The Warring States period	The Warring States period	From the Qin Dynasty to the Western Han Dynasty; or the Eastern Han Dynasty
Doctors and Medical Books (*Yi Jia Yu Yi Ji,*《医家与医籍》)	Liang Naigui, 1983	The Spring and Autumn Period and the Warring States period	Written by Bian Que	—

(continued)

Table 1.2 (continued)

Name of book	Author and publication year	*Huangdi's Internal Classic*	*Classic of Difficulties*	*Shennong's Classic of Materia Medica*
Summary of Chinese Medical Books (*Zhong Guo Yi Ji Ti Yao*, 《中国医籍提要》)	The editorial committee of *Summary of Chinese Medical Books*, 1984	The Spring and Autumn Period and the Warring States period	Written by Qin Yueren, later than *Huangdi's Internal Classics*	The Qin and Han Dynasties; or the Warring States period
Essays on reading ancient medical books (*Du Gu Yi Shu Sui Bi*, 《读古医书随笔》)	Li Jinyong, 1984	No earlier than the late Warring States period	The late Han Dynasty	—
Historical Materials of Chinese Pharmacy (*Zhong Guo Yao Xue Shi Liao*, 《中国药学史料》)	Xue Yu, the chief editor, 1984	—	—	1BC
Theories of Various Schools of Traditional Chinese Medicine in Different Dynasties (*Zhong Yi Li Dai Ge Jia Xue Shuo*, 《中医历代各家学说》)	Qiu Peiran, the chief editor, 1984	From the Spring and Autumn and the Warring States period to the Han Dynasty	—	Included in the description of three stages of social development in the *Book of Rites*
Compendium of Chinese Medical History (*Zhong Guo Yi Xue Shi Gang*, 《中国医学史纲》)	Kong Jianmin, 1988	*Plain Questions*: the first century *Miraculous Pivot*: from the end of the first century to the beginning of the second century	In the middle of the second century	The Western Jin Dynasty (AD 265–317)
Essentials of Ancient Medical History (*Gu Dai Yi Yao Shi Jing Hua Lu*, 《古代医药史精华录》)	Zhang Zhihan, 1989	The Spring and Autumn and the Warring States period	—	Finalized in the Han Dynasty, including the knowledge of medicinal herbs before the Han Dynasty

(continued)

Table 1.2 (continued)

Name of book	Author and publication year	*Huangdi's Internal Classic*	*Classic of Difficulties*	*Shennong's Classic of Materia Medica*
Academic History of Traditional Chinese Medicine (Zhong Yi Xue Shu Shi,《中医学术史》)*	Yan Shiyun, the chief editor, 1989	From the Warring States period to the Western Han Dynasty	—	—
A Brief Summary of History of Traditional Chinese Medicine (Xing Lin Shu Zhen – Zhong Yi Yao Shi Gai Yao, 《杏林述珍——中医药史概要》)	Fu Weikang, 1991	The Warring States period	—	From the first century to the second century
A Brief History of Chinese Medicine (Zhong Guo Yi Xue Shi Lve,《中国医学史略》)	Jia Dedao, 1993, 2nd edition	From the Warring States period to the Qin and Han Dynasties	The Eastern Han Dynasty	Approximately in the Eastern Han Dynasty

Note The symbol * indicates that the book is a textbook for medical colleges

dynasties. After using the statement that "*Shennong's Classic of Materia Medica* is a summary of drug knowledge since the Warring States, Qin and Han Dynasties" to gloss over the "blank" of several hundred years, it directly moves on to the two sections about "Zhang Zhongjing and *Treatise on Cold Damage and Miscellaneous Diseases*" and "the achievements of Hua Tuo" in the third century.[7]

Why does medicine exhibit advanced development? If there was no "horizontal comparison" in the study of the general history of science and technology, perhaps this question would not have arisen. Nevertheless, shortly after I joined the research team in 1983, I found myself frequently confronted with this inquiry. Today, I believe I can provide an answer: the notion of "advanced development" is an artificial construct.

[7] Du Shiran et al., *Science and Civilization in China*, Beijing: Science Press, 1982, pp.195–200.

Chapter 2
Legacy of Medical Excellence: The Qin Dynasty's Renowned Physicians and the Han Dynasty's Continuation of Its Medical System

It is obviously unreasonable that medicine prospered in the pre-Qin period (before 221BC) but failed to progress during the Han Dynasty (202 BC-AD 220), an important historical period of more than four hundred years when Chinese culture developed prosperously. The discovery of medical texts within the Mawangdui Han Tomb prompted researchers to propose the notion that "classics are not the sole creation of an individual within a specific timeframe", "they were compiled from the Spring and Autumn and Warring States period (770–221 BC) until the Han Dynasty", or "they represent a synthesis of medical knowledge since the pre-Qin era". Although these broad assertions aimed to reconcile the contrasting medical advancements between these two historical periods, their practical significance remains limited.[1]

Nevertheless, general "correction" is also of great importance. Some meticulous historians base their determination of the time period for the completion of *Huangdi's Internal Classic* (*Huang Di Nei Jing*, 《黄帝内经》) when there is no enough evidence to support that these classics were written in the pre-Qin period. For them, such corrections have the potential to challenge or reframe the belief that *Huangdi's Internal Classic* is completed in the pre-Qin period. What's more, it is widely acknowledged that certain elements within the book are indeed "inherited". However, the crucial issue lies in the absence of a thorough analysis regarding the precise mechanisms of this "inheritance". Therefore, discussing concepts such as inheritance, summarization, and further development, or asserting that works like the *Huangdi's Internal Classic* should have been completed within the Han Dynasty or even the Eastern Han Dynasty (AD 25–220), becomes meaningless without a specific examination of the process by which this transmission occurred.

Even if we refute the notions suggesting "no medical development in the Han Dynasty" or "advanced development in the pre-Qin period", and even if we adjust the estimated age of classics like *Huangdi's Internal Classic* to a time within the Han Dynasty, it remains imperative to shift our focus toward the concepts of "inheritance"

[1] Li Jinyong, *Essays on Reading Ancient Medical Books* (*Du Gu Yi Shu Sui Bi*, 《读古医书随笔》), Beijing: People's Health Publishing House, 1984, p. 1.

Y. Liao, *Reconstructing the Image of Medicine in the Qin and Han Dynasties*, Understanding China, https://doi.org/10.1007/978-981-95-6801-7_2

and "creation" in terms of medical advancements after the Western Han Dynasty (202 BC-AD 8). Though "inheritance" stresses the accumulation of quantitative changes, which is different from "creation" driven by a number of internal and external factors, it is important to recognize that these two aspects are not inherently separate. In fact, they often coexist and depend on each other's influences. Consequently, it is highly pertinent to conduct a comprehensive analysis of "time" and "region" as well as the role of inheritance and creation in different fields of medical knowledge. Let us begin by exploring the concept of "inheritance". In terms of "time", it is essential to thoroughly explore the extent to which the Western Han Dynasty inherited medical and cultural aspects from the pre-Qin period. Additionally, considering the element of "region", due consideration should be given to the influence of the "empire of Qin" and the "oriental areas" annexed by the Qin Empire.

Throughout history, when historians discuss the Han Dynasty, it is commonly believed that the political and cultural foundations of the Han Dynasty stemmed from the Qin Dynasty.[2] In other words, the remarkable civilization fostered by the Han Dynasty developed upon the groundwork laid by the unification of China by the Qin Dynasty (221–207 BC). Consequently, references to these two dynasties are often intertwined, as evident in phrases like "the civilization of the Qin and Han Dynasties" or "the Qin and Han Empire".[3] An intriguing story found in *The Commentary of Zuo* (*Zuo Zhuan*, 《左传》) tells the story of the king of the state of Jin sending envoys to the state of Qin on two occasions to seek exceptional doctors. This longstanding narrative has contributed to the belief that the Qin Dynasty possessed a wealth of skilled physicians, thereby serving as evidence supporting the notion that the Han Dynasty inherited comprehensive medical knowledge from the Qin Dynasty.

However, the Qin Empire, which was established by Ying Zheng (the first emperor of the Qin Dynasty) in 221 BC and ended with the failure of Ying Ziying (the last emperor of the Qin Dynasty) in 207 BC, lasted for just over ten years. In light of this, it is necessary to consider the following three questions:

(1) What aspects of the unified Qin Empire, which existed for over ten years, could be inherited by the Han Dynasty?
(2) Despite the short duration of the Qin Empire itself, the state of Qin had a history spanning several hundred years. As a formidable state that conquered the other six states and achieved the great unification of China, what was the extent of its cultural development during this extended period?
(3) Considering the strict laws and harsh punishment associated with the state of Qin, what level of expertise did the many "great doctors" possess? How advanced was the state of medicine at that time?

---

[2] 107 results will be shown after searching "the Han Dynasty inheriting the system of the Qin Dynasty" in *Complete Library in the Four Branches of Literature* (*Si Ku Quan Shu*, 《四库全书》), and 466 results for the keyword "the Han Dynasty inheriting from the Qin Dynasty".

[3] Ru Xin remarked, "The Qin-Han Dynasty in the East and Rome in the West stand as the epitomes of the two great ancient civilizations". This statement can be found on pages 8 to 12 of *Exhibition of the Qin-Han Dynasty and Rome* (*Qin Han Luo Ma Wen Ming Zhan*, 《秦汉-罗马文明展》) compiled by State Administration of Cultural Heritage in China and Ministry of Cultural Heritage and Art Activities in Italy.

Table 2.1 Lineage of the state of Qin: legendary era-1

Nyu Xiu: matriarchal era
Da Ye: son of Nyu Xiu
Da Fei: "Shun" giving him the surname "Ying"
Da Luo: the tenth generation of Da Fei's family
Fei Zi: son of Da Luo, building the city of Qin
Qin Hou: son of Fei Zi
Gong Bo: son of Qin Hou
Qin Zhong: son of Gong Bo, King Xuan of Zhou appointing him as "Da Fu", a senior official in feudal China
Duke Zhuang of Qin: son of Qin Zhong

Answers to these questions are very important for investigating the legacy of Han Dynasty medicine, as "the history of the Qin and Han Dynasties has already been extensively studied". Therefore, writing about medical history does not necessarily require a laborious exploration of firsthand literature to delve into the history of the state of Qin. However, it is still essential to provide readers with a brief introduction to key historical points, as it allows them to consider the possibility of the Han Dynasty inheriting rich medical knowledge from the Qin Dynasty.

2.1 A Brief History of the State of Qin

There exist differing opinions regarding the origin of the Qin people's ancestors. Some people propose they came from the east and others think they came from the west. It is also speculated that they might be descendants of "Nyu Xiu" or "Da Fei" (Bo Yi), who were offspring of Zhuan Xu, one of the five ancient emperors of China. The Qin people acquired the surname "Ying" for their involvement in assisting Yu with flood control and aiding Shun in livestock rearing.

Gradually, they settled in the vicinity of Tianshui in Gansu province, where they cultivated expertise in horse breeding and carriage driving. After more than ten generations of development, in the ninth century BC, Fei Zi, who served King Xiao of Zhou, was granted permission to establish "the city of Qin".[4] It was during this period that the term "Qin people" was adopted.

The lineage of subsequent rulers of the Qin regime began to be meticulously recorded in Chinese history following King Xuan of Zhou's (821 BC) proclamation of the title of Duke Zhuang of Qin to the eldest son of Qin Zhong, king of the state of Qin during the Zhou Dynasty (approximately 1046–256 BC). This clear lineage chronology of the Qin rulers can be seen in Table 2.1.

The history of Qin spans from Duke Zhuang of Qin to Ying Ziying, who governed for a mere 64 days, encompassing a total of 32 rulers. A concise overview featuring

[4] Lin Jianming said in the *Development History of the State of Qin* (*Qin Guo Fa Zhan Shi*, 《秦国发展史》), "Qin is located in the east of Qingshui county, Gansu province. Since then, people living there have been called 'Qin people'". p. 5.

Table 2.2 Lineage of the state of Qin: from "Duke" to "Emperor"[5]-2

Duke Xiang of Qin (Son of Duke Zhuang of Qin, 777–766 BC)	After the downfall of the Western Zhou Dynasty (around 1046–771 BC), King Ping of Zhou relocated the capital to Luo Yi in the east. Duke Xiang of Qin, who served as an escort during this move, was rewarded with the title of "Duke", officially establishing the Qin tribe as a vassal state. However, there are limited recorded accounts from this period, aside from information about the royal family.
Duke Wen of Qin (Son of Duke Xiang of Qin, 765–716 BC)	In the fourth year of his reign (762 BC), the capital of Qin was moved to the area where the Qian River and Wei River converged". During his reign, historical writing began to focus on events. For instance, there is a record mentioning "Dun", believed to be a civil official responsible for recording historical events. The political system gradually improved after the capital relocation.
	One notable political characteristic was the absence of a land allocation system, signifying that all land belonged to the ruler. This played a crucial role in the rapid growth of national strength.
→ Duke Xian of Qin (grandson of Duke Wen of Qin, also called "Duke Ning of Qin", 715–704 BC) → Chu Zi (younger son of Duke Xian of Qin, 703–698 BC) → Duke Wu of Qin (the eldest son of Duke Xian of Qin, 697–678 BC) →	
Duke De of Qin (the younger brother of Duke Wu of Qin, 677–676 BC)	He constructed a new capital called "Yong" and ushered in a new era of development. The flourishing of agriculture, handicraft industries, and commerce contributed to the county's prosperity.
→ Duke Xuan of Qin (the eldest son of Duke De of Qin, 675–664 BC) → Duke Cheng of Qin (the second son of Duke De of Qin, 663–660 BC) →	
Duke Mu of Qin (the youngest son of Duke De of Qin, 659–621 BC)	King Ping also conquered the Xirong tribes residing in the western part of ancient China. During this time, numerous notable individuals held various roles, including assistants, generals, civilian and military officers, and figures like Bole. Subsequently, the state of Qin gradually transitioned into a feudal society, although its national strength began to decline.

(continued)

[5] Tables 2.1 and 2.2 are mainly based on the 10th volume of the *History of Qin* (*Qin Shi*, 《秦史》) by Wang Quchang (pp. 59–99). *Development History of the State of Qin* by Lin Jianming and other scholars' opinions are also taken into consideration. The dates of each king's reign mentioned in this account are sourced from *A Chronology of Chinese History* (*Zhong Guo Li Shi Nian Dai Jian Biao*, 《中国历史年代简表》) (Beijing: Cultural Relics Publishing House, the second edition, 1994).

Table 2.2 (continued)

<table>
<tr><td colspan="2">→ Duke Kang of Qin (the eldest son of Duke Mu of Qin, 620–609 BC) → Duke Gong of Qin (son of Duke Kang of Qin, 608–604 BC) → Duke Huan of Qin (son of Duke Gong of Qin, 603–577 BC) → Duke Jing of Qin (son of Duke Huan of Qin, 576–537 BC) → Duke Ai of Qin (son of Duke Jing of Qin, 536–501 BC) → Duke Hui of Qin (grandson of Duke Ai of Qin, 500–491 BC) → Duke Dao of Qin (son of Duke Hui of Qin, 490–477 BC) → Duke Ligong of Qin (son of Duke Dao of Qin, 476–443 BC) → Duke Zao of Qin (the eldest son of Duke Ligong of Qin, 442–429 BC) → Duke Huai of Qin (the second son of Duke Ligong of Qin, 428–425 BC) → Duke Ling of Qin (grandson of Duke Huai of Qin, 424–415 BC) → Duke Jian of Qin (son of Duke Huai of Qin, 414–400 BC) → Duke Hui of Qin (son of Duke Jian of Qin, 399–387 BC) → Duke Chu of Qin (son of Duke Hui of Qin, 386–385 BC) → Duke Xian of Qin (son of Duke Ling of Qin, 384–362 BC) →</td></tr>
<tr><td>Duke Xiao of Qin
(Son of Duke Xian of Qin, 361–338 BC)</td><td>To counter this, Shang Yang was appointed as the prime minister to implement comprehensive reforms aimed at bolstering the country's power.</td></tr>
<tr><td colspan="2">→ King Huiwen of Qin (son of Duke Xiao of Qin, 337–311 BC) → King Daowu of Qin (son of King Huiwen of Qin, 310–307 BC) → King Zhaoxiang of Qin (son of King Huiwen of Qin, young half-brother of King Daowu of Qin, 306–251 BC) → King Xiaowen of Qin (son of King Zhaoxiang of Qin, 250–250 BC) → King Zhuangxiang of Qin (son of King Xiaowen of Qin, 249–247 BC) →</td></tr>
<tr><td>The first emperor of the Qin Dynasty (Ying Zheng)
(Son of King Zhuangxiang of Qin, 246–210 BC)</td><td>At the beginning of his reign, Qin had annexed Hanzhong and Bashu, a region including Sichuan province, Chongqing and the surrounding areas. It had Hedong, Taiyuan and Shangdang counties in the north and Yingyang in the east. The Zhou Dynasty ended, and Nan and Sanchuan counties were set up. In the 26th year of his reign (221 BC), he unified China and established the Qin Dynasty. He declared himself "the First Emperor".</td></tr>
<tr><td>The second emperor of the Qin Dynasty
(Son of the first emperor, 209–207 BC)</td><td>He usurped the throne with the help of Li Si and Zhao Gao and killed ministers and relatives who posed a threat. He was eventually put to death by Zhao Gao.</td></tr>
<tr><td>Ying Ziying[6]
(207 BC)</td><td>Zhao Gao believed that despite the re-establishment of six states and the territorial decline of Qin, it was crucial to preserve the reputation and legacy of the Qin Empire. To achieve this, he orchestrated the ascension of Zi Ying to the throne However, Zi Ying's reign was short-lived, lasting a mere 46 days. Following a defeat of the Qin military by Liu Bang, a general from the state of Chu, Zi Ying ultimately surrendered. This event marked the conquest of Xianyang by Liu Bang, leading to the downfall and conclusion of the Qin Dynasty.</td></tr>
</table>

[6] There are three opinions about the identity of Ziying: the son of the second emperor's brother, the son of the second emperor or the younger brother of Ying Zheng. Though the first one is the most popular one, scholars believe that the third one is more reasonable. More details can be seen in the article "Analysis of the Identity of Ziying" by Cui Shuting which has been included in the second volume of *Collection of Essays on History of the Qin and Han Dynasties* (*Qin Han Shi Lun Cong*, 《秦汉史论丛》) (Xi'an: Shaanxi People's Publishing House, 1983, pp. 203–213).

key individuals can be found in Table 2.2, providing a brief introduction to some of the significant figures in Qin's history.

2.2 The Culture of the State of Qin

In the book *History of Qin and Han Dynasties* (*Qin Han Shi*, 《秦汉史》), Qian Mu claimed that Qin people lacked their own distinctive culture.[7] His intention is to emphasize that the culture of Qin is formed through the gradual assimilation of merits from other cultures. While the opinions of other scholars may not align as unequivocally with Qian Mu's claim, they also acknowledge the eclectic nature of Qin's culture and attempt to explain it from various perspectives.

Geographic conditions are one of the factors contributing to these characteristics of Qin culture. Regarding the influence of these conditions on the Qin culture, there are basically two aspects. Firstly, the gradual eastward migration of the Qin regime basically followed the route of the Zhou people's eastward migration. Originating from "Xichui",[8] the Qin people moved eastward to "the confluence of Qian River and Wei River"[9] in the reign of Duke Xiang and Duke Wen, which was about 200 km in a straight-line distance. Eighty years later, Duke De established a new capital in the city of "Yong",[10] moving from the narrow Pingyang Valley[11] to the vast and flat place

[7] Qin Mu, *History of Qin and Han Dynasties*, this view is always repeated when it comes to the cultural characteristics of the state of Qin and its process of drawing cultural advantages from others p. 8.

[8] Xichui, also called "Xi Quanqiu", an ancient name for the western part of China, refers to the area in the northeast of Li county located in the southwest of the city of Tianshui in Gansu province. The ancestors of the Qin people lived here. However, some people hold that "Xichui" has already been a general term without corresponding places in today's China; Duke Xiang and Duke Wen lived in "Xichui" while Fei Zi and Duke Zhuang lived in "Xi Quanqiu", and Quanqiu should be divided into two parts-the west and the east.

[9] The confluence of the Qian River and Wei River: The Qian River is a tributary of the Wei River. Starting from the south of Liupan Mountain in Gansu province, it flows southeast through Long county and Qianyang and finally comes to the Wei River in Baoji.

[10] Yong refers to the county of "Fengxiang" (also called "Fufeng") in Shaanxi province. Both places are situated in the south of Qishan Mountain. The former is westward and the latter is eastward. *Study of History and Geography in the Qin and Han Dynasties* (*Qin Han Li Shi Di Li Yan Jiu*, 《秦汉历史地理研究》) (pp. 104–111) by Xu Weimin states that the palace of Qin is located in the south of Fengxiang county in Shaanxi province, with Yong River in the south and Qian Mountain in the north; and since 1962 thousands of cultural relics have been unearthed after the exploration and excavation of archaeologists, such as "No. 1 tomb of the King of Qin".

[11] It is mentioned in the fifth volume of *Records of the Grand Historian* (*Shi Ji*, 《史记》) that Duke Ning of Qin moved to Pingyang in the second year of his reign. "Duke Ning of Qin" is also called "Duke Xian of Qin" (reigned from 715 BC to 704 BC). *The History of Qin* by Wang Quchang states on page 101 that Pingyang is the Pingyang pavilion in Mei county. Accordingly, this place should be in the Shaanxi Province south of "Fufeng" and on the bank of Wei River, which is the same with the accounts of the position in *A Concise Historical Atlas of China* (*Jian Ming Zhong Guo Li Shi Di Tu Ji*, 《简明中国历史地图集》) (p. 12) edited by Tian Qixiang.

of "Zhouyuan",[12] the former site of the Zhou Dynasty. Later, the Qin people passed "Jingyang" north of Xianyang and "Liyang" east of Xianyang before ultimately settling in Xianyang in the era of Duke Xiao (350 BC).[13] Remarkably, the locations of Qin's capital were nearly the same as Fengdu and Haojing, two capitals of the Zhou Dynasty. Therefore, the Qin people undoubtedly had increasingly close contact with the "remaining people of Zhou", as well as the cultural relics, classics left after King Ping of Zhou moved eastward, and even "scholars like court histori ans".[14] This led to the saying that the cultures of Zhou and Qin were closely related: "Zhou and Qin were actually one".[15] On the other hand, since the Qin people lived in the western borderlands, they naturally came into contact with the "Xirong". This led to the inclusion of foreign elements in their cultural composition: "Relatively speaking, they adopted the most from the 'Rong and Di' cultures, to the extent that they were called 'Sinicized by the Rong' and regarded as a branch of the 'Xirong'".[16] Archaeological excavations confirm that Gansu was the first frontier of ancient China to accept and disseminate the culture of the East and the West. For example, the painted pottery and bronze "scepter heads" unearthed here, as special utensils indicating status and symbolizing authority, "were relatively concentrated in ancient Egypt, the Near East, Anatolia, and areas around the Black Sea and the Caspian Sea. In China, scepter heads are only distributed in Gansu, western Shaanxi, Xinjiang, and other places. Their forms are very similar to those found in the Near East and Central Asia. It can be basically determined that the scepter, a utensil with special functions, is not an inherent cultural characteristic of Chinese civilization and should be an external factor".[17]

The second characteristic of Qin culture is focusing on national prosperity, military strength, and the consolidation of centralized power.

---

[12] Zhouyuan includes part of the two counties of Qishan and Fufeng in Shaanxi province. King Tai of Zhou once built cities, palaces, ancestral temples, and houses in Zhouyuan, making it become the earliest capital of the Zhou Dynasty. Later, although King Wen of Zhou and King Wu of Zhou relocated the capital to Fengdu and Haojing respectively, Zhouyuan was still an important political center. Archeological work in Zhouyuan made many major discoveries in the later 1970s.

[13] Scholars hold varying opinions regarding the number of capitals that the Qin Dynasty possessed, with estimates ranging from six to nine. Xu Weimin's book *Study of History and Geography in the Qin and Han Dynasties* (pp.3-4) suggests that, in addition to the well-known capitals of Yong and Xianyang, which were equipped with large palaces, there were several smaller capitals. These included Xichui (Xi Quanqiu), situated at the confluence of the Qian River and Wei River, as well as Pingyang, Jingyang, Liyang, and other locations with relatively compact palace structures. The smaller palaces made it convenient for the Qin rulers to frequently relocate their capital as needed.

[14] Lin Jianmin, *Development History of the State of Qin*, p. 50.

[15] Han Xing, *A Combination of Confucianism and Law: Study of Politics and Culture of the Qin and Han Dynasties* (*Ru Fa Zheng He: Qin Han Zheng Zhi Wen Hua Lun*, 《儒法整合:秦汉政治文化论》), p. 63.

[16] Huang Liuzhu, *A Brief Introduction to the Culture of Qin* (*Qin Wen Hua Gai Shuo*, 《秦文化概说》). It is quoted from *A Combination of Confucianism and Law: Study of Politics and Culture of the Qin and Han Dynasties* by Hanxing, p. 62.

[17] *Civilization of Silk Road in Gansu Province* (*Gan Su Si Chou Zhi Lu Wen Ming*, 《甘肃丝绸之路文明》) edited by Gansu Provincial Museum, pp. 4–5.

Fig. 2.1 The place where the first emperor of Qin buried scholars of Confucianism. It is located in the southwest of Hongqingbao village in the district of Lintong, Shaanxi province

As a small state dwelling in ancient China's western region, the state of Qin managed to rise to prominence through rapid development, culminating in the remarkable achievement of unification. This success can be attributed to several factors, including the absence of burdensome traditional systems and an eagerness among the Qin people to learn and pursue innovation. Historians argue that the cultural progress of Qin was apparent in its early adoption of ironware and extensive cattle farming. However, it is essential to acknowledge that a ruling group striving for rapid economic development and national strength naturally places a strong emphasis on cultivating a culture that aligns with these objectives. In today's background, it is called pragmatic culture that serves national economic growth. As a result, historians describe the cultural history of the state of Qin as one where Confucianism did not hold significant influence until the reign of King Zhuangxiang of Qin (249 BC to 247 BC) (Fig. 2.1).[18] During the reign of King Zhuangxiang, the prime minister Lyu Buwei had about 3000 proteges whose writings were compiled into a book known as *Master Lyu's Spring and Autumn Annals* (*Lyu Shi Chun Qiu*, 《吕氏春秋》). This compilation showcases the diverse schools of thought studied by these proteges, including Confucianism, Mohism, Taoism, logical reasoning, and legalism, reflecting the evolving intellectual landscape of Qin during that era. Sadly, King Zhuangxiang's reign was short, lasting only three years. His son, Ying Zheng, ascended the throne at the age of 13, with Lyu Buwei assisting him in governing the state. However, as Ying Zheng grew older, his differences in governance philosophy with Lyu Buwei intensified. In the eighth year of his reign (239 BC), Ying Zheng aimed to assert his own ideas in governing the state, leading to an escalating conflict with Lyu Buwei. In a symbolic act of protest, Lyu Buwei publicly hung his book on the gate of Xianyang, declaring that anyone who could change even a single word would receive a substantial reward of one thousand gold. This "protest", however, backfired, resulting in Lyu Buwei's dismissal and subsequent exile to the region of Shu. Tragically, he ultimately chose to take his own life by consuming poisonous alcohol during his journey.

[18] Lin Jianming, *Development History of the State of Qin*, p. 126.

Even today, when we read *Master Lyu's Spring and Autumn Annals*, we can vividly sense the depth and richness of its content. The study of "history" requires a chronological understanding, but the brief reign of King Zhuangxiang limits our ability to thoroughly explore the dynamic academic environment of that time, where diverse doctrines coexisted and flourished. Ying Zheng, on the other hand, embraced the legalist doctrine, which aimed to consolidate central power and strengthen the state. However, this emphasis on legalism also fostered a political system that was characterized by its violent nature. "This violent disposition facilitated swift victories in the wars of unification but also contributed to the Qin Dynasty becoming akin to a prison, ultimately hastening its downfall".[19]

In short, the so-called "Qin culture" should refer to a regional culture that once existed during the Spring and Autumn Period and the Warring States Period and was owned by the Qin people (the Qin tribe). Its core area and distribution range were roughly in present-day Shaanxi and Gansu provinces in the northwest of China.[20]

It is not entirely accurate to label all the cultures within the empire's territory after Emperor Shi Huang unified China as "Qin culture". Furthermore, it is inappropriate to overemphasize the significance of Emperor Shi Huang's decision to "relocate 120,000 wealthy households from across the country to Xianyang", as well as similar actions taken by Emperor Gaozu of Han in terms of "cultural construction". Their focus was primarily on the wealth of powerful families and influential landowners. Their objective was to "strengthen the center and weaken the periphery to maintain control over the realm". Therefore, the ancients described the social features of Xianyang in the Qin Dynasty and Chang'an in the Han Dynasty as follows:

> As a result, people from different regions mixed together, and customs varied. The noble people advocated extravagance, while the common people despised benevolence and righteousness. When the society was prosperous, merchants pursued profits; when it was poor, thefts were out of control. In the neighborhoods, people attached great importance to money and goods in weddings, and the funerals were overly elaborate. So the capital area of the Han Dynasty was known as 'hard to govern', which was the same in ancient and modern times.[21]

Accordingly, the Han Dynasty mainly inherited the bureaucratic system from the Qin Dynasty. In terms of medicine, the Han Dynasty set up the imperial health administration and appointed its leader to manage everything about medicine, following the precedent set by the Qin Dynasty. However, such practices primarily pertained to the political system rather than the specific cultural elements that would foster the development of medical theories within the state of Qin.

Some may question whether medicine is considered a practical skill that the Qin people would value, given their emphasis on useful expertise. Indeed, medicine would

---

[19] Lin Jianming, *Development History of the State of Qin*, p. 134.

[20] Han Xing, *A Combination of Confucianism and Law: Study of Politics and Culture of the Qin and Han Dynasties*, p. 62.

[21] *Records of Geographical Conditions of the Capital City and its Surrounding Areas in the Qin and Han Dynasties* (*San Fu Huang Tu*, 《三辅黄图》), Volume 1, Customs of the Qin and Han Dynasties. The author is unclear. Historians believe that it is maybe written during the Southern Dynasty or the Tang Dynasty.

not have been overlooked by the Qin. However, the key point of discussion here is not the presence of doctors or specific understanding and treatment methods for certain diseases in the state of Qin. Instead, it revolves around whether the cultural system of Qin could contribute to the production of medical classics, such as the *Huangdi's Internal Classic*.

The reason why medical masterworks like the *Huangdi's Internal Classic* are still regarded as classics today lies in their predominantly "theoretical" content and highly logical theoretical frameworks. The interpretation of physiology and pathology in these works is not merely based on experiential summaries but rooted in abstract principles. However, during the pre-Qin period, the Yin-Yang School and the School of Huang Di and Lao Zi were primarily seen as "political theories" and were not embraced by the state of Qin, which instead advocated legalism and focused on promoting agriculture and strengthening military power. Following the conquest and unification of various states, the first emperor of Qin started to adopt the theory of the Five Elements. In his view, the Zhou Dynasty represented "fire" and the Qin Dynasty, having defeated the Zhou, symbolized "water" capable of extinguishing "fire".[22] This suggests that the Qin people began accepting the theory of the Five Elements, albeit belatedly. However, the cultural environment of Qin did not provide robust support for the development of medical theory within its borders, let alone establish a connection with the medical advancements and production of theoretical classics seen during the Han Dynasty.

2.3 Medicine of the State of Qin

Given that there has already been extensive research on the history of the Qin and Han Dynasties, it would be beneficial to explore the works of experts who have delved into the history of the Qin Dynasty specifically. By examining these materials, we can gain insights into the achievements in the field of medicine within the state of Qin during that time.

[22] *Records of the Grand Historian*, Volume 6. In the biography of the first emperor of Qin, it is mentioned that he extensively studied the principles of the Five Elements. Based on this knowledge, he concluded that the Zhou Dynasty represented the element of fire, while the Qin Dynasty should embody the element of water, as water has the power to extinguish fire. Additionally, the predominant color of clothing and flags was black, reflecting this symbolic association with the element of water. Furthermore, the Qin Dynasty had a preference for utilizing the number six as a unit of measurement or categorization.

2.3.1 *Limited Coverage of Medicine in Three Versions of* The History of the Qin Dynasty

Wang Quchang spent decades complementing the history of seventeen dynasties, enlisting the help of his relatives to gather a vast array of reference materials.[23] The *History of the Qin Dynasty* written by him includes two specific chapters "Prescriptions and Technique" and "Fangshi (occult masters)". Although it can be said that he searched exhaustively, what was the result? To begin with, let us delve into the chapter titled "Prescriptions and Technique". Within this chapter, a total of seventeen individuals have been documented, each with their own unique names and areas of expertise. Table 2.3 provides an overview of their names and skills.

The chapter "Fangshi" mentions a story that Xu Shi and Lu Sheng were asked to seek magical medicine for the first emperor of the Qin Dynasty.[25]

Additionally, Ma Feibai, who was determined fifty years ago to write a *History of the Qin Dynasty* to fill the gap in the *Twenty-Four Histories* and aimed to rival his hometown's worthy predecessor Ma Duanlin, wrote *Collected History of the Qin*. Though he devoted himself to collecting historical materials from the 1930s, his work was not finalized until 1979,[26] resulting in a comprehensive volume spanning nearly

Table 2.3 Individuals and their roles recorded in the chapter "Prescriptions and Technique" in the *History of the Qin Dynasty* by Wang Quchang[24]

Name	Roles
Bu Tufu	Be responsible for divination through turtle shells during the period of Duke Mu of Qin
Bo Le, Jiufang Gao	Experts at finding excellent horses
Huan, He, Qu, Xia Wuqie	Doctors
Cheng Miao, Wang Cizhong	Creating clerical script, an archaic style of Chinese calligraphy
You Zhan	A dwarf who excels at reasoning
Zhao Ping	Gaining renown for cultivating exceptionally delicious melons in the post-Qin Dynasty era
Liao Zhongyao, He Shehu, Qin Jing	Mastering the art of marksmanship with joined forces and successfully hunting down a formidable tiger
Zhuo Shi, Cheng Zheng, Wankong Shi	Becoming skilled in the art of iron smelting resulted in prosperity and wealth

[23] See the "postscript" written by Wang Qucahng's son in the *History of the Qin Dynasty*. pp. 308–310.

[24] Wang Quchang, The *History of the Qin Dynasty*, pp. 266–273.

[25] Wang Quchang, The *History of the Qin Dynasty*, pp. 273–275.

[26] See the "publishing notes" and "author's preface" in *History of the Qin Dynasty Based on a Collection of Documents* (*Qin Ji Shi*, 《秦集史》) by Ma Feibai, published by Zhonghua Book Company in 1982.

800,000 words. However, despite its extensive content, there are limited insights into medical matters within his book.

Therefore, the general process of my research on "medical development of Qin" can be divided into the following stages.

Initially, it was hypothesized that the Qin Dynasty had accumulated a wealth of medical knowledge over its long history, which may have been overlooked due to the sole association of "Qin" with Ying Zheng's establishment of the dynasty. Efforts were made to uncover evidence supporting this notion from various texts, but this endeavor proved futile. Subsequently, through an exploration of political, cultural, and economic materials concerning the Qin people, it was realized that a well-defined medical theory could not have flourished during a time when immediate interests such as national enrichment, military strength, and legal advocacy took precedence. Consequently, the conjectured conclusion shifted from "ignored substantial medical knowledge" to "lack of medical knowledge", awaiting verification. In other words, the focus of reading shifted from "discovering clues" to "verifying the non-existence". Generally speaking, it is relatively simple to confirm the presence of medical knowledge through the discovery of historical records. However, it is far more challenging to deny its existence, as no one can exhaustively study all historical materials. Lack of evidence does not definitively prove non-existence. Nevertheless, considering the efforts of historians dedicated to the study of the Qin people's history, it is highly plausible that the Qin populace did not advance significantly in the realm of medicine.

Furthermore, *History Manuscript of Qin* (*Qin Shi Gao*, 《秦史稿》) authored by Lin Jianming, a renowned expert in Qin history, contains a chapter titled "Development of Science and Technology". This segment emphasizes that Qin's remarkable scientific achievements were primarily concentrated in two domains: medicine and astronomical calendar.

> During ancient times, the governance and influence over medical practices in the state of Qin were firmly established under the direct control of the government itself. The noteworthy physicians were predominantly appointed and guided by the ruling authorities. In the Spring and Autumn period, the medical field in Qin had yet to exhibit distinct characteristics. However, following the reign of Duke Mu of Qin, the Qin State began to emerge as a medical powerhouse, ranking prominently among the various vassal states. Especially in the late Spring and Autumn period, the state of Qin became one of the most developed areas of medical science at that time, boasting numerous renowned doctors. The expertise of these Qin physicians was in such high demand that rulers from other vassal states sought their services to treat various ailments. The research and inventions pioneered by Qin doctors played a pivotal role in the history of medicine, with some of their discoveries still holding influence to this day, especially in the realm of pathology.

The book delves into the specific narratives surrounding He and Huan, two esteemed physicians of the Qin State who are lauded as embodying the pinnacle of medical expertise during that era.[27] Therefore, one might inquire about the proficiency level of the "good doctors" in the Qin State in terms of their medical expertise.

---

[27] Lin Jianming, *History Manuscript of Qin*, pp. 91–93.

2.3.2 *Doctors He and Huan Only Give Diagnoses Instead of Treatments*

The term "excellent doctors in Qin"[28] specifically refers to He and Huan. While those who study medical history are already acquainted with them, it is necessary to provide a concise introduction considering the diverse range of readers.

In the 23rd year of Duke Huan's reign (581 BC), Doctor Huan (缓) was dispatched to the state of Jin to provide medical treatment to Duke Jing of Jin. The process of this event was meticulously documented in the *Commentary of Zuo*, as follows:

> Duke Jing of Jin fell gravely ill, prompting the dispatch of Huan to provide medical care. Before Huan's arrival, Duke Jing dreamed of two children (two children here refer to the disease). One of them expressed concern, stating, "He is a skilled doctor. I fear he may harm us. What should we do?" The other child responded, "If we conceal ourselves above huang and below gao, he will be powerless against us." (two children here refer to the disease) Upon Huan's arrival, he declared, "The illness cannot be cured, as it resides in the area above huang and below gao, beyond the reach of medical intervention." Impressed, Duke Jing exclaimed, "Truly, you are an exceptional doctor." As Doctor Huan departed, he received generous rewards from Duke Jing in recognition of his services.[29]

In ancient China, the term "gao" referred to the fatty tissue connecting to the heart and the term "huang" denoted the region between the heart and the diaphragm according to *Explanation and Study of Principles of Composition of Characters* (*Shuo Wen Jie Zi*, 《说文解字》).[30] Later "gao huang" became a specific acupoint which is located in the posterior region of the human body. Positioned under the fourth thoracic vertebra and three inches lateral to the posterior midline. An idiom in China, known as "disease going into the area between gao and huang" emerged to describe the incurable diseases. Historical records indicate that in the 36th year of Duke Jing's reign in Qin (541 BC), Doctor "He" was sent to diagnose Duke Ping of Jin because of the proximity and marital ties between the Qin and Jin states.

---

[28] In the eighth volume of *Supplementary Notes on 'The Commentary of Zuo'* (*Zuo Shi Zhuan Xu Shuo*, 《左氏传续说》) written by Lyu Zuqian in the Song Dynasty (AD 960-1279), the concluding remark of seeking a doctor from the state of Qin is "Qin state boasting a number of good doctors". In the twentieth volume of *Investigative Reflections on Classics and Histories* (*Kun Xue Ji Wen*,《困学纪闻》) written in the Song Dynasty, Wang Yinglin, the author, firstly quoted from Lyu Zuqian's statements and then said, "It is mentioned in *Shi Zi* (《尸子》) that Yi Qu is an experienced doctor in the state of Qin". Yan Ruoqu in the Qing Dynasty added, "Bian Que was killed by Li Xi, the minister of the imperial health administration. It is worth noting that Han Fei once made a remark praising the doctors of Qin for their remarkable expertise in effectively treating". The sixth volume in *Collection of Wang Zhong* (*Wang Zhong Wen Ji*, 《王忠文集》) written by Wang Yi in the Ming Dynasty begins with the sentence that "Ancient people always say excellent doctors come from the state of Qin". All of this evidence serves to substantiate the notion that Qin is home to a multitude of exceptional doctors.

[29] *Annotation and Interpretation of 'The Commentary of Zuo'* (*Chun Qiu Zuo Zhuan Zhu Shu*, 《春秋左传注疏》), Volume 26, annotation made by Du Shi in the Jin Dynasty and interpretation given by Kong Yingda in the Tang Dynasty.

[30] *Annotation and Interpretation of 'The Commentary of Zuo'*, Volume 26, annotation made by Du Shi in the Jin Dynasty and interpretation given by Kong Yingda in the Tang Dynasty. While in the *Origin of Chinese Characters*, it pertains to the area above the heart and below the diaphragm.

> Duke Ping of Jin sought medical help from the state of Qin. Duke Jing of Qin sent Doctor "He" to examine him. Upon examining Duke Ping, Doctor "He" declared, "Your illness is a result of excessive sexual indulgence and cannot be cured." … Zhao Meng expressed admiration, saying, "You are truly an exceptional doctor." Then, Zhao Meng rewarded Doctor "He" generously and sent him back.[31]

Regarding these two significant historical medical accounts, there are three noteworthy points that merit attention:

(1) The two famous "good doctors" of Qin did not actually give patients any treatment, so I refer to their actions as "diagnosis" rather than "medical treatment".
(2) Both paragraphs are from *The Commentary of Zuo instead of Spring and Autumn Annals* (*Chun Qiu*, 《春秋》). They are not found in *The Commentary of Gongyang* (*Gong Yang Zhuan*, 《公羊传》) and *The Book of Gu Liang* (*Gu Liang Zhuan*, 《谷梁传》), two of the three books of commentaries on the Spring and Autumn Annals.
(3) Even in *The Commentary of Zuo*, the dialogue between doctor He and Zhao Meng[32] primarily focuses on Doctor He's admiration for Zhao Meng's governance. He praises Zhao Meng for successfully serving as the prime minister of the state of Jin for eight years and maintaining order. Additionally, Doctor He emphasizes the importance of setting limits on entertainment and sexual activities. However, the detailed explanations of key theoretical concepts such as "yin and yang" and "six qi" are provided in the later annotations by others. For example:

    > The natural world consists of six qi. [Note]: These include yin, yang, wind, rain, darkness and brightness. There are also five flavors. [Note]: Gold corresponds to pungency, wood to sourness, water to saltiness, fire to bitterness, and earth to sweetness. The five flavors originate from the interaction of yin, yang, wind, and rain.

(4) In *The Commentary of Zuo*, it is explained that engaging in sexual activities with women during nighttime may lead to mental disorders due to "the yang attribute of women and the generation of internal heat". This clearly shows that at that time there was no specific classification placing heaven, the sun, and men solely under "yang", and earth, the moon, and women solely under "yin".

Later, the annotators' interpretation that women are associated with yang is far-fetched. They argued that since women often follow men, women should be considered yang as well. Consequently, this perspective implies that there would only be "yang people" and no "yin people". Hence, this notion lacks coherence and does not make sense.

---

[31] *Annotation and Interpretation of 'The Commentary of Zuo'*, Volume 41, annotation made by Du Shi in the Jin Dynasty and interpretation given by Kong Yingda in the Tang Dynasty.

[32] Zhao Meng is another name for Zhao Dun, an important minister of the state of Jin, but it had passed away at that time. Thus, Zhao Meng in this article should be one descendants of Zhao Dun. As the historians say that all of Zhao Dun's offspring is called Zhao Meng.

Another event recorded in *The Commentary of Zuo* can account for why "women are considered to be yang" at that time.

> Duke Wen of Jin dreamed of fighting with Chu Zi. The King of Chu pressed on the Duke of Jin and then "salted" his brain. [Note]: "Salt" here means to sneeze. The Duke of Jin was frightened because of this. Zifan said, "It's auspicious! We have the support of heaven, and Chu will admit its guilt." [Note]: The Duke of Jin was facing upwards, thus having the support of heaven; the King of Chu was facing downwards to the ground, so he would admit his guilt.[33]

The reason why this was considered auspicious when the Duke of Jin was pressed to the ground and spat on is because he faced upward to the sky, signifying that he received assistance from heaven. Based on this observation, it becomes easier to understand why women were believed to possess the attribute of yang within the context of sexual activity. I mention this instead of quoting other contents more related to medical theory to prove that though yin-yang concept was utilized in reasoning, the theory of yin and yang was relatively simplistic compared to the more complex understanding found in *Huangdi's Internal Classic*. At that time, the concept of "simple dialectic" was not employed to delve into the intricate relationship between yin and yang, such as concepts like "mutual rooting", "unity of opposites", and "dynamic balance".

There is another famous doctor named Yi Qu in the state of Qin, who performed various treatments for notable individuals. He once cut off the acne for King Xuan of Qin, treated hemorrhoids for King Hui of Qin, and removed the dorsal carbuncle for Zhang Yi.

2.3.3 Applying the Medical Principles to State Governance

Shi Jiao, one of Shang Yang's proteges, recorded stories about doctor Qu in his now-lost book. Quoted statements from the *Readings of the Taiping Era* (*Tai Ping Yu Lan*, 《太平御览》) shed light on these accounts:

> According to *Shi Zi*, Qu was an excellent doctor in the state of Qin. He managed to cut off the acne for King Xuan of Qin and treat hemorrhoids for King Hui of Qin, both of whom made full recoveries. Zhang Yi, the prime minister of the state of Qin, sought Qu's help when he suffered from a dorsal carbuncle. Zhang Yi reassured Qu, saying, "Don't be afraid of my position. Do what you must. I entrust my back entirely to your care." Then Qu treated him without any hesitation and Zhang Yi was cured. Qu's expertise in treating illnesses and Zhang Yi's unwavering trust demonstrate the principle that extends to both medical treatment and state governance. There must be key controls in both self-cultivation and state governance before successful management can be achieved.[34]

---

[33] *Annotation and Interpretation of "The Commentary of Zuo"*, Volume 15, annotation made by Du Shi in the Jin Dynasty and interpretation given by Kong Yingda in the Tang Dynasty.

[34] *Readings of the Taiping Era* compiled by Li Fang in the Song Dynasty et al., Volume 371 and Volume 724.

When historians analyze these accounts, they often point out that there is no historical record of a King Xuan in the Qin dynasty, suggesting a potential mistake. They agree that King Hui refers to King Huiwen of Qin. However, it is not crucial since this is merely a fable. In *Zhuang Zi* (《庄子》), there is a similar tale where a man named Cao Shang from the state of Song receives generous rewards from the king of Qin. Zhuang Zi remarks, "I've heard that the king of Qin is seeking experienced doctors for his illness. For removing carbuncles and abscesses, one carriage will be given as a reward. For treating hemorrhoids, five carriages will be granted. The more challenging the position, the greater the reward. Did you treat the king's hemorrhoids? How did you get so many carriages? It's best for you to leave now".[35] This story satirizes those who are obsequious and fawning. It does not delve into specific names of doctors or kings. Shi Zi's retelling of the story regarding Zhang Yi's dorsal carbuncle suggests that while it may not depict actual historical events, it instead emphasizes the underlying concept that "buyers" create demand that drives "sellers" to produce their products. This concept also applies to Shang Yang's strategies for governing the state. No matter how correct, effective, and useful those strategies may be, they only hold meaning when there is a demand for them. Moreover, absolute trust and delegated authority are necessary. Doctors strive to provide the best treatments for their patients, mirroring the relationship between proteges who offer advice to their "masters".

This is the so-called "learning the principle of politics from medical stories". In fact, Huan's diagnosis of "disease goes into the area above huang and below gao" and He's suggestion of "restraining sexual desire" also contain some metaphorical preaching and warnings to the king. Another example is found in *Master Lyu's Spring and Autumn Annals.* In the pre-Qin period, it recounts the story of Wen Zhi, a renowned doctor from the state of Song, who cured the king of the state of Qi by infuriating him. However, after the king recovered, he cruelly executed Wen Zhi by boiling him alive.[36]

This story exemplifies the notion of unwavering loyalty, where one serves the leader even if death is the consequence. Some medical history books refer to Han Feizi's statement that "babies will have abdominal pain without shaving their heads" as medical knowledge.[37] Without delving into the accuracy of this statement or the interpretation of the ancient Chinese character "Fu" (abdomen, 腹), its significance lies in the context.[38] It conveys the idea that politicians should not govern solely based on public opinion because civilians often prioritize short-term gains over long-term benefits". Just like when a little child has his head shaved, he must be held by one person and his loving mother begins to shave regardless of whether he is crying or

---

[35] *Commentaries on 'Zhuang Zi'* (*Zhuang Zi Zhu*, 《庄子注》) written by Guo Xiang in the Jin Dynasty, Volume 10.

[36] Master Lyu's *Spring and Autumn Annals*, Volume 11.

[37] *Han Feizi,* Volume 19, annotated by He Huan in the Yuan Dynasty.

[38] *The History of the Warring States Period* (*Zhan Guo Shi*, 《战国史》) written by Yang Kuan (p. 224). Experts added notes that "disease will worsen if without immediate treatment". The Chinese character of "Fu (複)" was mistaken for "Fu (腹 abdomen)" in this sentence.

not. *Strategies of the Warring States* (*Zhan Guo Ce*, 《战国策》) has documented a story of Bian Que coming to Xianyang to diagnose King Wu of Qin, which also contains the wisdom of governance.

> In the encounter between Bian Que and King Wu of Qin, the king showed Bian Que his illness and Bian Que offered to treat it. However, the king's attendants said, "Your Majesty's illness is in front of the ear and below the eye. Treating it may not cure it completely, and it may make you deaf and blind." When the king told Bian Que about this, the doctor became enraged and threw down his medical instrument, the Bian stone. He exclaimed, "You consult those who know about the matter but then let those who know nothing spoil it. If this is how you handle the state affairs of Qin, then with one move you will lead the country to ruin!".[39]

Bian Que's story is referenced as an example in a discussion about political strategies between the king and ministers of the state of Han, which is also documented in the same book.

> The prime minister of the state of Han commented, "The reason people show goodwill towards Bian Que is because they seek his expertise in curing their ailments. Without any illnesses, no one would treat Bian Que with such kindness. Currently, you have managed to win over Lord Pingyuan due to your strained relationship with the state of Qin. However, your close association with Lord Pingyuan is precisely why the state of Qin harbors animosity towards you. I urge you to carefully consider this situation."[40]

Such examples were prevalent in ancient times, especially before the Han Dynasty. These events, often cited by later generations as "medical historical materials", are essentially "using medicine to illustrate politics". Otherwise, in an era when having a wealth of knowledge like "carrying five carts of books" only meant having a few bundles of bamboo slips with not many words on them, why would historians and scholars waste so much ink on recording such trivial matters about disease treatment when writing historical accounts or making arguments?

Furthermore, two noteworthy points should be highlighted. Firstly, in books like Shi Zi, there is basically no content related to "medical knowledge" such as human physiology, causes of diseases, or treatment strategies. However, the situation was different in the Han Dynasty. Confucian-court officials used medicine as a metaphor for politics, and this approach resembled what doctors did when analyzing a patient's condition.[41] These differences clearly demonstrate that theoretical medical knowledge progresses over time, leading to an increase in both the quantity and depth

---

[39] *Strategies of the Warring States*, Volume 4, annotated by Gao You in the Han Dynasty and Yao Hong in the Song Dynasty.

[40] *Strategies of the Warring States*, Volume 28, annotated by Gao You in the Han Dynasty and Yao Hong in the Song Dynasty.

[41] *The Discourses on Salt and Iron* (*Yan Tie Lun*, 《盐铁论》) written by Heng Kuan in the Western Han Dynasty documents that medical metaphors appear many times in the debate between ministers and recommended local officials. The typical one is the statements in the fourth volume. Recommended local officials said, "Bian Que knew the cause of disease as soon as he touched the pulse. If yang qi is abundant, yin should be supplemented. If cold qi is excessive, yang should be replenished. The harmony of qi and pulse can remove all evil qi. Inexperienced doctors do not know the principle of pulse and the differences between blood and qi. Improper acupuncture will hurt the skin without any benefits for treating the disease. It is reasonable to remove the excess and supplement the deficiency instead of making the rich richer and the poor poorer".

of "medical knowledge" in the examples cited. Secondly, it is worth considering why all the diseases mentioned in examples from *Zhuangzi*, *Shi Zi*, and *Strategies of the Warring States* are "surface" diseases. This actually reflects the general level of people's understanding of disease at that time. In the process of understanding the objective world, there is a gradual progression from obvious "external phenomena" to invisible and intangible "internal movements", from tangible "matter" to ethereal "spiritual realm", and from the "concrete" to the "abstract". To illustrate this, let us take a simple example for comparison. When discussing the disadvantages of amnesty and redemption, Wang Fu, a famous Confucian scholar in the Eastern Han Dynasty, began by saying, "Those who treat diseases must first know the deficiency or excess of the pulse and where the qi is stagnated, and then prescribe the treatment. Thus, the disease can be cured and the lifespan can be prolonged". Although he still used medicine as a metaphor, his purpose was to illustrate that "Those who govern a country must first know what the people suffer from and the root causes of disasters, and then set up prohibitions. Thus, evils can be curbed and the country can be secured".[42] Here, the "medical principle" he mentioned is no longer about treating surface sores and performing surgical operations, but rather about pulse-taking, diagnosing deficiency and excess, and prescribing medicine.

2.3.4 Authentic Medical Records of the State of Qin

In 1976, an astounding discovery took place within the tomb of Qin in Shui Hudi, Yunmeng county, Hubei Province, as over 1150 bamboo slips were unearthed. Among these remarkable findings, a significant portion is dedicated to the laws and legal practices of the ancient state of Qin. Notably, medical texts primarily feature in an article titled *Models for Sealing and Physical Examinations* (*Feng Zhen Shi*, 《封诊式》). Within this documentation, detailed accounts of on-site investigations, autopsies, and diverse examinations can be observed. For instance, medical practitioners were instructed to assess defendants suspected of leprosy, ensuring their placement in isolation areas if required. Additionally, examinations focused on determining the causes of vaginal bleeding in plaintiffs through the study of aborted fetuses resulting from altercations.[43]

Autopsies were carried out by people called "Lingshi", a specialized occupation, while the examination of aborted fetuses and vaginal bleeding involved both "Ling Shi" and "Liqie", job titles assigned to those with considerable experience in matters of pregnancy and childbirth. Traditional doctors, however, were solely tasked with diagnosing leprosy in defendants. One particular individual aroused suspicion due

[42] *Discourses on State Governance* (*Qian Fu Lun*, 《潜夫论》) written by Wang Fu in the Han Dynasty, Volume 4.

[43] *Bamboo Slips Unearthed from the Tomb of Qin in Shuihudi* (*Shui Hu Di Qin Mu Zhu Jian*, 《睡虎地秦墓竹简》) compiled by the research group of these bamboo slips, Beijing: Cultural Relics Publishing House, 1978, p. 263 and p. 275.

to his symptoms claimed, "At the age of three, I developed psoriasis, causing my eyebrows to disappear. Although I am unaware of my precise ailment". Consequently, a doctor was summoned to conduct a thorough diagnosis. After the examination, the doctor replied, "The defendant has no eyebrows and has a damaged nose. Damage to the nasal cavity is evident, as there is an absence of sneezing when the nose is irritated. Furthermore, festering can be observed on various regions of the defendant's hands and feet, and he can't make normal sounds when asked to shout. Collectively, these symptoms strongly indicate a diagnosis of leprosy".

2.3.5 Xia Wuqie, a Doctor Standing Next to Qin Shi Huang

The *Records of the Grand Historian* documents that during Jing Ke's assassination attempt on Qin Shi Huang (the First Emperor of the Qin Dynasty, 259–210 BC), Xia Wuqie took action by throwing a bag of medicine toward Jing Ke.[44] Sima Qian, the author, used the characters "Shi (attendant)" and "Yi (doctor)" to simply and clearly indicate Xia Wuqie's position and identity. It is difficult to tell whether "Shi Yi" (an attendant doctor) is a specific job title at that time.

Who was Xia Wuqie, the doctor attending on the King of Qin? Apart from the various meanings given in dictionaries, the Chinese character "qie" (且) has another meaning and usage resembling the shape of a penis with the pronunciation of "ju" (居).[45] This meaning and usage are often avoided in discussion. In ancient China, stone models of penises were utilized for reproductive worship and referred to as "Shi (stone) Qie" (石且) or "Shi (stone) Zu" (石祖) when made of stone, and as "Tao Qie" (陶且) or "Tao Zu" (陶祖) when made of ceramics. The copper shovel is known as "the shape of Zu".

Inspired by this, it is plausible to consider that "Wu Qie" is probably not the real name of the doctor, but rather a reference to his status as a eunuch. The *Book of the Later Han* (*Hou Han Shu*, 《后汉书》) states that "there was one chief of the medicine in the inner palace, with a salary of four hundred dan", and the annotation says "a eunuch".[46] This indicates a longstanding practice of employing eunuchs as doctors within the palace, at least in the court of the Qin Dynasty after it unified the world. Wang Quchang, adhering to the basic concept that "the Han Dynasty followed the systems of the Qin Dynasty", quotes the records in the *Book of the Former Han* (*Han Shu*, 《汉书》) and believes that this was also the situation in the Qin system. He also explains that "one was in charge of all doctors, and the other was in charge of the

[44] *Records of the Grand Historian*, Volume 86 (The Eighty-sixth Book of Assassins). In the Tang Dynasty, Zhang Shoujie's *Remarks on the Records of the Grand Historian* (*Shi Ji Zheng Yi*, 《史记正义》) notes that the word "ti" is pronounced as "toss" (zhi). The meaning is also the same as "throw".

[45] *Index of Records of the Grand Historian* (*Shi Ji Suo Yin*, 《史记索隐》) written by Sima Zhen in the Tang Dynasty annotates that the pronunciation of "Qie" should be "ju".

[46] *Book of the Later Han*, Volume 37.

king's inner palace". As mentioned in the *Strategies of the State of Yan* (*Yan Ce*, 《燕策》), the attendant doctor Xia Wuqie belonged to the Shao Fu.[47] It appears that the attendant doctors who "served inside the king's palace" and "belonged to the Shao Fu" had a different role compared to general physicians.

These are roughly the limited historical materials about medicine in the State of Qin. Judging from the theory and treatment techniques reflected in them, there is probably not much valuable knowledge for the people of the Han Dynasty to inherit.

[47] Wang Quchang, The History of the Qin Dynasty, pp. 128–129.

Chapter 3
Differences Between the East and West in Ancient China: The Inheritance of Medical Knowledge From Six States

Qin Shi Huang, the first emperor of the Qin Dynasty (221–207 BC), "burned books except that of medicine, divination and tree planting[1]".

When it comes to the inheritance of medical knowledge in the Han Dynasty (202 BC–AD 220), it should be noted that besides the culture and medicine of the Qin Dynasty, those vassal states and regions annexed by Qin should also be taken into account. While political regimes may change, cultures persist through time, and this is particularly evident in the case of medicine, which is a practical skill rooted in experience and transmission.

Certainly, the inheritance of knowledge during ancient times is undeniable. What matters is the degree and nature of the inheritance. During the middle of the Spring and Autumn period (770–476 BC), larger states gradually emerged through the annexation of smaller states, leading to intense competition among them. At the end of the Spring and Autumn period, the states of Han, Zhao, and Wei dismembered the state of Jin and the family of Tian took the control of the state of Qi. These events marked the onset of the Warring States period (475–221 BC), a time when seven powerful states vied for rulership over China.[2] The term "Warring States" not only described the political landscape but also the condition of vassal states during that era.[3] Unquestionably, every state found it imperative to enrich their resources and bolster military strength. Now, when it comes to the "prosperity of academic thoughts" in the Warring States period, it is common to refer to the flourishing philosophies and schools of

---

[1] The 6th chapter "A Biography of the First Emperor of Qin" in *Records of the Grand Historian* (*Shi Ji*, 《史记》), Volume 6.

[2] Book of Price Maintaining Chapter 8 in *Records of the Grand Historian*, Volume 30.

[3] The *History of the Warring States Period* (*Zhan Guo Shi*, 《战国史》) written by Yang Kuan (p.8, note 1) gives the example like Dun Ruo saying there are six warring states recorded in *Strategies of the Warring States* (*Zhan Guo Ce*, 《战国策》). Yang Kuan also points out that the term of "Warring States" is not used as the name of a historical era until Liu Xiang wrote *Strategies of the Warring States* in the late Western Han Dynasty (202 BC-AD 8).

Y. Liao, *Reconstructing the Image of Medicine in the Qin and Han Dynasties*, Understanding China, https://doi.org/10.1007/978-981-95-6801-7_3

thought and consider them influential in the scientific and technological advancements. However, it is crucial to recognize that all academic schools served political interests at that time.

3.1 Culture in the East

Qian Mu provides a precise analysis of the cultural characteristics in the pre-Qin period, particularly in distinguishing between the eastern and western regions. Essentially, he grouped the Qin in the west together with the scholars from the "Three Jin states" of Han, Zhao, and Wei, while regarding the states of Qi and Lu in the east, where Confucianism and Taoism originated, as a unified whole (Fig. 3.1).

Most of the people in the states of Qi and Lu set great store by culture and history. They took the whole society as the object of their academic pursuits and emphasized general human life, rather than starting from the narrow-minded goal of national prosperity. Therefore, their discussions and ideas often aimed at seeking thorough improvements for the entire society, which was common to both the Confucian and Mohist schools. Later, when the Taoist school emerged, its attitude toward learning was the same. As for the scholars from the Three Jin states, their vision and spirit were often confined to their own states, and they only aimed at seeking the prosperity of their own states. Their intention was just to make rough reforms based on the current

Fig. 3.1 Zhusi academy in Qufu. Originally referred to as the Lecture Hall, this site holds immense significance as the birthplace of Confucianism and the cultivation of Confucian culture. Unfortunately, it suffered destruction during the Yuan Dynasty (AD 1271–1368), but subsequent restoration efforts have been undertaken numerous times

situation and they couldn't pay attention to the whole society and all aspects of human life. Generally speaking, their thoughts were just about taking advantage of the situation and only pursuing immediate benefits. Therefore, their discussions often valued power over culture and emphasized reality while neglecting history. This was the case with the Legalist, Military, and Strategists of vertical and horizontal alliances schools. This is the general difference.

From the perspective of the people in the Qin state, culture was merely seen as a shortcut to achieve national prosperity and strength. They failed to recognize the beauty and profound meaning of the eastern culture, nor could they accept and assimilate it for their own use. Those whom the people in the Qin state learned from and trusted were just utilitarian scholars from the Three Jin states. Xun Kuang and Han Fei were both influenced by the practical and utilitarian trend in the Three Jin states.[4]

3.1.1 Jixia Academy and "Contention of a Hundred Schools of Thought"

Medical historians and even historians of science attach great importance to the "Jixia Academy" in the state of Qi, the academic atmosphere of the "contention of a hundred schools of thought" there, and the important role of the "theory of yin-yang and five elements" with natural-philosophy color advocated by people like Zou Yan (305 BC–240 BC) in the development of scientific fields such as medicine. However, the actual historical situation may not be like this (Fig. 3.2).

First of all, historians generally agree that the theory of yin-yang and five elements was put forward by Zou Yan, a noted scholar of the Jixia Academy[5] in the state of Qi. Recognizing that "the rulers became more and more licentious and extravagant, and failed to cultivate virtue, improve themselves and bring benefits to the common people", Zou Yan proposed a new theory of governing the world, which was different from the "benevolence and righteousness" advocated by Confucius and Mencius. It really impressed the rulers at that time: "When the rulers first saw his theory, they were deeply affected and intended to follow it". Therefore, "he was respected when traveling among the vassal states", which was definitely different from "Confucius suffering from hunger in the states of Chen and Cai, and Mencius being in trouble in the states of Qi and Liang". However, the result was still "it could not be put into practice later". Sima Qian (145 BC–90 BC) summarized the main points of Zou Yan's theory, saying that "in general, it must end with the practice of benevolence and righteousness, frugality, and the proper relationships between the sovereign and ministers, the superior and the inferior, and among relatives".[6] This was of course a

---

[4] Qian Mu, *History of the Qin and Han Dynasties* (*Qin Han Shi*, 《秦汉史》), pp. 5, 9, 21.

[5] Lin Jianming et al., *Social Civilization of the Qin and Han Dynasties* (*Qin Han She Hui Wen Ming*, 《秦汉社会文明》), p. 268.

[6] Quoted from *Records of the Grand Historian*, Volume 74.

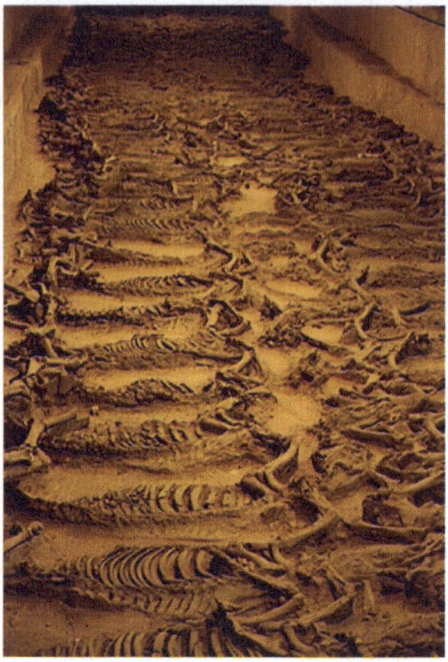

Fig. 3.2 The sacrificial horse pit at the ancient city of Linzi in the state of Qi. The extensive practice of burying horses with the deceased in Linzi, the ancient capital of the state of Qi, serves as a testament to the impressive capabilities of the Qi civilization. This grand-scale tradition showcases Qi's national capacity and its significant cultural and historical heritage

political theory. Nevertheless, it was only superficially accepted until Qin Shi Huang (259 BC–210 BC) made his eastern tour. Based on this theory, he deduced the cycle of the Five Virtues and simply changed the clothing color.

Regarding Jixia Academy and its academic atmosphere, Han Xing's concise opinions came to the point:

> Established in the capital of Qi by Duke Huan of Qi (Tian Wu, 374–357 BC), Jixia Academy resembled a modern-day equivalent of the Chinese Academy of Social Sciences. This institution served as a nexus for academic research, education, and political consultation. Scholars enjoyed a supportive environment as the state provided the necessary facilities and covered living expenses, giving them the freedom to delve into their studies, engage in discussions, conduct investigations, and even travel as they pleased. Undeniably, Jixia Academy held a position of paramount importance during the middle and late Warring States period, influencing academia, culture, thought, and education. However, this very aspect also presented a challenge. Given their dependence on the state's material support, scholars naturally found themselves considering national politics extensively during debates. Consequently, thinking from the ruler's perspective became an ingrained ideological inclination among later Chinese scholars.[7]

Therefore, the Jixia Academy should not be regarded as the place and time coordinate where medicine could develop in the "letting a hundred flowers bloom and a hundred schools of thought contend" atmosphere during the Warring States period.

[7] Han Xing, *A Combination of Confucianism and Law: Study of Politics and Culture of the Qin and Han Dynasties* (*Ru Fa Zheng He: Qin Han Zheng Zhi Wen Hua Lun*, 《儒法整合:秦汉政治文化论》), pp. 17–18.

Fig. 3.3 A picture of Mount Penglai. This painting, created by Yuan Jiang during the Qing Dynasty (AD 1616–1912), portrays the mythical dwelling place of immortals, a destination coveted by both Qin Shi Huang and Emperor Wu of the Han Dynasty

3.1.2 Fangshi From the States of Yan and Qi and the Search for Immortals

The state of Lu, situated to the south of Qi, and Yan, located to the north, played significant roles in the dissemination of Confucianism. And when it comes to immortals and Fangshi (occult masters), ever since Sima Qian mentioned "the Fangshi by the sea in Yan and Qi", the states of Qi and Yan have been renowned for their expertise in this field.

Sima Qian, a notable historian, believed that the search for immortals began with King Wei of Qi, King Xuan of Qi, and King Zhao of Yan, who dispatched explorers on a voyage in search of three enigmatic mountains. This narrative gained further prominence when Qin Shi Huang took an interest in the pursuit of immortals. Intrigued by these stories, he tasked Xu Shi with leading a group of boys and girls on a voyage to seek immortals, while Han Zhong, Hou Gong, and Shi Sheng were assigned to search for the elixir of life.[8] These are well-known stories, so there's no need to cite them one by one. However, from an academic perspective, some noteworthy points emerge (Fig. 3.3).

(1) It was only after Qin Shi Huang came into contact with the eastern culture that all his actions related to "seeking immortals" took place. Therefore, in terms of the so-called "Qin culture", there was originally no such content. In contrast, in the

[8] The 6th chapter "A Biography of the First Emperor of Qin" in *Records of the Grand Historian*, Volume 6

regions of Yan and Qi, the practice of seeking immortals had a long history and was an inherent part of their cultures.

(2) Among the "fangshi" (alchemists), there were those called "Xian Men":

> Song Wuji, Zheng Boqiao, Chong Shang, and Zi Gao were all from the state of Yan and called "Xian Men". They were proficient in dealing with ghosts and immortals.
>
> In an endeavor to fulfill his aspirations, Qin Shi Huang embarked on an eastward voyage, seeking the renowned mountains and eight immortals, along with the elusive "Xian Men".[9]
>
> In the 32nd year of his reign, Qin Shi Huang dispatched Lu Sheng to locate Gao Shi, one of the esteemed "Xian Men".[10]

What does the term "Xian Men" mentioned here mean? Commentators have interpreted "Xianmen Gao Shi" as the names of two ancient immortals respectively. For instance, in *Collected Annotations on Records of the Grand Historian*, it quotes "Wei Zhao said: an ancient immortal" under the entry of "Xian Men"; and in *Correct Meaning of Records of the Grand Historian*, it further comments under "Gao Shi" that "also an ancient immortal". However, if we take into account two other citations: the phrase "those like Xian Men" implies that it refers to "a certain group of people"; and "Xianmen Zigao" most likely directly refers to "Xianmen Gao Shi". Joseph Needham holds that "Xian Men" refers to "Shaman".[11] Considering that the territory of the Yan State was close to the Northeast, which has been the stronghold and an area where "Shamanism" has been widely spread from ancient times to the present, Needham's view is extremely worthy of attention. Nevertheless, there are also those who interpret "Xian Men" as "Sramana".[12] In this case, its origin would shift from the Northeast to the Southwest. It doesn't matter which interpretation readers find acceptable.

(3) Since what these alchemists said and did ultimately involves the issue of "life extension", it is indeed appropriate to include them in the category of "medicine". However, it should be noted that what these alchemists talked about did not contain the content of using health-preserving "techniques" to seek health and longevity as commonly mentioned. They only focused on external "immortality pills".

3.1.3 *Eastern Learning Spreading to the West and* Master Lyu's Spring and Autumn Annals

After the Spring and Autumn period when "in the realm of scholarship, political and educational affairs were not separated, and officials also served as teachers", not

9 *Records of the Grand Historian*, Volume 28.

10 The 6th chapter "A Biography of the First Emperor of Qin" in *Records of the Grand Historian*, Volume 6

11 Joseph Needham (UK), *Science and Civilization in China*, Volume 2, the version of Chinese translation, p148.

12 Ma FeiBai, *History of the Qin Dynasty Based on a Collection of Documents* (*Qin Ji Shi*, 《秦集史》), p.358

only did the social structure undergo a huge transformation, but also academically, the Warring States period witnessed a new situation of "the flourishing of social scholarship and the spread of the learning of court-officials into the thoughts of various philosophers".[13]

It is by no means difficult to find in the words of various philosophers some statements about the ways to maintain health and the principles of diseases and premature deaths. An exemplary representation of this is found in *Master Lyu's Spring and Autumn Annals* (*Lyu Shi Chun Qiu*,《吕氏春秋》), compiled by the proteges of Lyu Buwei (AD ?–235). Although it mainly focuses on "the art of governance", there are indeed many contents related to the causes and principles of health and diseases, as well as the correct lifestyle. Yang Kuan, in his work the *History of the Warring States Period*, specifically highlights the section on "the development of medicine and physiological hygiene", with *Master Lyu's Spring and Autumn* Annals serving as the primary reference:

> They believe that the human body has three hundred and sixty joints, nine orifices, Five Zang-organs and Six Fu-organs. Optimal health is characterized by resilient skin, unobstructed blood vessels, strong muscles and joints, a calm mind, and the smooth circulation of essential qi. If so, a man can be free from disease (The chapter "Removing Depression" in *Master Lyu's Spring and Autumn Annals*). In their mind, the method of governing the body for a longer life is to "use the new and abandon the old" and to "update essential qi every day and remove all evil qi" (The chapter "Beginning with Self" in *Master Lyu's Spring and Autumn Annals*). In terms of diet, it is advised to maintain regular eating habits and avoid both excessive hunger and overeating to keep Five Zang-organs healthy. It is also cautioned against the consumption of excessively oily and spicy foods as well as strong liquor, as these can lead to diseases.[14] Because fatty meat and strong drink hurt intestines (The chapter "Competing the Natural Span of Life and chapter of The Foundation of Life" in *Master Lyu's Spring and Autumn Annals*). Excessively indulging in delicious food without moderation is believed to block the flow of qi, hindering the possibility of achieving a longer life, as mentioned in the chapter "Attaching Importance to Self" of *Master Lyu's Spring and Autumn Annals*. When it comes to exercise, they think that physical inactivity would hinder the circulation of essence, finally leading to the qi stagnation and subsequent illness, as outlined in the chapter "Competing the Natural Span of Life" of *Master Lyu's Spring and Autumn Annals*. It is cautioned against relying too heavily on carriages for transportation, as this may compromise one's health, as stated in the chapter "The Foundation of Life" of *Master Lyu's Spring and Autumn Annals*. The chapter "Desires" in *Master Lyu's Spring and Autumn Annals* explains the potential consequences of excessive engagement in sexual activities, such as the development of diseases affecting the ears, eyes, mouth, and the occurrence of pitting edema, stiff muscles and joints, as well as blockages in blood vessels and the nine orifices. Additionally, the chapter "The Foundation of Life" in *Master Lyu's Spring and Autumn Annals* emphasizes that seductive dancing and folk songs have the potential to erode human nature. In terms of living spaces, they believe that a big room with excessive yin and a high stylobate can result in an imbalance of yin and yang, which in turn may lead to disease, as discussed in the chapter "Attaching Importance to Self" of *Master Lyu's Spring and Autumn Annals*.[15]

---

[13] Qian Mu, *History of the Qin and Han Dynasties*, p.4

[14] From my point of view, it is more proper to explain "the beginning of disease" as dizziness and headache.

[15] Yang Kuan, *The History of the Warring States Period*, pp. 233–234.

Though these statements are related to methods of life preservation and principles of disease in traditional Chinese medicine, most of them are just common sense. Some individuals may argue, "The early theories of life, disease cause and physiology are inherently simple. How can we evaluate them using modern standards?" To assess the accessibility of these ideas, we can refer to earlier medical texts.

Moreover, it is challenging to ascertain whether the aforementioned statements have been incorporated into the field of medicine, becoming an integral part of medical knowledge system. More importantly, *Master Lyu's Spring and Autumn Annals* do not originate from the state of Qin. This book was written by the proteges of Lyu Buwei, who hailed from various regions. Even Lyu Buwei himself was not a native of Qin but an immigrant who migrated from the east to the state of Qin. Therefore, relatively speaking, these are "foreign cultures" for the state of Qin.

In summary, before Dong Zhongshu reformed Confucianism with the theory of yin-yang and the five elements and before Taoism shifted its focus to physical health preservation after withdrawing from the political stage, these political theories that developed in the states of Qi and Lu in the East did not have the nature and function of directly promoting the germination of medical theories. Even during the reign of Emperor Wu of the Western Han Dynasty, the immortal-seeking arts in the states of Yan and Qi still simply and "externally" manifested as seeking immortals and elixirs, lacking the connotation of "medical health preservation" that focuses on the cultivation of both the mind and the body as defined.

3.2 Medicine in the East

The above primarily discusses the influence of Confucianism on medicine. If we only consider the medical situation in that area, there is an aspect that cannot be ignored. In the east, the state of Qi achieved the status of "the leader of the Five Hegemons" during the Spring and Autumn period under the governance of Guan Zhong, who held the position of "Senior Minister" and was respected as "Uncle Zhong" by Duke Huan of Qi. In Guan Zhong's so-called "Nine Benevolent Policies", matters related to health such as "caring for the sick" and "inquiring about the sick" can already be seen.

> The so-called "caring for the sick" means that in all capitals of the states, there are officials in charge of caring for the sick, the deaf, the blind, the mute, the lame, those with limited mobility, those with hemiplegia, and those who are so disabled that they can't take care of themselves. The ruler takes them in and provides for them (not only taking care of them but also treating their illnesses), and the government supplies them with food and clothing. The support stops only when they recover from their illnesses. This is what is called "caring for the sick".[16]

[16] *Guan Zi* (《管子》), Volume 18. "Nine policies to seek benefits for civilians" include "raising the old, loving children, caring about orphans, supporting the life of the sick and giving them treatment, helping windowed men and women get married, paying a visit to the sick, reporting the situation of the poor, helping people out of poverty and commemorating the people who died for the state".

Among the capitals of the states during the Warring States period, Linzi, the capital of the State of Qi, was the largest in scale.[17] There were 70,000 households in the city. On the streets, the wheel-hubs of carriages bumped into each other, and people rubbed shoulders. When they joined their collars, it would form a curtain; when they raised their sleeves, it would form a canopy; when they wiped off their sweat, it would be like rain. Moreover, families were well-off and people were high-spirited.[18] With such a high level of civilization and economic strength, there was naturally a necessary environment for the growth of medicine in this area.

3.2.1 Qin Yueren, also Known as Bian Que

In the Warring States period, there was a famous doctor called Qin Yueren who was praised as Bian Que. He practiced medicine in the northern era of the State of Qi, where was connected with the State of Yan and known for producing many "alchemists".

Bian Que was from Zheng in Bohai Commandery. His family name was Qin and his given name was Yueren. Sima Qian, in the chapter "Bian Que" of his work, made it clear that Bian Que and Qin Yueren were distinct individuals. But he later accounted, "Doctors treated patients in both the state of Qi and Zhao. One doctor in the state of Zhao was called Bian Que". This implies that the doctor in the state of Qi was not Bian Que. Since Sima Qian's book spans several centuries, it further underscores that Bian Que and Qin Yueren were indeed separate entities. Hence, annotations of *Records of the Grand Historian* identified Qin Yueren as an accomplished doctor who possessed skills akin to Bian Que. It is mentioned that Qin Yueren hailed from the state of Lu and, therefore, was also referred to as the doctor of Lu. (*Remarks on the Records of the Grand Historian* (*Zheng Yi*,《正义》). Lu, a small state during the Spring and Autumn period, was situated in the Changqing district of Jinan, Shandong province. The *Explanation of Records of the Grand Historian* (*Ji Jie*,《集解》) quotes that Xu Guang said the place of Zheng should be the county of Mao in Hejian, Hebei province. The *Index of Records of the Grand Historian* (*Suo Yin*,《索引》) states that there is no county of Zheng in Bohai.[19] All these pieces of evidence serve to establish that Bian Que and Qin Yueren were not the same people. In addition, the evidence validates that Qin Yueren was the doctor of the state of Qi. Consequently, it is appropriate to discuss Bian Que and Qin Yueren as separate individuals.

Besides the chapter "Bian Que", a similar description can also be seen in the chapter "Zhao State" in the *Records of the Grand Historian of China*. It can even

[17] Yang Kuan, *The History of the Warring States Period*, p.47.

[18] *Strategies of the Warring States*, Volume 8.

[19] All quotations are seen in Volume 150 of *Records of the Grand Historian*. Three famous books of annotation of *Records of the Grand Historian* are *Index of Records of the Grand Historian* by Sima Zhen in the Tang Dynasty (AD 618–907), *Remarks on Records of the Grand Historian* by Zhang Shoujie in the Tang Dynasty, and *Explanation of Records of the Grand Historian* by Pei Yin in the Song Dynasty.

be said that the accounts of Bian Que treating Zhao Jianzi, a scholar-official of the state of Jin, in the chapter "Bian Que" are based on the contents of the chapter "Zhao State". Though Bian Que enjoyed a high reputation, he just did the same thing as two famous doctors in the state of Qin did at that time. He only explained the cause of diseases and did not give any treatment. The story goes as follows: "Zhao Jianzi fell ill, remaining unconscious for five days, causing great concern among the doctors". At this critical moment, Bian Que visited and recounted a similar case. Duke Miu of Qin had experienced a comparable situation and emerged from a seven-day coma. Upon awakening, he joyfully spoke of his visit to where the immortals reside. He mentioned that his extended stay was necessary for learning purposes, as an immortal had revealed to him that the state of Jin would fall into turmoil. Bian Que then assured Zhao Jianzi's servant, saying, "His blood vessels are normal, and there is no need for panic. He will awaken in three days and speak". After two and a half days, Zhao Jianzi woke up as Bian Que said, and he told some political prophecies about the future of the state of Jin.[20]

Our book will not comment on the specific content of Zhao Jianzi's seven-day dream related to the politics of the State of Jin. Readers only need to know two points: Firstly, in early Chinese society, there were numerous instances of future prediction through dreams, and oneiromancy was more important than other techniques of divination like astrology. Although the story contains contents about the doctor's diagnosis and treatment, it is more closely related to politics. Secondly, the interpretation of dreams in *Huangdi's Internal Classic* (*Huang Di Nei Jing*, 《黄帝内经》) is based on the physiology and pathology of the human body, devoid of predictive elements. The differences between these two points should not be interpreted by the shifting from idealism to materialism or from the perspective of changes and development. This is because several centuries separate these two concepts, with the former concerning political prophecies and the latter focusing on the medical interpretation of physiology and pathology.

In the following accounts about treating the prince suffering from corpse-like syncope of the state of Guo, the chapter "Bian Que" in the *Records of the Grand Historian* mentions the diagnostic methods of inspection and palpation, the theory of meridian-related disease causes, and the use of acupuncture, hot medicated compress, and decoction. These diagnostic and treatment methods are totally different from the practice of "old Bian Que" and the famous doctors "He" and "Huan" in the state of Qin, who merely engaged in discussions without providing any treatment. At this time, the self-introduction is clearer: "My name is Qin Yueren and I'm living in the town of Mao in the county of Bohai of the state of Qi". Notably, he did not refer to himself as Bian Que. The specific relationship of inheritance in skills and succession of the name between Bian Que and Qin Yueren still remains unclear.

The chapter "Bian Que" consists of three stories with the final one revolving around the encounter between Bian Que and Duke Huan of Qi.

> When Bian Que passed the state of Qi, Duke Huan of Qi treated him as an important guest. As soon as Bian Que saw Duke Huan in the court, he said, "Your Majesty has a disease in

[20] *Records of the Grand Historian*, Volume 43.

the interstitial space. If not treated, it will get worse." However, Duke Huan denied being sick. After Bian Que left, he remarked to his attendants that doctors often treat those who are not ill in order to showcase their skills.

Five days later, Bian Que came to see Duke Huan again and stated that his disease had progressed to the blood vessels, emphasizing the need for treatment. Duke Huan, once again, denied being sick and displayed a hint of anger.

Another five days passed, Bian Que came to see Duke Huan again, informing him that his disease had now affected his stomach and intestines. He cautioned that without timely treatment, the condition would worsen. Duke Huan remained silent and appeared displeased.

In yet another five days, Bian Que visited Duke Huan once more. However, upon seeing him, Bian Que promptly turned around and left. Curious about this abrupt departure, Duke Huan sent people to inquire. Bian Que explained that a disease in the interstices could be treated with a hot medicated compress, a disease in the blood vessels with needling stones, and a disease in the stomach and intestines with medicinal wine. Yet, the disease had eventually reached the bone marrow, and the god of controlling life was powerless to cure it. Therefore, Bian Que chose not to recommend any treatment. Five days later, Duke Huan fell ill and urgently requested Bian Que's presence at the palace. Unfortunately, Bian Que had already fled, and soon after, Duke Huan was reported to have passed away.

Clearly, this tale serves as a political fable, emphasizing the importance for rulers to understand the principle of preventive measures. The description of disease progression in the account moves from the outermost layer of the body to the innermost: starting with the skin (interstices), then moving to the blood vessels, followed by the stomach and intestines, and finally reaching the bone marrow. Some readers may question the significance of this sequence, as diseases typically develop from the outside to the inside. However, the matter is not as straightforward as it seems. To gain a deeper understanding of this issue, let us compare the approaches of two renowned doctors of the late Eastern Han Dynasty (AD 25–220): Hua Tuo and Zhang Zhongjing. Zhang Zhongjing, known for his work *Treatise on Cold Damage and Miscellaneous Diseases* (*Shang Han Za Bing Lun*, 《伤寒杂病论》), follows the classification system outlined in the chapter "Discussion on Heat" in *Plain Questions* (*Su Wen*, 《素问》). He classifies diseases based on the "six meridians". In contrast, Hua Tuo explains the same types of diseases "from the exterior to the interior". Their perspectives on the location of pathogenic factors and treatment methods differ significantly (See Table 3.1).

I have used the chart for its visual impact, as it helps convey information directly. This is particularly helpful for general readers who may not require in-depth knowledge of the specific details and effects of the theory and treatment methods. Instead, readers should focus on several key points:

(1) The chapter "Discussion on Heat" in the *Plain Questions* and *Treatise on Cold Damage and Miscellaneous Diseases* are both based on the theoretical system of "six meridians". However, in the former, all six disease courses are attributed to excessive heat, advocating sweating and purgation as the means to remove the excessive heat evil. In the latter, symptoms from day four to six manifest as cold deficiency, while the first three days involve a mix of deficiency and excess. As a result, treatment methods vary across different stages.

Table 3.1 Comparison of the similarities and differences among the *Plain Questions*, Zhang Zhongjing, and Hua Tuo on febrile disease

		Chapter of discussion on heat in *Plain Questions*	*Treatise on Cold Damage* and *Miscellaneous Diseases* written by Zhang Zhongjing	Hua Tuo
Day One	Location of disease	Meridian of greater yang (excessive heat)	Meridian of greater yin (deficiency and excess)	Skin
	Treatment methods	Sweating	With sweat: expelling pathogenic factors from muscles; without sweat: inducing perspiration	Ointment rubbing and fire moxibustion
Day Two	Location of disease	Meridian of yang brightness (excessive heat)	Meridian of yang brightness (excessive heat)	Skin
	Treatment methods	Sweating	Hot but not excessive: clearing; Excessive symptoms of Fu-organs: purgation	Needling, releasing the flesh, and inducing perspiration
Day Three	Location of disease	Meridian of lesser yang (excessive heat)	Meridian of lesser yang (disharmony of yin and yang)	Muscle
	Treatment methods	Sweating	Harmonizing yin and yang	Inducing perspiration
Day Four	Location of disease	Meridian of greater yin (excessive heat)	Meridian of Greater yin (cold in a deficiency condition)	Chest
	Treatment methods	Purgation	Warming the middle	Vomiting
Day Five	Location of disease	Meridian of Lesser yin (excessive heat)	Meridian of Lesser yin (cold in a deficiency condition)	Abdomen
	Treatment methods	Purgation	Restoring yang to save from collapse	—
Day Six	Location of disease	Meridian of Reverting yin (excessive heat)	Meridian of Reverting yin (cold in a deficiency condition)	Stomach
	Treatment methods	Purgation	Restoring yang to save from collapse	Purgation

(2) Hua Tuo does not have an independent medical masterpiece that has been handed down, so even the traditional Chinese medicine practitioners do not care about the specific contents of his "medicine". The statements attributed to Hua Tuo in Table 3.1 are sourced from the 29th volume of *Essential Prescriptions Worth A Thousand Gold for Emergencies* (*Qian Jin Yao Fang*, 《千金要方》)

written by Sun Simiao in the Tang Dynasty (AD 618–907). The annotations indicate that these contents are credited to Hua Tuo. However, similar content can also be found in the *Treatise on the Pathogenesis and Manifestations of All Diseases* (*Zhu Bing Yuan Hou Zong Lun*, 《诸病源候论》) by Chao Yuanfang in the Sui Dynasty (AD 581–618) and *Arcane Essentials from the Imperial Library* (*Wai Tai Mi Yao*, 《外台秘要》) written by Wang Tao in the Tang Dynasty. Yet, the narrator remains unnamed. It is evident that during that era, this was an alternative system coexisting with the theory of the "six meridians".

(3) Although the story of Duke Huan of Qi carries a political fable nature, it also contains medical theory content. Several hundred years later, in the period of Hua Tuo and the following Sui and Tang Dynasties, this medical theory remained as significant as other prevailing theoretical systems. Shouldn't this fact prompt a re-examination of the diverse knowledge that independently developed in different regions but coexisted for an extended period?

(4) The chapter "Discussion on Heat" in the *Plain Questions* initially proposed the division of external-contraction febrile disease into six stages. In *Treatise on Cold Damage and Miscellaneous Diseases*, Zhang Zhongjing developed this idea into a classic theory in the late Eastern Han Dynasty. This process did not inherit any aspect from the state of Qi or eastern medicine. But on the other hand, the research and arguments on "Bian Que's theory of pulse" at the end of our book discuss in detail how the contents of medicine in the east were incorporated into *Huangdi's Internal Classic*.

The issue of inheritance and discarding is complex and requires a case-by-case analysis to fully understand its intricacies.

3.2.2 *Yang Qing, Gongsun Guang, and Chunyu Yi*

In *History of the Qin Dynasty Based on A Collection of Documents*, Ma Feibai, the author attempted to emulate the style of the "*Records of the Grand Historian*" but failed to include a biography of renowned doctors. However, he did include six doctors, namely Huan, He, Qu, Li Xi, Xia Wuqie, and Yang Qing, in the chapter titled "Biography XVII".[21] Different from other books on the history of the Qin Dynasty, it put Yangqin in the group of doctors of the state of Qin. In the fourth year of the reign of the emperor Wen (176 BC), Chunyu Yi, the supervisor of the granary in Linzi, was convicted of a crime and was supposed to be sent west to Chang'an. When his daughter Ti Ying learned of this, she went to the capital with her father and submitted a petition to the emperor, stating that she was willing to become an official maid to redeem her father's punishment. As a result, Chunyu Yi was freed from punishment. This incident not only left behind the touching story of "Ti Ying Saving Her Father" but also deeply moved the emperor, who abolished the corporal punishment that year. Later, when Sima Qian was writing the biographies of medical

[21] Ma Feibai, *History of the Qin Dynasty Based on a Collection of Documents*, pp. 348–350.

Fig. 3.4 Ti Ying saving her father. This has become one of the historical themes that are depicted by many people today

practitioners in *Records of the Grand Historian*, besides gathering old stories and combining the accounts of Bian Que and Qin Yueren, he compiled the "Biography of Cang Gong" based on Chunyu Yi's detailed responses to the emperor's inquiries about his teachers and the effects of his treatment. Chunyu Yi mentioned that he first learned medical knowledge, including the theory of transforming yin and yang and the grammar of medical transmission, from Gongsun Guang in Tangli, Zichuan. Through Gongsun Guang, he received further teachings from Yang Qing, Gongsun Guang's brother. Chunyu Yi's responses regarding the effects of his medical practices included detailed descriptions of 25 medical cases, which are regarded as "the earliest medical records in Chinese history" today (Fig. 3.4).

According to the record of *Records of the Grand Historian,* Yang Qing was from Linzi in the state of Qi. In 180 BC, when Yang Qing was over 70 years old, he bestowed upon Chunyu Yi several books on pulse theory derived from Huangdi and Bian Que, methods of diagnosing ailments based on one's complexion, and Chinese materia medica.[22] Based on this information, it is speculated that Yang Qing was born sometime between 260 and 250 BC. Several decades after his birth, Ying Zheng defeated the state of Qi and unified China, which explains why it is appropriate to say that both Yang Qing and his brother Gongsun Guang were physicians in the state of Qi, the state of Qin, and the Han Dynasty.

The specific contents of Gongsun Guang's "transformation between yin and yang and transmission grammar" remain unknown, and it is even difficult to make literal interpretation. Similarly, it is also not easy to accurately punctuate the medical books

[22] *Records of the Grand Historian*, Volume 150, Chapter 45.

passed down by Yang Qing to determine their titles and quantities. However, the focus shifts to where Yang Qing acquired his medical knowledge and books. When posed with this question, Chunyu Yi claimed ignorance of Yang Qing's sources. It is unclear whether he deliberately avoided answering or he really did not know it. Yang Qing, who held the title of "Gongsheng" (entitled to ride in the same carriage as the monarch),[23] was wealthy and good at medicine but was unwilling to treat patients and remained unknown to the nobles in the state of Qi. In my opinion, he could be a court doctor or an official who lived as an ordinary person after the fall of the Qin Dynasty. Despite Chunyu Yi's early interest in medicine and having studied under several teachers in his youth, learning their key teachings and mastering all their medical prescriptions, the treatments he tried often didn't work. It was only after he served Yang Qing diligently for many years and thus won Yang Qing's favor that Yang Qing asked him to completely discard his previous prescriptions and gave him all the secret prescriptions instead. This shows that his knowledge and books were different from those passed down in the state of Qi or among the common people. And the character "jin" (forbidden) itself often specifically refers to the palace, such as "Forbidden City" and "forbidden area". Also, when passing on the books to Chunyu Yi, Yang Qing told him, "Be careful not to let my descendants know that you are learning my prescriptions".[24] This might be out of disdain for "skills" to maintain the dignity of his "Gongcheng" status, or it might be an attempt to conceal the origin of the books.

However, the most crucial problem lies in the lack of satisfactory research and analysis of the contents of the sole medical record dating back to the early Western Han Dynasty (202 BC-AD 8). While the record offers extensive information, there persists a challenge in comprehending the terms, concepts, and doctrines presented due to inadequate knowledge derived from earlier classics like *Huangdi's Internal Classic*. There may be two explanations for this. First, it is plausible that medicine itself underwent significant transformations over such a long period. Second, the medical record may belong to a distinct system of medical knowledge altogether. Although studying medical works unearthed from the Mawangdui archaeological site during the same period as the record offers some connections to *Huangdi's*

[23] In the *Previous Rituals of Officials in the Han Dynasty* (*Han Guan Jiu Yi*, 《汉官旧仪》) written by Wei Hong, it is explained that the eight-level official can sit with the monarch in the carriage. In *A Collection of Essays on the History of Qin and Han Dynasties* (*Qin Han Shi Lun Cong*, 《秦汉史论丛》) written by Gao Min, there is an article discussing the system of awarding the title of nobility. It lists examples of "being rewarded with the right to sit with the monarch" in other books to prove that people always confuse the system of awarding the title of nobility established by Shang Yang and the system of titles with twenty levels in the Qin and Han Dynasties. This essentially indicates that certain titles among the twenty ranks are absent in the *Book of Shang Yang* (*Shang Jun Shu*, 《商君书》), but the Han people have already confused them. Nonetheless, the phase "sitting with the monarch in the carriage" is not included among those titles. It can be found in the *Book of Shang Yang* and the chapter "The List of Official Titles" in the *Book of the Former Han* (*Han Shu*, 《汉书》). In this way, it is impossible to ascertain whether the title bestowed upon Yang Qing originated from the state of Qin or the Han Dynasty.

[24] The quotation can be seen in *Records of the Grand Historian*, Volume 150, Chapter 45.

Internal Classic, it still fails to provide a comprehensive understanding of the record. Consequently, the second explanation appears to be the more plausible one.

3.3 Culture in the South

According to historians, the southern region of ancient China was predominantly occupied by the states of Chu and Yue.[25] A significant event occurred during the 23rd year of King Huai's reign (306 BC) when the state of Yue was conquered by the state of Chu and subsequently transformed into a prefecture under its rule. This effectively united Chu and Yue into a single state. In terms of cultural and medical practices, the eastern and western regions of Chu exhibited distinct characteristics. Initially, the local culture of Chu and Yue was overshadowed by the perception of them as barbarians. However, the field of archaeology has shed new light on this vast area, with numerous Neolithic relics being discovered. These artifacts have proven to be just as exquisite as those found in the Central Plain, challenging previous assumptions (Fig. 3.5).

3.3.1 The Culture of Chu and Qin Spreading Southward

During the reign of King Cheng of Zhou, the region of Chu emerged as an independent state with its capital in Danyang which is present-day Zhijiang in Hubei province. However, in 680 BC, King Wen of Chu relocated the capital from the Three Gorges Valley to Ying, now known as Jiangling in Hubei province. Ying was referred to as Jinan city due to its location south of Mount Ji. Ying served as the capital of the state of Chu for 400 years until Bai Qi, a general from the state of Qin, conquered this city in 278 BC. The book *New Discussion* (*Xin Lun*, 《新论》) written by Huan Tan vividly depicts the thriving vibrancy of Ying, describing crowded streets where people jostled and bumped each other, causing new clothes worn in the morning to be torn by evening due to the constant rubbing against others.[26]

During the early Spring and Autumn period, powerful states like Qin, Jin, and Chu would often transform those defeated states into their own counties. Therefore, with the continuous expansion of the state of Chu, an increasing number of counties were established, leading to the saying that Chu had "nine counties" (symbolic for a significant number).[27]

---

[25] Huang Liuzhu, *The Spread of the Culture of Qin to the South* (*Qin Wen Hua De Nan Bo*, 《秦文化的南播》), recorded in the sixth volume of *A Collection of Essays on the History of Qin and Han Dynasties* compiled by the Chinese Research Society of Qin-Han History, pp. 253–266.

[26] Li Fang (the Song Dynasty (AD 960–1279)) et. al., *Readings of the Taiping Era* (*Tai Ping Yu Lan*, 《太平御览》), Volume 776.

[27] Yang Kuan, *The History of the Warring States Period*, pp. 111.

Fig. 3.5 Relics of Zhanghua Palace in the state of Chu. It was found in the town of Longwan in Qianjiang, Hubei province. Plate tiles, semicircle-shaped tiles and eaves titles unearthed from here bear a striking resemblance to similar objects found in the city of Jinan, also in Hubei province. It is speculated that these findings could potentially be liked to the famous Zhanghua Palace according to the literature and geographical evidence

When discussing the development of southern culture, historians tend to focus more on the southward spread of Qin culture during the later Warring States period, which accompanied their victories and annexation. In the early period, most of the arguments emphasized the importance of "taking control of the states of Ba and Shu" on the basis of historical documents:

> In 316 BC, under the leadership of Zhang Yi and Sima Cuo, the Qin state successfully conquered the states of Ba and Shu. This conquest brought the area beyond the Qinling Mountains in the southwest of the Qin state, namely the area of the Sichuan Basin today, under the control of Qin. This achievement held immense significance for the development of the Qin state. The strategic positions of Ba and Shu, located on the flank of Qin, made them crucial territories. Moreover, these states possessed fertile lands abundant in resources, further bolstering their importance. This acquisition played a pivotal role in creating favorable material conditions for the ultimate unification of China.[28]

In addition to agriculture, historians especially pay attention to the matter that the people of the Qin state "got rich" by relying on their own skills and making full use of local mining and metallurgical resources. The Zhuo family, the Cheng family, and the Wan family, who were mentioned when talking about the "skilled craftsmen" of the Qin state before, all belong to this kind. For example, when the Zhuo family was relocated to Jiameng (an ancient county established by the Qin state, located in today's

[28] Lin Jianming, *Development History of the State of Qin* (*Qin Guo Fa Zhan Shi*, 《秦国发展史》), pp. 106–107.

Guangyuan, Sichuan), they "started iron-smelting on the iron-rich mountains" and "their wealth outshone all in Yunnan and Sichuan". "They were so rich that they owned a thousand servants, and enjoyed the pleasure of hunting in their fields and ponds, comparable to a monarch".[29]

Furthermore, after Ying Zheng (known as Qin Shi Huang) ascended to the throne, he orchestrated the relocation of two prominent families to the region of Shu.[30] The large-scale migration of powerful families inevitably facilitated the dissemination of knowledge and culture.

In recent years, with the increase of archaeological excavations, historians have broadened their perspectives beyond relying solely on historical records. For example, the article *The Spread of the Culture of Qin to the South* by Huang Liuzhu[31] argues that some elements of Qin culture exist in Chu culture by comparing the laws of the Qin and Chu states and the *Ri Shu* (《日书》). Archaeologists have also discovered evidence in excavated tombs and have reinforced these findings.

> In 223 BC, the state of Qin overpowered and destroyed the state of Chu, conquering its entire eastern section (including the present-day eastern part of Hubei Province). In 278 BC, the Qin general Bai Qi captured Ying city, allowing the state of Qin to extend its dominance over the western part of the state of Chu (including the western part of Hubei province now). Therefore, the eastern and western parts of Hubei province was respectively governed by the state of Qin and Chu during the 50 years from Bai Qi taking control of Ying to the state of Qin's triumph over the state of Chu. Archaeologists point out that the tombs in eastern Hubei show the features of the culture of Chu while the tombs in western Hubei display minimal influence from the Chu culture.[32]

However, in my opinion, these views still stem from an observation of issues rooted in court politics. When it comes to practical skills and knowledge, the richness of local culture and the interactions with the Central Plain are all worthy of concern.

3.3.2 *States of Chu, Wu, and Yue*

Wu and Yue were two ancient states located in the lower reaches of the Yangtze River. Toward the end of the Spring and Autumn period, Gou Jian, the king of Yue, successfully conquered the state of Wu and established dominance over the region. During the Warring States period, in 306 BC (the 23rd year of King Huai of Chu), the state of Chu destroyed the state of Yue and set up a prefecture there.

---

[29] *Records of the Grand Historian*, Volume 129, Chapter 69.

[30] Chapter 6 in the 6th volume of *Records of the Grand Historian* states that after Miu Du was killed, more than 4000 households of his family were deprived of the title of noble and moved to the region of Shu. Chapter 25 in the 85th Volume records that Ying Zheng ordered Lyu Buwei and his families to move to the region of Shu.

[31] It is recorded in the 6th volume of *A Collection of Essays on the History of the Qin and Han Dynasties* compiled by the Chinese Research Society of Qin-Han History, pp. 253–266.

[32] Tan Weisi, Wan Quanwen, "An Archaeological Discovery of Half a Century in Hubei Province". *The Unearthed Cultural Relics of Hubei Province* edited by Tan Weisi, pp. 1–10.

Because of its vast territory, some literati thought that King Huai of Chu was the only one worthy of the title of "hegemon" in the Warring States period. However, 80 years later, the state of Qin defeated the state of Chu and set up Chu prefecture. In the following year, the state of Qin captured the former territory of the state of Yue and set up the Kuaiji Prefecture. The regions of Chu and Yue were merged into the territory of the state of Qin. The eastern part of the state of Chu, including the original land of the states of Wu and Yue, is often called "the south of the Yangtze River" and sometimes it specially refers to the southern coast of the middle and lower reaches of the Yangtze River.

As the saying goes, "This region was characterized by its wet and humid climate, and there was a popular belief that the men here had shorter lifespans".[33] Wang Zijin proposed dividing the development of the south of the Yangtze River into several stages. Initially, the area served as a destination for sending criminals, much like the Ba and Shu regions. Later, during the early Han Dynasty, the south made considerable progress but still lagged behind economically and culturally when compared to the middle and lower reaches of the Yellow River. In the Eastern Han Dynasty, significant changes occurred. Records of south-to-north grain diversion began to appear in 107 AD (the first year of the Yongchu reign), while the traditional economic zones of the Guanzhong Plain and the North China Plain witnessed a decline. This transformation had a profound impact on Chinese history. Meteorologists such as Chu Kochen, studying the history of climate change in ancient China, suggested that the climate in the south of the Yangtze River was more conducive to human habitation. As a result, it led to large-scale immigration to the south of the Yangtze River during the Han Dynasty. Lingling county experienced the most significant growth, with a 906.5% increase in households and a 618.6% population increase from AD 2 to AD 140. Yuzhang county followed suit, with figures of 502.6% and 374.21% for households and population growth, respectively.[34]

Later, Wang Zijin substantiated his claims about the development of the culture in the south of the Yangtze River by referencing the *Book of the Later Han* (*Hou Han Shu*, 《后汉书》) and *Records of the Three Kingdoms* (*San Guo Zhi*, 《三国志》). These historical texts provided evidence that immigration and economic growth played pivotal roles in this region's cultural advancement. During the early Eastern Han Dynasty when the country had just been pacified and roads were not yet accessible, those who had fled to the south of the Yangtze River to avoid chaos found the poor road conditions deterring them from returning to the Central Plain. As a result, Kuaiji was known to have quite a number of talented people.[35] In the late Eastern Han Dynasty, after reading Yu Fan's *Annotation of Book of Changes* (*Yi Zhu*, 《易注》), Kong Rong remarked that "the outstanding figures in the southeast are not just the bamboo

[33] *Records of the Grand Historian*, Volume 129, Chapter 69.

[34] Wang Zijin, The Climate Changes in the Qin and Han Dynasties and the Economic and Cultural Progress in the South of the Yangtze River, recorded in the sixth volume of *A Collection of Essays on the History of Qin and Han Dynasties* compiled by the Chinese Research Society of Qin-Han History, pp. 14–31.

[35] *History of the Later Han*, Volume 160.

arrows of Kuaiji".[36] Moreover, when Sun Ce met with scholars from the Central Plain and heard them say "People in the east are talented, but it's a pity that their learning is not extensive enough and they may fall short in discussions", he wanted Yu Fan to go and meet the court officials in the Central Plain "to refute those arrogant speakers from the Central Plain".[37] These examples demonstrate the confidence and intellectual prowess of the talents in this region, which should not be underestimated.

However, solely examining the migration of scholars from the Central Plains to the regions south of the Yangtze River due to chaos and never returning, or from the perspective of the theory of immigration and economic development, overlooks the evolution of indigenous culture. The "Hemudu Site" discovered in Yuyao, Zhejiang Province in 1973 covers a total area of 50,000 square meters and has four superimposed cultural layers. The lowest layer dates back 7,000 years. A large number of items were unearthed here, including production and daily necessities made of bones, clay, and wood, remains of rice, components of stilt-style architecture, and especially delicate jade wares whose intricate crafting techniques still perplex researchers. These objects provide a comprehensive glimpse into the prosperity of the local culture at that time, leaving a lasting impression on the world. Another significant find that garnered global attention is the bronzes excavated from Shang Dynasty (1600–1100 BC) tombs in Dayangzhou Township, Xingan County, Jiangxi Province. This discovery ranks alongside the Unearthed Ruins of Yin in Anyang city, Henan province and the site of Sanxingdui in Guanghan city, Sichuan province. However, it also raises thought-provoking questions. Some esteemed archaeologists, starting from the Central Plains regime's demand for copper resources, suggest that the nature of these bronzes is "the Yin culture crossing the Yangtze River, bringing a new spring to the south of the river". Yet, most experts argue that the cultural relics unearthed from these tombs differ significantly from those found in the Central Plains. They infer that there was an indigenous bronze culture in the area of Ganjiang River developing in parallel with the Central Plain culture, indicating the existence of another regime coexisting with the Shang Dynasty. The owner of the tomb in Xingan is considered the supreme ruler of the indigenous people.[38] Although the true cultural form of the "region of Nanman" remains shrouded in mystery, it is evident that a local culture existed prior to the introduction of Central Plain culture. However, there is a considerable temporal gap between these archaeological finds and the gradual establishment of the Traditional Chinese Medicine (TCM) system. Therefore, these archaeological results can only serve as supporting evidence for the existence and development of local culture in the south of the Yangtze River.

---

[36] The 12th chapter "The state of Wu" in *Records of the Three Kingdoms*, Volume 12.

[37] The 12th chapter "The state of Wu" in *Records of the Three Kingdoms*, Volume 12.

[38] *National Museum of China and Department of Culture of Jiangxi Province, Jiangnan in the Shang Dynasty: Collection of Cultural Relics Unearthed in Dayangzhou Town, Xingan County, Jiangxi Province* (*Shang Dai Jiang Nan—Jiang Xi Xin Gan Da Yang Zhou Chu Tu Wen Wu Ji Cui*, 《商代江南——江西新干大洋洲出土文物辑粹》), Beijing: China Social Sciences Press, 2006, pp. 5, 15.

3.4 Medicine in the South

3.4.1 From Jade Turtle to Fuyang Bamboo Slips of the Han Dynasty

In 1987, during an excavation of a Neolithic tomb in Lingjiatan, Changgang Township, Hanshan County, Anhui Province, archaeologists discovered a jade document placed on the chest of the deceased, sandwiched between a jade tortoise back and a jade plastron. This tomb, dating back 5,000 years, revealed a wealth of relics and exemplified a high level of cultural development during its early stages.[39]

By now, many scholars have conducted intensive research on the jade document, speculating that its shape and depictions may be related to ancient astronomy, calendric systems, and Yi-ology. Unfortunately, without textual evidence, all interpretations based on ideology and culture remain speculative and can easily be subject to "over-interpretation". However, as a Neolithic relic, it is stunning and evokes curiosity about the origins of significant ideological elements in Chinese traditional culture. Looking further northeast from the excavation site of the Hanshan jade document which is located northeast of Chaohu Lake, we find that the shore of Huaihe River belonged to the state of Xu in the Spring and Autumn period. King Yan of the state of Xu was particularly fond of collecting natural goods, as evidenced by the records which described his eccentric nature: "King Xu Yan liked strange things. He ordered people to dive into the deep water to find peculiar fishes and to search unusual animals in remote mountains, which he then proudly displayed in his courtyard".[40] Based on this, Xu Xusheng, an ancient history researcher, remarked that:

> The records about King Yan in *Shi Zi* (《尸子》) were not intended as praise, but rather to highlight his eccentricities. However, the pursuit of knowledge often stems from curiosity. Just as King Alexander's collection of animals and plants laid the foundation of Western biology through Aristotle, it is intriguing to consider that had the state of Xu prioritized defending itself and preserving its collection, the foundation of Chinese biology might have been established two or three hundred years earlier than that of Greece.[41]

Mr. Xu may have overestimated the collection of King Xu Yan, but he recognized the correlation between curiosity and science. He did not adhere to the theory of "social development needs". Instead, he astutely discerned the clues within historical documents that emphasized the role of politics in shaping people's understanding of the natural world during its nascent stage. This perspective holds valuable lessons. As he pointed out, "The scarcity of historical materials from the Xu state does not

---

[39] Anhui Institute of Archaeology. *The Brief Report on A Neolithic Tomb Unearthed in Lingjiatan, Hanshan County, Anhui Province* (*An Hui Han Shan Lin Jia Tan Xin Shi Qi Shi Dai Mu Di Fa Jue Jian Bao*, 《安徽含山凌家滩新石器时代墓地发掘简报》), Cultural Relics, 1989, (4): pp. 1–9.

[40] Written by Li Kai in the Qing Dynasty, "King Xuyan" of the chapter "Biographies for Treacherous Ministers of Zhou State" in *Ancient History* (*Shang Shi*, 《尚史》), Volume 28. It quoted the content from *Shi Zi*.

[41] Xu Xusheng, *The Legendary Era of Ancient Chinese History* (*Zhong Guo Gu Shi De Chuan Shuo Shi Dai*, 《中国古史的传说时代》), Beijing: Cultural Relics Press, the new edition in 1985, p. 188.

stem from a lack of cultural sophistication, but rather from its deviation from the mainstream culture of the Central Plains. Consequently, historical accounts based on the orthodox culture offer only cursory insight into their accomplishments".[42]

Undoubtedly, what is referred to as "pharmaceutical knowledge" gradually accrued through the process of comprehending and exploring the uses of natural resources. A prominent example lies in the *Classic of Mountains and Seas* (*Shan Hai Jing*, 《山海经》), which chronicles over 100 species of animals, plants, and minerals, many of which possess medicinal properties still recognized today. The book's actual age remains undocumented; however, it is generally believed to have been completed between the end of the Spring and Autumn period and the early Han Dynasty. It is attributed to multiple authors and primarily covers the state of Chu, the Bashu region, and the state of Qi. This is the prevalent conclusion derived from five decades of research.[43] The contents of the book suggest that Chu culture occupies a central position. Interestingly, in Fuyang, Anhui province, west of the territory of King Xu Yan, recently unearthed documents exhibit striking similarities to the descriptions found in the *Classic of Mountains and Seas.*

The bamboo discovered in 1977 from the tomb of Xiahou Zao, Lord of Ruyin, during the Han Dynasty excavation in Fuyang, Anhui province initially bore the title *Miscellaneous Prescriptions* (*Za Fang*, 《杂方》). However, based on the content found in the first set of slips, researchers later reclassified them as *Everything* (*Wan Wu*, 《万物》). These slips primarily focus on the laws, nature, and applications of various objects. Within their pages, approximately 30 substances can be identified as potential "medicines" for treating diverse ailments. These substances are further categorized into groups such as jade, grass, wood, animals (beasts, poultry, insects, and fish), fruits, rice cereals, and vegetables. Some of the passages delve into uses that do not align with our modern knowledge system, including claims like "rabbit hair can be crafted into coats", "spiders enable swift movement", "a large mirror suspended above a basin helps us perceive our surroundings", and "the body becomes as light as a cloud over the mountain". However, it is these very descriptions that offer insight into the evolution of pharmacological knowledge. Believed to have been compiled in the early Western Han Dynasty, its principles closely resemble those found in the *Classic of Mountains and Seas*, positioning it as a foundational work in the field of herbal medicine.[44]

---

[42] Xu Xusheng, *The Legendary Era of Ancient Chinese History*, Beijing: Cultural Relics Press, the new edition in 1985, p195.

[43] Yuan Ke, *The Forword: Annotation of the Classic of Mountains and Rivers* (*Shan Hai Jing Quan Yi*, 《山海经全译》), Guiyang: Guizhou People's Publishing House, 1991, p. 1.

[44] Hu Pingsheng, Han Ziqiang, "A Few Words On Everything", *Cultural Relics*, 1988, (4): pp. 48–53.

3.4.2 *Vessel in Sichuan Province and Hunan Province*

In the conclusion of the chapter "Bian Que", Sima Qian asserts that the study of vessels today follows the principles established by Bian Que. However, it remains unclear what exactly these vessels refer to. Are they related to pulse diagnosis, or do they involve the meridians associated with qi-blood circulation in traditional Chinese medicine? These two possibilities share some similarities and dependencies, but they also have significant differences. Given the lack of thorough research and newly unearthed relics in the past, one can only speculate that the vessels mentioned by Bian Que serve as the foundation for pulse diagnosis and the meridian-collateral theory in the *Huangdi's Internal Classic*. However, after the discovery of primordial medical texts and cultural artifacts pertaining to pulse studies, those with a sharp eye can see the regional variations in the early understanding of vessels (Fig. 3.6).

This difference in the concept of meridians arises due to the depictions of meridian vessels found primarily in the medical texts discovered in the Mawangdui Han Tomb and the wooden sculpture of human figure discovered in Mianyang, Sichuan Province, both of which were situated within the Western Chu and Bashu regions.

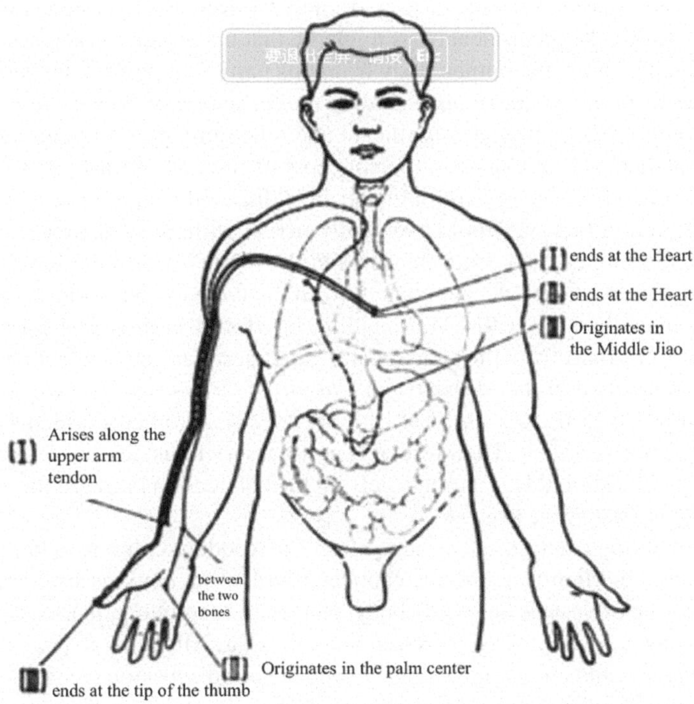

Fig. 3.6 The circulation of meridian vessels. Hand greater yin meridian is shown in the picture. (I) Cauterization Canon of the Eleven Vessels of the Foot and Forearm (II) Moxibustion Classic on the Eleven Yin and Yang Channels (III) Miraculous Pivot· Meridian Vessel

The two meridian-related books found in the Mawangdui Han Tomb describe eleven vessels that originate from the hands or feet, each with distinct lengths and endpoints. Therefore, they cannot form the so-called "endless channel for qi-blood circulation". This meridian vessel theory, which is also applicable to disease diagnosis and can also be called a form of "pulse-diagnosis method", contrasts with the vessel theory presented by Bian Que, which focuses on the theory of qi-blood circulation without specifying the meridian vessels. Additionally, variances exist between the lines depicted on the wooden sculpture of human figure discovered in Mianyang, Sichuan Province, and the meridian descriptions found in the medical texts unearthed from the Mawangdui Han Tomb.

(1) One end of each vessel depicted on the wooden sculpture of human figure is primarily located near the sense organs such as the eyes, ears, mouth, and nose. Even the vessels at the back of the body also reach here over the top of the head. Due to the absence of accompanying labels or text, it is challenging to determine which end represents the "start" and which end represents the "end" of the vessels. If we consider the end near the sense organs as the "end", a plausible theoretical explanation could be that the qi of the human body travels through the vessels to reach the eyes, ears, mouth, and nose, thus enabling the sensory functions of vision, hearing, taste, and smell. Conversely, if we consider this end as the "start", the theoretical basis might be that natural qi, the source of energy and vitality, enters the human body from this particular point. It is crucial not to ignore this point, as questions regarding the formation of sensory functions and the origin of life energy are significant and inherently present in all knowledge systems that seek to elucidate the concepts of life and disease. If we examine the source of life "energy" through the most direct life experience—such as the understanding that we would eventually perish without sustenance—concepts like "gathering essence from the sun and the moon" would not have emerged. A comparison of the relevant content from the two versions of meridian vessel books unearthed from the Mawangdui Han Tomb is provided for reference (Table 3.2). Apart from the hand greater yang meridian starting from behind the ear, as described in the *Moxibustion Classic of the Eleven Yin-yang Meridians* (*Yin Yang Shi Yi Mai Jiu Jing*, 《阳十一脉灸经》), all other meridians originate from the hands or feet. Therefore, it is unnecessary to list all the starting points, and attention should be focused solely on the connection between the end of the vessels and the sense organs.

(2) Meticulous distribution and arrangement: The wooden sculpture of human figure discovered in Mianyang initially captivates with its meticulous arrangement and distribution of vessels across its body. However, regrettably, it appears that this aspect has gone unnoticed by researchers thus far. Allow me to provide further explanations and encourage readers to engage in a discussion regarding whether this represents one of the differences between the two meridian vessel books discovered in Mawangdui and *Miraculous Pivot* (*Ling Shu*, 《灵枢》).

First, there is one meridian on each of the middle fingertips of the "wooden figure's" two hands. These two meridians converge at the top of the head along the midline of the upper limbs. In other words, if the "wooden figure" could stretch its hands horizontally, a connecting line running through the middle fingertips of both hands and the center of the top of the head would be seen from above. Regardless of what kind of "meridians" they are, what I want to point out first is their characteristic of running across the left and right ends and being located on the midlines of the upper limbs and the head.

Second, still looking downward along the sagittal plane of the body, if the body or lower limbs are cross-sectioned, six meridians will be seen. There are two in the front, two in the back, and one on each side, regularly distributed on the body and lower limbs (Fig. 3.7).

Table 3.2 The end of meridians and their connections with sense organs recorded in the two meridian vessel books of Mawangdui

Names of meridians	Cauterization Canon of the Eleven Vessels of the Foot and Forearm		Moxibustion Classic on the Eleven Yin and Yang Channels	
	Endpoints	elated organs	Endpoints (or start points)	Related organs
Foot greater yang meridian	Head	Ear, eye, nose	Head	Eye
Foot lesser yang meridian	Head	Ear, eye	? (Missing words)	? (Missing words)
Foot yang brightness meridian	Head	Mouth, nose	?(Missing words)	? (Missing words)
Foot lesser yin meridian	Head	Tongue	Head	Tongue
Foot greater yin meridian	Thigh	—	Heart	—
Foot reverting yin meridian	Thigh, chopped meat	—	Head	Eye
Hand greater yang meridian (shoulder meridian)	Shoulder	—	Head (start point)	Ear
Hand lesser yang meridian (ear meridian)	Head	Eye	Head	Ear
Hand yang brightness meridian (teeth meridian)	Head	Mouth	Head	Nose
Hand lesser yin meridian	Hypochondrium	—	Arm	—
Hand great yin meridian	Heart	—	Arm	—

Fig. 3.7 The distribution of meridian vessels on the truncus and lower limbs of the wooden sculpture of human figure

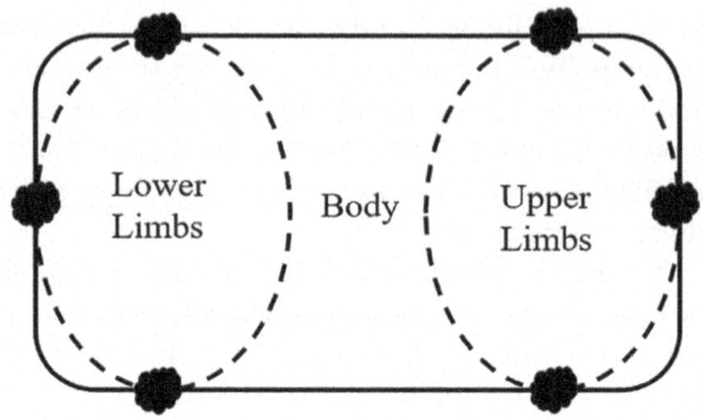

For readers who are interested, you may buy a replica of this "wooden figure" (available at Mianyang Museum) and have a try following this train of thought. You're bound to make more discoveries. Here I've just given two examples to show that it's different from the "meridians" in Hunan and those of Bian Que. As for what theoretical concepts are hidden behind this "standardized" arrangement, it remains unknown due to the lack of any written explanations.

3.4.3 Theory of "Five Elements" in the History of the Lost Yue Nation

> Yan Di initially governed China and later passed on his position to Huang Di. Huang Di, in turn, ruled over China by adhering to the laws of nature. Thus, Shaohao was in charge of the Western region and was assisted by Chi You, being responsible for the element of Metal; Xuan Ming was in charge of the Northern region and was assisted by Baibian, being responsible for the element of Water; Taihao was in charge of the Eastern region and was assisted by Yuan He, being responsible for the element of Wood; Zhu Rong was in charge of the Southern region and was assisted by Pu Cheng, being responsible for the element of Fire; Hou Tu was in charge of the Central region and was assisted by Hou Ji, being responsible for the element of Earth. With the governance of these five regions in place, a set of rules was established. Therefore, even if these officials were to exchange their positions, they could still assist in maintaining the normal order of all things. Your Majesty, if you carefully consider and adopt my advice, you can either achieve supreme kingship or become a powerful overlord.[45]

During the final period of the Spring and Autumn era, a war erupted between the Yue State and the Wu State. After suffering defeat, King Goujian of Yue surrendered to the Wu State, hoping to save both his life and his state. After his eventual release, King Goujian devoted himself to strengthening the Yue State and ultimately succeeded in destroying the state of Wu. The passage cited above is from the words of Ji Ni, a counselor of Goujian, who presented strategies for strengthening the country to him

[45] The chapter "Ji Ni's Internal Classic" in *The History of the Lost Yue Nation* (*Yue Jue Shu*, 《越绝书》) written by Yuan Kang of the Han Dynasty, Volume 4.

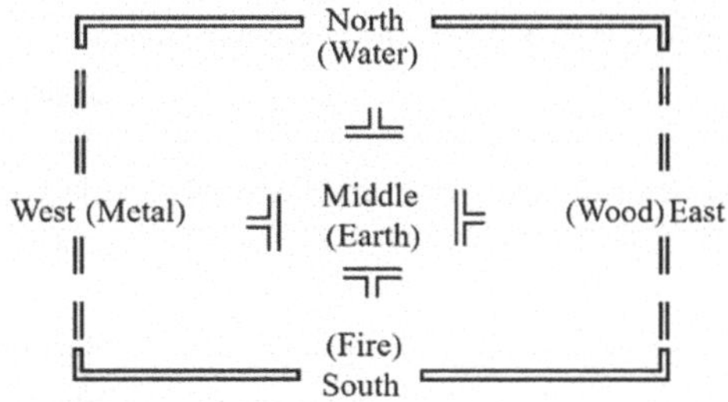

Fig. 3.8 The match of "five elements" and "five directions"

after Goujian's return to the Yue State, as recorded in *History of the Lost Yue Nation*, a historical book of the Yue State that includes this historical event (Fig. 3.8).

These excerpts are quoted because they contain significant content frequently found in the transmitted version of *Huangdi's Internal Classic*, namely the correspondence between the "Five Elements" and "Five Directions". The reason why it is said to be important is that when it comes to "directions", in people's general awareness, they still immediately think of the "east, south, west and north" (Four Directions); the same is true in traditional Chinese culture. The internal determining factor lies in that the "Yin-Yang School" emphasizes the "order of the four seasons", and its spring, summer, autumn, and winter (Four Seasons) can only be matched with the east, south, west, and north (Four Directions) Therefore, please never underestimate this slight change from the "Four Directions" to the "Five Directions", because it represents that when the "Five Elements" became the "guiding principles" in the theoretical system of natural philosophy at that time, in order to solve the problem of their coordination with important categories such as "directions" and "seasons", the concepts of the "Five Directions" and the "Five Seasons" were created.

By comparing it with *Guan Zi*,[46] which is regarded as the "academic journal of the Jixia Academy", it's not difficult to find the different characteristics in terms of "transformation" and "correspondence" respectively. Firstly, both Guanzi and Yuejueshu mention "up and down" and "the four directions"—that is, the spatial concept of what is commonly called "the six directions" and "the up, down, four directions are called the universe". However, the ways they handle these concepts are different. In *Guan Zi*, there is no attempt to transform them to match the "Five Elements" (as shown in Table 3.3). While the *History of the Lost Yue Nation*, although also referring to "the upper heaven" and "the lower earth", achieves the transformation from "the four directions" to "the five directions" by placing them on a higher level, namely the so-called "The Yellow Emperor served Heaven above and governed the Earth below".

In terms of the concept of time, *Guan Zi* mentions that Chi You, one of the six ministers of Huang Di, was in charge of the "Heavenly Way"—"Time". Then, by dividing the 360-day year into five periods, with wood, fire, earth, metal, and water

46 Han Xing, *A Combination of Confucianism and Law: Study of Politics and Culture of the Qin and Han Dynasties*, p. 26. It mentions that this statement was quoted from *The History of Chinese Philosophy* (new version) (*Zhong Guo Zhe Xue Shi Xin Bian*, 《中国哲学史新编》) by Feng Youlan and quotes a similar statement by Gu Xiegang.

Table 3.3 "Six-directions" system in *Guan Zi*

	Governor	Direction	The governed
Huang Di governed the world with the assistance of six ministers	Chi You	Natural laws	Seasons
	Da Chang	Favorable geographical position	Cereals
	She Long	East	Spring
	Zhu Rong	South	Summer
	Da Feng	West	Autumn
	Hou Tu	North	Winter

each governing 72 days, it realizes the correspondence between the "Five Elements and Heavenly Time", that is, the so-called "Using the Five Elements to regulate Heavenly Time".[47]

It is important to note that the concept of the "repeated circulation of the properties of five elements" proposed by Zou Yan of Jixia Academy is characterized by the cyclic nature of wood, fire, earth, metal, and water properties, which should be represented through seasons rather than directions. This explains why *Guan Zi* retained the notions of four and six directions in spatial coordinates but changed the four seasons to five time periods in the temporal dimension. Upon closer examination, we can identify another mode of transformation. The 18 days at the end of the four seasons are attributed to "earth", resulting in a total of 72 days. Consequently, there are 72 days in each of the "five seasons". This facilitates the correspondence between the four seasons and the five elements, but it fails to fully manifest the primary idea of the repeated circulation of the properties of the five elements. In other words, these two modes of transformation illustrate the different motivations behind them: one, driven by "political" consciousness, establishes a theoretical foundation for explaining how "everything must be based on natural law", while the other focuses solely on addressing the compatibility issue between the "four seasons" and the "five elements" from an academic perspective. Through this example and various similar instances, we can gain insights into the "integrated development of cultural diversity" during that era. This not only satisfies the need for research in medical history but also aids researchers seeking a deeper understanding of the core academic ideology of the Qin and Han Dynasties. However, in order to avoid digressing too far when delving into this topic, I must pause here and return to the main theme.

The History of the Lost Yue Nation was first recorded in the *History of the Sui Dynasty* (*Sui Shu*, 《隋书》) and the old and new versions of the *History of Tang Dynasty* (*Tang Shu*, 《唐书》). In these records, it was all stated that the author was Zigong. In the *Chongwen Comprehensive Bibliography* (*Chong Wen Zong Mu*, 《崇文总目》) of the Song Dynasty, "Zixu" was added under the author's name. However, Chen Zhengsun, also a scholar of the Song Dynasty, clearly disagreed with these

[47] *Guan Zi* noted by Fang Xuanling in the Tang Dynasty, Volume 14, Chapter 41, Wuxing.

claims. In his *Zhizhai Content Explanation* (*Zhi Zhai Shu Lu Jie Ti*, 《直斋书录解题》), he stated, "There is no author's name recorded. The legend that it was written by Zigong is incorrect", Nevertheless, we cannot deny the documentary value of this book in recording the historical facts of the State of Yue. As contemporary scholars mentioned in the "Preface to the Proof-read Edition" of this book, "Professor Bao Lüping, the director of the Department of Foreign Languages at Hangzhou University, translated the *History of the Lost Yue Nation* into English as *Lost History of Yue*. I think such a translation truly grasps the essence of the Chinese title". Thus, it is obvious that as a historical document, this book holds significant value.[48]

This leads us to the following speculations: Is it possible to find some kind of connection between the cultural system of the Yue State in the pre-Qin period and the chapters in *Huangdi's Internal Classic* that repeatedly discuss the correspondence between the "five directions" (east, south, west, north, and center) and the "five elements"? Or could this be evidence to support the view that *Huangdi's Internal Classic* was not written by one person at one time, but gradually developed from the Spring and Autumn period to the Warring States period? However, the matter is not as straightforward as it seems.

There are three different opinions on the authorship of the *History of the Lost Yue Nation* Nevertheless, some scholars from the Ming and Qing Dynasties discovered an implicit reference to the author's name at the end of the book (Volume 15):

> The text records and presents such claims. There was a person. Taking "qu" (去) as the base, when combined with "yi" (衣), it forms the surname. For the given name, it's composed of "mi" (米) covered by "geng" (庚). It's said that Yu came on an eastern expedition and died and was buried in this territory. The author didn't directly state his identity but used such analogies to make it clear. He expressed his insights and also showed his naivete, briefly classifying events to wait and tell future generations. After the text was composed and the diction was finalized, it was the work of a local sage. This sage had a surname formed by "kou" (口) followed by "tian" (天), and had the same given name as Qu Yuan, the minister of the Chu State.

The combination of "qu" (去) and "yi" (衣) forms the surname "yuan" (袁), and the combination of "mi" (米) covered by "geng" (庚) creates the given name "kang" (康). This "Kang Yuan" is the one who "recorded and presented those claims". As for the one who "composed the text and finalized the diction", it was Wu Ping, whose surname is formed by combining "kou" (口) and "tian" (天), and who shares the same given name as Qu Yuan, the minister of the Chu State. Their hometowns are also clearly indicated in it, which is Kuaiji (now Shaoxing, Zhejiang Province), the location of the Mausoleum of Dayu. Based on the statement in Volume 2 of the book that "from the time when Goujian moved the capital to Langya to the 28th year of the Jianwu era (52 AD), a total of 567 years had passed", it can be inferred that Yuan and Wu lived in the early Eastern Han Dynasty. In this way, when looking at the issue of the correspondence between the "five elements" and "directions" in the

---

[48] *The History of the Lost Yue Nation* (the version revised by Yue Zumou), Shanghai, Shanghai Chinese Classics Publishing House, 1985, "the preface of the revised edition of Lost History of Yue", p.13.

History of the Lost Yue Nation, we should realize that it represents the prevailing academic and ideological framework of the early Eastern Han Dynasty.

Furthermore, by comparing the records of Gou Jian seeking advice on strengthening the state and seeking revenge in the chapter "Ji Ni's Internal Classic" in *The History of the Lost Yue Nation* with related records in the *History of Wu and Yue in the Spring and Autumn Period* (*Wu Yue Chun Qiu*, 《吴越春秋》)[49] written in the Eastern Han Dynasty, we discover that both texts elaborate on the methods of agricultural development based on the changing seasons and yin-yang principles. However, the crucial difference is that there is no content associated with the relationship between "five elements and directions" in *History of Wu and Yue in the Spring and Autumn Period*.

[49] *History of Wu and Yue in the Spring and Autumn Period* written by Zhao Yu of the Han Dynasty, Volume 5.

Chapter 4
Innovation and Development: Diverse Medical Practices Throughout the Han Dynasty

Examining the extent to which subsequent generations could draw upon medical advancements from the Warring States Period (475–221 BC) to the Qin Dynasty (221–207 BC), it becomes evident that the medical developments during the Han Dynasty (202 BC–AD 220) were truly innovative. Furthermore, it is crucial to delve deeper into this "innovation" and explore its specific manifestations within different historical periods.

4.1 "Quiet Period" of the Medical Development at the Beginning of the Western Han Dynasty

During his battle against Qing Bu, Emperor Gaozu of the Han Dynasty (Liu Bang) was wounded by flying arrows and fell ill on his way back. As his illness became more and more severe, Empress Lyu invited an excellent doctor to treat him. Upon the doctor's arrival, Liu Bang inquired about his illness, to which the doctor assured him of a potential cure. However, Liu Bang berated the doctor, stating, "I, as a commoner, personally wielded a three-foot sword in battles, and now I have become the ruler of the empire. Isn't this destiny? God rules life. Even Bian Que (a famous doctor in the Spring and Autumn and the Warring States Period) (770 – 207 BC) won't change my fate". Therefore, the doctor was not allowed to treat him.[1]

It is widely known that medical services were once a privilege exclusive to the elites before becoming accessible to all. While Liu Bang's refusal of treatment could be attributed to his arrogance, it also reflected the society's limited reliance on medicine during that time. As the supreme ruler, he held a dismissive view of medical practices, which was likely influenced by both the competence of the physicians and

[1] *Book of the Former Han* (*Han Shu*,《汉书》), Volume 8.

Y. Liao, *Reconstructing the Image of Medicine in the Qin and Han Dynasties*, Understanding China, https://doi.org/10.1007/978-981-95-6801-7_4

their overall effectiveness. Thus, it is not surprising that a proverb during the Han Dynasty cautioned against seeking treatment from unqualified practitioners.[2]

When Liu Xiang and his son, Liu Xin, compiled the chapter "Catalog of Literary Works" in the *Book of the Former Han* (*Han Shu*,《汉书》) at the end of the Western Han Dynasty (202 BC–AD 8), they listed a series of books on medical classics, classical prescriptions, sexual healthcare, and health preservation in the chapter of "Category of Prescriptions and Techniques" (an academic field focusing on medicine and health preservation). Then they gave the following historical summary:

> Those in the field of "prescriptions and techniques" are all equipped with the means to preserve life and were once in charge of relevant affairs as court officials. In ancient times, there were Qi Bo and Yu Fu; in the middle era, there were Bian Que and Yi He. They could discuss diseases in relation to the state and infer political situations from diagnoses. After the Han Dynasty was founded, there was Cang Gong. Now, their techniques have become obscure. Therefore, we discuss these books to arrange and present the knowledge of prescriptions and techniques.[3]

In the past, I always regarded it as an expression of the "sage and hero view of history"—how could medicine be "worse now than in the past" when it is always progressing? However, as I gradually delved deeper into the study of the history of the Qin and Han dynasties, I actually felt that there was a great deal of information in these few dozen words and the definitions were accurate. Firstly, it gives the most concise and profound definition of the "arts of prescriptions and techniques". Whether it is the medical classics and classical formulas respected as "medicine" by modern people, or the sexual techniques and the pursuit of immortality that are looked down upon, in essence, they are all "the knowledge of life", so it is said to be "the means to preserve life". Treating illnesses and saving the dying are part of "benevolent governance", and this knowledge originally belonged to the official institutions, so it is said to be "one of the responsibilities of the royal officials". Secondly, it is about the definition of the medical remarks of Bian Que and Yi He recorded in historical books. The idea of "discussing diseases in relation to the state and inferring politics from diagnoses" is exactly the issue that has been emphasized in Chapter 2 of this book, so there is no need to elaborate further.

It should be noted here that it is an evaluation of Han-Dynasty medicine. That is, although there was "Cang Gong" (Chunyu Yi), a doctor who truly had medical "skills", it was still not enough to change the generally under-developed and "dim" situation. Regarding such a general description, positioning legendary figures like Qi Bo and Yu Fu as good doctors in ancient times does have the aspect of not breaking away from the stereotype of the "sage-hero view of history", but it also has the aspect of objectively evaluating the general picture of Han-Dynasty medicine. Looking at the overall reviews of other fields in this book, there are absolutely no such words that praise the past and belittle the present. For example, when evaluating the "Six Arts" (103 schools, 3,123 articles), it says that after the establishment of the Han Dynasty, "to explain a text of five characters, it would take as many as twenty or

[2] *Book of the Former Han*, Volume 30.

[3] *Book of the Former Han* (*Han Shu*,《汉书》), Volume 30.

thirty thousand words"—complaining that it was too complicated; when it comes to the "Hundred Schools of Thought" (189 schools, 4,324 articles), it states that "now different schools each promote their own strengths, exhaust their knowledge and thoughts to clarify their meanings"—they give full play to their abilities; and about the "Poetry and Fu" (106 schools, 1,318 articles), it says that "since Emperor Xiaowu established the Music Bureau and collected folk songs, there were the songs of Dai and Zhao, and the styles of Qin and Chu", all of which can be used for comparison.[4]

Therefore, it can be concluded that the medical development in the early Western Han Dynasty lacked the presence of medical classics or a well-established knowledge system inherited from the pre-Qin period, including the Qin Dynasty. Notably, there were no remarkable developments in this field in the cultural sphere of the central government either. From this perspective, the medical development in the early Western Han Dynasty can be regarded as a continuation of the actual situation where there was no "advanced development" since the pre-Qin period.

Theoretically, there is no need to investigate the "cause" for things that haven't happened—there is "cause" only when there is "effect". However, this does not prevent us from getting a better understanding of the general picture of academic development during this period through the comments of historians.

In the realm of political power transitions, a common trend emerges where the victorious party, having obtained power through military strength, often possesses a lower level of "culture" compared to the defeated faction. Furthermore, it is customary for a newly established regime to adhere to the existing political system. A notable example of this phenomenon can be observed in the Song Dynasty (AD 960–1279), a period characterized by strong national power and cultural prosperity similar to that of the Han Dynasty. After Emperor Zhao Kuangyin assumed power in the mid-tenth century (AD 960), until Emperor Shenzong of Song (reigned from AD 1068 to AD 1086) implemented the "Yuanfeng New Official System" over a hundred years later, there was not only no unified new system formulated, but "the chaos even deepened, resulting in a phenomenon of mixed and ever-changing situations". The root cause of this phenomenon can be traced to the fact that the monarchs of the Five Dynasties all came to power from vassal states. The founding monarch of the Song Dynasty, Zhao Kuangyin, was no exception, and his prime minister, Zhao Pu, also came from the staff of a vassal town.[5] In contrast, the backgrounds of the founding monarchs and ministers during the Han Dynasty were even less illustrious. Emperor Gaozu of Han (Liu Bang), who "did not engage in productive work and was fond

---

[4] "Six arts", "scholars", and "poems and fus" collect 3,123, 4,324, and 1,318 chapters from 103, 189, and 106 schools respectively. And "fu" actually is an ancient Chinese style of writing, which is concerned with literary style and rhythm, and has the nature of both poetry and prose.

[5] Qu Tui Yuan: "Overview of the Official System of Past Dynasties". As an introduction or commentary, it was published in the front volume of the new edition of *The Official List of Past Dynasties* (*Li Dai Zhi Guan Biao*,《历代职官表》) (written by Huang Benji, the Qing Dynasty), Shanghai Ancient Books Publishing House, 1980.

of wine and women", was actually an idle commoner.[6] At that time, the meritorious officials were so-called "commoner-turned-generals and ministers".[7] This is one of the main reasons why the Han Dynasty adopted the official system of the Qin Dynasty. Furthermore, due to the severe exploitation by the rulers of the Qin Dynasty, the social economy declined extremely. So the top priority at that time was to solve the problem of people living in destitution. In terms of ideology, the early Western Han rulers, who had little "cultural knowledge", like the emperors of the Qin Dynasty, also held a derogatory attitude toward Confucianism. Therefore, this period was not the specific time for the great development of science and culture in the Han Dynasty.

4.2 Prosperity of Vassal States and Various Schools in Different Regions

When it comes to "academic renaissance", historians largely focus on "the rule of Wen and Jing" primarily highlight the reign of Emperor Wen (179–157 BC) of the Han Dynasty and Emperor Jing (156–141 BC) of the Han Dynasty. This marked the prime time of the Han regime following the earlier stage of Emperor Gaozu of Han and Empress Lyu's rule. Additionally, it is crucial to focus on an important development during the reign of Emperor Hui of the Han Dynasty. In the fourth year of his rule (191 BC), Emperor Hui abolished the regulation that prohibited commoners from privately collecting books. This act restored the freedom of circulating books for the general populace, which offered a new opportunity for academic development.[8]

However, it should be noted that, in terms of the actual social situation at that time, the society obviously hadn't reached the level of "cultural popularization among the common people". Therefore, when thinking about the development of medicine, especially the issue of its "theorization" process, we must not take it for granted that once the "regulation on banning private possession of books" was abolished and the situation of "academic renaissance" emerged, medicine would rapidly grow among the common people like bamboo shoots after a spring rain. Moreover, the "academic learning" and "culture" that historians see and talk about usually basically refer to the "orthodoxy of Confucian classics" and "literature and art". Then, in this case, where was the soil for the growth of medicine? We have to pay attention to the differences between the "royal courts" and the "vassal states".

"Royal courts" refer to the place where the emperor was located, while "vassal states" were the places where the kings enfeoffed by the Han royal family were based. From Emperor Gaozu, Emperor Hui, and Empress Lyu to Emperors Wen and Jing in the Han Dynasty, the regime still adhered to a policy of simplicity and frugality. However, the vassal states were quite different. They didn't have to worry about the

[6] *Records of the Grand Historian* (*Shi Ji*,《史记》), Volume 8.

[7] Qian Mu, *History of Qin and Han Dynasties* (*Qin Han Shi*,《秦汉史》), p. 38.

[8] Qian Mu, *History of Qin and Han Dynasties*, p. 76.

livelihood of the people across the country like the emperor in the royal court, nor did they need to consider waging wars against neighboring areas to expand the territory. They could focus on the construction of their own palaces and tombs. Therefore, the *Book of the Former Han* describes the grandeur of the vassal states as "having the same systems of palaces and official positions as the capital". They could also recruit talents and discuss "miscellaneous learning" The *Writings of Prince Huainan* (*Huai Nan Zi*,《淮南子》), written by the guests of King Huainan Liu An, can be regarded as a representative. The pursuit of a high-quality, even luxurious present life, as well as the ways and techniques for health and longevity, were naturally included. Perhaps this was the fertile ground for the growth of medical knowledge during this period and the specific location of such soil. The saying "The situation of the world is like a person suffering from a severe swelling. The size of a shin is almost as big as the waist; the size of a finger is almost as big as the thigh" precisely refers to the situation at that time when the power of the vassal states exceeded that of the royal courts.

> King Liu De of Hejian was enfeoffed in the second year before the reign of Emperor Jing. He was devoted to learning and had a great passion for ancient knowledge, always seeking truth from facts. Whenever he obtained a good book from the common people, he would surely have it carefully copied and give the copy back to the owner, while keeping the original. He also offered gold and silk as rewards to attract more books. As a result, people with knowledge and skills from all directions traveled thousands of miles here. Some who had ancestral old books often presented them to the king. Therefore, the number of books he collected was almost as large as that in the imperial collection of the Han Dynasty[1]. At that time, Liu An, King of Huainan, also loved books. However, most of the people he recruited were good at empty arguments. All the books collected by King Liu De were ancient pre - Qin classics written in ancient scripts.[9]

The so-called "ancient pre-Qin books" mentioned later in the text refer to works such as *Rites of Zhou* (*Zhou Li*,《周礼》), *Book of Documents* (*Shang Shu*,《尚书》), and *The Book of Rites* (*Li Ji*,《礼记》), which were discussed by the disciples of Confucius and Mencius. These books played an important role in the revival of the "Six Arts" of Confucianism, all playing an important role in the revival of "six arts". However, in terms of what we today call scientific and philosophical thoughts and skills, it is probably the non-Confucian, so-called "mostly superficially argumentative" things under the patronage of Liu An, the King of Huainan, that deserve more attention.

These examples highlight the extensive book collections and literary pursuits among the kings of vassal states. Notably, Fujita Katsuhisa, a Japanese expert, observed that even the prime minister of the state of Changsha during the time of Emperor Wen of the Han Dynasty, predating Liu De and Liu An, possessed a rich compilation of books. These instances collectively indicate the widespread popularity of books within the vassal states of the Han Dynasty.[10]

Furthermore, in previous research, it came to my attention that in the archaeological excavations of the tombs of the vassal-state kings, those from the Western Han

---

[9] *Book of the Former Han*, Volume 53.

[10] Fujita Katsuhisa (Japan), *Research of Historical Records of the Warring States from "Records of the Grand Historian"*, Shanghai: Shanghai Ancient Books Publishing House, 2008 Chinese version, p. 25.

Dynasty seem to account for a relatively large proportion. Moreover, the unearthed cultural relics related to medicine in these tombs are apparently more numerous than those in tombs of other eras. Examples include the "Tomb Group of the King of Chu" in Xuzhou, Jiangsu; the "Tomb of Liu Sheng, the Prince of Zhongshan" in Mancheng, Hebei; the "Tomb of Liu Bo, the Prince of Changyi" in Hongtushan, Juye, Shandong; the "Tomb of the Marquis of Dai at Mawangdui" in Changsha, Hunan; and the "Tomb of the King of Nanyue" in Guangzhou. Here, just taking several "pestles and mortars" as an example, we can get a glimpse of the situation. Therefore, I can't help but wonder: if this is really the case, then what's the reason? Although I've consulted some professionals about this phenomenon, I haven't received a clear answer. I can only leave it as a question for further thinking and research. When reading *History of the Qin and Han Dynasties* by Qian Mu, there is a subsection titled "The Vassal States' Promotion of Learning". I think his summary that "The central court was thrifty while the vassal states were extravagant"[11] is precisely the best answer to this question.

Because of this perspective, what Sima Qian wrote in the chapter of "Cang Gong" in *Records of the Grand Historian* (*Shi Ji*,《史记》) not only serves as an early record of medical cases but also provides us with a glimpse into the life of a doctor in the court of vassal states. Several notable cases are recorded:

> The wife of the official Beigong Sikong in the Qi State fell ill. All the doctors thought it was caused by wind invading the body and the main problem was in the lungs.
>
> Han, a maid of the King of Jibei, suffered from back pain, aversion to cold and fever. All the doctors thought it was a feverish illness.
>
> Chunyu Sima in the Qi State fell ill. At that time, the doctor Qin Xin was beside him.
>
> Sui, the personal physician of the King of Qi, fell ill and self-prepared and took the "Five Minerals Medicine".
>
> When the King of Qi was still the Marquis of Yangxu, he was seriously ill. All the doctors diagnosed it as syncope.
>
> The King of Jibei sent the imperial physicians Gao Qi and Wang Yu to learn from me. I taught them about the pathways of the meridians.
>
> The King of Zichuan often sent Feng Xin, the supervisor of horses in the Grand Granary, to learn the correct medical prescriptions from me. I taught him the theories of diagnosis methods and the principles of using medicine.
>
> Du Xin from the family of the Marquis of Gaoyong was interested in pulse-taking and came to learn from me. I taught him about the upper and lower meridians.
>
> Tang An from Zhaoli in Linzi came to learn from me. I taught him the five diagnostic methods and the knowledge of upper and lower meridians. Later, he was appointed as the personal physician of the King of Qi.[12]

The occult masters of ancient times were highly educated individuals who possessed adept knowledge in medicine. They sought fortune and favor by attaching themselves to the kings of vassal states, going to great lengths to please their royal patrons. This led to significant advancements in the four main categories of medical knowledge:

---

[11] Qian Mu, *History of Qin and Han Dynasties*, p. 78.

[12] *Records of the Grand Historian*, Volume 150.

"theory", "practical skills", "sexual healthcare", and "health preservation". These advancements surpassed the level of understanding seen in the pre-Qin era and introduced innovations to the field. It's worth noting that the theories and methods of occult masters diverged due to their regional influences, resulting in the independent growth of various distinct schools across different areas.

For example, after the discovery of numerous medical books in the Western Han Dynasty tomb at Mawangdui, in 1984, two medical works titled the *Book on Meridians* (*Mai Shu*,《脉书》) and the *Book of Daoyin* (*Yin Shu*,《引书》) were found in a tomb of the early Western Han Dynasty in Zhangjiashan, Jiangling County, Hubei Province.

Despite the potential dating back to the pre-Qin period, the medical knowledge contained in these books remained prevalent during the early Western Han Dynasty, showcasing the continuous circulation of this knowledge. This serves as a basis for evaluating the state of medical development during the Western Han Dynasty. *The Book of Daoyin* comprises five parts, with the second part outlining 38 Daoyin (conduction exercise) methods, and the fourth part describing 24 Daoyin methods. However, only six methods bear the same name in both sections, indicating that *the Book of Daoyin* incorporated a variety of techniques from different schools. A comparison between these texts and the depicted illustrations of Daoyin from the Mawangdui tomb reveals that some expressions align, while others deviate entirely. This illustrates the diverse development of Daoyin practices during the Western Han Dynasty. To elaborate further, it's important to mention that the meridians and collaterals inscribed on the wooden sculpture of human figure unearthed in Mianyang, Sichuan Province in 1993 differ significantly from those recorded in the medical books from the Mawangdui tomb or *Huangdi's Internal Classic* (*Huang Di Nei Jing*,《黄帝内经》). This highlights the varying interpretations and explanations of the theory of meridians and collaterals. Considering that this theory describes the existence of invisible tissues within the human body, it is not surprising that different systems of expression emerged in ancient times when information exchange was extremely challenging.

In essence, the aforementioned examples not only present specific knowledge but also depict a picture of coexisting schools of thought. However, it is difficult to find traces of such knowledge in direct records and circumstantial evidence from the pre-Qin period. Therefore, it is more reasonable to attribute this innovation to the Han Dynasty.

4.3 Knowledge Collection and Theories Unity

Although theoretical "thoughts" and practical "skills" are inherently interconnected, people often tend to approach them as separate realms of knowledge. In contemporary society, "science" and "technology" are perceived as interdependent and inseparable, forming a close relationship over time. In the field of medicine, there is still a significant tendency for science and technology to exist independently, a phenomenon that was even more pronounced in ancient times. For instance, it was commonly believed that this kind of "knowledge" about what diseases a certain drug or acupoint can treat contained a great deal of "experience" elements. Hence, when different "interpretations" emerged, they were often consolidated with an inclusive and eclectic approach, leaving it to future generations to select and verify. Some individuals would advocate for specific theories by adhering solely to one school, while others would combine ideas, adopting what is useful and discarding what is deemed useless. These are the three approaches taken to reconcile divergent thoughts. Let's take pharmacology as an example. In *Wu Pu's Materia Medica* (*Wu Pu Ben Cao*,《吴普本草》) during the Three Kingdoms era (AD 220–280), one herbal medicine could have up to eight different descriptions. *Shennong's Classic of Materia Medica* (*Shen Nong Ben Cao Jing*,《神农本草经》), on the other hand, was guided solely by Shen Nong's theory. Huangfu Mi in the Western Jin Dynasty (AD 266–316) merged two types of acupuncture treatments. Until now, when it comes to theories of herbal medicine and acupuncture therapy from different schools, some prudent TCM scholars still choose to harmonize and integrate them.

However, the process of unifying "theoretical ideas" has persisted from ancient times to the present day. Thus, the so-called classic theories of twelve meridians and five Zang-organs and six Fu-organs are formed on the basis of the different interpretations of the core theory like meridians and Zang-Fu organs. This unification work is exemplified in the texts of *Plain Questions* (*Su Wen*,《素问》) and *Miraculous Pivot* (*Ling Shu*, 《灵枢》). Let me present an example.

> Huang Di asked, "I have heard differing opinions among fangshi (alchemists or those skilled in esoteric arts). Some consider the brain and the marrow as Zang-organs, while others consider the intestines and the stomach as Zang-organs; and there are those who view them as Fu-organs. I am perplexed by these contradictions. Each claims their perspective to be correct. I am unaware of the foundation of their reasoning and seek clarification on this matter."
>
> In response, Qi Bo explained, "The brain, marrow, bones, vessels, gallbladder, and female uterus are all engendered by the qi of the earth. They correspond to the storage aspects of yin and bear the image of the earth. Therefore, they store without draining, earning their designation as extraordinary organs. On the other hand, the stomach, large intestine, small intestine, triple burner, and urinary bladder are generated by the qi of heaven. Their qi resembles that of heaven, enabling them to drain without storing. They receive and transform the turbid qi from the five depots, earning their title as organs of transmission and transformation. These are locations where nothing can remain for an extended period, as they serve as conduits for transportation and drainage."
>
> Then Qi Bo added, "as for the so-called five Zang-organs, they store the essence qi and do not drain it. Hence, even if they are full, they cannot be replete. As for the six Fu-organs, they transmit and transform things, but do not store them. Hence, they may be replete, but they cannot be full."

Tables 4.1 and 4.2 summarize the main points for better comprehension. Upon examining Qi Bo's statements, we can observe that despite using different names, such as extraordinary organs and organs of transmission and transformation (Table 4.1) and the five Zang-organs and six Fu-organs (Table 4.2), there are similarities between the two systems. Both systems include two types of organs, namely yin and yang. Additionally, the number of organs in both systems reflects this duality, with five representing yang and six representing yin. Furthermore, one system originates from the qi of earth, while the other is derived from the qi of heaven. Lastly, one system emphasizes storage without discharge, whereas the other focuses on discharge without storage. In essence, these two theoretical systems represent distinct approaches to understanding the Zang-Fu organs.

The contemporary textbook of *Outline of Chinese Medici*ne (*Zhong Yi Xue Gai Lun*,《中医学概论》) presents the theory of Zang-Fu organs, as demonstrated in Table 4.3, which was formulated by doctors in later generations when they organized them into a system.

However, it is noteworthy that professors have not provided explanations for the recurring appearance of the gallbladder and the overlapping boundaries of organ functions. Therefore, in order to truly understand traditional Chinese medicine, it is crucial to adopt a historical perspective and employ logical analysis. Only in this way

Table 4.1 Comparison of extraordinary organs and organs of transmission and transformation

Category	Number	Zang-Fu	Earth or Heaven	Function
extraordinary organs	6	brain, marrow, bone, pulse, gallbladder, uterus	qi of earth	storage without discharge
organs of transmission and transformation	5	stomach, large intestine, small intestine, triple energizer, bladder	qi of heaven	discharge without storage

Table 4.2 Comparison of five Zang-organs and six fu-organs

Category	Number	Zang-Fu	Earth or Heaven	Function
five Zang-organs	5	heart, liver, spleen, lung, kidney	qi of earth	storage without discharge
six Fu-organs	6	stomach, large intestine, small intestine, triple energizer, bladder, gallbladder	qi of heaven	discharge without storage

Table 4.3 Theory of Zang-fu organs in contemporary traditional Chinese medicine

Category	Zang-Fu organs	Functions
five Zang-organs	heart, liver, spleen, lung, kidney	storage without discharge
six Fu-organs	stomach, large intestine, small intestine, triple energizer, bladder, gallbladder	discharge without storage
extraordinary organs	brain, marrow, bone, pulse, gallbladder, uterus	storage without discharge

can we understand the essence of "construction" and comprehend the ancient practice of "unifying different thoughts" and further "modernization" in today's society.

It is important to highlight that the unity of different Zang-Fu theories, pharmaceutics, and the evolution of meridian theory as represented in the *Miraculous Pivot* are all products of the phenomenon where "occult masters followed theories they deemed right". If we consider this phenomenon as a form of "innovation", then the integration of different theories can be seen as another form of innovation. Because besides the generation of new knowledge, reorganization of concepts, changes in connotation and denotation, and the establishment of comprehensive systems are also examples of innovation that operate at a higher level.

4.4 The Illustrious Figures of the Eastern Han Dynasty: Zhang Zhongjing, Hua Tuo, and Beyond

Zhang Zhongjing (around AD 150–219) and Hua Tuo (around AD 145–208) are often referred to as the "Gemini" of medical science in the Han Dynasty by some medical historians.[13] This metaphor accurately captures their significance as they both lived during the same period, with Zhang specializing in internal medicine and Hua in surgery. Jian Bozan, a renowned historian, recognized the contributions of Zhang Zhongjing and Hua Tuo as representative of medical advancements in the Han Dynasty.[14] However, the medical innovation and development in the Eastern Han Dynasty (AD 25–220) go beyond the accomplishments of Zhang Zhongjing and Hua Tuo. For example, *Shennong's Classic of Materia Medica*, a classic book of pharmacology, *Classic of Difficulties* (*Nan Jing*,《难经》), *Huangdi's Classic of the Bright Hall* (*Ming Tang Jing*,《明堂经》), an acupuncture book focusing on the role of acupoints instead of meridians, were all completed around the early and middle Eastern Han Dynasty (Figs. 4.1 and 4.2).

When it comes to the history of the Qin and Han Dynasties, historians often consider them as a whole due to the title of emperor being introduced by Qin Shi Huang, the first emperor of the Qin Dynasty, representing both supreme power and a new national system.[15]

Consequently, earlier scholars tended to focus on the relationship between the Qin and Western Han Dynasties, neglecting the Eastern Han Dynasty's significance. They prefer to examine the changes in different stages of the Western Han Dynasty and pay less attention to the Eastern Han Dynasty. For example, the *History of Qin*

---

[13] Zhang Zhihan, *Essentials of ancient medical records* (*Gu Dai Yi Yao Shi Jing Hua Lu*,《古代医药史精华录》). Ji'nan: Shangdong Education Press, 1989, p. 28.

[14] Jian Boza n, *History of the Qin and Han Dynasty* (*Qin Han Shi*,《秦汉史》). Beijing: Beijing University Press, 1983, pp. 510–511.

[15] Gan Liming, Liu Xinguang: *The Reform of the Qin and Han Dynasties and Its Reasons, Results, Success, and Failure* (*Hong Ji Chu Dian——Qin Han Gai Ge Ji Yin Guo Cheng Bai*,《宏基初奠——秦汉改革及其因果成败》). Nanjing: Nanjing University Press, 2000, pp. 1, 61.

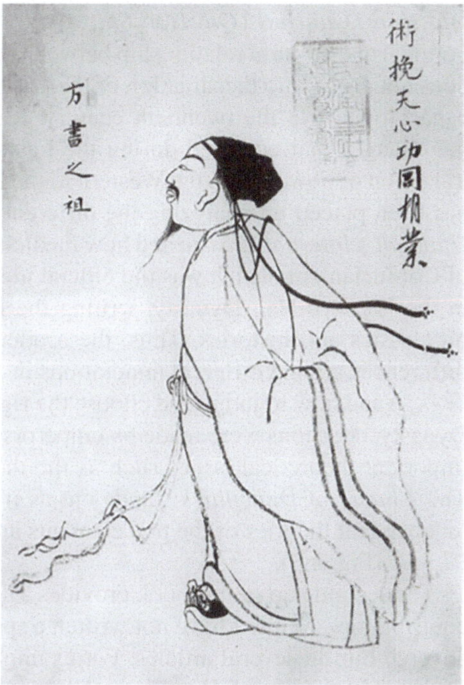

Fig. 4.1 Portrait of Zhang Zhongjing. During the Qing Dynasty, a significant cultural belief was portrayed in a picture, where Chinese characters indicated that one's contribution to a nation is equally as crucial as that of a prime minister

Fig. 4.2 An ancestral temple for Zhang Zhongjing in Nanyang. It was built in 1632. The picture shows the outside after the renovation in 1981

and Han Dynasties (*Qin Han Shi*,《秦汉史》) by Qian Mu discusses in detail the political and cultural relationship between the Qin and Western Han Dynasties, but it does not probe into the situation of the Eastern Han Dynasty. However, perspectives began to shift in the twentieth century, leading scholars to pay more attention to the changes that occurred during the Eastern Han Dynasty, especially around the transition of power from the Western to the Eastern Han. Nevertheless, limited focus has been placed on analyzing the different statements within medical classics like *Huangdi's Internal Classic* and how medicine evolved to align with the development of Confucianism, which was the official ideology of the Han Dynasty. For example, in the Western Han Dynasty, writing the law and making judgments were based on classics and histories. Thus, the academic community concentrated on telling differences among different annotations in *Spring and Autumn Annals* (*Chun Qiu*,《春秋》) and how to judge and choose the right opinion. However, in the Eastern Han Dynasty, decisions were made by emperors, shifting the academic discourse toward important theoretical issues such as the theory of the five elements and yin-yang. The *Classic of Difficulties* stands out as it harmoniously combines the previously independent theories of the five elements and yin-yang found in *Huangdi's Internal Classic* (Fig. 4.3).

The second part of this book provides a detailed discussion on Zhang Zhongjing's contributions. While I have not written a specific research paper on Hua Tuo, I have covered him in several articles. For example, a general analysis has been made on 16 treatment cases of Hua Tuo released by Shanghai Jiao Tong University Press contained in *Book of the Later Han* (*Hou Han Shu*,《后汉书》) and *Records of the Three Kingdoms* (*San Guo Zhi*,《三国志》). It is important to note that Hua Tuo, often recognized as a "surgical expert", was actually a versatile "general practitioner" who treated a wide range of diseases, including internal, external, gynecological, and pediatric conditions. He employed a combination of acupuncture and medicine

Fig. 4.3 The picture of Hua Tuo. It has been collected by the National Palace Museum and is widely used today

Fig. 4.4 The temple built for Hua Tuo. It is located in the hometown of Hua Tuo—Haozhou, Anhui

in his treatments. Records indicate that Hua Tuo possessed the ability to predict whether patients would survive or not. This foresight, along with the wise decision to abstain from further treatment for terminally ill patients, was considered an important criterion for assessing one's medical expertise in history. As mentioned in Chapter Three of the first part, both Bian Que and Hua Tuo believed that diseases progressed from the exterior to the interior. In the second part of this book, which explores pulse-taking diagnosis, the lineage of the *Classic of Difficulties*, and early materia medica texts, Hua Tuo and his apprentices are also included in the discussion (Fig. 4.4).

Hua Tuo's prominence as a renowned ancient doctor, particularly in the field of surgery, can be attributed to various factors. From my point of view, it is largely influenced by the story depicted in the *Romance of the Three Kingdoms*, where Hua Tuo successfully helps Guan Yu remove poison by scraping it off the bone. Additionally, as Western "surgical therapy" gained acceptance in China, medical historians sought corresponding examples within our own history. However, it is unsurprising that Hua Tuo faced criticism in ancient times, as medical practitioners of the era generally held a negative stance toward surgery. For example, in the Song Dynasty, Zhuge Xing wrote a postscript to *A record of the life of various physicians* (*Yi Shuo*,《医说》) written by Zhang Gao. In it, he expressed a negative evaluation of Hua Tuo's "methods of dissecting the abdomen and reconnecting tendons", considering them "techniques obtained through special means, not in line with the normal principles of herbal medicine in *Shennong's Classic of Materia Medica*". He also said that only Zhang Zhongjing's works were "the origin of all prescriptions, and scholars should follow them".[16] Similar statements are also found in the preface to the *Classified Materia Medica* (*Zheng Lei Ben Cao*,《证类本草》). Yu Tuan, a physician in the

[16] Zhang Gao (the Song Dynasty), *A record of life of various physicians* (*Yi Shuo*,《医说》), Shanghai: Shanghai Science and Technology Press, 1984.

Ming Dynasty (AD 1368–1644), wrote *The True Story of Medicine* (*Yi Xue Zheng Zhuan*,《医学正传》) in which he praised *Huangdi's Internal Classic* and *Classic of Difficulties* as "the classics for physicians"; he considered Zhang Zhongjing's *Treatise on Cold Damage Diseases* (*Shang Han Lun*,《伤寒论》) in the Eastern Han Dynasty to be an unalterable and wonderful classic through the ages. However, he blamed Hua Tuo's surgical method of "cutting the abdomen and back, washing the intestines and stomach to get rid of diseases".[17] Similarly, Yu Chang, who wrote *The Rules of Medicine* (*Yi Men Fa Lyu*,《医门法律》) in the Qing Dynasty (AD 1616–1912), accused Hua Tuo of "getting involved in superstition and heresy, and it was actually the Confucian scholars who first cut off the correct medical lineage".[18]

Laparotomy is a surgery that involves various technical conditions in all aspects and is not something that can be accomplished merely with an "anesthesia" method. In *Treatise on Causes and Manifestations of Various Diseases* (*Zhu Bing Yuan Hou Lun*,《诸病源候论》), a famous medical book in the Sui Dynasty (AD 581–618), there is a statement about the intestinal rupture caused by external injuries:

> In cases where a patient has a broken intestine, if only one end of the intestine is visible, it is impossible to reconnect it. However, if both ends are visible, a prompt reconnection is necessary. Needles are used to join the broken segments of the intestine, followed by the application of chicken blood on the junction to prevent the leakage of qi. Finally, the intestine is carefully pushed back into the abdomen.[19]

It shows that though hundreds of years have passed from the period from the Three Kingdoms (AD 220–280) to the Sui Dynasty, medical techniques had not advanced sufficiently to perform laparotomy when one end of the intestine was still located within the abdomen. Therefore, even if Hua Tuo had conducted abdominal operations, they would likely have been limited to the outside of the peritoneum at most. It is worth noting that the historical records of Hua Tuo's apprentices primarily focus on traditional medical skills, such as acupuncture, herbal medicine, and practicing the "Wu Qin Xi" exercises for health and longevity. As a result, the exceptional nature of Hua Tuo's surgical abilities warrants further investigation and scrutiny.

---

[17] Chen Menglei et al. (the Qing Dynasty), *Ancient and Modern Books Integrating the Complete Records of the Ministry of Medicine* (*Gu Jin Tu Shu Ji Cheng Yi Bu Quan Lu*,《古今图书集成医部全录》), Volume 52, Beijing: People's Health Publishing House, 1991 Edition, Volume 12, p. 38.

[18] Yu Chang (the Qing Dynasty), *The Rules of Medicine* (*Yi Men Fa Lyu*,《医门法律》), Volume 2.

[19] Chao Yuanfang (the Sui Dyansty), *Treatise on Causes and Manifestations of Various Diseases* (*Zhu Bing Yuan Hou Lun*,《诸病源候论》), Volume 36.

Chapter 5
Analysis of Forces That Drive the Medical Development in the Pre-Qin Period and Han Dynasty

The development of a branch of knowledge, in general, depends on both internal and external forces and their interaction. Internal forces stem from human intelligence and the desire for exploration and the pursuit of truth. They manifest in two aspects of understanding and utilizing nature. Understanding nature entails the continual introduction of explanatory viewpoints and the gradual construction of explanatory systems, such as science, discovery, theory, and doctrine. Utilizing nature involves the application of technology and invention. (The initial use of technology is known as invention, while its ongoing utilization is referred to as technology.[1]) Scientific and technological advancements are acknowledged if they provide effective explanations or practical skills that unveil the inherent laws of the world. Conversely, explanations or skills that fail to meet these criteria are considered superstition or witchcraft. It's important to note, however, that both science and superstition stem from human intelligence. As knowledge accumulates or existing theories face insurmountable challenges, individuals are driven to seek new explanations. This process often leads to the emergence of new theories or technologies through cross-disciplinary learning and the integration of diverse knowledge fields or foreign cultures. These phenomena generally represent the basic manifestations of the so-called "internal driving forces".

External forces play a role in knowledge development, driven by factors such as social needs and financial incentives. However, regardless of whether the motivation is to fulfill objective needs or pursue profitability, the advancement of knowledge ultimately relies on individual intelligence. The only distinction lies in the absence of intrinsic interest, which is unrelated to reality.

"Human wisdom" only makes sense when compared with that of other living beings. When it comes to humans themselves, it turns into an empty phrase with a certain "-ism" nature. Although there are differences in intelligence among people, they are all human beings. Therefore, from this perspective, the "sage-centered view of history" has a very reasonable aspect. All inventions and creations come from

[1] This concise way of description is put forward by Zhang Kaixun.

Y. Liao, *Reconstructing the Image of Medicine in the Qin and Han Dynasties*, Understanding China, https://doi.org/10.1007/978-981-95-6801-7_5

the outstanding individuals within humanity. Among the masses, the smart ones can only understand and inherit the "theories" advocated by the outstanding or use their inventions and creations. Those who are good-for-nothing or clumsy can't even mimic others. Therefore, regarding the "sage-centered view of history", we should criticize its stance that the present is inferior to the past and its ignorance of the fact that the knowledge system has gradually developed and changed, as well as its specific application in constructing the historical interpretation system.

Under the interaction of internal and external forces, the growth of a body of knowledge involves a complex process, including independent inventions and discoveries, the gradual quantitative growth through accumulation, and the significant changes caused by major inventions and discoveries and system reorganization. However, overall, it still shows a linear development trend. Regarding the development of medicine from the Pre-Qin period to the Han Dynasty, from the most macroscopic level, it is of course a process of accumulation and gradual change. The fundamental difference between the so-called two "images" lies in this: when the medical classics represented by *Huangdi's Internal Classic* (*Huang Di Nei Jing,*《黄帝内经》) are placed in a certain period of the Pre-Qin era, the statement of "accumulation and gradual change" becomes an empty phrase. It only abstractly affirms the existence of a "development process" but fails to provide a specific explanation of how such high-level "classics" were formed and what specific developments occurred in the hundreds of years after their birth. The key point of the "reconstructed" image is to divide this historical period into three major stages and present a relatively specific "development process".

In the first chapter, reference is made to the book *Science and Civilization in China* (*Zhong Guo Ke Xue Ji Shu Shi Gao,*《中国科学技术史稿》), authored by the research team led by Du Shiran around 1980. One of the book's important objectives is to address the issue of the "parallel development" and the lack of integration between science-technology research and social development. It aims to achieve this by elucidating the interactive relationship between these two aspects. However, many years after this masterpiece in the general history of science and technology was published, Du repeatedly mentioned that this problem remained unresolved, and there still existed a "two-skin" phenomenon, referring to the disconnection between S&T and social development. Just like the question of whether medicine in the Qin and Han Dynasties experienced advanced development, this issue has long haunted those in the field of S&T history research. Consequently, the following analysis of the driving forces of medical development during this period serves as a concrete attempt to bridge the gap of the "two-skin" problem.

5.1 Absence of Forces in the Pre-Qin Period

At any time, there exists the light of wisdom as the "internal driving force" for the growth of knowledge in human society. However, it remains a matter for specific analysis as to where this light of wisdom converges. Historians have undoubtedly centered their discussions on the politics, economy, law, education, and culture of the pre-Qin states and the two great empires of the Qin and Han Dynasties. Their discussions revolve around how outstanding figures, within a specific historical environment and under certain needs, demonstrated their intelligence, put forward new doctrines, or implemented new policies, thus promoting social development and progress. Even for the Qin state (770–207 BC), which some historians regarded as "uncultured", we can still find some achievements in literature and art. We can even assert that around the reign of Duke Mu of Qin (?–621 BC), the Qin state was not only not backward in economy and culture among the vassal states at that time but also led in some aspects and was one of the advanced slave-owning states then.[2] However, why is it quite difficult to see the "fruits of medicine" in this well-explored "familiar land" they have repeatedly delved into? The fundamental reason lies precisely in what they call "culture", which usually mainly refers to the prosperity of agriculture and handicrafts, especially emphasizing the use of ox-ploughing and iron tools. Of course, it also includes language, characters, poetry, songs, music, dance, and even astronomical calendars, divination, and sacrifice. But all these are not sufficient to prove that "medicine" would necessarily develop synchronously.

5.1.1 Marriage Is the Key to Population Growth

Throughout history, the population has consistently stood as the paramount productive force, predating the ascendancy of science and technology as the primary productive force. Similarly, the number of soldiers was a crucial determinant of a country's military might before the advent of long-range weapons of mass destruction. Consequently, rulers of both past and present have always recognized the significance of population. However, while leaders of the past focused on increasing their population, contemporary rulers strive to limit its growth.

Medicine, as a tool for treating diseases and reducing mortality, directly contributes to population growth or, at the very least, the maintenance of the current population size. On this basis, ancient rulers are supposed to pay close attention to the important role of medicine in promoting a country's productivity and military power. However, this perspective, although seemingly reasonable today, was rarely entertained by rulers over the course of ancient history. The prevailing mindset dictated that engaging in warfare represented a quicker and more efficient method of acquiring land and population. With land and population, a country could rapidly increase its productivity, secure abundant agricultural and artisanal resources, and bolster its military forces. In addition to incentivizing farming and military service, encouraging childbirth emerged as another pivotal policy for fostering national prosperity and military strength.

King Gou Jian of Yue state (520–464 BC) endured hardships as he meticulously planned his retaliation against the state of Wu during the Spring and Autumn

---

[2] Lin Jianmin, *The History of the Qin state* (*Qin Guo Fa Zhan Shi*,《秦国发展史》), p. 42.

period (770–476 BC). This endeavor involved implementing various policies, such as population growth, wealth accumulation, and education.[3] Among these initiatives, promoting childbirth held a particularly vital role.

> King Gou Jian implemented a series of measures to boost the population swiftly in preparation for his retaliatory campaign against the state of Wu. One such measure was prohibiting young adults from marrying the elderly and vice versa. Parents of unmarried girls at the age of 17 would be held accountable, while parents of unmarried men at the age of 20 would also face consequences. Additionally, pregnant women were required to report their condition to the authorities, who would dispatch a doctor for their care. Upon giving birth to a boy, the family would be rewarded with two bottles of wine and a dog; for a girl, two bottles of wine and a piglet. If three babies were born, a wet nurse would be provided, and if two babies were born, the king would cover the upbringing costs of one baby. King Gou Jian further decreed that the children of widows, widowers, the sick, and the poor should be taken in and raised by the government.[4]

All these measures were aimed at rapidly increasing the population for the purpose of amassing troops to exact revenge on the state of Wu. Guan Zhong (about 723–645 BC), the chancellor and reformer of the state of Qi (1046–221 BC), put forward nine policies of benefiting people. One of these policies encouraged the coupling of widows and widowers, providing them with farmland and a house, and requiring them to serve the state after three years. This policy not only demonstrated a sense of humanistic care but also aimed to mobilize more people for state service.[5] Similarly, the wife of Cao Pi (emperor Wen of Wei, AD 187–226) during the Three Kingdoms era (AD 220–280) recognized the need to strengthen their military forces and proposed prioritizing the marriage of women to soldiers due to their scarcity.[6] Subsequently, Cao Rui, Emperor Mingdi of Wei, issued a decree between AD 233 and 237 that all women who had married non-soldiers should be taken away and remarried to soldiers.[7]

In a word, during those times, the connection between medicine and population growth had not yet been established.

5.1.2 The Status of Physicians and Medicine

According to the first Chinese dictionary, *Explanation and Study of Principles of Composition of Characters* (*Shuo Wen Jie Zi,*《说文解字》), which was compiled

[3] Du Yu (the Jin Dynasty), Kong Yingda (the Tang Dynasty), *Annotation and Interpretation of the Commentary of Zuo* (*Chun Qiu Zuo Zhuan Zhu Shu,*《春秋左传注疏》), Volume 57.

[4] Zhao Yu (the Han Dynasty), *The Spring and Autumn Annals of Wu and Yue* (*Wu Yue Chun Qiu,*《吴越春秋》), Volume 6.

[5] Fang Xuanling (the Tang Dynasty), *Guan Zi* (《管子》), Volume 18.

[6] The Chapter "History of the Wei Dynasty" in *Records of the Three Kingdoms* (*San Guo Zhi,*《三国志·魏志》), Volume 5.

[7] He Zhuo (the Qing Dynasty), *Yimen's Reading Notes* (*Yi Men Du Shu Ji,*《义门读书记》, Volume 26.

by Xu Shen (AD 58–147) in the Eastern Han Dynasty (AD 25–220), "医Yi" originally referred to workmen who treated diseases. This ancient definition implies that physicians and their skills were considered a type of workmanship. This differs significantly from contemporary views on the medical profession, medical science, and medical technology, which many may find hard to agree with. Therefore, it is important to remind readers not to analyze ancient concepts solely through a modern lens. On the other hand, it is evident that medical workmen, unlike other skilled laborers who create physical objects directly, possess unique abilities such as diagnosing diseases through observing the complexion and feeling the pulse, treating diseases through acupuncture and medicinal practices, dispelling the pathogenic factors and preserving health. They devote their diligent efforts and wisdom to helping infertile couples conceive, aiding patients in their recovery, and strengthening the weak. Therefore, the chapter "Catalog of Literary Works" in the *Book of the Former Han* (*Han Shu*,《汉书》) describes the knowledge and techniques of medical workmen as the way of prolonging life. Additionally, the techniques utilized by medical workmen are deeply intertwined with ancient Chinese philosophy and human ethics, which diminishes their status as mere skilled laborers. However, these observations were made by later generations. The status of medical workmen is very low in a longer period of history, even in the Han Dynasty and the Three Kingdoms Period if you think of the persons mentioned earlier in this book, like Gongcheng Yangqing, Chunyu Yi, Hua Tuo, or behaviors and manners of medical workmen like "refusing to treat patients", "not letting their descendants learn their skills", "feeling ashamed of being a physician". So, Han Yu, a great writer in the Tang Dynasty (AD 618–907), expressed a disdainful attitude toward witch doctors, musicians, and various craftsmen.[8] Despite the existence of doctors, assistants, and officials responsible for medical education in the Imperial Medical Bureau, physicians were still called medical workmen. However, in the Song Dynasty (AD 960–1279), intellectuals regarded medical research as one of the Confucian ways of deepening their understanding of the world. They viewed it as knowledge and skills necessary for serving the monarch and parents, preserving health, aiding those in distress, cultivating moral character, regulating the family, and practicing benevolence. This shift in perspective provided theoretical support for elevating the status of medicine and physicians. Consequently, medical theoretical research started to be highly valued and developed. Understanding these historical dynamics is crucial in comprehending why Zhang Congzheng (AD 1156–1228), one of the four renowned physicians in the Jin and Yuan Dynasties, titled his medical book *Confucians' Duties to Their Parents* (*Ru Men Shi Qin*). It also enables us to grasp the knowledge and medical ethics of the so-called "Confucian physicians". Furthermore, this understanding can serve as an explanation for satisfying our psychological needs. After all, Li Shizhen (AD 1518–1593), a highly influential figure in Chinese medicine, turned to the study of medicine after failing the Imperial Examination multiple times. Even during the Meiji Restoration in Japan, which introduced western science and technology and

[8] Yao Xuan (the Song Dynasty), *The Literary Highlights in the Tang Dynasty* (*Tang Wen Cui*,《唐文粹》), Volume 47.

politics in an all-round way, the proportion of medical graduates from the imperial university in 1890, the year of the first Imperial Diet, was the lowest. Meanwhile, the proportions of graduates in agriculture, law, arts, science, and engineering had all increased. The number of students majoring in arts, science, and engineering was nearly or more than twice as high as those majoring in medicine. It is only in recent years that the social status of medicine and physicians has begun to approach that of other scientific and technological fields.

In a word, when the status of physicians and medicine was low in the pre-Qin period, it was impossible for elites to choose medicine as their careers. Thus, the development of medical theories would not be promoted.

5.1.3 The Contention of the Hundred Schools of Thought

When delineating the identity and status of physicians and medicine, medicine needs to be distinguished from the hundred schools of thought. There was an assertion that medicine was once thought to be included in the hundred schools of thought in the pre-Qin period (a period in which different ideas are put forward and exchanged from divergent schools) or the trend of free talk in Jixia Academy. However, this assertion lacks factual evidence, as confirmed by both historians' specific commentaries on the "hundred schools of thought" and the account of Philosophers listed in the "Catalog of Literary Works" chapter of the *Book of the Former Han*. Since this point has been reiterated numerous times, further elaboration is unnecessary at this juncture.

5.2 The Situation in the Han Dynasty

5.2.1 Confucian Physicians

After the Song Dynasty, the emergence of Confucian physicians sparked interest in exploring the relationship between Confucianism and medicine within the field of medical history. However, the existing research predominantly focuses on the moral values and cultural literacy of Confucian physicians, providing limited insight into the influence of Confucianism on medical theory. While this book does not encompass the subsequent issues that arise in later periods, it is essential to examine the impact of Confucianism on the development of medical theories during the Han Dynasty. This analysis warrants a two-fold approach.

Firstly, it is crucial to acknowledge that the advancements in Confucianism during the Han Dynasty, such as the refinement of yin-yang and five-element theories and the expansion of imperial college education, were driven by political motives and necessities, lacking direct correlation with medical development and education. Historians have extensively discussed this aspect, highlighting the four significant transformations that occurred regarding the integration of Confucianism with political governance during the Han Dynasty[9]:

(1) The founder and first emperor of the Han Dynasty, Liu Bang (BC 256–195), underwent a transformation in his perception of Confucianism, shifting from disdain to respect.
(2) Emperor Wu of the Han Dynasty (BC 156–87) established Confucianism as the predominant official ideology.
(3) Emperor Yuan of the Han Dynasty (BC 74–33), a devout follower of Confucianism, modified the laws of the Han Dynasty to appoint Confucian scholars to influential positions, setting a precedent for Confucian governance.
(4) The monarchs of the early Eastern Han Dynasty were well-versed in Confucian classics and advocated governing the country based on Confucian principles.

Regarding imperial and private education:

> In the context of the Eastern Han Dynasty, the establishment of the imperial college by Emperor Guangwu (6 BC-AD 57) led to a significant expansion of both imperial and private education. By the reign of emperor Shun and emperor Zhi, the imperial college had 250 buildings with 1850 rooms, covering 30,000 students. The imperial court offered preferential treatment to the students of the imperial college and stipulated that the children of the officials from the general to the petty official should be enrolled in the college. Eminent Confucian scholars were employed to teach Confucian classics at the imperial college, and the emperor himself delivered lectures there to elevate its status. This flourishing of the imperial college system in turn promoted the development of local sinology, with schools established throughout the county to meet the widespread demand for studying classics and pursuing official positions.
>
> At the same time, local private schools also saw a period of prosperity. Many Confucian scholars, particularly those from families with a strong Confucian background, established schools that adhered to the principles of the patriarchal clan system and taught Confucianism. These schools attracted large numbers of students, sometimes reaching hundreds or even thousands. Some students would travel great distances and bring food to seek instruction from accomplished Confucian scholars temporarily residing in fine houses. (the chapter of "Confucian Scholars" in the *Book of the Former Han*). This was one of the characteristics of the landlords of the Eastern Han Dynasty in the ideological field.[10]

Education focused on Confucianism with the purpose of cultivating officials. Therefore, Fangji (methods and techniques) did not hold a prominent place in education like today. Huang Liuzhu, an expert who studied the history of the Qin and Han Dynasties, mentioned that people who were skilled in Fangji could be an official when he discussed the official promotion system, and it was commonly seen in the

[9] Liu Houqin, *Confucianism and the Society of the Han Dynasty* (*Ru Xue Yu Han Dai She Hui*,《儒学与汉代社会》), Jinan: Qilu Press, 2002, pp. 1–7.

[10] Li Konghuai, "The Formation and Characteristics of Landlords in the Eastern Han Dynasty" in *A Collection of Essays on the History of Qin and Han Dynasties* (*Qin Han Shi Lun Cong*,《秦汉史论丛》), Series 1, Xi'an: Shaanxi People's Press, 1981, pp. 190–207.

Western Han Dynasty (202 BC-AD 8). For example, according to the chapter "The Biography of Zhou Ren" in the *Book of the Former Han*, Zhou Ren was summoned to emperor Jingdi because of his excellent medical skills. Zhou had served as the emperor's retainer when he was still the crown prince.[11] However, it is worth noting that this is the only known case related to medicine mentioned in the book.

Nevertheless, the development of medicine was significantly affected by the social environment, as discussed in the relevant thematic studies of this book. Scholars believe that Traditional Chinese Medicine (TCM), which emphasizes the harmony between heaven and humanity, reflects the societal influence on medicine.[12] In particular, the authors of the current edition of *Huangdi's Internal Classic* and *Classic of Difficulties* (*Nan Jing*, 《难经》) played a significant role in constructing a theoretical system based on the concepts of yin-yang and the five elements, separate from clinical practice. Obviously, both their writing proficiency and cultural literacy show that they are far from being craftsman-like physicians who make a living with practical skills; instead, they are Confucian physicians. This emphasis is crucial because Confucianism greatly influenced the development and systematization of medical theories during the Eastern and Western Han Dynasties. While it was common for Confucian scholars to possess knowledge of medicine during ancient times, these Confucian physicians were distinct from ordinary Confucians in several ways. First of all, they were not concerned about li (礼, rites or rituals), the social factor of Confucianism, but instead the philosophical cosmology and methodology serving as tools for reasoning and explaining rituals. Secondly, although ordinary Confucian scholars were familiar with medical theories (including physiology, pathology, treatment, pharmacology, etc.), they usually applied these well-known superficial and simple truths to discuss politics and state affairs. ("Eminent doctors are able to eliminate the country's calamity and defects".) On the contrary, Confucian physicians applied grand doctrines such as the principles of yin-yang and five elements, the principle of the unity of heaven and humanity, and even military doctrines to medicine which was scorned by gentlemen. It is the contributions of these Confucian physicians that enabled the various self-claimed doctrines of Fangshi to form a theoretical system full of "construction" characteristics. This system, characterized by the integration of philosophical and cosmological principles into medicine, forms the fundamental nature and distinctive features of TCM.

The relationship between Japanese Kampo medicine and Confucianism has long been a topic of debate. Many modern practitioners of Kampo medicine have a strong foundation in Confucianism and place great emphasis on empirical evidence. As a result, there is an ongoing discussion about whether the ancient Chinese medical classics were actually written by Confucians. For example, Yamagata Daini (AD 1725–1767), a Japanese Confucian physician, who was murdered by the shogunate

[11] Huang Liuzhu, *the Official Selection System in the Qin and Han Dynasty* (*Qin Han Shi Jin Zhi Du*,《秦汉仕进制度》), Xi'an: Northwest University Press, 1985, pp. 221–222.

[12] Liu Houqin, *Confucianism and the Society of the Han Dynasty* (*Ru Xue Yu Han Dai She Hui*,《儒学与汉代社会》), Jinan: Qilu Press, 2002, pp. 465–466.

for conspiracy, commented on *Plain Questions* (*Su Wen*,《素问》), *Miraculous Pivot* (*Ling Shu*,《灵枢》) and *Classic of Difficulties*:

> Although they claim to be from the time of the Yellow Emperor (Xuan Yuan) and Qi Bo, in fact, they were written by Confucian scholars of the Han Dynasty. There are indeed many aspects that cannot be relied upon. However, since that time was not long after ancient times, there are still some ancient medical principles preserved in them.
>
> The *Huangdi's Classic of Eighty-one Difficulties* (*Ba Shi Yi Nan*,《八十一难》) is also somewhat ancient and should not be completely discarded. Therefore, by carefully selecting and studying the content of these two books, we can gain some assistance in medical skills. As for the theories of the five - element phases and six - qi (yunqi theories), they have no relation to medical practice and can all be discarded.[13]

Yoshimasu Todo (AD 1702–1773), a representative of ancient formula sect of Chinese medicine in Japan, offered a similar perspective:

> In ancient times, there were three classes of doctors, namely physicians, yin-yang doctors and Daoist doctors. According to *The Rites of Zhou* (*Zhou Li*,《周礼》), physicians, like Bian Que and Zhang Zhongjing, identified diseases based on their specific manifestations and treated them with prescriptions. Yin-yang doctors, exemplified by Chunyu Yi (Taicang Gong), solely relied on the theory of yin-yang and five elements for treating diseases, showing little regard for practicality, which often resulted in ineffective treatments. Daoist doctors, on the other hand, aimed to attain immortality through the refinement of qi and the consumption of elixirs. Though fewer in number, representatives of this group include Ge Hong, Tao Hongjing, and Sun Simiao. After the deaths of Bian Que and Zhang Zhongjing, there was a noticeable absence of discussions about the physician's way of treating diseases, and Chunyu Yi was blamed for this.[14]
>
> Despite the intriguing pulse theory and etiology found in the *Plain Questions*, *Miraculous Pivot* and *Classic of Difficulties*, the absence of treatment prescriptions renders them purely theoretical. This calls into question the value of merely discussing the theory without offering practical remedies.[15]

While these medical classics written by Confucian physicians demonstrate a high level of theoretical knowledge, much of their content, such as the theory of five circuits and six qi, may be incomprehensible and impractical. Even the practical aspects may be inaccessible to ordinary doctors with limited medical skills. As a result, for a significant period after the Han Dynasty, medicine seemed to prioritize practicality over theory. It was only during the Song and Ming Dynasties when theoretical doctrines reached their peak resurgence.

---

[13] Syanken Datugu (Japan), *Correcting Medical misconceptions*(『醫事撥亂』), Kano Library of Tohoku University, Japan, No. 9 - 31,963–1.

[14] Yoshimasu Todo (Japan), *Iji wakumon* (『医事或問』), Kyoto: Shibunkaku, 1970 reprint, pp. 50–51.

[15] Yoshimasu Todo (Japan), *Toudousenseimondou* (『東洞先生問答』), Kyoto: Shibunkaku, 1970 reprint, p. 63.

5.2.2 *The Evolution of the Huang-Lao School (the Teachings of Huang Di and Lao Zi)*

Although many people, including well-known science historians like Joseph Needham, believe that Taoist doctrines are most closely related to immortal arts and the study of life, in fact, there are fundamental differences between them. The basic nature of Huang-Lao learning and Taoist theories is also political theory. It just emphasizes another worldly strategy of letting nature take its course and governing by non-action. This can be illustrated by the fable of "Lord Wen Hui (King Hui of the Wei)", where the cook Pao Ding shares his wisdom on cutting up an ox. Just as the fable in "The Secret of Caring for Life" of *Zhuang Zi* (《庄子》) expresses, when Lord Wenhui learned about nourishing life from the story of the cook Ding dissecting an ox: Only by cutting the ox (the object) along the bone gaps and muscle textures can the knife (the tool) be undamaged. This is the best way to govern society and the people—what is meant by "nourishing life" (here, "nourishing" means governing, and "life" refers to the people). Therefore, when historians discuss the "Huang-Lao learning", they all emphasize that in essence, it is also a kind of political theory to remedy the current social problems.

However, in the Warring States Period (475–221 BC), when the principalities were competing for hegemony, this political strategy could neither be appreciated by the state of Qin and the Three Jin (the states of Han, Wei, and Zhao), nor be widely adopted in the eastern cultural circle where it originated. The early Han Dynasty marked a turning point in the acceptance of the Huang-Lao School's principles. The fall of the oppressive Qin Dynasty allowed people to recuperate and embrace a simpler way of life. Figures like Prime Minister Xiao He and Cao Can became exemplars of generosity and peaceful working styles, which aligned with the principles of the Huang-Lao School.[16] Emperor Hui of the Han introduced amnesty and reduced laws and regulations that harmed common people,[17] further demonstrating adherence to the principles of the Huang-Lao School. Interestingly, the Huang-Lao School also created a favorable environment for the revival of Confucianism. The relaxed and inclusive nature of the Huang-Lao School's philosophy allowed for the resurgence of Confucian ideals during this period.[18]

Then the Huang-Lao School began to retire from politics when the Emperor Wu of the Han adopted Confucianism as the official ideology while rejecting all other schools of thought. However, the thought of "conforming to nature and being indifferent to fame or gain", along with the political theories of "granting reward in spring and summer but punishment in autumn and winter", and "giving priority to

[16] *Book of the Former Han*, Volume 89.

[17] *Book of the Former Han*, Volume 2.

[18] Hanxing, *A Combination of Confucianism and Law: Study of Politics and Culture of the Qin and Han Dynasties* (*Ru Fa Zheng He: Qin Han Zheng Zhi Wen Hua Lun*,《儒法整合:秦汉政治文化论》), p. 124.

morality over penalty",[19] which were explained based on the law of nature, were integrated with the thought of "the order of the four seasons" advocated by yin-yang theorists. Thus, the aforementioned thoughts of the Huang-Lao school were gradually developed into a scientific way of preserving health. For example, the meaning of the noun "生sheng", which originally referred to the people, has gradually changed into the verbal form "生长grow". This change is reflected in the concept of "the order of the four seasons" of the Yin-yang theorists and the health preservation principle of "birth in spring, growth in summer, harvest in autumn, and storage in winter" in *Huangdi's Internal Classic*.

The gap between the political theory of the Huang-Lao School and the art of "seeking longevity and becoming an immortal" in Fangshu was thus bridged. In this way, these two seemingly unrelated doctrines gradually merged. On the other hand, the "way of immortality" that involved sending Fangshi (occult masters) to the sea to search for immortals and magic pills the pre-Qin period and the reign of Emperor Wu of the Han Dynasty (156–87 BC) seems to have been replaced by the techniques of health preservation and longevity that contain more "medical" content. These techniques gradually enhanced their theoretical content and status by relying on the names and doctrines of the Yellow Emperor and Laozi, leaving later generations with the impression that they originated from the Huang-Lao School, or that the Yellow Emperor and Laozi were originally the founders of the art of immortality and health preservation. Regarding the reasons for this "evolution" process, Jian Bozan believed it was to counter the influence of the incoming Buddhism.

> In order to counter the influence of Sakyamuni, Fangshi (occult masters) exalted Lao Zi. Fearing that the influence of Laozi alone might not be sufficient, they also exalted the Yellow Emperor. Consequently, by the early Eastern Han Dynasty, the main deities worshiped by Fangshi were no longer the immortals of Penglai, but the Yellow Emperor and Laozi. According to "The eight princes of emperor Ming" in the *Book of the Later Han* (*Hou Han Shu*,《后汉书》), Liu Chong, the Prince Min of Chen, "offered sacrifices to the Yellow Emperor and Laozi, praying for longevity and blessings." This is evidence that the Yellow Emperor and Laozi had been elevated to the altar by Fangshi around the time of Emperor Ming.[20]

However, in my opinion, rather than "resistance", it's more like "imitation" and "association". You have Sakyamuni as the founder, and I have the Yellow Emperor and Laozi; you have Buddhist scriptures, and I have Laozi's *Tao Te Ching* (*Dao De Jing*,《道德经》; also known as the *Five Thousand Words*, Wu Qian Yan,《五千言》). This characteristic seems to be more prominently manifested when Taoism came into being, as various Taoist "ritual implements" are similar to, or even identical to, those used in Indian religious practices.

[19] There are four silk books buried in Mawangdui Han Dynasty Tomb. And it is generally believed that these are the *Huangdi's Four Canons* (*Huang Di Si Jing*,《黄帝四经》) that were mentioned in the chapter "Catalog of Literary Works" in *Book of the Former Han*.

[20] Jian Bozan, *History of Qin and Han Dynasties*(*Qin Han Shi*,《秦汉史》), Beijing: Peking University Press,
1983, pp. 567–568.

Chapter 6
The Influence of Foreign Cultures

From a broader perspective, the influence of foreign cultures can also be regarded as an external factor or momentum that drives knowledge development and reform. However, historical research cannot be content with macroscopic theoretical generalizations like science or philosophy. It must conduct in-depth analysis of specific details, so I will discuss it as a separate chapter.

When it comes to the exchanges between China and foreign countries during the Qin and Han Dynasties (221 BC-AD 220), "Zhang Qian' s Western Expedition" and the "the Silk Road" have attracted worldwide attention. (Fig. 6.1)

Zhang Qian (164–114 BC) was sent by Emperor Wu of the Han (156–87 BC) to the western regions, thereby opening a new era of communication between China and the countries along the Silk Road. The civilization of the western regions quickly made its way into the heart of China through this international trade route, which later became known as the Silk Road. In just a few decades, numerous goods from the western regions, including sesame, flax, fig, pomegranate, mung bean, cucumber, green onion, carrot, garlic, saffron, coriander, jiulin vine, glass, woolen cloth, gemstone, potion, and glue, were introduced to the Central Plains.[1]

The development of medicine during the Qin and Han Dynasties also benefited from the Silk Road's trade, as various medicinal substances, previously unknown in China, made their way through this route. While these medicines are of great importance, I will not delve into them further, assuming that there is a general understanding of them. Instead, let us briefly review the regional issues of early civilizations in Eurasia. (Fig. 6.2)

[1] Lin Jianming, et al., *Social Civilization of the Qin and Han Dynasties* (*Qin Han She Hui Wen Ming*, 《秦汉社会文明》), Xi'an: Northwestern University Press, 1985, p.28.

Y. Liao, *Reconstructing the Image of Medicine in the Qin and Han Dynasties*, Understanding China, https://doi.org/10.1007/978-981-95-6801-7_6

Fig. 6.1 The oil painting of *Zhang Qian on a mission to the Western regions* by Wang Kewei

Fig. 6.2 The inheritance of Ayurveda of Indian traditional medicine

Experts who study the world's oldest Mesopotamia culture have put forward various theories about the origins of the Sumerians. Some point out that the similarities between the Sumerian language and Chinese suggest an eastern origin[2]; while others argue that although the Sumerian language has no known connections to any existing language, their culture bears resemblance to the ancient civilization of India.[3] The question of where exactly the Sumerians migrated from to settle in the delta of the Tigris and Euphrates rivers remains a mystery. Considering they had a good knowledge of architecture and astronomy and created cuneiform writing

[2] L. S. Stavrianos (USA), *World Since 1500: A Global History* (*Quan Qiu Tong Shi-1500 Nian Yi Qian De Shi Jie*,《全球通史——1500年以前的世界》), Shanghai: Shanghai Academy of Social Sciences Press, 1988, Chinese translation, p.119.

[3] Edward McNall Burns, Philip Lee Ralph (USA) et al., *World Civilizations* (*Shi Jie Wen Ming Shi*,《世界文明史》), Beijing: The Commercial Press, 1987, Chinese translation, Volume I, p.68.

and plenty of artworks. These factors, along with their self-designated term "black-headed people", lead us to speculate that their origin might indeed be somewhere in the East.[4] It is worth noting that, although we often refer to ancient Egypt and ancient Greece as Western civilizations, they are actually located in the eastern part of Europe. In terms of early civilization, historians are generally concerned with the influence of tribal migration from the east to the west, aggression and conquest, resource plundering and business activities on the cultural formation of this region. By the first century, a civilization belt had formed across the Eurasian continent from the Roman empire in the west to ancient China in the east.

Ancient China, located in the eastern end of this civilization belt, existed for a long period of history in a relatively independent and isolated environment due to limited transportation access. Many scholars attribute this geographical isolation as an important factor contributing to the emergence and growth of Chinese civilization. Chu Kochen (AD 1890–1974), a former principal of Zhejiang University, once remarked that cultures do not develop overnight, but rather over extended periods of time. However, the prospect of cultural development can be hindered by invasions from neighboring barbarian nations. In fact, the birth of ancient cultures around the world is often associated with isolated regions that are well-protected. A splendid culture can only be bred in such an area".[5] Nevertheless, the exchange of knowledge, skills, and communication between the Chinese nation and other civilizations has never been completely cut off. Moreover, those interactions commenced early, covered a wide range of domains, and exerted a great impact that likely extended beyond what is documented in literature.

When it comes to medicine, the spread and communication of knowledge and skills can be summarized as follows:

1) Knowledge and skills can be spread through people and book. The specific forms can be manifested in people's migration, official diplomacy, scholars' exchanges, missionary work by monks, and business trade. However, what is often overlooked is the role of the "wandering scholars" (those who carry skills) in the folk.
2) Since traditional medicines share the same essence, when identical or similar theories and treatment methods are seen in different regions or cultural systems, it is actually very difficult to judge whether they occurred independently in multiple ways or originated from communication and dissemination.
3) China and West Asian countries, having developed their own relatively mature cultural systems, primarily engage in medical and pharmaceutical exchanges at the technical level. This includes the exchange of new drugs, therapeutic techniques, and other similar advancements. However, foreign theories are rarely

---

[4] Gerhard Venzmer (Germany), *5000 Years of Medicine* (*Shi Jie Yi Xue Wu Qian Nian*,《世界医学五千年》), p.5.

[5] Chu Kochen, the Chapter "The Relation of Climate to Human Life and Other Living Things" in the *Collected Works of Chu Kochen* (*Zhu Ke Zhen Wen Ji*,《竺可桢文集》), Beijing: Science Press, 1979, p.218.

embraced outright. Rather, if foreign theories are considered, they are carefully selected and adapted to fit existing practices.

4) Chinese medical knowledge, being an integral part of the overall culture, was later introduced to neighboring countries alongside the spread of Confucianism. However, its primary dissemination was toward eastern countries like the Korean Peninsula and Japan. Conversely, medical knowledge introduced to China originated from the West, with early influences coming from India and later influences from Europe and America during modern times.

6.1 The Footprints of Ashvins (Indo-European Divine Twin Horsemen)

The *Susruta Samhita* is one of the three medical classics of Ayurveda, which constitutes a system of medicine with historical roots in the Indian subcontinent. At the beginning of this book, Divodasa Dhanvantari, the king of Kasi, explains the origin and inheritance of medical knowledge, stating "The medical knowledge of Ayurveda was created by Brahma, passed on to Prajapati,[6] then to Ashvins and finally to Indra". He further adds, "I'll impart this knowledge to those who are eager to study for the betterment of mankind".

Similarly, the beginning of the Charaka Samhita, another revered medical classic of Ayurveda, also describes the origin of Ayurveda as attributed to Brahma and passed down to Prajapati, Ashvins, and Indra. Importantly, it was not the mortal Dhanvantari who approached Indra to acquire this knowledge, but rather the immortal Bharadvaja.

The transmission of Ayurveda from Brahma to Indra is depicted in a divine realm, although it is a mythical account. Notably, Ashvins hold significance in this narrative. They are among the oldest gods in the ancient religious system of the Indo-Europeans and first appeared in the Yamnaya culture on the shores of the Caspian Sea and Black Sea between 3200 and 2200 BC. "*Rigveda*, an ancient Indian classic of Brahmanism, calls it Nāsatya or Aśvinau. According to Aryan religious legends, Ashvins are twin messengers of God who often come at dawn to bring wealth to mankind and ward off disasters and diseases[7]". Ashvins have been eulogized as the God of medicine in *Rigveda*. While explaining why general surgery ranked first among the eight medical specialties, Dhanvantari also told a fairy tale about Ashvins. During the war of the Gods, Rudra severed the head of Yajna (the god of sacrifice), and the Gods beseeched: "Ashvins! You are the bravest among us, please restore Yajna's head to its original

[6] Free translation "Lord of Creatures". It is a title for the god of creation in Hindu mythology. The *Vedas* refers to Indra and other gods. *Manava-Dharma-Sastra* (*The Laws of Manu*) refers to Brahma and sometimes Manu. It is said that there are 7 lords, but some say 21, and the names are not the same.

[7] The introduction and pictures of Ashvins are all taken from Lin Meicun's *A Chinese Porcelain Jar Associated with Marco Polo: A Discussion from an Archaeological Perspective* (*Gu Dao Xi Feng-Kao Gu Xin Fa Xian Suo Jian Zhong Xi Wen Hua Jiao Liu*,《古道西风——考古新发现所见中西文化交流》) pp. 3–32.

place". Ashvins agreed and, as a token of gratitude, the Gods offered gifts to them. Ultimately, Ashvins successfully reattached Yajna's head.[8]

It is noteworthy that as one of the polytheistic Gods worshiped by Indo-Europeans (including Aryans, Iranians, Scythians, and Sakas), Ashvins not only holds a prominent place in the primitive religious system of these groups but also finds representation in various artistic artifacts. Statues of Ashvins can be observed in Persian stone carvings, the Saka Burial Mound in Kazakhstan, and Iranian bronze wares. Besides, the image of Ashvins can also be found in rock paintings in the Tianshan and Yinshan Mountains, on a Gold Medal unearthed in Warring States period tombs (475–221 BC) in Inner Mongolia, on eaves tiles unearthed in the tombs of the Lower Capital of the Yan State (1044–222 BC), and on clan emblems on bronze ware from the late Shang Dynasty (1600–1046 BC).

Ashvins, born on the northern shore of the Caspian Sea and the Black Sea, was brought to India, Persia, Iran, and northern China along with Ayurveda, the traditional medicine of ancient India. Ayurveda, influenced by the western regions, expanded and assimilated into other cultures as an independent system. Regrettably, they remain uncertain to this day whether Ashvins brought any specific medical knowledge or medicinal substances with them.

6.2 The Image Change of Xi Wang Mu (the Queen Mother of the West).

Although King Mu of the Western Zhou Dynasty also traveled far on a thousand-li horse, his precious steed obviously couldn't match the divine horse "Ashvins" in terms of stamina. So he only reached the Kunlun Mountains in the Western Regions and paid a visit to the "Queen Mother of the West".[9]

Consequently, contemporary scholars researching the history of Sino-foreign medical exchanges have brought up the issue of the "Queen Mother of the West":

> In the native Chinese myths and legends about the origin of medicine, there are few references to the elixir of life, which is believed to have come from outside … originating from foreign countries… The earliest legends of the elixir of life all came from foreign lands, at least from the area around the Kunlun Mountains … The "Land of Immortality", "Immortal Trees", "Immortal Mountains" and "Immortal People" in the *Classic of Mountains and Seas* (*Shan Hai Jing*,《山海经》), *Classic of Regions Beyond the Seas* (*Hai Wai Jing*,《海外经》) and

[8] Yano Michio (Japan), *Indo igaku gairon* (『インド医学概论』) or its Chinese translation *Ayurveda: Traditional Indian Medicine* (*A Shu Fei Tuo-Yin Du De Chuan Tong Yi Xue*,《阿输吠陀——印度的传统医学》), Shenyang, Liaoning Education Press, 2002, pp.34, 78.

[9] Volume 16 of *Classic of Mountains and Seas* (*Shan Hai Jing*,《山海经》): "There is a mountain named Kunlun. According to Volume 43 of the *Records of the Grand Historian* (*Shi Ji*,《史记》), the Queen Mother of the West, a woman who wore a jade hairpin with tiger teeth and a leopard tail, lived in a cave on this mountain. During the Western Zhou Dynasty, King Mu ordered Zaofu to travel to the west and encountered the Queen Mother of the West. The king was so delighted by this encounter that he even forgot to return".

> *Classic of the Great Wilderness* (*Da Huang Jing*,《大荒经》), which serves as proof.... It is probably also the place under the jurisdiction of the Queen Mother of the West.[10]

In the long-term life contact, it is reasonable and credible that the Huaxia nation has absorbed some medical knowledge from the cultures of the Western Regions. However, when it comes to the legend of the Queen Mother of the West, specific analysis is necessary because the image of the Queen Mother of the West actually varies in records from different periods. The purpose of the analysis is to illustrate that the content of myths and legends is modified because different eras have different cultural needs. As emphasized in the first chapter of this book, "legends" are not only useful but also very important. The most basic principle for using them is to "analyze the cultural issues of a specific time and region based on their time and place of origin".

The earliest record about the Queen Mother of the West probably dates back to the second year of the Taikang era (281 AD) of the Western Jin Dynasty, when a man named Bu Zhun from Jixian County looted the tomb of King Xiang of Wei and obtained the bamboo-slip book *Tale of King Mu, Son of Heaven* (*Mu Tian Zi Zhuan*,《穆天子传》). It mentions that King Mu traveled to the land of the Queen Mother of the West and drank and sang with her[11] Sima Qian's *Records of the Grand Historian* (*Shi Ji*,《史记》) has a similar account, while the *Classic of Mountains and Seas* mainly describes its geographical location and customs. Since the *Writings of Prince Huainan* (*Huai Nan Zi*,《淮南子》) states that "Yi asked the Queen Mother of the West for the elixir of immortality", she has been associated with "immortality" and "elixirs of immortality[12]". Subsequently, in the "rumored stories" about Emperor Wu of the Han Dynasty, namely the *Biography of Emperor Wu of the Han Dynasty* (*Han Wu Di Nei Zhuan*,《汉武内传》), which is traditionally attributed to Ban Gu but the real author is unknown, the name "Queen Mother of the West" appears 39 times, and she becomes a veritable immortal. Later generations cited it, for example:

According to *the Biography of Emperor Wu of the Han Dynasty*, among the superior celestial medicines of the Queen Mother of the West is the Kongtong spiritual melon. It also records that the Queen Mother of the West said to the Lady of Highest Prime, "We went to the Huozhu Mountain together and ate the spiritual melons. They tasted really good. I've been remembering that taste for as long as 7,000 years."[13]".

---

[10] Ma Boying et al., *A Chinese-Foreign Exchange History of Medicine* (*Zhong Wai Yi Xue Wen Hua Jiao Liu Shi*,《中外医学文化交流史》), Shanghai: Wenhui Press, 1993, pp. 5–6.

[11] *Tale of King Mu, Son of Heaven*, Volume 2: King Mu conquered the west and "passed by the land of the Queen Mother of the West"; Volume 3: "the King Mu and the Queen Mother of the West drank in the Yaochi and she sang a song for the King Mu".

[12] Gao You (the Eastern Han Dynasty), *An Overall Interpretation of Philosophical Work of Master Liu An* (*Huai Nan Hong Lie Jie*,《淮南鸿烈解》), Volume 6.

[13] Ouyang Xun (the Tang Dynasty), *Classified Excerpts from Ancient Writers* (*Yi Wen Lei Ju*,《艺文类聚》), Volume 87.

The Biography of Emperor Wu of the Han Dynasty also states that among the superior celestial medicines of the Queen Mother of the West is the Great True Red Ganoderma.[14]

It further mentions that there is a miraculous divine sparrow from Mount Kunlun among the Queen Mother of the West's celestial medicines.[15]

In *the Biography of Emperor Wu of the Han Dynasty*, the Queen Mother of the West said that among the celestial medicines, there are the Mysterious Frost and Crimson Snow.[16]

In addition, in the works of Sun Simiao, a medical expert in the Tang Dynasty, we can still find citations from the *Immortal Scripture of the Queen Mother of the West* (*Xi Wang Mu Xian Jing*,《西王母仙经》).[17] The *Record of Heretofore Lost Works* (*Shi Yi Ji*,《拾遗记》) states that King Zhao of Zhou asked the Queen Mother of the West for the "divine moth" to make the Nine-Revolving Divine Elixir, but "the Queen Mother didn't give it to him".[18]

As is well-known, Liu An, the King of Huainan, was fond of esoteric arts, and Emperor Wu of the Han Dynasty sought immortality just like the First Emperor of Qin. Therefore, at that time, the image change of the Queen Mother of the West, who was originally described as "looking like a human, with a leopard's tail, tiger's teeth, good at roaring and with disheveled hair", was a result of the trend of the times.[19] In later books, although such accounts are often seen, they are just continuous transcriptions of the above texts.

6.3 Brahmanism and Buddhism Entering the Qin Dynasty

The people of the Qin state originally lived in the western border areas and were in close contact with the Western Rong tribes. Therefore, their culture was greatly influenced by the latter. This point has already been explained when introducing the characteristics of "Qin culture" earlier. Ma Feibai's *History of the Qin Dynasty Based on a Collection of Documents* (*Qin Ji Shi*,《秦集史》) elaborates on this issue with the following three points.[20] Among them, it seems to also involve another figure

---

[14] Ibid. Volume 98.

[15] Li Fang et al. (the Song Dynasty), *Readings of Taiping Era* (*Tai Ping Yu Lan*,《太平御览》), Volume 922.

[16] *The Valley of Numerous Beautiful Flowers: An Encyclopedia for the Folk in the Song Dynasty* (*Jin Xiu Wan Hua Gu Qian Ji*,《锦绣万花谷前集》), Volume 2.

[17] Sun Simiao (the Tang Dynasty), *Essential Prescriptions Worth a Thousand Gold for Emergencies* (*Bei Ji Qian Jin Yao Fang*,《备急千金要方》), Volume 82.

[18] Wang Jia (the Jin Dynasty), *Record of Heretofore Lost Works*, Volume 4.

[19] *Classic of Mountains and Seas*: Volume 2.

[20] Ma Feibai, *History of the Qin Dynasty Based on a Collection of Documents*, pp. 358–360; pp. 361–363; pp. 364–366; pp. 714–715.

named "Agnivesa"[21] in the process of traditional Indian medicine being passed from the "divine realm" to the "human world".

1. The induction of Indian Agni to China: Agni, Wuji, etc., are the transliterations of Agni in the Indian Agni cult.
2. Duke Mu of Qin is said to have made Buddha statues (according to *Extensive Records of the Taiping Era* (*Tai Ping Guang Ji*,《太平广记》) quoting *Forest of Gems in the Garden of the Dharma* (*Fa Yuan Zhu Lin*,《法苑珠林》).
3. During the time of Emperor Shi of Qin, eighteen Samanas including Shi Lifang (释利房, or written as室利防) came to the East (according to *Collection on the Propagation and Clarification of Buddhism* (*Guang Hong Ming Ji*,《广弘明集》), etc., "释" can also be written as "室"; "房" can also be written as "防").

It's entirely up to the readers to judge whether to accept Ma's first point, that is, whether they agree with the view that "Anqi and Wuji are the transliterations of Agni in the Indian Agni cult". Personally, I not only deeply admire those who can analyze the meaning of words from the aspects of pronunciation and rhyme, but also think that they can connect many seemingly unrelated or unsolvable things and give unexpected explanations. The main basis for Ma's view is as follows:

> According to *The Index of Historical Records* (*Shi Ji Suo Yin*,《索隐》), quoting Yue Yan, "Song Wuji is an immortal in the moon"; it also quotes *Illustrated Book of Bai Ze* (*Bai Ze Tu*,《白泽图》) as saying, "The essence of fire is called Song Wuji. Presumably this person is the God of Fire." And Anqi and Anqisheng are the Kitchen Gods. Therefore, *Treatise about the Sacrifice for Heaven and Earth* (*Feng Shan Shu*,《封禅书》) states, "Sacrificing to the Kitchen God can attract spiritual beings. With the presence of spiritual beings, cinnabar can be transformed into gold…" Both of them are similar to the God of Fire (Agni) recorded in the *Indian Vedas*. In a family, Agni is the main god of the kitchen, so it is also called the master of the household.

Furthermore, it is discussed that the pronunciations of "An (安)" and "Wu (毋)" are similar and both can be read as "Ah (阿)".

> Since the Six Dynasties, Agni has been transliterated as "A Qi Ni (阿耆尼)". Anqi and Wuji are just the forms with the suffix "ni" omitted. In ancient China, Song Wuji was not the only person named Wuji. During the Spring and Autumn period (770–476 BC), there was Fei Wuji in the State of Chu and Han Wuji in the State of Jin. During the Warring States Period, there was Wei Wuji, the Lord of Xinling in the State of Wei. Perhaps these names were all inspired by the fire god Agni.[22]

It should be noted that while it's definitely okay to call Agni the God of fire, whether it's appropriate to refer to it as the "Agni cult" is debatable. After all, Agni is just one of the deities in ancient Indian "Vedism" and "Brahmanism". As for the differences between them, "from the perspective of historical development, Vedism, Brahmanism, and Hinduism are just three stages of the evolution of the same religion.

---

[21] Agnivesa is a human rather than a god and that's why he got the name. Indian medical historians said: Since the knowledge in Charaka's work comes from this, it is better to call it "*Agnivesa Collection*" rather than "*Charaka Collection*".

[22] Ma Feibai, *History of the Qin Dynasty Based on a Collection of Documents*, pp. 364–366.

Hinduism, as one of the oldest religions in the world, originated from ancient Vedism, went through the Brahmanism period, and gradually evolved into what it is today".[23]

Regarding Ma's claim that Buddhism was introduced to China during the reign of Duke Mu of the Qin, the old saying still holds true: it's easy to make a general statement, but difficult to provide a detailed argument. In general, no one will argue with you if you say that there were influences of foreign cultures in the pre-Qin period. However, if you specifically claim that Buddhism was introduced to China as early as the times of Duke Mu of Qin and Emperor Qin Shi Huang, you need to check whether the evidence is substantial and reliable.

The basis for Ma's second and third points is recorded in the literature as follows: In the *Forest of Gems in the Garden of the Dharma* compiled by the Buddhist monk Daoshi in the Tang Dynasty, it is stated that during the time of Duke Mu of Qin, a stone Buddha was discovered in Fufeng. Not recognizing what it was, Duke Mu discarded it in the horse-stable, defiling the statue. The god who protected the statue was angry and made the duke fall ill.[24]

In the Song Dynasty (AD 960–1279), Li Fang (AD 926–996) was ordered by the imperial court to supervise the compilation of *Extensive Records of the Taiping Era*. In addition to the above-mentioned content, it also added the content of "Asked about the stone statue that emerged from the mountain of Xianji Temple in Fangzhou, which dynasty was it erected in? The answer was: The statue was made by Duke Mu of Qin". And the story about "Eighteen Buddhist monks including Shilifang coming from the west" was also first mentioned by Buddhist monks in the Tang Dynasty.

According to sutra catalogues by people like Dao'an (a Buddhist monk) and Zhu Shixing (a Buddhist monk), during the reign of the First Emperor of the Qin Dynasty, eighteen Samanas including Shi Lifang came with Buddhist scriptures to convert the emperor. The emperor didn't accept their preaching and imprisoned them. At night, sixteen Vajras came and broke into the prison to set them free. Frightened, the emperor kowtowed and apologized.

However, when asked "Although there is such a saying, we don't know the exact time. What evidence is there?", the responders' answers appear rather feeble.[25]

> In *the Biography of Liu Xiang* (*Liu Xiang Zhuan*,《刘向传》), it is mentioned that Liu Xiang (77–6 BC) observed historical records widely and read the classics thoroughly." He often claimed, "As I searched through the ancient books, I often came across Buddhist scriptures." When he wrote the *Biographies of Various Immortals* (*Lie Xian Zhuan*,《列仙传》), he stated, "I rummaged through the collected books… and 74 of them have already appeared in the Buddhist scriptures." When Liu Xiang mentioned "the collected books", it probably referred to the books stored among the common people during the time of Emperor Qin Shi Huang; or it was said to be the books stored in Confucius' residence. Based on this, isn't it true that Buddhism had already spread in China before the Qin and Han Dynasties.[26]

---

[23] Zhu Mingzhong, "Some Questions Concerning Hinduism and Buddhism", *South Asia Studies* (*Nan Ya Yan Jiu*,《南亚研究》), 1991(1): 34.

[24] Shi Daoshi (the Tang Dynasty), *Forest of Gems in the Garden of the Dharma*, Volume 22.

[25] Li Fang (the Song Dynasty), *Extensive Records of the Taiping Era*, Volume 93.

[26] Shi Daoxuan (the Tang Dynasty), *Collection on the Propagation and Clarification of Buddhism*, Volume 11.

Despite the speculative nature of the answer, there is no direct evidence to confirm the arrival of Shi Lifang and the other eighteen Samanas in China. The *Biographies of Various Immortals*, long attributed to Liu Xiang, fails to provide evidence of the 74 individuals having appeared in sutras. Although the comments from monks in the Tang Dynasty do not sufficiently demonstrate the introduction of Buddhism during the reign of Duke Mu of Qin and the first emperor of the Qin Dynasty, it does not imply a denial of communication between China and Western Asia during that time. Furthermore, religion is not the exclusive means of transmitting technology and secular knowledge. Whenever there are exchanges between people, there will inevitably be exchanges of ideas, culture, technology, and commerce. What's more, Agni plays a crucial role in traditional Indian medicine. The principal classic of Ayurveda—*Charaka Samhita* (the current edition was completed not later than the second century)—shows that almost all medical knowledge is derived from his teachings. Just as Ma said, "Agni and his two followers - Anqi Sheng and Song Wuji - were not Buddhists and it was Brahmanists instead of Buddhists who introduced the Jinzhen Bozhang (金针拨障), a form of cataract surgery, also known as gold needle couching, into China during the Tang Dynasty". This is supported by the poem "Dedication to the Eye Doctor Who Is a Brahmanist"[27] written by a Chinese poet, Liu Yuxi (AD 772–842) in the Tang Dynasty.

In the autumn, my eyesight was afflicted by a cataract, causing distress in my life. My vision became blurred, making me appear much older than my years. I began to confuse red for green, and my eyes became sensitive to sunlight and susceptible to even the slightest breeze. The doctor has the skill of gold needle couching. How can he enlighten me (cure my eye problem)?

6.4 The Controversy About Birthplace of Traditional Medicine

A few years ago, a Chinese-American doctor in the United States carried out a study in the Aboriginal tribes during his spare time. Upon observing therapeutic techniques that resembled those used in traditional Chinese medicine (TCM), particularly the use of needles to stimulate specific points in the body and change fetal position, he hypothesized that these techniques might have been introduced to China through the Bering Strait centuries ago. Then he completed his doctoral thesis on this topic in a research institute in China.

Similarly, those who have some understanding of Indian culture and its traditional medicine will also notice the similarities between Chinese and Indian cultures. Several years earlier, Chinese and Japanese scholars argued whether Hua Tuo (AD 145–208) came from the western regions, which sparked intense criticism. However,

[27] *Yu Ding Quan Tang Shi* (《御定全唐诗》) in 46th year of Kangxi (1707) Volume 357.

most individuals refrained from commenting. Based solely on knowledge, it is challenging to determine whether these similarities are due to communication or independent occurrence. As we all know, *Manava-Dharma-Sastra* (*The Laws of Manu*) in the ancient Indian contains teachings comparable to Confucius's "three obediences and four virtues", yet no one has commented on a potential relationship between them. Similar phenomena and issues can be observed in the field of medicine as well. For example, the occurrence of the number "7" in witchcraft therapy, as mentioned in the second part of the book, and the use of "three Yin and three Yang" by doctors to divide time into six parts share great similarities with foreign cultures, but differ from Chinese culture as a whole. However, it cannot be concluded that these similarities are a result of communication.

In addition, even if a similarity is affected by communication, questions arise regarding which side originated the similarity and whether it was directly derived from specific knowledge or underwent intermediary processes during communication. For example, when encountering the statement, "babies born at seven months can survive, while those born at eight months cannot" in Arabic and Indian literature, it becomes difficult to determine its origin or when and how the Chinese became acquainted with this concept.

Nevertheless, I believe that discovering these similarities and differences between ancient cultures and traditional medicine of foreign countries is where the true fascination lies. It allows us to understand not only what is close to us but also the intriguing disparities and resemblances that exist. Exploring these unfamiliar yet familiar understandings and thoughts can be a captivating experience.

Chapter 7
Crafting the Big Picture from In-Depth Topic Studies

This chapter summarizes the conclusions of my previous researches in various aspects and redraws the basic picture of the medical history of the Western and Eastern Han Dynasties, with specific examples as the supporting points for constructing the overall picture. Here, the verification process and others' research work are not detailed. Those who are interested in the details can refer to the relevant articles in the "Special Topics".

In fact, each article in the special-topic research has two-fold significance. In terms of the "disciplinary history", that is, the so-called "internal history" research, it contains specific analyses of the growth process of various aspects of medical knowledge, making the concept of "gradual development" no longer an empty phrase. It can be said that it interprets from the perspective of understanding the "reasons why". On the other hand, each article also has significance in the aspect of "social history". It makes the so-called "external history" research no longer superficial. It can be said that it also reaches the level of understanding the "reasons why" in bridging the gap between "science" and "society".

7.1 The Core of the Reconstructing: The Age of the Current Version of *Huangdi's Internal Classic*

One of the fundamental "coordinate points" and "elements" for the overall picture of the medical development history from the Pre-Qin to the Han Dynasties is the production era of several epoch-making important works. And among them, the most important one is naturally the world-famous classic, *Huangdi's Internal Classic* (*Huang Di Nei Jing*, 《黄帝内经》). In his 1979 article "*The Formation of Huang Di's Internal Classic*", Keiji Yamada pointed out based on the characteristics of the Mawangdui medical manuscripts: Firstly, *Huangdi's Internal Classic* should

Y. Liao, *Reconstructing the Image of Medicine in the Qin and Han Dynasties*, Understanding China, https://doi.org/10.1007/978-981-95-6801-7_7

be regarded as "a microcosm of the contemporary medicine". Secondly, he criticized Chinese researchers for "deducing the compilation time of the silk-manuscript medical books based on the unproven assumption that *Huangdi's Internal Classic* was a work of the Warring States Period. On the contrary, one should deduce the compilation process and time of the former from the established facts about the latter".[1]

However, relying solely on the Mawangdui medical manuscripts cannot resolve the issue of the latest possible compilation time of the transmitted version of *Huangdi's Internal Classic*. Although the first bibliographic catalog in Chinese history, *Seven Categories of Writings* (*Qi Lve*,《七略》), which was compiled at the end of the Western Han Dynasty (202 BC-AD 8), has long been lost, when Ban Gu of the Eastern Han Dynasty compiled the "Catalog of Literary Works" in the *Book of the Former Han* (*Han Shu*,《汉书》), he basically copied its content. Since *Huangdi's Internal Classic* is recorded in it, it indicates that at the latest, by the time *Seven Categories of Writings* was compiled at the end of the Western Han Dynasty, this book must have already existed. Therefore, many researchers, including Keiji Yamada, although they strongly argue that the upper-limit of the compilation time of the transmitted version of *Huangdi's Internal Classic* should be after the burial time of the Mawangdui tombs due to the discovery of the Mawangdui medical manuscripts, have never had any doubts that its lower-limit is at the latest before the compilation of *Seven Categories of Writings*. However, the key to the problem lies precisely in the fact that the current version of *Huangdi's Internal Classic*, which consists of two independent works, *Plain Questions* (*Su Wen*,《素问》) and *Miraculous Pivot* (*Ling Shu*,《灵枢》), is not the 18 volumes of *Huangdi's Internal Classic* recorded in "*Catalog of Literary Works*". In other words, these two works, despite sharing the same title, are not identical. Therefore, the mention of *Huangdi's Internal Classic* in *Seven Categories of Writings* does not provide evidence for the latest completion time of the current version of *Huangdi's Internal Classic*. The main grounds for reaching this important conclusion are all detailed in the discussion of this special-topic research. Additionally, it is worth noting the significance of the period when Wang Mang (45 BC–AD 23) was in power. He summoned scholars from all over the country to write treatises in the court and unified "divergent theories". The traditional Chinese medicine theory of Zang-Fu organs, which is found in the transmitted version of *Huangdi's Internal Classic* and is still in use today, is not only the result of the process of "unifying divergent theories" but also an excellent example to observe this transformation process. Wang Mang unprecedentedly ordered the imperial physicians to conduct dissections and explicitly stated that it was for medical purposes. There may be some connection between this and the records of the morphology and dimensions of Zang-Fu organs and meridians in the transmitted version of *Huangdi's Internal Classic*. During the Eastern Han Dynasty (AD 25–220), when Ban Gu compiled the "*Catalog of Literary Works*" of the *Seven Categories of Writings* by Liu Xiang (77–6 BC) and Liu Xin (50 BC–AD 23), he clearly recorded the deletions and additions of books. However,

---

[1] Keiji Yamada (Japan), *The Philosophical and Scientific Culture of the Ancient East Asia* (*Gu Dai Dong Ya Zhe Xue Yu Ke Ji Wen Hua*,《古代东亚哲学与科技文化》), pp. 244, 259.

there is no record of adding any works by Liu Xiang and Liu Xin, who supported Wang Mang's usurpation, or any works that emerged during this period. Therefore, although we can infer from the absence of records in the *Seven Categories of Writings* that there was no *Plain Questions* and *Miraculous Pivot* before that time, we cannot deny the possibility that they were compiled during Wang Mang's era just because they are not recorded in the "*Catalog of Literary Works*".

After that, Keiji Yamada made a significant revision to his previous views based on my above-mentioned analysis and research on the compilation time of the present-day *Huangdi's Internal Classic*. He stated, "This is a sharp critique hitting the 'Achilles' heel' of all researchers, myself included. It's a remarkable discovery of an issue".[2] In his subsequent report, he further mentioned, "In my current view, among the existing *Huangdi's Internal Classic*, no more than 20 chapters were written in the Western Han Dynasty, and the rest were penned from the Xin Dynasty under Wang Mang to the early Eastern Han Dynasty".[3] However, as people worldwide are accustomed to calling *Plain Questions* and *Miraculous Pivot* the *Huangdi's Internal Classic*, and to clarify that it's not the one recorded in the *Catalog of Literary Works*, I refer to it as the "present-day *Huangdi's Internal Classic*" or the "transmitted *Huangdi's Internal Classic*".

With the clear chronological reference provided by the Mawangdui medical manuscripts and their close content-related connection with the current version of *Huangdi's Internal Classic*, and after a series of analyses and verifications on the logical soundness as mentioned above, it can be said that there are quite sufficient reasons to move the compilation era of the current version of *Huangdi's Internal Classic* from the pre-Qin period to the transition between the Western and Eastern Han dynasties. At least, it is much more reasonable than blindly adhering to the sage-based historical view without any evidence and simply claiming that the book was written in the pre-Qin era. Subsequently, as research on medical issues in the Han dynasties has continued to expand and deepen, many pieces of evidence that can serve as "collateral evidence" have been discovered. Therefore, in the whole "reconstruction" process, it is better to downgrade this important "conclusion" to the status of a "working hypothesis". Fortunately, subsequent research has not only failed to refute this "working hypothesis" but also further "corroborated" its rationality and credibility. Moreover, since many subsequent studies are closely and directly related to this, it can be said that determining the compilation era of the current version of *Huangdi's Internal Classic* plays a core role and holds great value in the entire "reconstruction" task.

---

[2] Keiji Yamada (Japan), *Bogao's Thoughts on Metrological Anatomy and Body Measurement* (Haku Kō ha no keiryō kaibōgaku to jintai keisoku no shisō). See *A Continuation of the Scientific History of Ancient China* (Chūgoku kodai kagaku shiron. Zokuhen) by Keiji Yamada, Tan Tanaka. Kyoto: Institute for Research in Humanities, Kyoto University, 1991, p.482.

[3] Keiji Yamada (Japan), *The Philosophical and Scientific Culture of the Ancient East Asia*, p.259.

7.2 Beyond the "Story"—Presenting the Theories of Bian Que

Although general medical history works all highly praise the pre-Qin divine doctor Bian Que, they basically just retell several "stories" in the *Biographies of Bian Que and Cang Gong* in *Records of the Grand Historian* (*Shi Ji*, 《史记》), without any "academic research". The reason is that although "Bian Que's Internal and External Classics" were recorded in the *Seven Categories of Writings*, modern people have no way to get even a glimpse of them. Since the view that the current version of *Huangdi's Internal Classic* is the "18 volumes of *Huangdi's Internal Classic*" recorded in the *Seven Categories of Writings* has been negated, the parallel relationship between the current version of *Huangdi's Internal Classic* and other "medical classics" recorded in the *Seven Categories of Writings* or "*Catalog of Literary Works*" in The History of the Han Dynasty has also been denied.[4] Since *Plain Questions* and *Miraculous Pivot* were compiled after the collation and recording of these medical books in the late Western Han Dynasty, there is a possibility that they absorbed the contents of other medical classics such as those of "Bian Que" and "Bai Shi" recorded in the *Seven Categories of Writings*. When carefully reviewing the existing materials from this perspective, the following unexpected results were obtained: In the current version of *Huangdi's Internal Classic* and the *Classic of Difficulties*, we can indeed find the contents of "Bian Que's Pulse Theory" recorded in *Pulse Classic* (*Mai Jing*,《脉经》) written by Wang Shuhe of the Western Jin Dynasty. The most typical example is that the entire chapter "Discourse on Very Strange Diseases" in *Plain Questions* appears word-for-word in "Bian Que's Rules of Pulse-taking" in *Pulse Classic*.

It is an obvious truth that the *Pulse Classic*, which was completed in the Western Jin Dynasty, could not have copied the text from the current version of *Huangdi's Internal Classic* and called it "Bianque's Rules of Pulse-taking". Therefore, we can only assume that Wang Shuhe, who served as the imperial medical officer, could still find the remains of Bian Que's works in the official library when he was collecting various old theories of "pulse diagnosis" from ancient times and compiling the *Pulse Classic*. Based on this, we can not only study the specific content of "Bian Que's Pulse Theory", which was praised by Sima Qian as "Even today, those who talk about pulse diagnosis in the world all follow Bian Que", but also provide strong evidence for the above-mentioned view that the current version of *Huangdi's Internal Classic* is not the one recorded in "*Catalog of Literary Works*" and for the analysis of its compilation time. This is exactly the issue discussed in the article "Research on Bian Que's Pulse Theory".

---

[4] Seven Works included in Catalog of Literary Works are: *Huangdi's Internal Classic* (*Huang Di Nei Jing*,《黄帝内经》), *Huangdi's External Classic* (*Huang Di Wai Jing*, 《黄帝外经》), *Bianque's Internal Classic* (*Bian Que Nei Jing*, 《扁鹊内经》), *Bianque's External Classic* (*Bian Que Wai Jing*, 《扁鹊外经》), *Family Bai's Internal Classic* (*Bai Shi Nei Jing*, 《白氏内经》), *Family Bai's External Classic* (*Bai Shi Wai Jing*, 《白氏外经》), *Parallel Chapters* (*Pang Pian*,《旁篇》).

7.3 The Vital Historical Materials—Medical Works from Mawangdui

In 1973, a significant archaeological discovery that caused a sensation both at home and abroad was made at Tomb No. 3 of Mawangdui Han Tomb in Changsha, Hunan province, where a large number of silk manuscripts and bamboo slips were unearthed. Among the unearthed bamboo and silk documents, there were more than ten medical documents that originally had no titles. The collation team named them respectively according to the content of each book as follows:

Moxibustion Classic of the Eleven Foot-arm Meridians (Zu Bi Shi Yi Mai Jiu Jing,《足臂十一脉灸经》);

Moxibustion Classic of the Eleven Yin-yang Meridians · Version A (Yin Yang Shi Yi Mai Jiu Jing,《阴阳十一脉灸经》甲本);

Rules of the Meridians (Mai Fa,《脉法》);

Death Signs of the Yin-yang Pulses (Yin Yang Mai Si Hou,《阴阳脉死候》);

Prescriptions for Fifty-two Diseases (Wu Shi Er Bing Fang,《五十二病方》) (The above 5 are on the same piece of silk);

Inedia and Breatharianism (Que Gu Shi Qi,《却谷食气》);

Moxibustion Classic of the Eleven Yin-yang Meridians · Version B (Yin Yang Shi Yi Mai Jiu Jing,《阴阳十一脉灸经》乙本);

Diagram of Daoyin (Dao Yin Tu,《导引图》) (The above 3 are on the same piece of silk);

Health-preservation Prescriptions (Yang Sheng Fang,《养生方》);

Miscellaneous Prescriptions (Za Liao Fang,《杂疗方》);

Book on Obstetrics (Tai Chan Shu,《胎产书》) (The above 3 are on the same piece of silk);

Among the above 11 items, since the texts of the two versions of the *Moxibustion Classic of the Eleven Yin-yang Meridians* are basically the same, there are actually ten works.

Additionally, among the bamboo and wooden slips unearthed from this tomb, there are four medical books, and they are respectively named as follows:

Ten Questions (*Shi Wen*,《十问》) (bamboo slips);

Methods of Intercourse between Yin and Yang (*He Yin Yang*,《合阴阳》) (bamboo slips);

Prayer-healing Prescriptions (*Za Jin Fang*,《杂禁方》) (wooden slips);

Discussion of the Perfect Way of Sexual Healthcare (*Tian Xia Zhi Dao Tan*,《天下至道谈》) (bamboo slips).

The reason why the unearthed medical books from Mawangdui are of inestimable significance to the study of the medical history of the Qin and Han Dynasties is as follows. Firstly, before the unearthed of these medical books, there was almost no direct medical historical material for the entire Western Han Dynasty, except for the life experience of Chunyu Yi (about 215–140 BC) and his 25 "medical cases" described in the chapter "Biographies of Bian Que and Cang Gong" in the *Records of the Grand Historian.* Secondly, the unearthed medical books from Mawangdui are not only large in quantity, which can provide relatively rich first-hand materials for the research on specific details in many aspects of the medical field, but also generally cover the four aspects of "medical classics, classical prescriptions, sexual techniques, and longevity art" in the "Category of Prescriptions and Technique" of *Seven Categories of Writings* compiled by Liu Xiang and Liu Xin when they sorted out ancient books in the late Western Han Dynasty. As "*Catalog of Literary Works*" roughly defined the connotation of "fangji": these are all "tools for maintaining life and the responsibility of an official position". More precisely, they should be called works related to the "study of life", not just about the prevention and treatment of diseases. These four aspects are the extension of the concept of "fangji" in that era. Thirdly, since they are the "responsibility of an official position", these works on the study of life should not be regarded as "cultural relics" for burial, but as carriers of knowledge in use at that time. Therefore, even though those who hold the view that the current version of *Huangdi's Internal Classic* was written before the Qin Dynasty date the works of the Mawangdui medical books back to the period between the Spring and Autumn (770–476 BC) and the Warring States Period, or even earlier, the most reliable evidence is the burial date of the tomb—168 BC, which can only be regarded as an objective manifestation of the actual level of medical development from the pre-Qin era to the early Western Han Dynasty. If we look at the issue from this perspective, then it is not purely accidental that these medical works coincide with the scope of the "Category of Prescriptions and Techniques" in *Seven Categories of Writings*.

Some scholars with an archaeological background have criticized the collation team for the "naming" of these unearthed medical books, which they believe violates the naming principles of archaeology. This criticism is also of great reference value. For example, regarding the so-called two versions of "*Moxibustion Classics of the Eleven Meridians*", if the people at that time had given them a title, they would probably have simply called them "*Book on Meridians* (*Mai Shu*,《脉书》)" or "*Meridians* (*Jing Mai*,《经脉》)". There is no way they would have used the words "eleven" and "moxibustion". The reason is that such perspectives come from the comparison made by modern people between these books and the classicized "meridian theory", and the close relationship between the meridian theory and acupuncture-moxibustion therapy. In an era when there was no classicized "twelve-meridian" system and no treatment method of needling according to meridians, there was obviously no question of "only eleven meridians" or "only moxibustion method". Just as the term "Traditional Chinese Medicine" emerged only when Western medicine was introduced and two different medical knowledge systems coexisted, before that, "medicine" was simply "medicine", and there could be no distinction between Chinese and Western medicine. In addition, it is also quite possible that *Death Signs of the Yin-yang Pulses* and *Rules of the Meridians*, which are written on the same silk scroll, are parts of a single work with consecutive content. Even all the texts recorded on this silk scroll might originally have formed a single medical book that covers "from theory to treatment". Of course, these are still just speculations and have little bearing on the overall situation. To sum up, the Mawangdui medical books are of crucial value in studying the transition of traditional medicine from empirical medicine to theoretical medicine in the early Western Han Dynasty, the establishment of the early meridian theory system, the use of moxibustion and acupuncture, the evolution of drug therapy, and the development of health preservation, providing the most precious and reliable materials. Therefore, apart from the special-topic studies in the following aspects, these unearthed medical books are often mentioned in discussions of other issues, such as the methods of "acupuncture and moxibustion" and "pulse diagnosis", especially the "way of health preservation".

7.3.1 Moxibustion Classic of Eleven Yin-Yang Meridians

When analyzing the dating of the literature from the perspective of the font, the Mawangdui Silk Manuscripts Collation Group noted that "the silk manuscripts containing five medical texts, including the *Moxibustion Classic of the Eleven Foot-arm Meridians*, feature a font similar to seal script. Among the Mawangdui silk manuscripts, this represents an earlier-style font, suggesting that the copying likely took place around the transition from the Qin to the Han Dynasty".[5] However, two points merit attention. First, the term "copying date" implies that the actual time of the texts' creation could be much earlier. Second, the conclusion that these texts are early works is based solely on the resemblance of the font to seal script. It is well-known that small seal script remained widely used during the Western Han Dynasty, especially in its early stages. It wasn't until the late Western Han Dynasty that clerical script gradually became more prevalent. As for the time span between the "copying date" and the "formation date", it's truly impossible to determine. It could have been a long interval or just a short while. In conclusion, considering the pre-Qin states' medical development levels and the medical knowledge recorded in *The Rites of Zhou* (*Zhou Li*,《周礼》) (widely regarded as a work from the early Western Han), we can't claim that these texts were merely burial items in the early Western Han Dynasty.

There are two types of works related to the meridian theory discovered at Mawangdui. These works record the 11 meridians in two different orders: one with the foot meridians followed by the hand meridians and the other with the yin meridians followed by the yang meridians. This is why the working group named them as *Moxibustion Classic of Eleven Foot-arm Meridians* and *Moxibustion Classic of Eleven Yin-yang Meridians*. Despite the differences between the two works, they both provide information about the running courses, indications of diseases and therapeutic principles of the 11 meridians. The *Moxibustion Classic of the Eleven Yin-yang Meridians* is similar to Chapter 10 "Meridians" in *Miraculous Pivot*. Therefore, I will analyze this classic which integrates yin-yang theory and is more comparable to today's 12-meridian theory. Different origins of characters, how this early-stage literature was further modified in *Huangdi's Internal Classic*, as well as the theoretical

[5] Working Group on Silk Manuscripts from Mawangdui Han Tomb. *Silk Manuscripts from Mawangdui Han Tomb* [Part Four] (*Ma Wang Dui Han Mu Bo Shu*,《马王堆汉墓帛书》). "Publication Note". p.2.

foundation of the "11 meridians" will be analyzed. For medical history researchers, studying this topic has two implications. First, when compiling various works, ancient people did not have the awareness of "copyright", so they had the freedom to delete or add any information. Second, it is important to consider the chronological order in order to examine the evolution of ideas, concepts, and techniques. It is taboo to explain what existed in the past based on later connotations and usages.

7.3.2 Therapy of Incantation

Former generations have generally explained witchcraft therapy, including incantation therapy, as either superstition or psychotherapy. Both criticism and praise were made from the standpoint of science. In his research, Keiji Yamada noted that in the works from Mawangdui, although words or language are used to treat diseases, the treating target is not "mental disorders". Instead of digging deeper on this issue, he analyzed in detail the whole process of one witchcraft case in *Prescriptions for Fifty-two Diseases*. I was invited to translate this article,[6] and the process was challenging. This was because the long-term "conventional education" had led me to overlook witchcraft, recognizing it as something that can be disregarded. After translating, as well as reading *Golden Bough* (*Jin Zhi*, 《金枝》), *Primitive Mentality* (*Yuan Shi Si Wei*, 《原始思维》), and other related works, my academic horizons have been significantly expanded. Back to the story, although Zhu You (祝由, prayer-healing, a witchcraft therapy involving the usage of words or language) is mentioned in the current version of *Huangdi's Internal Classic*, there is in fact no practical content. The unearthed medical works from Mawangdui fill the gap in our understanding of witchcraft therapy during the Han Dynasty. The findings of this special topic are as follows:

(1) According to the works from Mawangdui, there is no commonality in the diseases treated by incantation. However, after rearranging the "keywords", it was found that the disease cause is "anything that can understand language (from animals to elves that are alive)". From this, it can be inferred that the target of the incantation operator's language is the "disease cause", as implied in the term "Zhu You (祝由)"—to curse the disease cause. The character 祝 (zhu) or 咒 (zhou) signifies the act of cursing, and 由 (you) means "causes" in Chinese. Since the words are not directed toward the patient, interpreting them as "psychotherapy" is incorrect.

(2) The content of the curse consists of understandable threats instead of mysterious and incomprehensible spells from a wizard. The goal is to frighten away the disease causes that can understand language.

---

[6] Keiji Yamada (Japan). Twittering Bird at Night. Also included in *Selected Translation of Works on Chinese History by Japanese Scholars* (*Ri Ben Xue Zhe Yan Jiu Zhong Guo Shi Lun Zhu Xuan Yi*, 《日本学者研究中国史论著选译》). Beijing: Zhonghua Book Company, 1992, pp. 231-269.

(3) The incantation therapy mentioned in the works from Mawangdui relies solely on the words of the performer and does not involve external forces. Therefore, it can only be effective against disease causes that can understand language. However, the utilization of external forces can be seen in the chapter "Incantation Classic" in the *Supplement to the Essential Prescriptions Worth a Thousand Gold* (*Qian Jin Yi Fang*, 《千金翼方》) in the Tang Dynasty (AD 618–907), where methods such as gazing at celestial phenomena like thunder, lightning, and swallows, or calling upon gods, are employed to harness energy. This shows that witchcraft therapy was also developing and not limited to primitive societies with underdeveloped medicine, as is frequently described in works on medical history.

7.3.3 The Change of the Preparation Form of Peroral Medicine

In the chapter "Discourse on Decoctions and Wines" of the current version of *Huangdi's Internal Classic*, several preparation forms are mentioned, such as Tang Ye, medicinal liquor and medicinal materials. As for what Tang Ye (汤液) exactly is, there has been no specific explanation. Historically, it is falsely defined as "decoction" (汤剂). This special research starts with what Tang Ye is and discusses the vital change of preparation forms from solid to liquid in the Han Dynasty, based on the primitive prescriptions in the *Prescriptions for Fifty-two Diseases*.

It should be pointed out that although food is the primary ingredient of Tang Ye, it would be too "modern" to define it as food therapy. Apart from ingredients different from general herbs, the reason why Tang Ye is special lies in the methods and auxiliary materials used during processing, all of which can contribute to its efficacy. For example, Tang Ye that uses running water from thousands of miles away and is heated with hollow wood has the function of smoothing yin, yang, qi, blood, and meridians. This is the so-called "prescription of water and fire". Neglecting the materials involved in processing, such as the fuel used, and the methods employed would prevent a comprehensive understanding of how Tang Ye is utilized in disease treatment. Moreover, it could be mistakenly conflated with culinary techniques prioritizing "taste".

7.4 The Significance of the Eastern Han Dynasty When Compared with the Western Han Dynasty

The significance of the medical history in the Eastern Han Dynasty cannot be ignored, which can be supported by the following three topics.

7.4.1 *Before and After the Beginning of the Study of Materia Medica*

In Chinese, the name "Ben Cao", namely Materia Medica, first appeared as a 31BC official title in the chapter "Records of Outskirts Sacrifices" in the *Book of the Former Han*. In AD 5, Wang Mang summoned talents proficient in esoteric arts and Materia Medica to the capital of the time. In this way, the name "Ben Cao", which is equivalent to the concept of pharmacology, emerged around the turn of the Common Era. The research on Materia Medica during the Han Dynasty covered various aspects:

(1) The establishment of the name "Ben Cao" was roughly in the late Western Han Dynasty, suggesting that *Shennong's Classic of Materia Medica* (*Shen Nong Ben Cao Jing*, 《神农本草经》) could not have been written prior to the Qin Dynasty.
(2) *Wu Pu's Materia Medica* (*Wu Pu Ben Cao*, 《吴普本草》), written in the Three Kingdoms Period (220–280), introduces eight schools of Materia Medica, including Shen Nong, Huang Di, Qi Bo, Bian Que, Physician He, Tong Jun, Lei Gong, and Li Dangzhi. Therefore, *Shennong's Classic of Materia Medica*, incorporating only theories of Shen Nong, should not be regarded as the "summary of Materia Medica before the Han Dynasty".
(3) In the Liang Dynasty (AD 502–557), Tao Hongjing (AD 456–536) combined *Shennong's Classic of Materia Medica* with other supplementary content from various schools, giving rise to a more comprehensive *Classic of Materia Medica* (*Ben Cao Jing*,《本草经》). Prior to this, the study of Materia Medica had been developing in multiple ways. Earlier, it was *Wu Pu's Materia Medica* (*Wu Pu Ben Cao*, 《吴普本草》) that had the nature of "collecting" the theories of various schools. While the work truly worthy of being called an "integration"—extensively collecting and systematically organizing knowledge—was Tao Hongjing's *Variorum of the Classic of Materia Medica* (*Shen Nong Ben Cao Jing Ji Zhu*,《神农本草经集注》).

7.4.2 *Study on* Classic of Difficulties

Before the Tang Dynasty, the *Classic of Difficulties* was relatively less prominent. In early references, it was referred to as "Eighty-one Questions" (*Ba Shi Yi Wen*, 《八十一问》), or "Eighty-one Difficult Issues" (*Ba Shi Yi Nan*, 《八十一难》), without the titles "Classic" or "Huang Di". It was written in the form of self-questioning and self-answering with 81 questions in total. Interestingly, this number indirectly signifies the independence between the *Plain Questions* and *Miraculous Pivot*, both of which contain 9 volumes (81 chapters) each. So far, there have been two viewpoints on this work: first, the author was Bian Que (the Spring and Autumn period), or Qin Yueren (the Warring States period) who was praised as Bian Que. Second, it is an interpretation of *Huangdi's Internal Classic*. The discussions about *Classic of Difficulties* mainly focus on its theoretical framework:

(1) Many of the questions discussed in the *Classic of Difficulties* cannot be found in *Huangdi's Internal Classic*. This suggests that the *Classic of Difficulties* involved other lost literary works. If we consider the *Classic of Difficulties* as an annotation to the medical works headed by the existing version of *Huangdi's Internal Classic*, then it should be accomplished after them. Its integration of yin-yang and five-phase theories reflects characteristics of the Eastern Han Dynasty, and its writing style may also show influences from Buddhism.
(2) Instead of addressing theoretical issues in previous medical literature, the *Classic of Difficulties* used the topics to expound its own views and thus introduced a series of new "concepts". An example can be found in the current version of *Huangdi's Internal Classic*, where there are five Zang-organs and six Fu-organs, and the yin-yang nature of the numbers "five" and "six" does not align with that of the Zang-Fu organs. To solve this problem, in the *Classic of Difficulties*, the kidneys are interpreted as "the left kidney and the right life gate", making six Zang-organs. It also presents the "triple energizers" as having only a conceptual existence without a concrete physical entity, resulting in the presence of only five Fu-organs. In this way, the "Zang" (yin in nature) and "Fu" (yang in nature) can perfectly match the odd-yang and even-yin numerical properties. Consequently, not only does the new concept of the "life gate" emerge, but it also sparks endless debates about whether the "triple energizers", one of the six Fu-organs, are tangible or intangible. Moreover, on the basis of the "twelve meridians" in the *Huangdi's Internal Classic*, the concept of "eight meridians" which are analogous to lakes is proposed. There is also a new pulse-taking method of "only taking the pulse at the Cunkou" and dividing it into the "Chi" and "Cun" parts.
(3) An outstanding characteristic of the *Classic of Difficulties* is its integration of the yin-yang and five-phase theories, which are treated separately in the *Huangdi's Internal Classic*.

7.4.3 *Studies on* Treatise on Cold Damage and Miscellaneous Diseases

The research on *Treatise on Cold Damage and Miscellaneous Diseases* (*Shang Han Za Bing Lun*,《伤寒杂病论》) written by Zhang Zhongjing in the late Eastern Han Dynasty is extensive. However, few in-depth studies have been conducted on how it transformed the "six-meridian pattern identification" in the chapter "Discourse on Heat" in the *Plain Questions* (dividing exogenous febrile diseases into six stages according to the three yin and three yang, with different treatment methods for each stage) and the relationship between the "six meridians" here and the "six meridians" in the meridian theory. The special research in this regard points out that the six-meridian syndromes in "*Treatise on Febrile Diseases*" of *Plain Questions* are basically consistent with the records of the syndromes of each meridian at that time, while *Treatise on Cold Damage and Miscellaneous Diseases* is not restricted by the meridian theory and mainly focuses on the stage characteristics of the clinical course of the disease. This shows that from the turn of the Western and Eastern Han Dynasties to the late Eastern Han Dynasty, the "six-meridian syndrome differentiation" system for exogenous diseases, which was originally rooted in the meridian theory, had, to a certain extent, deviated from the meridian theory and become the names of various syndromes of exogenous diseases. Thus, it has become a relatively independent concept and theoretical system.

Another significant corresponding change is as follows: In "Treatise on Febrile Diseases" of *Plain Questions*, all the "diseases of the three yin" are "heat syndromes". Therefore, its treatment principle is to use diaphoresis when the disease is in the three yang (external heat) and purgation when the disease is in the three yin (internal heat). However, in the *Treatise on Cold Damage Diseases* (*Shang Han Lun*,《伤寒论》), the "diseases of the three yin" are all deficiency-cold patterns, so the treatment methods have changed to mainly using warm natured herbs. In the Qing Dynasty (AD 1616–1912), the so-called "new learning of warm-febrile diseases" emerged, which can actually be regarded as a return to the original appearance of the "six-meridian syndromes" in *Plain Questions*.

As for the reason why there was a shift from the view in "Treatise on Febrile Diseases" of *Plain Questions* that all the six meridians were in the state of "excess heat" to the view in *Treatise on Cold-Damage Disorders* that the three yin were in the state of "deficiency–cold" in the late eastern Han Dynasty, it should be noted that this shift may be related to the low-temperature climate caused by abnormal cold snaps in the late Eastern Han Dynasty and the harms (such as famine) caused by this climate change.

7.5 Theoretical Studies

7.5.1 The Connection Between the Establishment of Basic Theories and Morphological Perceptions (Anatomy)

Former generations have affirmed the anatomical achievements in *Huangdi's Internal Classic*, but none have addressed the correlation between the basic theories and morphological achievements. This approach aligns with the typical standard of modern medicine, which seeks scientific aspects in ancient materials. Some argue that traditional Chinese medicine (TCM) is considered philosophical medicine, leading to the belief that its theories are unrelated to anatomy. However, studies on this topic highlight that many physiological theories in TCM are based on limited morphological knowledge and speculation. Examples include:

(1) The ancients could see the shape of the gallbladder and its connection with the duodenum, but they couldn't know, as contemporary doctors (including both Chinese and Western medicine doctors) do, that bile comes from the liver. Moreover, they couldn't know its biochemical function of activating trypsin after being secreted into the duodenum to complete protein digestion. So they speculated that the essence of food flowed from the common bile duct into the gallbladder, and thus called it a "Fu-organ" that "stores essence without discharging".
(2) During a time when the function of the kidney was unclear, people observed the connection between the small intestines and the bladder through the "Jiao" (namely the omentum tissue), and they speculated that urine flowed into the bladder from the small intestines. This is also how the perplexing term "San Jiao" (namely triple energizers) originated.
(3) Meridians were explained based on visible superficial veins. With the introduction of Western medicine, people began to recognize the distinctions between the vascular system and the meridian system, raising the question of "What is a meridian?"

The origin and function of bile, the function of the kidney, and the formation of urine have become fundamental knowledge for modern medical researchers and TCM practitioners. Consequently, it becomes challenging for them to comprehend certain theories in ancient medicine. Similar issues arise when examining the false explanations for respiratory and circulatory mechanisms in ancient China.

7.5.2 The Circulatory and Respiratory Systems

In works on medical history at home and abroad, the theory of blood circulation expounded in the *Huangdi's Internal Classic*, written thousands of years ago, is widely praised. However, given the complexity and experimental evidence behind the recognition of the circulatory and respiratory systems in modern medicine, there are doubts about whether the Chinese theory of "blood circulation" existed before the Qin Dynasty. The findings on this topic are as follows:

(1) Ancient people did not have concepts like the heartbeat, and the criteria for determining death were solely based on body temperature and breathing. The twelve meridians were believed to originate from the "stomach", and the pulsation felt in the apex area was referred to as the "large stomach collateral", associated with the flow of "pectoral qi". These ideas suggest that during the compilation of the current version of *Huangdi's Internal Classic*, people believed that the generation and flow of blood were dependent on the process of "digestion".
(2) Traditional Chinese Medicine (TCM) views the pulse as the "movement of the vessel itself", and there is a saying that describes the pulse movement at a specific point as either fast or slow. This belief persisted even in works like *Bin Hu's*

Sphygmology (*Bin Hu Mai Xue*, 《濒湖脉学》) by Li Shizhen (AD 1518–1593) from the Ming Dynasty.

(3) The circulation of Qi was believed to be dependent on the lung. However, due to the lack of knowledge about "gas exchange" during that time, it was thought that Qi should circulate throughout the entire body.

7.5.3 Theory of Yin and Yang

Anyone engaged in the study of Chinese philosophy or culture recognizes the significance of yin-yang and five-phrase theories. Similarly, researchers in medical history and the entire Traditional Chinese Medicine (TCM) community regard these theories as fundamental to TCM. However, there has been limited exploration of how these concepts were applied in medical works during the Han Dynasty. The major points of this topic are as follows:

(1) Prior to the Qin Dynasty, yin-yang and five-phase theories existed as separate entities. This distinction continued until the Western Han Dynasty, and hence, in the current version of *Huangdi's Internal Classic*, certain chapters still maintain the separate treatment of yin-yang and five-phase theories.
(2) Dating back to the pre-Qin period, yin-yang theory focuses on "the circulation of yin and yang in the orderly four seasons". Thus, in the current version of *Huangdi's Internal Classic*, acupuncture-moxibustion therapy that lays emphasis on the circulation of qi and blood is mainly based on yin-yang instead of the five-phase theory.
(3) "Three yin and three yang", namely dividing yin and yang into three parts respectively, is a special form that only appears in medicine. However, even in the current version of *Huangdi's Internal Classic*, the most important classic that emphasizes basic theories, there is no explanation provided for its origins. Its application is confined to the realm of "time" circulation, including the year, month, and day.
(4) Over time, the significance of yin-yang theory gradually diminished. Its emphasis on "the order of the four seasons" gave way to the dialectical "opposition" of the two forces, transforming it into a tool of philosophical thinking.
(5) Yin-yang theory found acceptance in both the study of the *Books of Changes* (*Zhou Yi*,《周易》) and medicine, indicating that the development of these two domains was parallel. The statement "if you do not study *Book of Changes*, then you are not a great physician" originated only during the Ming Dynasty (AD 1368–1644). The *Book of Changes* serves as a vehicle for conveying the yin-yang theory rather than being its origin.

7.5.4 *Pulse-Taking Methods*

This special research mainly puts forward the concept of "ancient pulse-taking methods" based on the record in the *Classic of Difficulties*, which says "The pulse has three sections and nine manifestations, with yin and yang, light and heavy qualities, and sixty indications. One pulse changes with the four seasons. As it has been a long time since the sages' era, everyone adheres to their own methods. How can we distinguish them?", and then explores various pulse-taking methods.

(1) The analysis of several pulse-taking methods used in ancient times is presented (Fig. 7.1). These include:

Pulse-taking method based on three positions and nine indicators: This method relies on the theoretical foundation of heaven, earth, and human, with each divided into three indicators. However, its practical value is limited.

Pulse-taking method based on four seasons: This method focuses on the different pulses observed in each season, based on the subtypes of yin and yang. However, its clinical value is not significant.

Pulse-taking method based on the strength used: This method interprets the strength used in pulse-taking in relation to yin, yang, the four seasons, and the twelve months. Its theoretical foundation is similar to the previous method.

Pulse-taking method at Renying and Cunkou: This method takes yin-yang theory as its foundation, defining Renying (carotid pulse) as yang and external, Cunkou (radial pulse on the thumb side of the wrist) as yin and internal.

(2) The development of the modern pulse-taking method is closely associated with the approach used at Renying and Cunkou in the current version of *Huangdi's Internal Classic*. Initially, the *Classic of Difficulties* focused solely on taking

Fig. 7.1 The arts of medicine From *One Hundred Figures of the Folk Customs in Beijing* (*Bei Jing Min Jian Feng Su Bai Tu*, 《北京民间风俗百图》). This figure takes pulse-taking as the arts of medicine, embodying its function as the "cultural symbol" of TCM

the pulse at Cunkou, using the radial styloid process as Guan, the area on the wrist forward to the palm as Cun, representing conditions of yang, and the area backward as Chi, representing conditions of yin. Zhang Zhongjing and Hua Tuo (around 145–208) expanded on this method by taking pulses at Cun, Guan, and Chi, associating them with the upper, middle, and lower parts of the human body, respectively. Wang Shuhe in the Jin Dynasty (266–420) continued this approach in the *Pulse Classic*, establishing the correlation between Cun, Guan, and Chi on the wrists and the Zang-fu organs, which is still utilized today.

7.5.5 System of Acupuncture and Moxibustion

Studies on the theory of acupuncture and moxibustion in the Han Dynasty should focus on the following problems.

Firstly, the definition of acupuncture and moxibustion in the Han Dynasty differed from the modern understanding.[7] At that time, all the external treatments could fall into the category of acupuncture and moxibustion, for example, to burn the "verruca" and to cut the "carbuncle", just as described in the sentence "treatments employing fire and herbs are used to treat the internal, and acupuncture and moxibustion are used to treat the external".

Secondly, during the theorization of acupuncture-moxibustion therapy, the current version of *Huangdi's Internal Classic* emphasized meridians and the circulation of qi and blood, hence "reinforcing and reducing based on yin and yang" and the theory of "four seasons" originating from yin-yang theory abound in it. The *Classic of Difficulties* was the first to mention the correlation between the five elements and acupoints. When comparing the current version of *Huangdi's Internal Classic* with *Huangdi's Classic of the Bright Hall* (*Ming Tang Jing*,《明堂经》) written during the Eastern Han Dynasty, several distinct differences emerge:

(1) The former emphasizes reasoning and performing treatments based on "meridians", while the latter specifically focuses on the therapeutic effects of individual "acupoints".
(2) The former mostly selects acupoints on the "limbs", while the number of "trunk" acupoints in the latter increases.
(3) The former increases the stimulation amount by performing multiple shallow insertions at one location, while the latter is characterized by changes in the depth of needle insertion.

Huangfu Mi, a prominent figure during the Western Jin Dynasty (AD 215–282), played a significant role in the development of acupuncture and moxibustion.

---

[7] The modern concept of acupuncture and moxibustion refers to acupuncture and moxibustion therapies that are used to adjust the function of the Zang-fu organs, qi, and blood through the meridians and acupoints, thereby achieving the purpose of curing diseases. [Nanjing University of Traditional Chinese Medicine: *The Preface of Lecture of Acupuncture and Moxibustion* (*Zhen Jiu Xue Jiang Yi*,《针灸学讲义》), Shanghai: Shanghai Scientific and Technical Publishers, 1964].

He meticulously extracted key information from *Huangdi's Internal Classic* and *Huangdi's Classic of the Bright Hall*, merging them into a comprehensive treatise known as the *A-B Classic of Acupuncture and Moxibustion* (*Zhen Jiu Jia Yi Jing*,《针灸甲乙经》). This seminal work became the first monograph solely dedicated to the field of acupuncture and moxibustion. As a compilation, the subsequent evaluation of the *A-B Classic of Acupuncture and Moxibustion* primarily focused on its philological significance. Scholars recognized the potential benefits of having an additional version that could aid in the collation and preservation of these classical texts. However, it is important to note that the true value of this work lies in its integration of the two distinct acupuncture-moxibustion systems found in the aforementioned classics. Remarkably, these integrated theories continue to be employed in contemporary practice, forming the fundamental framework of acupuncture and moxibustion theory.

7.6 Theory of Five Circuits and Six Qi

The theory of five circuits and six qi is widely regarded as one of the most enigmatic and intricate aspects of traditional Chinese medicine (TCM), characterized by concepts such as "dominant circuit", "guest circuit", "year circuit", "dominant qi", "guest qi", "controlling heaven", and "terrestrial effect". These elaborate ideas, coupled with the complex calculation process, have contributed to the aura of mystery surrounding this theory within the realm of TCM fundamentals. From a scientific standpoint and in support of TCM, some contemporary scholars interpret the theory of five circuits and six qi as a form of "meteorological medicine". According to this viewpoint, diseases are influenced by yearly "circuits" and "qi", which, in turn, are determined by the yin-yang and five-phase nature of the heavenly stems and earthly branches. This raises the question of how this theory relates to the field of meteorology. This explanation aims to shed light on this connection by considering the following aspects.

(1) The seven chapters on the theory of five circuits and six qi in *Plain Questions* are believed to be supplemented by later generations, thus suggesting that this theory may represent a newer facet or a specific development in medical understanding during the Eastern Han Dynasty.
(2) There are still other chapters in *Huangdi's Internal Classic* that focus on health preservation, diseases, and treatment in terms of seasonal changes. Although these are closely associated with "the circulation of yin and yang in the orderly four seasons", and although the four seasons in them are transformed into five and endowed with the nature of five elements to correspond to the five-element theory, based on which diseases and treatments are discussed, these texts do indeed encompass elements of "meteorological medicine". This stems from the fact that the changes observed in the four seasons consistently adhere to the same objective pattern year after year. By applying this theory within a 12- or

60-year cycle to analyze the occurrence of specific diseases, a consistent and well-defined theory emerges. Simultaneously, a boundary is formed between this theory and the broader concept of "natural law". This boundary represents the essence of the theory of five circuits and six qi.

(3) The calculation methods employed in each of the seven chapters in the *Plain Questions* differ from one another, indicating that each author constructed their own system based on their unique understanding of "circuits" and "qi", influenced by the prevailing ideological trends of their time. Consequently, each calculation presents a distinct concept of "circuit" and "qi". In the Song Dynasty, due to a particular emphasis on this theory, an effort was made to unify these diverse concepts, giving rise to subsidiary concepts such as "dominant circuit", "guest circuit", "year circuit", "dominant qi", and "guest qi". In this way, a once simple theory gradually evolved into a complicated and mysterious one.

(4) Despite the apparent differences between each theory, they share a common objective of aligning circuits with qi (or matching heaven with earth, yin with yang, and heavenly stems with earthly branches). To achieve this, each circuit is further divided into excessive and deficient categories, while fire is differentiated into two types: sovereign fire and ministerial fire, to correspond with the six qi. Similar patterns have been observed outside the medical field as well. *Comprehensive Discussions in the White Tiger Hall* (*Bai Hu Tong Yi,*《白虎通义》) completed in AD 72 discusses why two of the five elements belong to yang while the remaining three belong to yin. This parallel reflects the amalgamation of yin-yang and five-phase theories evident in both the theory of five circuits and six qi and the *Classic of Difficulties*.

7.7 Way of Health Preservation

In most works on medical history, even including *China Great Encyclopedia* (*Zhong Guo Da Bai Ke Quan Shu,*《中国大百科全书》), the concept of "Yang Sheng (养生, health preservation)" in the sense of the study of life is associated with "Yang Sheng Zhu (养生主)" in *Zhuang Zi* (*Zhuang Zi,*《庄子》), and it is regarded as the origin of the "way of health preservation" with such connotations. Actually, the fable "Lord Wenhui learned about health preservation from the cook dissecting an ox" in *Zhuang Zi* expounds a purely political theory—"Yang (养)" means to govern, and "Sheng (生)" means the people. This is the first major fallacy that must be recognized when studying the development of health preservation science in the Qin and Han dynasties. The development process of "health preservation science" with the inherent definitions understood in medicine and commonly can be roughly described as follows.

(1) The chapter "On Purpose" in *Zhuang Zi* rebukes the techniques of "physique preservation" such as hanging like a bear and stretching like a bird, as well as breathing exercises, considering them as "deliberate" behaviors that go against "nature" and are not advisable. However, this indicates that such "physique

preservation" techniques did exist in the era of *Zhuang Zi*. The basic tenet of Taoist view of life is to follow nature, that is, to live out one's natural lifespan without premature death.

(2) Although Qin Shi Huang, the first emperor of the Qin Dynasty (259–210 BC) once ordered to lead virgin boys and girls to sail east on the sea in search of "immortal medicine", and in the Han Dynasty, "among the coastal areas of Yan and Qi, all the fangshi (square-masters, a group of people claiming to have supernatural powers and secret recipes in ancient China) clenched their wrists and claimed to have secret prescriptions and the ability to achieve immortality",[8] these records do not allow us to have specific knowledge of the then supernatural arts and the way of health preservation. However, the medical books unearthed from Western Han tombs such as those in Mawangdui and Zhangjiashan contain detailed records of the specific operation methods of practices like inedia, breathing exercises, therapeutic guiding exercises, and the art of sexual intercourse which combines sexual skills and sexual art. This is exactly what was classified as "immortality-seeking" and "sexual arts" in the "Catalog of Literary Works" of the *Book of the Former Han*, and together with "medical classics" and "effective prescriptions", they constitute the "means of nourishing life"—the "way of health preservation".

(3) The "school of Song and Yin", one of the Daoist schools during the Warring States Period, had many discussions on cultivating one's nature. The Daoist concept of following nature was also absorbed by medical classics like the current version of *Huangdi's Internal Classic* and transformed into the theory of cultivating one's nature of "being calm and void, and keeping the spirit within". On the other hand, during the Han Dynasty, when all other schools of thought were excluded and Confucianism was honored as the dominant ideology, the schools of Huang Di and Laozi had to withdraw from the political arena and turned to explaining the theories of Laozi and Zhuangzi from the perspective of self-cultivation. The *Annotation of Laozi by Heshanggong* (*Lao Zi He Shang Gong Zhu*, 《老子河上公注》) in the Eastern Han Dynasty is a typical example in this regard.

(4) Discussions on "cultivating one's nature" such as the relationship between the body and the spirit, moral cultivation, and moderating desires have existed in various schools of thought since the Pre-Qin Period. "Whether it was Laozi and Zhuangzi of Daoism, Confucius and Xunzi of Confucianism or Han Fei of Legalism, their views on nurturing the spirit, protecting the body, moderating one's desires, harmonizing emotions, and regulating diet were basically the same".[9] During the Three Kingdoms period, Ji Kang (around AD 224–263) wrote the *Discourse on Health Preservation* (*Yang Sheng Lun*, 《养生论》). Starting from the interdependent relationship that "the body relies on the spirit

[8] From the chapter "Annals of Emperor Wu", *Records of the Historian* (*Shi Ji*, 《史记》), Volume 12.

[9] Yan Shiyun. *The Academic History of Traditional Chinese Medicine* (*Zhong Yi Xue Shu Shi*, 《中医学术史》), Shanghai: Shanghai University of Traditional Chinese Medicine Press, 1989, pp. 44-45.

to exist, and the spirit needs the body to survive", he explained the principle that health preservation requires "cultivating both the body and the spirit", thus constructing the principle of health preservation like an ultimate truth.

Last but not least, when researchers examine medical history through a scientific lens, any content deemed "unscientific" tends to be filtered out. As a result, out of the four aspects of "medical classics, classical prescriptions, health preservation, and sexual healthcare" recorded in the *Catalog of Literary Works*, only the first two would typically receive attention. Even in recent years, when there has been a renewed focus on studying these areas, it predominantly falls under the domain of "cultural history" as a mere "phenomenon", rather than being integrated into the comprehensive framework of medicine during that era. As far as I'm concerned, the reason for this could largely be attributed to the trend of thought and education since the twentieth century, which made people consciously or unconsciously wear a pair of "glasses". These glasses turn people's inherent "stereoscopic vision" into two non-intersecting "parallel rays of light": one can see "science" on one side, and on the other side, one can see what was called "superstition" in the past and is now referred to as "culture". Unfortunately, these two perspectives rarely converge or interact with each other.

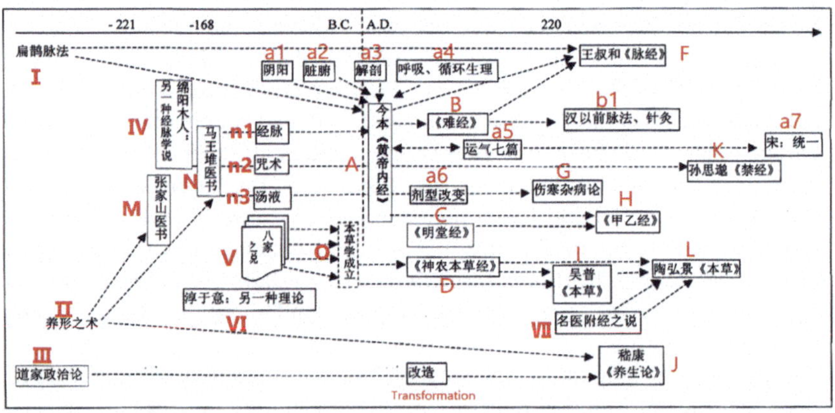

Fig. 7.2 Reconstructing the image of medicine in the Han dynasty (I) Bian Que's Rules of Pulse-taking (II) Ways to Preserve the Physique (III) Daoism Politics (IV) Wooden Figure with Meridians: Another Theory of Meridians (V) Eight Schools of Materia Medica (VI) Another Theory by Chunyu Yi (VII) Combined Theories of Various Schools (A) Current Version of Huangdi's Internal Classic (a1) Yin and Yang (a2) Zang-fu Organs (a3) Anatomy (a4) Respiratory and Circulatory Physiology (a5) Seven Chapters Supplemented by Wang Bing (a6) Changes of the Preparation Forms (a7) The Unification of the Song Dynasty (B) Classic of Difficulties (C) Classic of the Bright Hall (D) Shen Nong's Classic of Materia Medica (F) Pulse Classic by Wang Shuhe (G) Treatise on Cold Damage and Miscellaneous Diseases (H) A-B Classic of Acupuncture and Moxibustion (I) Wu Pu's Materia Medica (J) Discourse on Health Preservation by Ji Kang K. Incantation Classic by Sun Simiao (L) Variorum of the Classic of Materia Medica by Tao Hongjing (M) Medical Works from Zhangjiashan (N) Medical Works from Mawangdui n1. Meridians n2. Incantation n3. Tang Ye (O) Beginning of the Study of Materia Medica

While it is entirely necessary to make such a selection when traditional medicine is taught and learned as a practical skill to meet the health needs of the public, historical research must remove the "glasses" to restore its original appearance. This is also one of the meanings of the so-called "overall image". Based on the results of the above-mentioned thematic studies, the reconstructed development of medicine in the Western and Eastern Han Dynasties is shown in the Fig. 7.2.

Chapter 8
Discussions

8.1 Going down the Altar—*Huangdi's Internal Classic*

> The *Huangdi's Internal Classic* (*Huang Di Nei Jing*,《黄帝内经》), a renowned monograph, has served as the cornerstone of the theoretical framework of traditional Chinese medicine. Over the past two millennia, TCM has made significant strides by incorporating the academic principles presented in the *Huangdi's Internal Classic* and applying them in clinical practice. Even in modern times, TCM continues to adhere to these principles.
>
> —Notice of the Academic Seminar on *Huangdi's Internal Classic* held in Hong Kong in June, 2008.

It was agreed upon by the majority of TCM practitioners and traditional culture enthusiasts that the quoted statement accurately summarizes the status, role, value, and current state of the *Huangdi's Internal Classic* in the realm of traditional medicine. I share this viewpoint. However, behind such commendable recognition, the organizers of the conference expressed their purpose to undertake an analysis of relevant research to "chart a new path, new method, and new version for examining the *Huangdi's Internal Classic* in the 21st century". Such a goal can be attributed to two factors:

Firstly, since the appearance of the *Outline of Chinese Medicine* (*Zhong Yi Xue Gai Lun*,《中医学概论》)[1] in the middle of the twentieth century, there have been two distinct perspectives on "how to study TCM"—one stemming from the classical

[1] *Outline of Chinese Medicine* is one of the textbooks for medical colleges and universities compiled by Nanjing University of Traditional Chinese Medicine under the commission from the Ministry of Health of China. The "second volume" of its 1958 edition was "Selected Readings of the Classics", including introductions to *Huangdi's Internal Classic*, *Treatise on Cold Damage Diseases* (*Shang Han Lun*,《伤寒论》), *Synopsis of the Golden Chamber* (*Jin Gui Yao Lve*,《金匮要略》), and *Systematized Identification of Warm Diseases* (*Wen Bing Gai Shu*,《温病概述》). When reprinted in 1959, the introductions were changed to synopsis of "five circuits and six qi", internal medicine, ophthalmology, qigong and so on.

Y. Liao, *Reconstructing the Image of Medicine in the Qin and Han Dynasties*, Understanding China, https://doi.org/10.1007/978-981-95-6801-7_8

approach and the other from the *Outline*.[2] Consequently, this has raised questions regarding the significance and necessity of the "classics" in both theoretical studies and clinical application. Even within the "classical" camp, there exist diverging opinions on whether to meticulously examine every word of the classics or to extract the essence for profound comprehension.

Secondly, historical research in this field has been continuously advancing. As the late twentieth century approached, contrasting views on the age and authorship of the *Huangdi's Internal Classic* emerged, challenging previously held beliefs. Furthermore, there has been a focused analysis of the foundational theories and treatments, as well as a reinterpretation of important concepts in light of cultural and historical realities, which has shed new light on their original and evolving meanings.

The fact that the *Huangdi's Internal Classic* has come down from the "altar of divinity" makes it both possible and necessary for us to face the question of "how to read the *Huangdi's Internal Classic* in the 21st century". If this question can be solved, the problem of how to understand traditional Chinese medicine will also be resolved. Because the *Huangdi's Internal Classic* is worthy of being the epitome and representative of the characteristics of traditional medicine such as "longstanding history", "still surviving to date", and "the integration of culture and technology" in the coordinates of "time" and "space", thus forming a specific "fence". However, what is its practical value after all?

The key elements that form the foundation of the TCM theoretical system can be found throughout the various chapters of the *Huangdi's Internal Classic*. Its status as a classic is evident in its significant role in guiding medical practice. However, from a practical standpoint, there are certain challenges to consider.

Firstly, the nature of the classic means that the "key elements" are repeatedly mentioned in different chapters, as various authors interpret and apply them differently. Additionally, there are a significant number of irrelevant words and passages that merely mimic the classic without adding any real value. Secondly, as medicine has progressed in terms of theory and technique, it has become necessary to re-evaluate aspects such as etiology and treatments that were once considered relevant but are no longer applicable.

As early as the Western Jin Dynasty (266–316), Huangfu Mi recognized the shortcomings of the *Internal Classic*, such as its repetitive content and lack of substance. In response, he edited the *A-B Classic of Acupuncture and Moxibustion*, removing exaggerated expressions and repetitive content, and reorganized the classification. This led to the emergence of works like the *Great Simplicity of Huangdi's Internal Classic* (*Huang Di Nei Jing Tai Su*,《黄帝内经太素》), *Classified Classic* (*Lei Jing*,《类经》), and selected readings from the *Essentials of the Internal Classic* (*Nei Jing Zhi Yao*,《内经知要》). In the *Great Simplicity of Huangdi's Internal Classic*, *Huangdi's Internal Classic* was rearranged under 19 topics. Taking this as an example, we can measure the value of modern application of the theories and techniques in *Huangdi's*

[2] The "classic" camp emphasizes that learning TCM must start with "classics", and the various versions of *Outline* are destroying TCM. The "Introduction" camp maintains that the "classics" are unnecessary, it is enough to absorb the essence from the *Outline*.

Internal Classic (See Table 8.1). According to this table, there is not a significant amount of crucial content. If the important content is refined with systematical organization and expressed in modern languages, it involves such chapters as "yin-yang and five elements", "Zang-Fu organs", "meridians and acupoints", "causes and mechanisms of diseases", "diagnosis and therapeutic principles", which is actually consistent with the structure in the *Outline of Chinese Medicine.* It should also be noted that the *Outline* transcends *Huangdi's Internal Classic* both systematically and integrally. This results from the continuous development of TCM. If TCM is regarded as a practical skill utilized for the treatment of diseases, then the original texts of the *Huangdi's Internal Classic* may be set aside. However, if a deeper understanding of its origins and underlying principles is desired, studying the classic becomes necessary.

Why should we adopt a new approach, methodology, and version when studying *Huangdi's Internal Classic*? This is because we have realized various problems in reality, and desire a deeper understanding of this classic and "TCM". So, what do we mean by a new approach, methodology, and version? In my view, there are two aspects to consider. On the one hand, for practitioners who only use the skills (namely clinical doctors), it is beneficial to extract relevant content from the classic and compile it into textbooks, creating a new "version". This is precisely the purpose behind the development of works like the *Outline of Chinese Medicine.* On the other hand, for historians and individuals interested in the origins, development, and underlying concepts of *Huangdi's Internal Classic*, a cultural perspective within the historical context is necessary. It is crucial to move away from the traditional "sage-dominated historical view" and adopt an "evolutionary historical view".

8.2 The Academic Idol out of Psychological Needs

TCM classics can be likened to antique porcelain, as their value increases over time without being tied to practical use. Just like people still appreciate and utilize porcelain made using ancient techniques for their superior appearance and quality, TCM learns from the classics to gain essential philosophical insights. The revised textbook *Outline of Chinese Medicine* in 1959 deleted the chapters "Introductions to Huangdi's Internal Classic", "Treatise on Cold Damage Diseases", "Synopsis of the Golden Chamber", and "Systematized Identification of Warm Diseases" in the source text. This deletion demonstrated the distillation of the classics into this revised "Introduction". Therefore, rather than saying that TCM classics merely "resemble" antiques, they truly "are" antiques due to the following reasons:

1. TCM classics have a longstanding history;
2. Understanding TCM classics requires specialized knowledge and archaeological methods;
3. Successive generations have inherited and developed the techniques in TCM classics;

Table 8.1 The Subjects, Content and Remarks in *Great Simplicity of Huangdi's Internal Classic*

Subjects	Volume and whether it was lost	Chapters	Content and Remarks
Health preservation	Volume 1 (lost)	[note: The title of volume 2 is "second volume of health preservation", so volume 1 should be "first volume of health preservation".]	1. "Prevention" and "healthcare" are part of medicine at all times. 2. Theoretically, both TCM and modern medicine have similar concepts of "preventative treatment" and "healthcare". 3. Both traditional and modern medicine lay emphasis on the psychological and physical aspects as well as their mutual impacts.
	Volume 2 (the last chapter was lost.) search *Great Simplicity of Huangdi's Internal Classic* in classics publishing house in China for references.	Normalizing health preservation, six qi, nine qi, diet regulation, life span Yin and Yang	
Yin and Yang	Volume 3 (the first chapter was lost.) search *Great Simplicity of Huangdi's Internal Classic* in classics publishing house in China for references.	Comprehensive discourse on yin and Yang, yin-yang regulation, miscellaneous theories on yin and Yang	1. It is the core theory of TCM. 2. It is of vital importance in both theory and treatment. But there are great differences in conceptual interpretation and practical application between ancient and modern TCM.
	Volume 4 (lost)		
Unity of the human	Volume 5 (the first chapter was lost.)	Unity of yin and Yang, Unity of the four seas, twelve waters	1.The "unity between the heaven and the human" is explained concretely in it. 2.The explanations lack practical value.

(continued)

Table 8.1 (continued)

Subjects	Volume and whether it was lost	Chapters	Content and Remarks
Zang-fu organs	Volume 6 (the first chapter was lost.)	Assigning the division of the five Zang-organs, correspondences and manifestations of the Zang-fu organs, qi and fluids of the Zang-fu organs	1. It is an important theory of TCM. 2. Concepts such as "physique and spirit", "internal and external", "symptoms and signs" centered on five Zang- and six fu-organs are still of practical value in clinic.
	Volume 7 (lost)	(note: The title of volume 6 is "first volume of Zang-fu organs", so volume 7 should be "second volume of Zang-fu organs".)	
Meridians	Volume 8 (the first chapter was lost.)	Meridians like a circulatory ring, explanations on diseases of different meridians, explanations on Yang brightness meridians	1. It is an important theory of TCM. 2. Concepts such as "internal and external", "organs", "symptoms and signs" centered on meridians are still of practical use in clinic.
	Volume 9	Main meridians and their divergences, similarities and differences in the running courses of the meridians, different divergences of the meridians, fifteen collaterals, meridians and their cutaneous regions	
	Volume 10	Governor vessel, belt vessel, yin and Yang heel vessels, conception vessel, thoroughfare vessel, yin and Yang meeting vessels, branch and root of the meridians, root and connection of the meridians	
Acupoints	Volume 11	Fundamental transport points, five modifications and the transport points, treating diseases in the fu organs by sea points, qi holes, qi palaces, bone hollows	1. It is an indispensable part of TCM. 2. It is closely related to the theory of meridians with its own independence. 3. The acupoints and their indications changed dramatically in later generations.

(continued)

Table 8.1 (continued)

Subjects	Volume and whether it was lost	Chapters	Content and Remarks
Qi of nutrient and defense	Volume 12 (the first chapter was lost.)	Differences between nutrient and defense qi, movements of nutrient and defense qi, fifty circulations of nutrient qi, fifty circulations of defense qi	1. It explains physiologically the "circulation of qi and blood" in TCM. 2. It lacks practical value.
Measurements of the body	Volume 13	Meridian sinews, measurements of the bones, measurements of the intestines, measurements of the meridians	1.Those measurements of the body are not the actual value. 2"Proportionalized" measurements of the bones are still applied to acupoints location.
Diagnostic signs	Volume 14 (the first chapter was lost.)	Pulse manifestations in the four seasons, pulse manifestations of the true Zang-organs, pulse diagnosis in the four seasons, pulse diagnosis at Renying	1. Diagnosis is a common part in all medical systems. 2. "Four examinations" is summarized by later generations. 3. Modern pulse-taking method does not exist in *Huangdi's internal classic*. The methods employed in the book were phased out. 4. In *Huangdi's internal classic*, the main method of inspection is to inspect the complexion instead of the tongue.
	Volume 15	Diagnosis based on complexion and pulse, diagnosis based on complexion, Cunkou and Chizhong, pulse diagnosis at chi, pulse diagnosis at chi and Cun, diagnosis based on movements in the meridians of the five Zang-organs	
	Volume 16 (lost)	Diagnosis based on deficient and excessive pulses, miscellaneous diagnoses, discourse on meridians and pulses	
Symptoms and signs	Volume 17 (only the last chapter was preserved.)		
	Volume 18 (lost)		

(continued)

Table 8.1 (continued)

Subjects	Volume and whether it was lost	Chapters	Content and Remarks
Making-up prescriptions	Volume 19 Volume 20 (lost)	The ancient and contemporary, the essential ways, prescriptions and geography, suitable treatments for diseases of the physique and mind, prayer-healing, needles and stones, decoctions, organs and functions	1. Some content has fundamental significance of guidance. 2. Specific explanation on herbal therapy is very little, and has long been phased out. 3. The theory of "property and flavor" based on the attribute of "yin and yang", "five elements" is closely related to the basic theory of "Materia medica". It is not the basis for determining the efficacy of the herbs.
Nine needles	Volume 21 (lost)	Essential ways of the nine needles, essential explanations on the nine needles, various source points, likeness of the nine needles	1. Therapy of acupuncture has its own characteristics and system, but its modern connotation and denotation are different from the ancient ones. 2. The "nine needles" is actually a general term for all external treatment tools, including instruments to press and to cut abscesses and expel pus. 3. Most of the piercing skills were phased out in later generations.
	Volume 22	Rules of piercing, indications of the nine needles, triple piercing, three changes of piercing, five piercing, piercing of diseases in the five Zang-organs, five regulations of piercing, piercing the five evils, nine piercing, twelve piercing	
	Volume 23	Examining misleading piercing, examining qi and piercing, examining piercing in accordance with the norms, piercing abscesses contrary to and in accordance with the norms, examining collateral piercing, miscellaneous piercings	

(continued)

Table 8.1 (continued)

Subjects	Volume and whether it was lost	Chapters	Content and Remarks
Reinforcing and reducing	Volume 24	Prohibitions in accordance with the heaven, discourse on the fundamental Spirit, healthy qi, pathogenic qi, reinforcing, and reducing, deficiency, excess, reinforcing, and reducing, deficiency and excess	1. Therapeutic principles of reinforcing and reducing are based on the deficiency-excess nature of the disease. 2. The deficiency-excess theory experienced a gradual evolution from concrete to abstract. Correspondingly, specific treatments underwent the same transition.
Cold damage	Volume 25	Determining the febrile diseases, explanations on the febrile diseases, febrile diseases of the five Zang-organs, limpness of the five Zang-organs, explanations on malaria, three malaria, twelve malaria	1. Only the system of "six meridians" in chapter 31 "discourse on heat" of *plain questions* (*Su wen*,《素问》) be absorbed into the system of external contraction by later generations. 2. The actual application covers the theories, methods, prescriptions, and herbs of the two systems of "cold damage" and "warm diseases".

(continued)

Table 8.1 (continued)

Subjects	Volume and whether it was lost	Chapters	Content and Remarks
Cold and heat	Volume 26	Cold and heat Syncopes, syncope of the meridians, mutual movements of cold and heat, headache resulting from syncope, heart pain resulting from syncope, miscellaneous theories on cold and heat, abscesses, abscesses resulting from Worms, scrofula with sensations of alternating cold and heat, Laws of cauterizing cold and heat	1. Only the concepts of "cold syncope", "heat syncope", can be used as a tool of reasoning. 2. Contents applied to clinical practice are very limited.
Discourse on evils	Volume 27	Seven evils, twelve evils, staying evils, striking of the evils, transmission of the evils	1. The concept of "evil" can be applied as a tool of reasoning. 2. Contents for guidance in clinical practice are very limited.
Wind	Volume 28	Winds of various types, manifestations of various winds, miscellaneous theories on various winds, nine mansions and eight winds, wind at the eight seasonal turning points, discourse on impediment	1. The concept of "evil" is applied as a tool of reasoning. 2. Contents for guidance in clinical practice are very limited.
Discourse on qi	Volume 29 (the first chapter was lost.)	Triple qi, fluids, discourse on water, discourse on swelling, discourse on wind edema, discourse on cough	1. It contains several "key elements" that constitutes the theoretical system and physiological foundation. 2. It still serves as the guidance for clinical practice.

(continued)

Table 8.1 (continued)

Subjects	Volume and whether it was lost	Chapters	Content and Remarks
Miscellaneous diseases	Volume 30	Diseases during pregnancy, warm and Summerheat diseases, changes of the four seasons, disease of breath accumulation, disease of hidden beams, pain resulting from heat, spleen heat and wasting-thirst, gallbladder heat, headache with toothache, pain in the Chin, pain in the nape, throat impediment, throat pain, pain in the eyes, deafness, nosebleeds, susceptibility to anger, Zhen-diseased sinews, blood decay, vexation resulting from heat, cold body, melting flesh, panting while lying down, diminished qi, counterflow and fullness of qi, treating retching, lower Back pain, diseases of the thighs, pain in the knees, limpness and syncope, difficult urination and hematuria, seemingly affected by Worms and menstruation blockage, epilepsy, madness, reversal cold of the extremities, death resulting from reversal qi, reversal Yang qi, wind and qi contrary to the norms, body rigidity resulting from wind, wine wind, explanations on the meridians, measurements of the body, deficiency and excess of the meridians, prohibiting extreme deficiency, in accordance with the four seasons, regulations of the piercing of malaria, piercings of abdominal fullness, piercings of cholera, piercings of convulsions resulting from fright, piercings of abscesses in the armpits, explanations on diseases, diseases resulting from long-term qi counterflow, diseases resulting from six fu-organs, diseases resulting from the intestines and stomach, indications of the transport points of the meridians	1. The title "miscellaneous diseases" as it is, it actually covers "miscellaneous theories". Most of the titles and their respective content are not about "diseases". 2. The contents covered are related to "symptoms", hence "treatment based on symptoms" is also a feature of TCM. 3. Most of the contents are not in use in modern clinical treatment.

4. TCM classics possess immeasurable academic, historical, and cultural value. For those who seek to "revive the antiquity", these classics also provide valuable "spiritual support".

It's unclear when exactly the trend of "respecting the ancient and despising the present" started, but at least in the works of the Han Dynasty (approximately 202 BC–AD 220), discussions about this ideology can be found:

> Ordinary people in the world mostly respect the ancient and despise the present. As a result, those who expound doctrines must attribute them to Shen Nong and Huang Di, before they can persuade the chaotic and ignorant rulers. They make the origin of their doctrines seem remote and lofty, and thus people value them. Scholars are blinded by such theories and respect what they've heard. They sit upright seriously to praise these doctrines and straighten their necks to recite them.[3]

When criticizing the trend of "respecting the ancient and despising the present", the author of *Writings of Prince Huainan* (*Huai Nan Zi*,《淮南子》) accused those who "put forward their ideas under the guise of the ancients" of aiming to persuade the muddled rulers in troubled times, revealing the real purpose of those who "respected the ancient and despised the present" at that time. However, in the field of medicine, a craft that "gentlemen disdain", the situation is completely different—there is probably no blame of influencing muddled rulers in troubled times. At most, it's just "making the origin seem more ancient and lofty, and thus people value it", making "scholars be blinded by the theories and respect what they've heard, sitting upright and praising them, straightening their collars and reciting them".

There are mainly two forms of "writing in the name of ancient sages". The first involves directly attributing works to ancient sages, while the second form incorporates ancient theories to bolster one's own works. Both phenomena can also be observed in the field of medicine. Examples of the first form include *Essence on the Silvery Sea* (*Yin Hai Jing Wei*,《银海精微》), attributed to Sun Simiao (around 541–682) but written after the Song Dynasty (960–1279), and *Esoteric Teachings of Huatuo* (*Hua Tuo Shen Yi Mi Chuan*,《华佗神医秘传》) issued in 1920 by the Shanghai Ancient Books Preservation Society. The second form is more complex and encompasses various situations. It is natural in academia to build upon, supplement, and propagate knowledge based on the previous generations' understanding. The authors may not have intended to feign authorship by ancient sages or sought to garner praise and serious study from future generations. For example, Chapter 54 "Explanations on the Needles" in *Plain Questions* and Chapter 3 "Explanatory Remarks on the Small Needles" in *Miraculous Pivot* (*Ling Shu*,《灵枢》) are annotations of Chapter 1 "The Nine Needles and Twelve Origin Openings" in *Miraculous Pivot*.

---

[3] Gao You (the Han Dynasty), *An Overall Interpretation of Philosophical Work of Master Liu An* (*Huai Nan Hong Lie Jie,*《淮南鸿烈解》), Volume 19.

The authors of these "annotated" chapters offered their own interpretations of the original texts based on personal experiences. However, their explanations were later regarded as "the sayings of the sages" by subsequent generations.

The inclination toward "respecting the ancient" is influenced by various factors. One primary factor is the inherent human psychological need for "reverence". People often seek fulfillment in worshiping fictional gods, ancient sages, clay statues of Bodhisattvas, and living idols, as these entities resonate with their psychological needs. In the realm of medicine, individuals not only revere medical sages like Bian Que., Hua Tuo, and Zhang Zhongjing (around 150–219), but also the abstract idols of ancient "studies" or "classics". This mindset is not unique to China but is also prevalent in neighboring countries. For instance, after coming into contact with Traditional Chinese Medicine (TCM) during the Song (960–1279) and Ming Dynasties (1368–1644), Japan's Kampo Medicine began exclusively venerating Zhang Zhongjing's *Treatise on Cold Damage Diseases* (*Shang Han Lun*,《伤寒论》), abandoning the medical practices of the Song and Ming Dynasties, similar to the critique of Zhu Xi's (1130–1200) theories and the promotion of Confucianism in society. India provides another example, with the belief that all things originated within its borders and Ayurveda being the world's oldest medical system. Newly introduced cultures would be "sanskritized" and assimilated into literature, presenting them as ancient practices. Revivalists assert the absolute superiority of Ayurveda, attributing its decline to Islamic and British rule. They believe that by restoring its original purity, it can surpass Western medicine.[4]

The trend of "respecting the antiquity" inevitably leads to a loss of confidence among scholars, who dare not fathom the possibility of creating new and improved theories or practical skills. From their perspective, ancient medicine is an impeccably "finished body" since its inception, leaving no room for progress. All new things are just the "process of attaining the Dao", and their understanding stems either from studying under ancestors who "truly comprehend the system" or from their own "ability to comprehend", uncovering the hidden mysteries and practical applications within the minds of ancient sages. This approach prevents the formation of new knowledge from being seen as criticism but rather as "elucidation" of old theories or systems. Consequently, this perpetual cycle continuously amplifies the brilliance and reverence for "ancient academic achievements".

Understanding this unique way of "innovation" embedded in the development of traditional medicine leads to a fundamental realization: the surviving traditional medicine is not an exact "clone" of the ancient practice. While modern science openly embraces "progress" and "surpassing predecessors" as benchmarks for belief, goals, and judgment, Traditional Chinese Medicine (TCM) is sometimes hindered by a preoccupation with "respecting the antiquity". Researchers are often hesitant to discuss or even contemplate "innovation". In reality, whether it involves "respecting

[4] A translation by Yano Michio (Japan), *Indo igaku gairon* (『印度医学概论』); The Chinese version can be referred to *Ayurveda: Traditional Indian Medicine* (*A Shu Fei Tuo -Yin Du De Chuan Tong Yi Xue*,《阿输吠陀——印度的传统医学》), Shenyang: Liaoning Education Press, 2002, p. 51.

the antiquity" or "reviving the antiquity", it is merely a mindset rather than an objective fact.

8.3 The Mainstream and Tradition of Studies on the Medical History in the Qin and Han Dynasties

There have been doubts among scholars with a strong background in history regarding certain "unreasonable claims" about medical development in the Qin and Han Dynasties. However, rather than correcting these claims, they are often regarded as "differences in opinion" and frequently included in the "Literature Review" section of modern dissertations. There are mainly two reasons: First, from an academic standpoint, while these "doubts" may have been insightful, they primarily existed in literary and historical works and did not pose a significant threat to the established "tradition". Second, it is relatively convenient and safe to carry on with the "traditional" theories. From the perspective of sociology, since the establishment of the "History of Science" in China in the early 1950s, the main goals and tasks were focused on "unearthing and organizing the ancient heritage of Chinese science, technology, and culture". This aligns with the perspective of G.Sarton (1884–1956), considered the father of the history of science, who believed that "the history of science is the only history that can illustrate the progress of mankind". Such a value orientation has long dominated, leading society to explore ancient scientific and technological achievements within the framework of modern science and technology. Consequently, the research path of "emphasizing ancient accomplishments" has become the academic trend within this field. Therefore, suspicion of ancient knowledge and the pursuit of historical truth naturally went against this historical trend. It was not until the 1980s, with improvements in the research environment (including reduced political control and increased international academic contact), the emergence of new researchers, and a re-examination of the concept of "Whig history"[5] in Western society, that this situation began to change significantly. In this process, many works on humanity emerged. And these works no longer evaluate the history with the standards of whether it was "promoting" or "obstructing", "science" or "superstition", "correct" or "wrong", but focus on the overall description of the historical truth to explore its cultural features and evolution. However, maybe because of the powerful "traditional aspect" of TCM, although China has entered a historical period of "new historiography" since the early twentieth century, the research pace in medical history has lagged behind in comparison.

[5] Or "the Whig interpretation of history". The British Historian H. Butterfield (1900–1979) put forward this concept in *The Whig Interpretation of History* (1931). The main idea is that some historians study the past by referring to the present, simply dividing historical figures into two types, namely those who made and those who hindered progress. They used this rough method to select the right type to justify their own arguments. In the 1980s, the western world once again reflected on the interpretation of the "Whig history" and criticized the extreme anti-Whig tendency at the same time.

The historical expounding on medical works and important figures in the pre-Qin Period and the Han Dynasty can be seen in medical or general literary works of different ages. Since the twentieth century, medical history has gradually developed into a specialized discipline, fostering professional researchers and producing noteworthy works. The study of the history of the pre-Qin Period and the Han Dynasty, when TCM theories originated and flourished, naturally became more extensive and nuanced, exploring the interconnection between medicine and social politics, ideology, and culture.

In the twentieth century, a few academic works appeared such as *Medical History and Healthcare Organization* (*Yi Xue Shi Yu Bao Jian Zu Zhi*,《医学史与保健组织》), *History of Chinese Medicine*[6] co-authored by Wang Jimin (1889–1972) and Wu Lien-teh (1879–1960), and *History of Chinese Medicine* (*Zhong Guo Yi Xue Shi*,《中国医学史》) by Chen Bangxian (1889–1976) et al. However, the discipline was still in its early stages of development. It was not until the late 1970s that significant changes took place. However, if we take a cursory look at the various works (including textbooks) or articles on medical history during this period, we can feel the seemingly "default tone" which also led to the general agreement on many important issues. Tracing back to its source, this "default tone", which was observed automatically by later works, actually exists in the textbook *Outline of Chinese Medicine*. Although not primarily focused on medical history, the "Preface" of this textbook encapsulates the core content found in popular medical history works. One key aspect addressed in the "Preface" is the evaluation of TCM.

TCM is considered a vast treasure house, encompassing thousands of years of experience in combating diseases by the Chinese people. Under the guidance of its unique theoretical system, the principle of "pattern identification and treatment" was established. Beyond herbal therapy, TCM includes other specialized treatments such as acupuncture, qigong, and massage. Additionally, a relatively sophisticated system of medical management and education has been developed. TCM has played a vital role in maintaining the health of the Chinese population for thousands of years and has also contributed to the medical development in other countries.

Although the above evaluation is concise, it outlines the basic nature and features of TCM with its own innovation:

1. TCM, based on accumulated experience, distinguishes itself from pseudo-scientific and pseudo-arts practices like "witchcraft";
2. TCM possesses a distinct theoretical system that goes beyond empirical medicine;
3. TCM adopts the principle of "pattern identification and treatment", which fundamentally differentiates it from modern medicine;
4. This book approves of herbal, acupuncture, qigong, and massage therapies, expanding the treatment methods of TCM based on their unique features,[7] which can be seen as an innovative approach;

[6] The first edition was published in 1932, and a new edition has been published recently.

[7] "Qigong" was originally not part of TCM. In addition, it was also from this book that "cupping" become one therapy of acupuncture listed in "appendix".

5. Throughout history, TCM has developed a mature system of management and education, earning recognition at the Science, Technology, and Society (STS) level;
6. TCM has played an important role in people's health and the reproduction of the nation—as mentioned above;
7. TCM has its influence in the world, affirming its contribution to mankind.

In the following decades, both the TCM community and the government largely adhered to the aforementioned points, shaping so-called TCM policies, academic research, and research methods. For example, based on the first point, everything that does not conform to contemporary science is branded as "feudal superstition and dross". Consequently, there has been little investigation into the origins and development of these so-called "feudal superstition and dross" and the "constructive" nature of the basic theory, and few attempts to apply anthropological methods to analyze TCM history.

Similarly, though "pattern identification and treatment" has long been perceived as the core distinction between Western and traditional medicine, as well as a hallmark theory of TCM spanning thousands of years, very few have thoroughly examined its historical development.

Meanwhile, due to limited knowledge about traditional medicine outside of China, people often boast about TCM's global influence. This phenomenon, influenced not only by a self-centered perspective but also contemporary bias, predicts that the twenty-first century will be the century of TCM. However, few scholars have studied the impact of ancient foreign medicine on TCM.

The "historical review" that follows the "evaluation" has set the tone for TCM history development for decades. In other words, with time passing by, most of the medical history works have generally followed the "Preface" of the initial TCM textbook, offering scarce revolutionary insights. The key points in it are as follows (see Table 8.2).

I have emphasized the importance of "textbooks" on various occasions. In reform of teaching materials, the establishment of TCM basic theories and the stipulation of the specific content, *Outline of Chinese Medicine* did have its unique historical significance. However, it is important to note another significance of this textbook. It also serves as the important guidance in the overall image in depicting epoch-making representative figures, works and events that shaped the development of medicine. It summarizes the previous research as it was compiled in the middle of the twentieth century. Objectively speaking, subsequent works bear similarities in structure and content, even if they do not entirely follow this pattern. In other words, the historical descriptions in this book echo those before and after it, indicating that discussions on TCM history throughout the twentieth century were relatively consistent. Here, we do not have to dwell on the problems arising from this "invariability in the 100 years", nor do we need to pay much attention to the figures, works, and events after the Jin Dynasty (266–420). Just take a look at the first two historical periods in Table 13: *Huangdi's Internal Classic* was accomplished in the pre-Qin Period, followed by Zhang Zhongjing's *Treatise on Cold Damage and Miscellaneous Diseases* and Hua

Table 8.2 Key points about medical development through past dynasties in the "Preface" of *Introduction to Traditional Chinese Medicine*

Dynasties	Key Points
8th–third century BC	1. *Huangdi's Internal Classic* was accomplished. It takes the simple materialism of yin-yang and five-element as its medical theories... And provides the theoretical foundation for TCM 2. Bian Que. and his theories on pulse
Han dynasty (around 202 BC-AD 220)	1. Zhang Zhongjing further developed "pattern identification and treatment"; *Treatise on Cold Damage and Miscellaneous Diseases* made up for the loss of the "external classic" 2. Hua Tuo, surgery, and anesthesia (first in the world)
Jin dynasty (266–420)	1. *Pulse Classic* (*Mai Jing*,《脉经》) by wang Shuhe (210–280), the first monograph on pulse 2. Ge Hong's (284–364) alchemy promoted the development of pharmaceutical chemistry; simple and effective *Handbook of Prescriptions for Emergencies* (*Zhou Hou Bei ji fang*,《肘后备急方》); records of a variety of infectious diseases
Sui and tang dynasties (581–907)	1. *Treatise on the Pathogenesis and Manifestations of all Diseases* (*Zhu Bing yuan Hou Lun*,《诸病源候论》) records 1720 patterns and a variety of infectious diseases. It is the first monograph on causes of diseases and pathology. 2. *Essential Prescriptions Worth a Thousand Gold for Emergencies* (*Bei ji Qian Jin Yao fang*,《备急千金要方》) by Sun Simiao puts gynecology and pediatrics first 3. There are 1104 patterns in *Arcane Essentials from the Imperial Library* (*Wai tai mi Yao*, 《外台秘要》) by Wang Tao (670–755)
Song dynasty (960–1279)	1. Medical works were able to circulate due to the invention of the movable type printing 2. Medical education 3. Large works on prescriptions 4. The first monograph on forensic medicine - *Collected Cases of Injustice Rectified* (*xi yuan ji Lu*, 《洗冤集录》)
Jin and yuan dynasties (around 1115–1368)	Four schools of the Jin and yuan dynasties formed
Ming Dynasty (1368–1644)	1. Large works on prescriptions 2. School of warm tonification 3. Li Shizhen (1518–1593) and his *Compendium of Materia Medica* (*Ben Cao Gang Mu*, 《本草纲目》)
Qing Dynasty (1616–1912)	1. Literature collation 2. School of warm diseases 3. Materia medica, treatment methods, mature medical system, medical education, the influence on the world
Modern Times (1912-)	The reaction of Kuomintang and the wisdom of the Communist Party

Tuo's surgery in the Han Dynasty. These examples alone suffice to demonstrate the authority of this textbook.

I remember that a Japanese medical historian, Nakagawa Yonezo, criticized the danger of "fixed theory", "Once a theory is established, it becomes widely accepted and gradually transforms into an inflexible truth".

The reason why this incorrect view on the medical development in the pre-Qin and Han Dynasties "set the tone" is that this tone was already formed in the Republic of China (1912–1949) in *Treatise on the Origin and Development of Medicine in China* (*Zhong Guo Yi Xue Yuan Liu Lun*, 《中国医学源流论》) by Xie Guan (1880–1950):

The medicine of our country sprouted in ancient times, matured in the Zhou Dynasty (1046–256 BC), and was passed down during the Han Dynasty. Later, from the Jin (266–420) to the Tang Dynasty (618–907), relevant materials were collected, and new theories emerged during the Song and Ming Dynasties. The idea of "reviving the past" emerged in the Qing Dynasty and continued until the Republic of China. The introduction of Western medical methods marked a period of transformation for traditional Chinese medicine.[8]

This "tone" not only "standardized" the *Outline of Chinese Medicine*, but also has led to the mis-dating of important cultural relics unearthed from Han tombs. As a result of this "tone", there has been speculation that these relics may belong to an earlier dynasty instead of filling the medical image gap in the Han Dynasty.

Up until recently, the younger generation of researchers has continued to reference Xie Guan's theory, which states "three changes of TCM from the Zhou to the Qin Dynasties, from the Han to the Wei, and in the Republic of China". The first period signifies the "breakthrough of prescriptions and techniques", the second involves "revising the medical works", and from the end of the second period until the Republic of China, it was just "a continuous inheritance".[9]

In addition to overlooking the element of "time", the neglect of "space" is also a problem worthy of attention. Due to the limitations of transportation in ancient society, communication and knowledge-sharing differed greatly from modern times. Regions had distinctive dietary practices, clothing, architecture, written language, and spoken dialects. While the carriers of medical knowledge, such as cultural relics, documents, or records, shared similarities in content, researchers should always consider the possibility of cultural pluralism, independence, and coexistence. Hence, this book emphasizes the disparities among the "wooden figure with meridians" unearthed in Mianyang, Sichuan Province, the *Book on Meridians* (*Mai Shu*, 《脉书》) discovered in Hunan and Hubei Provinces, and the "theories of pulse" of Bian Que. and Chunyu Yi (approximately 215–140 BC) in eastern China. Therefore, the research on regional and cultural differences has been one of the key issues neglected

[8] Xie Guan, *Treatise on the Origin and Development of Medicine in China· Conclusion* (*Zhong Guo Yi Xue Yuan Liu Lun · Jie Lun*, 《中国医学源流论·结论》), Chengzhai Medical Publisher, 1936, third edition.

[9] Li Jianmin, *The Domain of Death and Life: The Origin and Development of Pulse Studies in the Zhou, Qin and Han Dynasties* (*Si Sheng Zhi Yu: Zhou Qin Han Mai Xue Zhi Yuan Liu*, 《死生之域: 周秦汉脉学之源流》), Taipei: Institute of History and Philology, Academia Sinica, agency Lexue Book Bureau, 2000, p. 49.

by historians. For researchers, their self-conscious observation on those differences depends on two aspects. First, they gradually acknowledge the commonalities in human thinking and understanding of nature through a broad understanding of different cultures worldwide. Second, they embrace this perspective through education, which enables them to observe and analyze problems continuously and apply them practically. For students, acquiring knowledge through practical experience and subsequently experiencing enlightenment is a gradual process. Therefore, the implementation of knowledge from top to bottom, known as "knowledge infusion", is of paramount importance. Hence, I consistently emphasize the importance of the teaching staff and textbooks. These aspects are crucial for students to gain comprehensive comprehension and enter the field of TCM, an essential component of general education.

One historian has pointed out, "The basic feature of Confucianism is its respect for ancient classics, on which everything is based".[10] In my opinion, for thousands of years and continuing to this day, this sentiment still holds true for numerous devout followers in the TCM community. Analyzing the impact of Confucian culture on TCM from this perspective may yield more profound insights compared to the superficial aspects of "moral cultivation" or "literary refinement".

On the other hand, the *Outline of Chinese Medicine* published in 1958 undoubtedly represents a milestone of epoch-making significance. As a textbook that is not a medical history textbook, it is quite normal for it to select and outline the most important aspects of the development history of this discipline in its "Preface", just like other textbooks. However, the problem lies in that it misplaces the era when the most important classic works were produced to the eighth to third centuries BC, and what's more, this statement is actually a longstanding and widespread view. I have chosen this book as because the target. On one hand, the short "Preface" indeed summarizes the tone and key points of the evaluation of TCM and the overview of its history from multiple aspects. It's hard to find such a concise and comprehensive written expression elsewhere. At the same time, in my heart, I actually want to use this to show my praise for the historical value of this book. If readers get this hint and skim through this book, they'll find that, except for the "Preface", the main text basically abandons the time coordinate and presents the theoretical system and practical skills of this discipline in a two-dimensional plane space. For example, it was this book that first officially incorporated the "ear acupoint chart" and "ear acupuncture therapy" (which did not originate from Chinese doctors) and the "cupping therapy" into the acupuncture and moxibustion therapy system. This will inevitably make learners mistakenly think that these are things that "have existed since ancient times" in China. Such a handling method is not bad for those who learn traditional Chinese medicine for practical purposes and only want to "know how it works". But for those who try to truly understand this knowledge and hope to figure out "why it works", it may cause some misleading. As many years ago, when someone asked Professor

[10] Han Xing: *A Combination of Confucianism and Law: Study of Politics and Culture of the Qin and Han Dynasties* (*Ru Fa Zheng He: Qin Han Zheng Zhi Wen Hua Lun*,《儒法整合:秦汉政治文化论》), p. 122.

Paul U. Unschuld, the former director of the Medical History Research Center at the University of Munich in Germany, "How many medical students choose your medical history course?" he replied, "About 10 percent", and then explained, "It's OK -- that's enough. Because not learning the history of medicine will never prevent them from prescribing medicine and performing surgeries for patients in the future. But if they want to become real doctors in the true sense, rather than just being satisfied with being craftsmen, they must learn the history of medicine".

Since the introduction of the *Outline of Chinese Medicine*, the Traditional Chinese Medicine (TCM) community has been divided into two contrasting camps. The reason is very simple. At the heart of this division lies a fundamental disagreement regarding the use of modern teaching materials and their impact on the preservation of TCM. While proponents argue that the *Outline* provides a comprehensive and accessible overview of TCM principles and treatments, critics fear that it undermines the foundations of TCM, insisting on the primacy of the "classics" as the starting point for learning. In my view, considering the practical needs of the majority of students, a middle ground can be found. It is advisable to further modify the *Outline of Chinese Medicine* by incorporating concise introductions to the origins and development of key components. By doing so, learners not only gain an understanding of the underlying rationales but also recognize that TCM, as an ancient knowledge system and therapeutic practice, continues to evolve. This approach acknowledges the importance of historical roots while making TCM more accessible for practical purposes. It is a strategy that promises optimal learning outcomes with minimal effort.

8.4 The Mystery of Medical Theories

Two prominent treatment methods in TCM, namely "acupuncture-moxibustion" and "herbal therapy", not only represent major therapeutic approaches but are also intricately linked to the foundational theories of TCM. These techniques and their associated theories emerged during the Han Dynasty. When elaborating on the medicine in the Qin and Han Dynasties, it is reasonable to give a certain explanation on its origin and development. However, it is worth noting that this field poses numerous puzzles, a rarity in other branches of historical research during this period.

The "acupuncture-moxibustion" therapy is related to the well-known mystery of the meridian theory. In this descriptive system about the "connections" of various parts of the human body, there are not only some contents that can be explained by the nerve theory, but also some that cannot be explained by the existing scientific knowledge. The latter is related to the understanding of the properties of medicinal herbs. Is it based on experience or the theory of medicinal properties? It is somewhat reasonable to say that the understanding of the efficacy of a certain drug comes from experience or analogical thinking. However, in actual use, we have no idea how the classic "prescriptions" that have been repeatedly proven effective were determined, especially how the knowledge of compatibility that has been proven to have "synergistic effects" by modern experimental science was obtained. In fact, we know as little about this as we do about the meridian theory. These are both two important medical achievements that took shape in the Han Dynasty and two important issues whose reasons we still don't know to this day. But imagine that even if one day scientific experiments clarify the reasons, we still can't explain how the ancient doctors at that time acquired this knowledge. In fact, for all the inventions whose original reasons have been lost over time, although it may be possible to reproduce their results using "scientific" methods, the process of their invention cannot be "reproduced".

This enigmatic aspect of medical history during the Qin and Han Dynasties adds to its allure. Much like the wonders of other early civilizations, we can marvel at the ingenuity of the ancient healers while acknowledging that their methods defy full comprehension.

Part II
Special Topics

A great man once said, "If you want to see the whole picture clearly, you must understand every part of it." In recent years, with the replacement of the research team and the influence of academic trends from abroad, the hot-spots of medical history research have gradually shifted to "cultural history", "social history", "disease history", etc. It seems that only in this way can there be innovation. However, there is still much to be done in "internal history research". To make a breakthrough, it is necessary to base on "detailed case studies". The content of this part includes some of my special works on the research of the medical history of the Qin (221-207 BC) and Han (around 202 BC-AD 220) Dynasties, which have been appropriately revised and compiled here for those who want to study it in detail. The main contents are as follows:

1. *The Position and Function of Ancient Anatomical Knowledge in Medical Theory* (*Gu Dai Jie Pou Zhi Shi Zai Yi Xue Li Lun Zhong De Di Wei Yu Zuo Yong*,《古代解剖知识在医学理论中的地位与作用》), *Studies in the History of Natural Sciences*(*Zi Ran Ke Xue Shi Yan Jiu*,《自然科学史研究》), 1987, (3).
2. *A Study of Huangdi's Internal Classic* (*Jin Ben Huang Di Nei Jing Yan Jiu*,《今本〈黄帝内经〉研究》), *Studies in the History of Natural Sciences*, 1988, (4).
3. *Studies on Bian Que's Theories on Pulse* (*Bian Que Mai Xue Yan Jiu*,《扁鹊脉学研究》), *Chinese Journal of Medical History* (*Zhong Hua Yi Shi Za Zhi*,《中华医史杂志》), 1988, (2).
4. *Studies on the Moxibustion Classic of 11 Yin-yang Meridians* (*Yin Yang Shi Yi Mai Jiu Jing Yan Jiu*,《〈阴阳十一脉灸经〉研究》), *Chinese Journal of Medical History*, 1989, (1).
5. *On the Origin and Development of Pulsological Techniques in the Pre-Han Period* (*Han Yi Qian Mai Fa Fa Zhan Yan Bian Zhi Yuan Liu*,《汉以前脉法发展演变之源流》), *Chinese Journal of Medical History*, 1990, (4).
6. *The Evolution of the Type of Medicament for Oral Administration in the Han Dynasty and a Study of Tang Ye* (*Han Dai Nei Fu Yao De Ji Xing Yan Bian Ji Tang Ye Yan Jiu*,《汉代内服药的剂型演变及"汤液"研究》), *Studies in the History of Natural Sciences*, 1990, (2).

7. *The Old Treatise on Cold Damage Diseases and the New Studies on Warm Diseases* (*Shang Han Jiu Lun Yu Wen Bing Xin Xue*,《伤寒旧论与温病新学》), *The Medical Arts of Qi Bo and Huang Di* (*Qi Huang Yi Dao*,《岐黄医道》), Shenyang: Liaoning Education Press, 1991.
8. *The Formation of Acupuncture in the Qin and Han Dynasties* (*Qin Han Zhi Ji Zhen Jiu Liao Fa Li Lun De Jian Li*,《秦汉之际针灸疗法理论的建立》), *Studies in the History of Natural Sciences*, 1991, (3).
9. *Research on Miscellaneous Records of Famous Physicians and its Relationship with Variorum of the Classic of Materia Medica* (*Kao Ding Ming Yi Bie Lu Ji Qi Yu Tao Hongjing De Guan Xi*,《考订〈名医别录〉及其与陶弘景的关系》), *Studies in the History of Natural Sciences*, 1992, (3).
10. *An Investigation into the Problems in Tao Hongjing's Herbal Works* (*Tao Hongjing Ben Cao Zhu Zuo Zhong Zhu Wen Ti De Kao Cha*,《陶弘景本草著作中诸问题的考察》), *Chinese Journal of Medical History*, 1992, (2).
11. *Time Features of the Medical Theory in the Classic of Difficulties* (*Nan Jing Yi Xue Li Lun De Shi Dai Te Zheng*,《〈难经〉医学理论的时代特征》), *Chinese Journal of Medical History*, 1993, (1).
12. *A Research on the Therapy of Incantation in Ancient China* (*Zhong Guo Gu Dai Zhou Jin Liao Fa Yan Jiu*,《中国古代咒禁疗法研究》), *Studies in the History of Natural Sciences*, 1993, (4).

13. *The Theory of Five Circuits and Six Qi in Traditional Chinese Medicine* (*Zhong Guo Chuan Tong Yi Xue De Yun Qi Xue Shuo*,《中国传统医学的运气学说》), *Journal of Dialectics of Nature*(*Zi Ran Bian Zheng Fa Tong Xun*,《自然辩证法通讯》), 1993, (2).

14. *Analysis on Various Deduction Patterns of Theory of Five Circuits and Six Qi in the Seven Major Discourses in Plain Questions* (*Su Wen Qi Pian Da Lun Yun Qi Bu Tong Tui Suan Fang Shi Zhi Fen Xi*,《〈素问〉"七篇大论"运气不同推算方式之分析》), *Chinese Journal of Medical History*, 1994, (2).
15. *The Relationship Between Daoism and Health Preservation* (*Dao Jia Si Xiang Yu Yang Sheng Xue De Guan Xi*,《道家思想与养生学的关系》), *Chinese Culture: Tradition and Modernization*(*Chuan Tong Wen Hua Yu Xian Dai Hua*,《传统文化与现代化》), 1993, (5).
16. *The Misunderstandings of the Respiratory and Circulation Mechanism in Ancient Chinese Medicine* (*Zhong Guo Gu Dai Yi Xue Dui Hu Xi Xun Huan Ji Li Ren Shi Zhi Wu*,《中国古代医学对呼吸、循环机理认识之误》), *Journal of Dialectics of Nature*, 1994, (1).
17. *The Yin-yang School, Yin-yang Theory, and Traditional Chinese Medicine* (*Yin Yang Jia Yin Yang Xue Shuo Yu Zhong Guo Chuan Tong Yi Xue*,《阴阳家、阴阳学说与中国传统医学》), *Chinese Culture: Tradition and Modernization*, 1995, (5).

Chapter 9
Research on the Current Version of *Huangdi's Internal Classic*

During the reign of Emperor Cheng in the Western Han Dynasty (32–6 BC), an initiative was undertaken to preserve and retrieve scattered works from various disciplines. The Emperor appointed Chen Nong, the Receptionist (Ye Zhe), to search for these scattered works among the common people. Liu Xiang, the Grand Master for Splendid Happiness (Guang Lu Da Fu), was summoned to proofread works from different schools and disciplines. Additionally, Li Zhuguo, the Attending Physician (Shi Yi), was tasked with proofreading medical works. After completing the proofreading of each work, Liu Xiang diligently organized the chapters, extracted key content, compiled a comprehensive catalog, and presented it to the Emperor. Following Liu Xiang's passing, Emperor Ai (reigning from 6–1 BC) enjoined Liu Xiang's son, Liu Xin (50 BC-AD 23), to continue the unfinished project. Liu Xin consolidated all the works and authored the *Seven Categories of Writings* (*Qi Lve*, 《七略》), marking the creation of the first book catalog in China.[1]

The *Seven Categories of Writings* records seven medical classic works with a total of 216 volumes, which are categorized as follows[2]:

> *Huangdi's Internal Classic* (*Huang Di Nei Jing*, 《黄帝内经》), 18 volumes;
> *Huangdi's External Classic* (*Huang Di Wai Jing*, 《黄帝外经》), 37 volumes;
> *Bian Que's Internal Classic* (*Bian Que Nei Jing*, 《扁鹊内经》), 9 volumes;
> *Bian Que's External Classic* (*Bian Que Wai Jing*, 《扁鹊外经》), 12 volumes;
> *Family Bai's Internal Classic* (*Bai Shi Nei Jing*, 《白氏内经》), 38 volumes;
> *Family Bai's External Classic* (*Bai Shi Wai Jing*, 《白氏外经》), 36 volumes;
> *Parallel Chapters* (*Pang Pian*, 《旁篇》), 25 volumes.

Among the revered classics mentioned earlier, it is the 18 volumes of *Huangdi's Internal Classic* that have been preserved and hold the highest authority. These

---

[1] Volume 30 of the *Book of the Former Han* (*Han Shu*, 《汉书》).

[2] *Seven Categories of Writings* was lost, but the content was preserved in the chapter "Catalog of Literary Works".

Y. Liao, *Reconstructing the Image of Medicine in the Qin and Han Dynasties*, Understanding China, https://doi.org/10.1007/978-981-95-6801-7_9

volumes are to TCM theory what the works of Confucius and Mencius are to Confucianism, *Tao Te Ching* (*Dao De Jing*, 《道德经》; also known as the *Five Thousand Words*, Wu Qian Yan, 《五千言》) to Daoism. This significant classic laid the foundation for TCM's basic theories and was considered a standard by later dynasties. However, the exact time of its compilation remains a subject of debate. It is speculated to have been completed during the Warring States Period (475–221 BC), at the transition from the Warring States Period to the Qin Dynasty (221–207 BC), or during the early Western Han Dynasty (202 BC-AD 8), among other possibilities. However, since its title is included in the *Seven Categories of Writings*, it must have been completed no later than the end of the Western Han Dynasty.

The current version of *Huangdi's Internal Classic* includes two parts, namely *Plain Questions* (*Su Wen*,《素问》) and *Miraculous Pivot* (*Ling Shu*, 《灵枢》). There are different views on whether they should be regarded as one or two separate works. An authoritative account of the origin and development of these two works exists:

Originally, *Huangdi's Internal Classic* incorporated 18 volumes, nine of which were under the title of *Plain Questions*, and the other nine volumes left untitled. However, in the Han and Jin (AD 266–420) Dynasties, the untitled nine volumes were referred to as *Nine Volumes* (*Jiu Juan*, 《九卷》), as mentioned in the prefaces of *Treatise on Cold Damage and Miscellaneous Disease* (*Shang Han Za Bing Lun*,《伤寒杂病论》) by Zhang Zhongjing (around AD 150–219) and *A-B Classic of Acupuncture and Moxibustion* (*Zhen Jiu Jia Yi Jing*, 《针灸甲乙经》) by Huangfu Mi (AD 215–282). The starting lines of the Nine Volumes indicate phrases like "The Ancient classic on Needle Therapy is to be established first", thus earning it the alternate title of the *Ancient classic on Needle Therapy* (*Zhen Jing*, 《针经》) … Over time, *Nine Volumes* expanded through additions by later generations, resulting in various versions. Gradually, *Plain Questions* and *Nine Volumes* diverged and eventually became the two distinct parts of *Huangdi's Internal Classic*. In the Tang Dynasty (AD 618–907), *Nine Volumes* was first named *Miraculous Pivot* by Wang Bing (AD 710–804).[3]

The above is the most common, orthodox, and classic introduction to *Huangdi's Internal Classic*. From it, the overall impression of the development of traditional medicine from the Warring States Period to the Qin and Han Dynasty is that the basic theoretical system of traditional Chinese medicine had been perfected in the pre-Qin period (before 221 BC), significantly ahead of the development in other fields of ancient Chinese science, technology, and culture. However, why do the medical books unearthed from the Mawangdui Han Tomb, whose burial date is known to be 168 BC, and the 25 medical cases of Chunyu Yi, a famous physician in the Western Han Dynasty (around 215–140 BC), reflect a much lower level of medical knowledge than *Huangdi's Internal Classic*. This is obviously puzzling. Additionally, during the Han Dynasty when science, technology, and culture witnessed great development, new progress in the medical field was not seen until the late Eastern Han Dynasty, represented by works such as *Shennong's Classic of Materia Medica* (*Shen Nong*

[3] Editorial Board of China Encyclopedia of Medicine: the chapter "Medical History" in *China Encyclopedia of Medicine* (*Zhong Guo Yi Xue Bai Ke Quan Shu*,《中国医学百科全书》), p.165.

Ben Cao Jing, 《神农本草经》) and *Treatise on Cold Damage and Miscellaneous Diseases*. Why did it seem like there was almost a blank in medical development during those several hundred years?

There are numerous such doubts. Therefore, in the study of the history of medicine in the Qin and Han dynasties, *Huangdi's Internal Classic* is of the utmost importance. However, the above-mentioned most orthodox and classic views on this book have always been the key factors hindering in-depth research and dispelling these doubts. It seems that everything has to start from scratch.

9.1 *Huangdi's Internal Classic* and Its Current Version

Regarding the division of *Huangdi's Internal Classic* into two parts (*Plain Questions* and *Miraculous Pivot*), the reason can be traced back to Huangfu Mi, a writer, historian, and physician from the Jin Dynasty (AD 265–317). In his preface to *A-B Classic of Acupuncture and Moxibustion*, he stated that according to the *Seven Categories of Writings* and the "Catalog of Literary Works", *Huangdi's Internal Classic* consisted of 18 volumes, with nine volumes on the *Ancient Classic of Needle Therapy* and nine volumes titled *Plain Questions*. This statement was considered true by later generations, as Huangfu Mi lived shortly after the Han Dynasty (202 BC-AD 220). Even today, this view is widely accepted and followed. However, there are reasons to question its accuracy based on the following facts:

(1) The *Catalog of Literary Works* incorporated 4 medical categories, namely medical classics, classical formula, sexual healthcare, and health preservation, covering 868 volumes of 36 schools. Unfortunately, apart from the 18 volumes of *Huangdi's Internal Classic*, all of the rest were lost. The loss of these works was so thorough that for thousands of years there has been no clue to some of the important terms like "Tang Ye (the earliest form of decoction,汤液)". Therefore, it raises the question of how *Huangdi's Internal Classic* managed to survive intact and remain popular throughout history.
(2) In *Catalog of Literary Works*, the terms "chapter" and "volume" are equivalent units. For example, the "Six Schools of Geomancy and Phrenology" consists of 122 volumes in total, with the *Classic of Mountains and Seas* (*Shan Hai Jing*,《山海经》) comprising 13 chapters. Adding up the numbers of chapters from various works results in a total of 122, indicating that 13 chapters are equal to 13 volumes. In "the school of rite", there are 56 volumes of *Ancient Classic of Rites* (*Li Gu Jing*, 《礼古经》), which is again equal to 56 chapters. Both are convincing examples. Therefore, *Huangdi's Internal Classic* should have only 18 chapters (equivalent to 18 volumes), which is significantly different from the current version containing 162 chapters.
(3) It is necessary to compare *Huangdi's Internal Classic* with other medical books recorded in the *Catalog of Literary Works*. For example, the *Prescriptions for Twelve Diseases on Impediment* (*Wu Zang Liu Fu Bi Shi Er Bing Fang*, 《五

藏六府痹十二病方》) has 30 volumes, the *Prescriptions for Sixteen Diseases on Hernia* (*Wu Zang Liu Fu Shan Shi Liu Bing Fang,*《五藏六府疝十六病方》) has 40 volumes, and the *Prescriptions for Twelve Diseases on Pure Heat* (*Wu Zang Liu Fu Dan Shi Er Bing Fang,* 《五藏六府瘅十二病方》) also has 40 volumes. All of these greatly exceed the number of volumes of *Huangdi's Internal Classic*. Can we believe that the prescription books only recording specific diseases like "Bi-syndrome", "hernia", and "Dan-syndrome" can be more than twice as long as the current version of *Huangdi's Internal Classic*?

(4) Another example is the *Ten Questions* (*Shi Wen,* 《十问》) unearthed from Mawangdui. Both domestic and foreign scholars believe that the two sections "Yao's Inquiry of Shun" and "Emperor Pan Geng's Inquiry of the Elder" in it are respectively related to the 23-volume *The Teachings of Yin Between Yao and Shun* (*Yao Shun Yin Dao,* 《尧舜阴道》) and the 20-volume *The Teachings of Yin Between Tang and Pan Geng* (*Tang Pan Geng Yin Dao,* 《汤盘庚阴道》) among the "Eight Works on Sexual Techniques" in the *Catalog of Literary Works*.[4] However, the section "Yao's Inquiry of Shun" only has 151 characters, and the section "Emperor Pan Geng's Inquiry of the Elder" only has 136 characters. No matter how these contents on sexual techniques developed, they could not reach the word-count of the current version of *Huangdi's Internal Classic*. Obviously, the 18-volume *Huangdi's Internal Classic* at that time had a very limited amount of text, approximately equivalent to 18 chapters. That is to say, it was only about one-ninth of the current version of *Huangdi's Internal Classic*.

(5) The current version of *Huangdi's Internal Classic* consists of two independent works, *Plain Questions* and *Miraculous Pivot*, but there is no such record in *Catalog of Literary Works*. Moreover, there are many differences between the two books in terms of terminology usage and theoretical selection. It can be clearly seen that they should originally be two independent works unrelated to each other.

(6) The names of the two books, *Plain Questions* and *Miraculous Pivot* (also known as *Nine Volumes, Ancient classic on Needle Therapy*) have always been used separately in the bibliographies of official histories of successive dynasties (see Table 9.1). Actually, it was only a speculation by Huangfu Mi, an official of the Jin Dynasty, to refer to them as *Huangdi's Internal Classic*. Wang Bing of the Tang Dynasty followed this view, and then later generations of medical scholars spread it through oral and written means up to now, but it has never been recognized by the bibliographies of successive dynasties.

Based on the inconsistencies mentioned above, it is clear that *Plain Questions* and *Miraculous Pivot* cannot be the 18 volumes of *Huangdi's Internal Classic* as recorded in the *Catalog of Literary Works*. This point was directly stated by Yao Jiheng (AD

---

[4] Working Group on Silk Manuscripts from Mawangdui Han Tomb. *Silk Manuscripts from Mawangdui Han Tomb [Part Four]* (*Ma Wang Dui Han Mu Bo Shu [Si]*, 《马王堆汉墓帛书》[肆]). pp. 148-149. Keiji Yamada [Japan], *Research on Newly Discovered Chinese History of Science · Translation and Notes* (*Shinhatsugen Chūgoku kagakushi shiryō no kenkyū · Yakutyo hen*). Kyoto: Institute for Research in Humanities, Kyoto University, 1985, p.297.

Table 9.1 Appellations of *Huangdi's Internal Classic*, *Plain Questions,* and *Miraculous Pivot* in the official catalogs and other works

Related catalogs or works	Appellations used or cited		
The chapter "Catalog of Literary Works" in *Book of the Former Han* (*Han Shu*,《汉书》)	18 volumes of *Huangdi's Internal Classic*		
Treatise on Cold Damage and Miscellaneous Diseases		*Plain Questions*	*Nine Volumes*
Pulse Classic (*Mai Jing*,《脉经》)		*Plain Questions*	*Ancient classic on Needle Therapy*
A-B Classic of Acupuncture and Moxibustion		*Plain Questions*	*Ancient classic on Needle Therapy*
The chapter "Catalog of Classics and Books" in *History of the Sui Dynasty* (*Sui Shu*,《隋书·经籍志》)		*Huangdi's Plain Questions*	*Huangdi's Acupuncture Classic* (*Huang Di Zhen Jing*,《黄帝针经》)
The chapter "Catalog of Classics and Books" in *The Old Tang's Historical Book* (*Jiu Tang Shu*,《旧唐书》)		*Huangdi's Plain Questions*	*Huangdi's Acupuncture Classic*
The chapter "Catalog of Literary Works" in *New History of the Tang Dynasty* (*Xin Tang Shu*,《新唐书》)		*Huangdi's Plain Questions*	*Huangdi's Acupuncture Classic*
The chapter "Category of Literary Works" in *Comprehensive Accounts* (*Tong Zhi*,《通志》)		*Huangdi's Plain Questions*	*Huangdi's Acupuncture Classic* *Internal Classic and Miraculous Pivot Classic*
The chapter "Catalog of Literary Works" in *History of the Song Dynasty* (*Song Shi*,《宋史》)		*Plain Questions* *Huangdi's Internal Classic · Plain Questions**	*Huangdi's Acupuncture Classic* *Huangdi's Miraculous Pivot Classic*
List of Writings Currently Held in the Nation of Japan (*nihon kokuken zaisuo mokuroku*)		*Huangdi's Plain Questions*	*Huangdi's Acupuncture Classic*

Note "*" is the annotated version by Wang Bing. The name *Huangdi's Internal Classic · Plain Questions* derives from here

1647–1715) in the Qing Dynasty (AD 1636–1912). In *Verification Between Real- and Pseudo-graphs of Ancient and Modern Times* (*Gu Jin Wei Shu Kao*,《古今伪书考》) he wrote, "The name of '*Huangdi's Plain Questions*' first appeared in the *History of the Sui Dynasty · Catalog of Classics and Books.* Wang Bing from the Tang Dynasty provided annotations, suggesting that the respective nine volumes of *Plain Questions* and *Miraculous Pivot* were actually *Huangdi's Internal Classic* as recorded in the *Catalog of Literary Works*. However, this is a false assumption". There were others who shared this view as well,[5] but their arguments lacked systematic elaboration and therefore held little impact.

9.2 Analysis of the Dialogs in *Huangdi's Internal Classic*

Although *Plain Questions* and *Miraculous Pivot* are two independent works, they share similarities in terms of structure and content. For example, they are both composed of nine volumes (81 chapters), both written in the form of questions and answers between Huang Di and his ministers. Additionally, they both quote some other medical works, among other similarities. These similarities may objectively lead people to accept Huangfu Mi's idea, yet these actually only reflect the time period in which the works were created. However, upon deeper analysis, differences can be found among these similarities. For instance, Huang Di plays different roles as either a teacher or a disciple, and the content involved in the questions and answers also has its own characteristics due to the different respondents.

In addition to Huang Di, *Plain Questions* also involves Qi Bo, Wei Yuqu, and Lei Gong. Since the seven chapters (Chapters 66–71 and Chapter 74) where the name Wei Yuqu appears were supplemented by Wang Bing of the Tang Dynasty when annotating *Plain Questions* with the "previously stored volumes" to fill in the missing parts, in fact, there are only discussions between Huang Di, Qi Bo and Lei Gong in *Plain Questions*. As for *Miraculous Pivot*, it involves five figures: Qi Bo, Bo Gao, Shao Yu, Shao Shi, and Lei Gong. Among the names of these ministers of Huang Di, only Qi Bo is recorded in the *Name Table of People from Antiquity* (*Gu Jin Ren Biao*,《古今人表》) in the *Book of the Former Han*. The origins of the other figures are difficult to trace. However, there are two points that predecessors didn't notice. First, in the Q&A sessions between Huang Di and his ministers, only in the "Lei Gong-Huang Di" Q&A mode does Lei Gong ask questions and Huang Di teach medical knowledge. In the rest, it's Huang Di who poses questions and the ministers give answers. Second, the medical principles described by the ministers are fundamentally different.[6]

---

[5] Wu Kaopan. *Verification of Huangdi's Internal Classic · Plain Questions · Miraculous Pivot* (*Huang Di Nei Jing · Su Wen · Ling Shu Kao*,《黄帝内经·素问·灵枢》考), *Chinese Journal of Medical History* (*Zhong Hua Yi Shi Za Zhi*,《中华医史杂志》), 1983, (2): 85; Fan Xingzhun. *A Brief History Traditional of Chinese Medicine* (*Zhong Guo Yi Xue Shi Lve*,《中国医学史略》), p.27.

[6] Japanese scholar Keiji Yamada has long realized the importance of this, "I noticed that there are traces of several schools in *Huangdi's Internal Classic* between questions and answers". (See *The*

In addition, there are some chapters in *Plain Questions* and *Miraculous Pivot* that do not adopt the question–answer format. In terms of content and writing style, these chapters show the characteristics of relatively primitive medical records on bamboo slips or silk. For example, Chapters 20 to 26 in *Miraculous Pivot* are written in the way of "For a certain disease, needling should be done at a certain place", so it is very likely that they are some kind of primitive medical records directly incorporated when *Miraculous Pivot* was compiled. Another example is Chapter 48 "Great Discussion on the Marvellous" in *Plain Questions*, which does not use the question–answer format. The whole text of this chapter is derived from the medical works of Bian Que in the Warring States Period. Therefore, the analysis of the Q&A relationship and content will contribute to clarifying the compilation sources of these two works.

It should be noted that when Wang Bing of the Tang Dynasty annotated *Plain Questions*, he "examined and compared the hierarchy of status to add content and make the meaning more explicit" regarding "the improper etiquette in the questions and answers between the sovereign and ministers". He also reorganized the content of each chapter, stating that "for the misplaced and fragmented texts with overlapping parts before and after, I studied their implications in detail, removed the redundancies, and retained the essentials".[7] Therefore, in the question–answer formats listed in Table 9.2, *Plain Questions* appears neater and more orderly than *Miraculous Pivot*. This is actually the biggest obstacle to the research on the compilation sources, but even in this situation, some valuable clues can still be found.

(1) Dialogs between Lei Gong and Huang Di (Chapters 75 to 81 in *Plain Questions*, Chapters 10, 48, 49 and one paragraph in Chapter 73 in *Miraculous Pivot*).

In the chapters where Lei Gong poses questions and Huang Di provides answers, there is a notable difference in writing style compared to other chapters in the book. For example, Lei Gong is referred to as "a junior disciple", and the opening phrases and word-using habits such as "Huang Di sitting in the Hall of bright", "today at high noon", "women and girls", "as the proverb says", "studying miscellaneous learning", "establishing the celestial measurements", "Shen Nong", "the two emperors" are not found in other dialogues with different question–answer relationships. Most of the other chapters directly start with expressions like "Huang Di asked", "Qi Bo replied", and when it comes to woman, the term used is "female". This indicates that these chapters marked with "Lei Gong's questions" may have other sources.

In terms of content, the first characteristic of the chapters with the "Lei Gong-Huang Di" dialogue is the discussion of the "Renying-Cunkou pulse diagnostic method". That is, the disease in the context of the "three yin and three yang" is determined by comparing the magnitudes of the "Renying" (carotid pulse) and the "Cunkou" (radial pulse at the wrist). This is an ancient pulse-taking method different from the current traditional Chinese medicine pulse-taking methods. If the modifier

---

Philosophical and Scientific Culture of the Ancient East Asia, p.251.) In my book, this is explained from a different perspective.

[7] The preface written by Wang Bing of *Huangdi's Internal Classic · Plain Questions*.

Table 9.2 Questioners or respondents in each chapter

Plain Questions	(01)Qi Bo(13)Qi Bo(25)Qi Bo(37)Qi Bo(49)-(61)Qi Bo(73)Qi Bo(02)-(14)Qi Bo(26)Qi Bo(38)Qi Bo(50)Qi Bo(62)Qi Bo(74)Qi Bo (03)Qi Bo(15)Qi Bo(27)Qi Bo(39)Qi Bo(51)Qi Bo(63)Qi Bo(75)Lei Gong(04)Qi Bo(16)Qi Bo(28)Qi Bo(40)Qi Bo(52)Qi Bo(64)Qi Bo(76)Lei Gong(05)Qi Bo(17)Qi Bo(29)Qi Bo(41)-(53)Qi Bo(55)Qi Bo(77)Lei Gong(06)Qi Bo(18)Qi Bo(30)Qi Bo(42)Qi Bo(54)Qi Bo(66)Wei Yuqu(78)Lei Gong(07)Qi Bo(19)Qi Bo(31)Qi Bo(43)Qi Bo(55)-(67)Wei Yuqu(79)Lei Gong
Plain Questions	(08)Qi Bo(09)Qi Bo(10)-(11)Qi Bo(12)-(20)Qi Bo(21)Qi Bo(22)Qi Bo(23)-(24)-(32)-(33)Qi Bo(34)Qi Bo(35)Qi Bo(36)-(44)Qi Bo(45)Qi Bo(46)Qi Bo(47)Qi Bo(48)-(56)Qi Bo(57)Qi Bo(58)Qi Bo(59)-(60)Qi Bo(68)Qi Bo(69)Qi Bo(70)Qi Bo(71)Qi Bo(72)Qi Bo(80)Lei Gong(81)Lei Gong
Miraculous Pivot	(01)Qi Bo(13)-(25)-(38)Qi Bo(51)Qi Bo(64)Bo Gao(76)Qi Bo(02)Qi Bo(14)Bo Gao(26)-(39)Qi Bo(52)Qi Bo(65)Qi Bo(77)-(03)-(15)Qi Bo(27)Qi Bo(40)Qi Bo(53)Shao Yu(66)Qi Bo(78)Qi Bo(04)Qi Bo(16)Huang Di(28)Qi Bo(41)Qi Bo(54)Qi Bo(67)Qi Bo(79)Qi Bo(05)Qi Bo self-teaching(29)Qi Bo(42)Qi Bo(55)Bo Gao(68)Qi Bo Shao Shi (06)Shao Shi(17)Qi Bo(30)Qi Bo(43)Qi Bo(56)Bo Gao(69)Shao Shi(80)Qi Bo Bo Gao(18)Qi Bo(31)Bo Gao(44)Qi Bo(57)Qi Bo(70)Qi Bo(81)Qi Bo(07)-(19)Qi Bo(32)Bo Gao(45)Qi Bo(58)Qi Bo(71)Bo Gao(08)Qi Bo(20)-(33)Qi Bo(46)Shao Yu(59)Bo Gao(72)Shao Shi(09)-(21)-(34)Qi Bo(47)Qi Bo(60)Qi Bo(73)Qi Bo(10)Lei Gong(22)-(35)Qi Bo(48)Lei Gong(61)Qi Bo Lei Gong(11)Qi Bo(23)-(36)Qi Bo(49)Lei Gong(62)Qi Bo(74)Qi Bo(12)Qi Bo(24)-(37)Qi Bo(50)Shao Yu(63)Shao Yu(75)Qi Bo

Note The numbers in parentheses are the chapter number in the current version; "-" means there is no form of questions and answers

words such as the questions and answers between the sovereign and his ministers are removed, it can be seen that its original form consists of short, mnemonic-like sentences, which quite feature the early medical books written on bamboo slips and silk. Chapters 48 and 49 of *Miraculous Pivot* preserve a relatively complete version of the "Renying-Cunkou pulse-taking method", and when other chapters involve this pulse-taking method, especially the meaning of "guan ge" (blockage and counter-flow), they are obviously based on this (see Table 9.3).

Feature two is related to meridians. Chapter 10 in *Miraculous Pivot* is the famous chapter on meridians, on which the current meridian models, wall-charts are based. It is recognized by the academia that this chapter was developed on the basis of *Moxibustion Classic of Eleven Yin-yang Meridians* (*Yin Yang Shi Yi Mai Jiu Jing*, 《阴阳十一脉灸经》) unearthed from Mawangdui. Moreover, it is more noteworthy that this chapter also takes the form of a dialogs between Lei Gong and Huang Di.

At the beginning of this chapter, it is stated that "Lei Gong asked Huang Di, 'in the words of the *Prohibition on Meridians* (*Jin Mai*,《禁脉》), for all the principles of acupuncture, the meridians come first…'". Since these words are found in *Miraculous Pivot - Chapter 48: Jin Fu*, annotators think that "*Prohibition on Meridians* (Jin Mai)" should be a mistake for "*Prohibition and Appropriation* (Jin Fu)". Actually, it is more reasonable to revise the title of Chapter 48 to "*Prohibition on Meridians* (Jin Mai)" based on this. Because when the monograph on meridians is called *Prohibition on*

Table 9.3 The content and citation of "Renying-Cunkou Pulse Diagnostic Method"

Chapters 48 "Prohibition and Appropriation" and 49 "The Five Complexions" in *Miraculous Pivot*	Other chapters in *Plain Questions* and *Miraculous Pivot*
▲ The movement in the vessels at the inch opening is controlled by the center. The movement in the vessels at the ren ying opening is controlled by the exterior. Both correspond to each other; they leave together and they come together. This is as if one were to pull a rope that has the same strength all along ▲ In spring and summer the movement in the vessels at the ren ying opening is somewhat more massive than at the inch opening. In autumn and winter, it is somewhat more massive at the inch opening. Those who are like this, they are called "healthy persons" ▲ When the movement in the vessels is at the surface and in the depth, and when a minimal or massive movement in the vessels at the ren ying and at the vessel opening is identical, the disease will be difficult to cure	Chapter 19 "The Four Seasonal Qi" in *Miraculous Pivot*: "At the qi opening the condition of the yin qi shows. At the ren ying opening that of the yang qi shows" Chapter 74 "Discussing Illness; Examining the Foot-long Section" in *Miraculous Pivot*: "When a person has a disease and the movements in the vessels at the inch opening and the ren ying opening are equally minimal or massive, and if they are identical in being at the surface or in the depth, it will be difficult to cure the disease"
▲ When the movement in the vessels at the ren ying opening is twice as massive than that at the inch opening, the disease is in the foot minor yang[8] conduit. When it is twice as massive and races, the disease is in the hand minor yang conduit ▲ When the movement in the vessels at the ren ying opening is twice as massive on top of the movement at the inch opening, the disease is in the foot greater yang[9] conduit. When it is twice as massive on top of the movement at the inch opening and races, the disease is in the hand greater yang conduit ▲ When the movement in the vessels at the ren ying opening is three times as massive on top of the movement at the inch opening, the disease is in the foot yang brilliance[10] conduit. When it is three times as massive on top of the movement at the inch opening and races, the disease is in the hand yang brilliance conduit ▲ When the movement in the vessels at the inch opening is twice as massive than that at the ren ying opening, the disease is in the foot ceasing yin[11] qi conduit. When it is twice as massive and races, the disease is in the hand heart ruler conduit ▲ When at the inch opening is twice as massive on top of the movement at the ren ying, the disease is in the foot minor yin[12] conduit. When it is twice as massive on top of the movement at the ren ying and races, the disease is in the hand minor yin conduit ▲ When the movement in the vessels at the inch opening is three times as massive on top of the movement at the ren ying opening, the disease is in the foot major yin conduit. When it is three times as massive on top of the movement at the ren ying opening and races, the disease is in the hand major yin conduit	Chapter 9 "Discourse on the Six Terms of a Year and on Phenomena associated with the Condition of the Depots[13]" in *Plain Questions*: "Hence, when the movement in the vessels at man's facing is once over normal fullness, the disease is in the minor yang. When it is twice over normal fullness, the disease is in the major yang. When it is three times over normal fullness, the disease is in the yang brilliance. When it is four times over normal fullness or more, the disease is 'obstructed yang[14]' When the movement in the vessels at the inch opening is once over normal fullness, the disease is in the ceasing yin; When it is twice over normal fullness, the disease is in the minor yin[15]; When it is three times over normal fullness, the disease is in the major yin[16]; When it is four times over normal fullness or more, the disease is 'yin closure' When the movement in the vessels that can be felt at man's facing[17] and at the inch opening are both four times over normal fullness, or more, this is 'closure and obstruction' {In case of 'closure and obstruction', the vessels are filled to extreme. The patient cannot live to the limits of the essence qi allotted to him by heaven and earth and will die.}"

(continued)

[8] "Minor yang" refers to "lesser yang" in Paul U. Unschuld's translation version of *Plain Questions* and *Miraculous Pivot*.

[9] "Major yang" refers to "greater yang" in Paul U. Unschuld's translation version of *Plain Questions* and *Miraculous Pivot*.

[10] "Yang brilliance" refers to "yang brightness" in Paul U. Unschuld's translation version of *Plain Questions* and *Miraculous Pivot*.

[11] "Ceasing yin" refers to "reverting yin" in Paul U. Unschuld's translation version of *Plain Questions* and *Miraculous Pivot*.

[12] "Minor yin" refers to "lesser yin" in Paul U. Unschuld's translation version of *Plain Questions* and *Miraculous Pivot*.

[13] 4 "Depots" refer to "Zang-organs" in Paul U. Unschuld's translation version of *Plain Questions*. He also uses "palaces" to explain "Fu-organs".

[14] 5 "Obstructed yang" refers to "repelling yang" in Paul U. Unschuld's translation version of *Plain Questions* and *Miraculous Pivot*.

[15] 6 "Minor yin" refers to "lesser yin" in Paul U. Unschuld's translation version of *Plain Questions* and *Miraculous Pivot*.

[16] 7 "Major yin" refers to "greater yin" in Paul U. Unschuld's translation version of *Plain Questions* and *Miraculous Pivot*.

[17] 8 "Man's facing" refers to "ren ying" in Paul U. Unschuld's translation version of *Plain Questions* and *Miraculous Pivot*.

Table 9.3 (continued)

▲ A movement in the vessels at the ren ying opening that is four times as massive on top of the movement at the inch opening, and that is both massive and accelerated, is called "overflowing yang qi". Overflowing yang qi constitute an external barrier. The patient will die; there is no cure ▲ When the movement in the vessels at the inch opening is four times as massive on top of that at the ren ying opening, that is called "internal closure". In the case of an internal closure, the movement in the vessels is both massive and accelerated. The patient will die; there is no cure[18]	Chapter 17 "Discourse on the Essentials of Vessels and the Subtleties of the Essence" in *Plain Questions*: "When yin and yang qi do not correspond to each other, the name of the disease is closure and obstruction" Chapter 17 "The Measurements of the Vessels" in *Miraculous Pivot*: "When the yin qi abound excessively, then the yang qi are unable to circulate. Hence that is called 'closure' When the yang qi abound excessively, then the yin qi are unable to circulate. Hence that is called 'barrier' When both the yin and yang qi abound, they cannot circulate through each other's sections. Hence that is called 'closure and barrier' Persons experiencing 'closure and barrier' cannot exhaust their time of life and die" Chapter 9 "End and Beginning" in *Miraculous Pivot*: "If the qi abound at the ren ying opening an additional four times more than normal ... that is called 'overflowing yang'. Overflowing yang generates an external barrier... If the qi abound at the vessel opening an additional four times more than normal, and if their movement is both massive and increased in speed, that is called 'overflowing yin' Overflowing yin generates internal block. If such a closure is not penetrable, the patient must die and cannot be cured... If the qi abound an additional four times more than normal ... that is called 'closure and barrier' In the case of closure and barrier the period by which life is shortened can be predicted"[19]

Meridians (Jin Mai), the two characters clearly hit the theme "meridian" and the academic background of the Qin and Han dynasties respectively—either it comes from the "Forbidden Palace" or it implies the meaning of "being prohibited from being spread".[20,21]

In *Plain Questions* and *Miraculous Pivot*, apart from the "classical" names, circulations, and directions of the meridians in the dialogues between Lei Gong and Huang Di, the meridian-related contents in other chapters are actually a collection of different and miscellaneous theories. Here are some examples:

[18] The Chapters 48 "Prohibition and Appropriation" and 49 "The Five Complexions" in *Plain Questions* of Paul U. Unschuld's translation version.

[19] Other Chapters in *Plain Questions* and *Miraculous Pivot* of Paul U. Unschuld's translation version.

[20] Volume 150 of the *Records of the Grand Historian* (*Shi Ji*,《史记》), there are "prohibited prescriptions", or "secret prescriptions". These terms should be referred to each other to explain "prohibition".

[21] In Chapter 48 "Prohibition and Appropriation" of *Miraculous Pivot*, "an oath is required" in order to transmit the knowledge. In Chapter 4 "Discourse on the True Words in the Golden Chest" of *Plain Questions*, it is said that "if it is not this kind of a person, do not teach him. If it is not this kind of truth, do not confer it".

In Chapter 2 of *Miraculous Pivot*, titled "To Consider the Transport Points as the Foundation", the number of meridians mentioned is the same as that in the medical books unearthed from Mawangdui, which is 11, and there is no pericardium meridian of hand-jueyin.

In Chapter 5 of *Miraculous Pivot*, called "Root and Connection", it is stated that the 12 meridians all originate from the extremities and terminate in the head, chest, and abdomen. However, there are actually only nine meridians described in this chapter. The described meridian pathways not only differ from those in Chapter 10, but also fail to form a complete circular pathway.

In Chapter 38 of *Miraculous Pivot*, titled "Movements Contrary to and in Accordance with the Norms, Being Well Nourished and Being Malnourished", it is mentioned that the lesser yin meridians only run downward, while in Chapter 10, they run upward.

In Chapter 7 of *Plain Questions*, titled "Further Discourse on Yin and Yang", it is stated that "People have four main meridians and twelve subordinate ones".

In Chapter 18 of *Plain Questions*, titled "Discourse on Phenomena Reflecting the Status of Qi in a Normal Person", "Xu Li" and "neck vessel" are mentioned. The movement of "Xu Li" is beneath the left breast, so it should refer to the cardiac pulsation that can be felt by hand. "Neck vessels" are also named after their locations and should refer to "Renying".

In Chapter 41 of *Plain Questions*, titled "To Pierce Lower Back Pain", there are various vessel names mentioned, such as the "separator vessel", "yin companion vessel", "transverse network vessel", "yang rope vessel", "yin meeting vessel", "yin rope vessel", "shining yang vessel", "dispersing vessel", "vessel of the flesh structures".

In chapter 28 of *Plain Questions*, titled "Discourse Thoroughly Deliberating upon Depletion and Repletion", there is a vessel referred to as the "capstring vessel".

In chapter 33 of *Plain Questions*, titled "Discourse Deliberating upon Heat Disease", there is "uterine vessel".

In chapter 52 of *Plain Questions*, titled "Discourse on Prohibitions in Piercing", there are mentions of the "stream vessel" and "large vessel".

In chapter 58 of *Plain Questions*, titled "Discourse on Qi Holes", the "sacrum vessel" is discussed.

Most of these meridian names are based on anatomical locations. Some of them are hard to understand and lack a systematic pattern, different from the system in *Miraculous Pivot - Chapter: Meridians.*

The noteworthy feature is the presence of theoretical inconsistencies in the dialogues between Lei Gong and Huang Di. For example, "Differing rules are presented regarding the diagnosis process based on the patient's age and social status. The patient must be asked whether he is of noble or low rank; whether he was a feudal lord who has been destroyed or harmed".[22] And also, "When the patient is of old age, search for the disease in the Fu-organs. When the patient is of young age, search for it in the meridians. When the patient is in the prime of his life, search for it in

[22] Chapter 77 "Discourse on Expounding the Five Faults" of *Plain Questions*.

the Zang-organs".[23] These follow another transmission order of diseases, which is clearly a theory of Huang Di and contradicts the one of Bian Que (skin and hair → flesh → vessels → sinews → bones → marrow).[24]

These phenomena indicate that the sections with the "Lei Gong-Huang Di" dialogues originally belonged to a certain school of thought and were incorporated into *Suwen* and *Lingshu* when these books were compiled. The same goes for the chapters of dialogues between Huang Di and other ministers, which should all be regarded as works of different schools. However, why is it that in other dialogues, it's always Huang Di asking questions and the ministers answering, while here it's Huang Di explaining medical theories? Considering that the core content of these sections is the theory of "meridians" and an ancient pulse-taking method (Renying-Cunkou pulse-taking method)—they can be said to be "self-contained systems". Then, can we assume that the core content of these sections is part of what was recorded as *Huangdi's Internal Classic* (and possibly even *Huangdi's External Classic*) recorded in the chapter "Catalog of Literary Works" of *Book of the Former Han*? In *Catalog of Literary Works*, it is named as "*Huangdi's Internal Classic*", implying that Huang Di delivered lectures rather than asking questions, as seen in other works such as *Discourse on Yin and Yang by Huangdi's Ministers* (*Huang Di Zhu Zi Lun Yin Yang*, 《黄帝诸子论阴阳》) and *Discourse on Buyin by Huangdi's Ministers* (*Huang Di Za Zi Bu Yin*, 《黄帝杂子步引》).

(2) Dialogues between Huang Di and Bo Gao (see Chapters 6, 14, 31, 32, 55, 56, 59, 64, 71, 77, etc.in *Miraculous Pivot*).

The writing style in the "Huang Di-Bo Gao" dialogues is not as prominent as the ones between Lei Gong and Huang Di, but there is still a significant use of conjunctions. The dialogues between Huang Di and Bo Gao mainly focus on the topic of "intestines and stomach".

The lengths of esophagus, stomach, and large and small intestines are described in Chapter 31 "Intestines and Stomach", the values of which are approximately equal to those measured in modern anatomy. In Chapter 32 titled "A Healthy Person Ends the Ingestion of Grain", these measurements are repeated to deduce further theories. The volume of the gastrointestinal tract is used to calculate the duration a person can survive without intake. The anatomy of the gastrointestinal tract is also described, including the absorption pathway of nutrients and the circulation pathway of nutrients and defense. The dialogues also list the combination of various foods and treatments, such as the "appropriation" and "prohibition" of five grains, five vegetables, five fruits, and five animals. Additionally, the diagnostic methods and muscle systems discussed in Bo Gao's dialogues are notably different from those in other chapters.

---

[23] Chapter 76 "Discourse on Demonstrating a Natural Approach" of *Plain Questions*.

[24] In the chapter "Biographies of Bian Que and Cang Gong" in *Records of the Grand Historian*. This is mentioned respectively in Chapter 56 "Discourse on Skin Sections", Chapter 62 "Discourse on Regulating the Conduits", Chapter 63 "Discourse on Misleading Piercing", Chapter 50 "Discourse on Prohibitions in Piercing" of *Plain Questions*.

(3) Dialogs between Huang Di and Shao Shi (see Chapters 6, 69, 72, and 79 in *Miraculous Pivot*).

"Huang Di-Shao Shi" dialogs feature in yin-yang theory, which differs from the theory of "three yin and three yang" in other chapters of *Plain Questions* and *Miraculous Pivot*. This yin-yang theory focuses on the dichotomy of yin and yang, namely "yin, yang, yin within yin, yang within yin, yin within yang, yang within yang". This theory is recorded as follows:

> In yin is yin; in yang is yang. If one has examined and therefore knows the conditions of the yin and yang qi, piercing them will be based on norms. In the interior, the five Zang-organs are yin and the six Fu-organs are yang. In the exterior, the sinews and the bones are yin and the skin is yang.
>
> If the disease is situated in the yin realm of the yin realm, the brook and stream points of the yin meridians are to be pierced. If the disease is located in the yang realm of the yang section, the sea points of the yang meridians are to be pierced. If the disease is situated in the yin realm of the yang section, the river points of the yin meridians are to be pierced. If the disease is situated in the yang realm of the yin realm, the collateral vessels are to be pierced.
>
> Diseases situated in the yang realm, they are called "wind". Diseases situated in the yin realm, they are called "impediment". If both yin and yang are affected by a disease, it is called "wind impediment".
>
> When the disease has manifested with physical appearances but without pain, it is of the yang type. When there are no physical appearances but there is pain, it is of the yin type.[25]

Regarding the classification of human types, unlike Bo Gao's division based on the five elements,[26] Shao Shi based it on yin-yang theory, dividing people into greater yin person, lesser yin person, greater yang person, lesser yang person, and person with harmonious yin and yang. Although both classifications consist of five categories on the surface, they differ in essence, with one being based on the five-element theory and the other on yin-yang theory. In Shao Shi's theories, there is no mention of the five-element theory.

(4) Dialogs between Huang Di and Shao Yu (see Chapters 46, 50, 53, and 64 in *Miraculous Pivot*).

Dialogs between Huang Di and Shao Yu mainly cover five-flavor theory, which is different from the ones between Huang Di and Bo Gao. There are two chapters covering five-flavor theory, namely Chapter 56 containing dialogs of Bo Gao and Chapter 63 containing dialogs of Shao Yu. The difference lies in the fact that Bo Gao's "five flavors" are based on the five-element theory, correlating sourness with the liver, bitterness with the heart, sweetness with the spleen, acridity with the lungs, and saltiness with the kidneys. On the other hand, Shao Yu does not adhere to the five-element theory. For example, he states that "Salty flavor enters the blood. If one consumes too much of it, this will cause thirst... Bitter flavor enters the bones. If one consumes too much of it, this will cause vomiting".[27]

---

[25] Chapter 6 "Longevity, Early Death, Hardness and Softness" in *Miraculous Pivot*.

[26] In his theories, Bo Gao divided different human types based on five phases, with each type further divided into another five sub-types according to five musical notes.

[27] Chapter 63 "On Flavors" in *Miraculous Pivot*.

In addition, when discussing the causes of various diseases, Shao Yu mostly attributes them to innate endowment. For instance, he states that "those with small bones and weak flesh easily suffer from alternating cold and fevers; when the flesh is not firm internally and the interstices are wide open, they may easily develop wind diseases; when all five Zang-organs are soft and weak, wasting-thirst easily develops".[28] Another example: "Those with a yellow complexion, a thin skin and weak flesh, they will not be able to overcome the wind in spring. Those with a white complexion, a thin skin and weak flesh, they will not be able to overcome the wind in the summer. Those with a greenish complexion, a thin skin and weak flesh, they will not be able to overcome the wind in autumn. Those with a red complexion, a thin skin and weak flesh, they will not be able to overcome the wind in winter".[29]

By examining the specific content involved in different dialogs mentioned above, it can be found that they have different standpoints throughout the discussion. This indicates that the *Miraculous Pivot* and *Plain Questions* do not have a unified theoretical core that runs through the whole work. Instead, they serve as collections of works containing varied views from different schools, adapted and compiled by individuals or groups over a specific period in history.

9.3 Disparities Existing in the Age of Different Chapters in *Plain Questions* and *Miraculous Pivot*

Since both *Plain Questions* and *Miraculous Pivot* are compilations in nature, it is determined that their contents must have a distinction between ancient and recent ones. As for the ancient parts, one can even find words like "Disease without other [damage]",[30] which are similar to the language used in oracle bone inscriptions. As for the recent parts, there is the official title "Zhong Zheng Zhi Guan"[31] that first appeared in the Cao Wei Dynasty, as well as the musical note names such as "upper jue tone", "major jue tone", "left jue tone", "iron hard jue tone", and "judicious jue tone"[32] which are not found in earlier literature. It can be said that a common feature of the two books is that on the one hand, they inherit and accommodate ancient medical theories and treatment experiences, and on the other hand, they are being processed and improved. Consequently, it is difficult to say which chapter is an earlier document and which is a later work. Instead, one can only analyze their relative chronology from the development levels of various branches of medicine. In the aspect of "meridians", as mentioned above, there is the comprehensive and systematic

---

[28] Chapter 46 "The Five Modifications" in *Miraculous Pivot*.

[29] Chapter 50 "On Courage" in *Miraculous Pivot*.

[30] Chapter 18 "Discourse on Phenomena Reflecting the Status of Qi in a Normal Person" in *Plain Questions*.

[31] Chapter 8 "Discourse on the Hidden Canons in the Numerous Orchid Chambers" in *Plain Questions*.

[32] Chapter 65 "Five Tones, Five Substances" in *Miraculous Pivot*.

"classical way" represented by "Meridians" in *Miraculous Pivot*, and there are also many meridian theories that name meridians by location and are consistent with the silk manuscripts unearthed from Mawangdui.

The same applies to the theory of Zang-Fu organs. Such concepts exist as "nine Zang-organs" (five spiritual and four physical Zang-organs), "eleven Zang-organs", "twelve Zang-organs", "the spleen being a solitary Zang-organ", "the kidney being a solitary Zang-organ", "the head being the house of bright essence, the back being the house of chest, the lower back being the house of the kidney, the knee being the house of the sinew, the bone being the house of the marrow", or the spleen belonging to one of the six Fu-organs.[33] As *Plain Questions* explains, there are differing viewpoints among "occult masters" regarding the classification of certain organs as Zang or Fu. "May I ask about these contradictions; all say of themselves they are right".[34] This suggests that the unification of the theory of Zang-Fu organs, an essential aspect of TCM, took time to develop. During the reign of Emperor Wu of the Han Dynasty (156–87 BC), "occult masters" were famous for their individual theories, but it was under the reign of Wang Mang (45 BC-AD 23) that these theories began to be unified.[35] Notably, *Comprehensive Discussions in the White Tiger Hall* (*Bai Hu Tong Yi*, 《白虎通义》), written in AD 79 during the early Eastern Han Dynasty, collated and established various theories, providing precise definitions for the five Zang-organs and six Fu-organs. Among them, the description of "triple energizers", one of the six Fu-organs that has been a subject of ongoing debate, closely aligns with the records found in *Miraculous Pivot*.[36]

Among these theories on Zang-Fu organs, the "nine Zang-organs" theory may have its origin in *The Rites of Zhou* (*Zhou Li*, 《周礼》), where it is stated that "(diseases) should be examined according to the activities of the nine Zang-organs". The significance of the "eleven Zang-organs" lies in the number "eleven". In medical works from Mawangdui, it is mentioned that there are 11 meridians. Additionally, in the chapter "Catalog of Temperament and Calendar" in the *Book of the Former Han*, a similar saying states that "the total number is eleven, and the laws of nature are all in it, which cycles endlessly". Therefore, "eleven Zang-organs" should also be one of the early theories.

---

[33] See Chapters 9, 30, 8, 19, 34, 17, 76 of *Plain Questions* respectively.

[34] Chapter 11 "Further Discourse on the Five Depots" in *Plain Questions*.

[35] Volume 12 in *Records of the Grand Historian* says, "In areas of Yan and Qi, occult masters all claimed in excitement that they had secret prescriptions, through which they could ascend to the class of the immortals". Volume 69 in *Book of the Former Han* says, "Wang Mang proposed to build tens of thousands of dormitories for learners", and "Thousands of learners all over the country with special talents were summoned and asked to compile works in order to correct the absurd remarks and unify different opinions".

[36] The chapter "Emotions and Temperament" in *Discourse on Virtues* (*De Lun Xia*, 《德论下》) in *Comprehensive Discussions in the White Tiger Hall* by Ban Gu (AD 32–92, the Han Dynasty) says, "The triple energizers are the pathway of the water and food, involving the process of the beginning and terminal of the qi", and "The upper energizer is like a hole, the middle energizer is like a knit, the lower energizer is like a ditch".

In terms of treatment methods, *Plain Questions* and *Miraculous Pivot* both focus on acupuncture. After the excavation of Mawangdui, people have gradually realized that in early times people employed moxibustion first, with acupuncture utilized much later. Therefore, the mention of the "nine needles" in the current version of *Huangdi's Internal Classic* should be attributed to the Han Dynasty. In the current version of *Huangdi's Internal Classic*, the gradual development of acupuncture and moxibustion is clearly evident. Some other treatment methods may have been already eliminated the Han Dynasty. For example, in Chapter 60 "Discourse on Bone Hollows" in *Plain Questions*, "At the location where one was bitten by a dog, apply three cauterizations. For it is in accordance with the laws to treat diseases resulting from harm caused by dogs that one applies the cauterization...". The level of this treatment method is basically the same as that of *Prescriptions for Fifty-two Diseases* (*Wu Shi Er Bing Fang*,《五十二病方》). To treat edema, "There are five lines above the sacrum, each line has five holes. There are two lines above the crouching rabbit, and each line has five holes ...There is one line above each ankle. Each line has six holes".[37] These are precisely the locations for "wind-water-illness with a swollen skin requiring 57 punctures".[38] Such treatments, not based on the pattern identification of the Zang-Fu organs, can only be considered more primitive methods. Similarly, the treatment of cold and febrile diseases in early literature involved applying acupuncture and moxibustion to the whole body in a general manner:

> First, perform moxibustion on Dazhui (GV14) at the nape, with the number of moxa-cones corresponding to the patient's age in years. Next, perform moxibustion on the coccyx, also with the number of moxa-cones based on the patient's age. Identify the depressions at the back Shu-points and perform moxibustion there. Raise the arm and perform moxibustion on the depression on the shoulder. Perform moxibustion between the two flanks. Perform moxibustion on the tip of Xuanzhong (GB39) above the lateral malleolus. Perform moxibustion between the little toe and the second toe. Perform moxibustion on the depression below the calf. Perform moxibustion behind the lateral malleolus. Perform moxibustion on the area above the clavicle where it feels hard and painful like a tendon when pressed. Perform moxibustion in the depression between the sternum and ribs in the center of the chest. Perform moxibustion below the metacarpal bones of the palm. Perform moxibustion at Guanyuan (CV4), three inches below the umbilicus. Perform moxibustion on the pulsating artery at the pubic hairline. Perform moxibustion at the space three inches below the knee. Perform moxibustion on the pulsating artery of the Stomach Meridian of Foot-Yangming on the dorsum of the foot. Perform one round of moxibustion on the vertex of the head.[39]

This treatment method clearly dates back to a time before the name of the most common and frequently used acupoint like "Zusanli (ST36)", emerged. When compared with the theory centered around "six-meridian pattern identification" in chapters such as "Discourse on Heat" in *Plain Questions*, where different methods are used at different stages of diseases, it is obvious that there is a difference between the ancient and the more recent. In short, if we simply regard the current version of *Huangdi's Internal Classic* as a classic work of basic TCM theories and examine

[37] Chapter 60 "Discourse on Bone Hollows" in *Plain Questions*.

[38] Chapter 19 "The Four Seasonal Qi" in *Miraculous Pivot*.

[39] Chapter 60 "Discourse on Bone Hollows" in *Plain Questions*.

these primitive and outdated treatment methods on the same level as the overall theoretical and therapeutic level achieved in the current version of *Huangdi's Internal Classic*, we will by no means be able to grasp the structure of the whole book.

In addition, whether the articles in different chapters and from different schools in the book are based on the five-element theory is also one of the important clues to distinguish their writing sequence. Although the five-element theory pervades the current version of *Huangdi's Internal Classic*, there are also schools like the "Huangdi–Shaoshi" Q&A in various chapters of Spiritual Pivot that only discuss yin-yang and do not involve the five-element theory. This is in line with the historical background in the early Western Han Dynasty. On the one hand, Confucianism was gradually developing toward orthodoxy; on the other hand, the traditions of the ancient philosophers still had a certain influence, especially the schools of Taoism, Legalism, and Yin-Yang, which were still relatively prominent among the principalities.[40] From the records about the "four seasons" in the book, we can see the evolution of medical theory from a non-five-element system to a five-element system. Simply put, in order to conform to the five-element norms, a "long summer" season was added between the four seasons to achieve a perfect match among the five elements, four seasons, and five internal organs. Therefore, those texts that use the concept of "five seasons" must have been written later than those that use the "four seasons" theory. The "five-element theory" was first advocated by Zou Yan and others during the Warring States Period. Generally speaking, until the early Western Han Dynasty, the "five-element theory" mainly referred to the "mutual restraint of the five elements". The theory of "mutual generation of the five elements" was not considered complete until Dong Zhongshu put it forward in his *The Luxuriant Dew of the Spring and Autumn Annals* (*Chun Qiu Fan Lu*,《春秋繁露》). Therefore, many chapters in the current version of *Huangdi's Internal Classic* that extensively discuss the mutual generation of the five elements should have been completed after that.

9.4 The Age of the Current Version of *Huangdi's Internal Classic*

Previous scholars analyzed and concluded that *Huangdi's Internal Classic* was written during the Warring States Period rather than in ancient times. However, later generations speculated that the book was likely completed during the transition from the Warring States Period to the Qin Dynasty, based on the usage of certain expressions such as "common people" and "the first month is greater yang, is yin", which appeared during the Qin or Western Han Dynasty.[41] Except for the seven chapters

---

[40] *General History of Chinese Thought* (*Zhong Guo Si Xing Tong Shi*,《中国思想通史》) by Hou Wailu, et al., Volume II, pp. 56–61.

[41] Okanisi Tameto (Japan). *Verification on Medical Works Before the Song Dynasty*, pp. 1-10, 32-37, 228. Long Bojian. *The Age of Huangdi's Internal Classic* (*Huang Di Nei Jing De Zhu Zuo Shi Dai*, 《黄帝内经的著作时代》). *Medical History and Healthcare Organization*, 1957, (2):106. Many

(namely Chapters 66 to 71, Chapter 74) which were supplemented by Wang Bing in the Tang Dynasty, and Chapters 72–73 which were indeed supplemented by later generations, the rest of the current version of *Huangdi's Internal Classic* is estimated by most to be completed before the late Western Han Dynasty. These assumptions are undoubtedly influenced by the chapter "Catalog of Literary Works". As to the age of *Miraculous Pivot*, some maintain it was accomplished later than *Plain Questions*. However, as there are quotations from *Miraculous Pivot* in *Plain Questions*, others believe *Miraculous Pivot* should be the earlier one.[42]

Now that *Plain Questions* and *Miraculous Pivot* do not match the 18 volumes of *Huangdi's Internal Classic* recorded in the chapter "Catalog of Literary Works", it's becoming evident that the current version of *Huangdi's Internal Classic* might have been completed no later than the compilation of the *Seven Categories of Writings*. Based on this conclusion, it is necessary to reconsider the exact completion dates of the two works. According to *Seven Categories of Writings* and the preface of *Treatise on Cold Damage and Miscellaneous Diseases*, the completion of the two works should be no later than *Treatise on Cold Damage and Miscellaneous Diseases* and no earlier than *Seven Categories of Writings*. This is because, under the historical conditions of the time, only the government had the capacity to collect various works and carry out the collation and compilation job. If *Plain Questions* and *Miraculous Pivot* were finished before *Seven Categories of Writings*, then these two large medical works would have been officially included into it. Works that predate the *Seven Categories of Writings*, even if not included, would have likely been lost, as was the case historically.[43] Therefore, it is highly unlikely that they could have been preserved and passed down intact.

However, although neither the *Seven Categories of Writings* nor the "Catalog of Literary Works" recorded the names of *Plain Questions* and the *Ancient classic on Needle Therapy*, it should not be asserted that the two did not exist at the end of the Western Han Dynasty. This is because in the compilation process, Ban Gu still based his work on *Seven Categories of Writings* without incorporating works during Wang Mang's rule. Ban Gu had made notes on places where he added or deleted anything, but there is no record of incorporating works of Liu Xin or Wang Mang.[44]

---

other works on TCM history all rest on this assumption. Only *A Brief History of Chinese Medicine* by Fan Xingzhun holds that *Huangdi's Internal Classic* should have been accomplished in the late Western Han Dynasty.

[42] Wei Yaoxi. *The Age of Miraculous Pivot* (*Ling Shu Cheng Shu Shi Dai*,《<灵枢>成书时代》). *Chinese Journal of Medical History*, 1983, (2):88.

[43] Many ancient medical works mentioned in *Plain Questions* and *Miraculous Pivot* such as *Upper Classic* (*Shang Jing*,《上经》), *Lower Classic* (*Xia Jing*,《下经》), *To Estimate and Measure* (*Kui Duo*,《揆度》), *The Strange and The Normal* (*Qi Heng*,《奇恒》), *Laws of Piercing* (*Ci Fa*,《刺法》), were also not included in *Seven Categories of Writings* and lost.

[44] According to *Biography of Ban Gu* (*Ban Gu Zhuan*,《班固传》) in *Book of the Later Han* (*Hou Han Shu*,《后汉书》), Zhang Tao discusses the political environment Ban Biao (3-54) had experienced in *History and Various Schools of the Eastern Han Dynasty* (*Dong Han De Jing Xue Yu Shi Xue*,《东汉的经学与史学》), "Ban Biao's continuing records to the *Records of the Historian* were not detailed enough, so Ban Gu concentrated on refining them, hoping to achieve his father's aspirations. Shortly afterwards, someone reported to the emperor, claiming Ban Gu was secretly compiling the history of

Wang Mang once proposed (AD 4) to "summon thousands of students with special talents all over the country to compile works in order to correct the absurd remarks and unify different opinions".[45] Therefore, special attention should be paid to the scientific and technological works that may have appeared during Wang Mang's reign. Wang Mang even had the imperial physician perform anatomy on human bodies and measure five Zang-organs and vessels for medical purposes, which had no precedence.[46] The current version of *Huangdi's Internal Classic* not only records the length of the internal digestive system, but also makes bold statements such as "Once he has died, he may be dissected for us to observe his interior appearance".[47] This shows that many chapters involving anatomy in the current version of *Huangdi's Internal Classic* were most likely produced under such environment.

Another reason why the current version of *Huangdi's Internal Classic* is closely related to Wang Mang's period lies in the rise of the name "Huang Di". The connection between medical works and "Huang Di" came into being in the early Western Han Dynasty. As stated in the chapter "Biographies of Bian Que and Cang Gong", Chunyu Yi received "Medical works on pulse of Huang Di and Bian Que" from his master Yang Qing. At this time, Huang Di and Bian Que were of the same importance, and there was no overwhelming advantage of Huang Di. Meanwhile, Empress Dowager Dou (?-135 BC) praised the studies of Huang Di and Laozi, so the emperor and princes had to study the works on Huang Di and Laozi.[48] As a result, the "study of Huang Di" during this time developed just as a way to rule the country with significant features of Legalism. Emperor Wu promoted Confucianism, leading to the abolition of many projects related to Huang Di, such as the "bright hall".[49] In *Seven Categories of Writings*, medical works of various schools are still equally important. The medical classics under the name of "Huang Di" were not the largest in terms of quantity. However, in *Plain Questions* and *Miraculous Pivot*, Bian Que's works were rewritten and incorporated into the theories of Huang Di, with figures like Shao Shi, Shao Yu, and Bo Gao appearing in dialogues with Huang Di. This shift indicates that Huang Di had gained absolute prominence during this period,

---

the dynasty without permission. Emperor Ming (AD 28-75) ordered the local official to detain Ban Gu in prison, and took all the books from Ban Gu's residence. However, after examining, Emperor Ming understood Ban's ambition was to eulogize the Han Court, so he did not punish him. Instead, Ban Gu was summoned to proofread books, and later was promoted. He then was finally allowed to finish his work". See China Research Association on the History of the Qin and Han Dynasties: *A Collection of Essays on History of the Qin and Han Dynasties* (*Qin Han Shi Lun Cong*,《秦汉史论丛》), Volume 6, pp. 343-354.

[45] Volume 99 in *Book of the Former Han*.

[46] Volume 99 in *Book of the Former Han*: Wang Mang let the imperial physicians, "directorate for imperial manufactories (Shang Fang)" and skillful butcher dissect him, measure his five Zang-organs, make bamboo branches run within his vessels to find out the circulation of them, so as to apply this in the treatment of diseases.

[47] Chapter 12 "Conduit/Stream Waters" in *Miraculous Pivot*.

[48] Volume 49 in *Records of the Grand Historian*.

[49] Volume 12 in *Records of the Grand Historian*.

Fig. 9.1 The hermit Fu (His real name is unknown)

possibly due to Wang Mang's claim as a descendant of Huang Di.[50] As Wang Mang identified himself as a descendant of Huang Di, the society began to venerate the name. Under these circumstances, various ancient medical works might have been compiled together and given the name of Huang Di (Fig. 9.1).

In addition, in *Plain Questions* and *Miraculous Pivot*, the names of "The Ancient classic on Needle Therapy" or "Nine Needles" are repeatedly mentioned (see Table 9.4). This suggests that there may have been another work titled "The Ancient classic on Needle Therapy" existing prior to *Plain Questions* and *Miraculous Pivot*. Upon examining the official historical records, evidence of *The Ancient classic on Needle Therapy* does exist in the *Book of the Later Han*:

> At first, there was an unknown elderly man who frequently fished by the Fu River, earning him the nickname "the old man Fu". He begged for food among the people. When he saw those who were ill, he would sometimes use acupuncture and stone needles for treatment, and the patients would recover immediately. Later, the old man wrote the *Ancient classic on Needle Therapy* and *Rules of Pulse-taking* (*Zhen Mai Fa*,《诊脉法》). Cheng Gao had sought to learn from him for many years, and then the old man taught him. Cheng Gao chose to live in seclusion and did not serve the imperial court. Guo Yu studied under Cheng Gao when he was young. Eventually, during the reign of Emperor He (AD 89-106), Guo Yu rose to the position of "aide to the imperial physician (Tai Yi Cheng)".[51]

If Guo Yu, who held the position of "aide to the imperial physician" during Emperor He's reign, was indeed an apprentice of Cheng Gao, it would suggest that Fu's works

[50] Volume 98 in *Book of the Former Han.*

[51] Volume 112 of *Book of the Later Han.*

Table 9.4 Examples of the citation of *the Ancient classic on Needle Therapy* or *Nine Needles* in *Plain Questions* or *Miraculous Pivot*

Books	Chapters	Content Cited
Plain Questions	Chapter 20 "Discourse on the Three Positions and Nine Indicators" Chapter 26 "Discourse on the Eight Cardinal Turning Points and on Spirit Brilliance" Chapter 27 "Discourse on the Division and Union of True Qi and Evil Qi" Chapter 54 "Explanations on the Needles"	"I have heard about the art of the nine needles from you, Sir…" "The nine indicators in the three sections: treat them as the origin. The discourse on the nine needles: it is not necessary to keep it…"; "For 'to take the past as law', one should first know the *Needle Classic*…" "I have heard of the art of the nine needles outlined in nine chapters. You, Sir, have subsequently multiplied them by nine. Nine times nine is 81 chapters…" "I should like to hear an explanation of the 'nine needles'…"[52]
Miraculous Pivot	Chapter 1 "The Nine Needles and the Twelve Origin Openings" Chapter 28 "Oral Inquiry" Chapter 42 "The Transmission of Diseases" Chapter 45 "The Assessment from Outside" Chapter 48 "Prohibition and Appropriation" Chapter 61 "The Five Prohibitions" Chapter 67 "The Application of the Needles" Chapter 73 "Function and Competence" Chapter 78 "On the Nine Needles" Chapter 73 "Function and Competence"	"A 'needle classic' is to be established first…" "I have been informed of the classics on the nine needles, and we have discussed movements of the yin and yang qi contrary to and in accordance with the norms…" "I have received from your instructions on the nine needles…" "I have been informed of the *Nine Treatises on the Nine Needles*…" "I have gone through the 60 chapters on the nine needles…" "To clearly know the discourse on the nine needles…" "I have been informed by you of the nine needles…" "I have been informed by you of the nine needles…" "I have been informed by you of the *Nine Needles*…" "Lei Gong asked Huang Di: The *Discourse on the Needles* (*Zhen Lun*,《针论》) says…"[53]

were likely completed during the late Western Han Dynasty to the early Eastern Han Dynasty. These works, however, could not have been included in the *Seven Categories of Writings*. And the current version of *Huangdi's Internal Classic* which repeatedly cited *The Ancient classic on Needle Therapy* and *Nine Needles* may be the works of Fu. There are very few records about medicine in the *Book of the Later Han* by Fan Ye (AD 398–445), but he wrote that *Needle Classic* and *Rules*

[52] Chapters 20, 26, 27, 54 in *Plain Questions* of Paul U. Unschuld's translation version respectively.

[53] Chapters 1, 28, 42, 45, 48, 61, 67, 73, 78 in *Miraculous Pivot* of Paul U. Unschuld's translation version respectively.

of Pulse-taking by Fu were popular at the time,[54] to which we should give special attention. Originally, *Miraculous Pivot* had no title but "Nine Volumes". Later on, it was titled "The Ancient classic on Needle Therapy", the content of which probably developed on the basis of Fu's *The Ancient classic on Needle Therapy*. As a result, it was unlikely for it to be incorporated in *Seven Categories of Writings*. We cannot assert that *Miraculous Pivot* appeared earlier simply because some of the content appeared earlier than that of other works. For example, the content on a divination tool (or "Liu Ren Shi Pan") from Tomb of Xia Houying (?-172 BC, who had the title of Ruyin Marquis) is similar to that in Chapter 77 "The Nine Mansions, the Eight Winds" of *Miraculous Pivot*. However, this similarity does not serve as evidence for the earlier appearance of *Miraculous Pivot*.[55]

Compared with *Plain Questions*, *Miraculous Pivot* is generally considered easier to understand, leading to the belief that *Plain Questions* is the older of the two. However, references to *Miraculous Pivot* can be found within *Plain Questions*. This is actually because both works refer to the same pieces of literature, such as Fu's *The Ancient classic on Needle Therapy* and Bian Que's works on pulse. Table 9.5 takes Chapter 54 "Explanations on the Needles" in *Plain Questions* and Chapter 3 "Explanatory Remarks on the Small Needles" in *Miraculous Pivot* as an example to illustrate how the two works separately annotate the same original literature preserved in *Miraculous Pivot*. It is worth noting that the two interpretations are completely different.

Table 9.5 illustrates that the original literature possesses a higher theoretical level compared to the unearthed medical works from Mawangdui. The original literature provides clear instructions on manipulation methods, suggesting that it was probably created at a later period. Only after this was it possible to generate different annotations of the original literature, resulting in two different analyses found in Chapter 3 in *Miraculous Pivot* and Chapter 54 in *Plain Questions*. It was only then that someone could compile these annotations into a single work. In fact, there are no preserved examples of medical works predating the Western Han Dynasty entirely. The medical texts that have been passed down to us are from the Eastern Han Dynasty, such as the *Classic of Difficulties* (*Nan Jing*, 《难经》), *Shennong's Classic of Materia Medica*, *Treatise on Cold Damage Diseases*. Now *Plain Questions* and *Miraculous Pivot* should be added to this list.

Although this book places the age of *Plain Questions* and *Miraculous Pivot* after *Seven Categories of Writings*, it emphasizes that the age of different part in them may differ. In this way, isn't this statement consistent with other general statements—"in the Qin and Han Dynasties" or "from the Warring States Period to the Qin and Han Dynasties"? In fact, the main difference lies between "big beginning and small ending" and "small beginning and big ending". In the first case, most of the content in the two works was written before the Qin Dynasty with later generations

54 Volume 112 of *Book of the Later Han*.

55 Anhui Heritage Working Group, et al. *Briefing on the Excavation of the Ruyinhou Tomb of the Western Han Dynasty* (*Fu Yang Shuang Gu Dui Xi Han Ru Yin Hou Mu Fa Jue Jian Bao*, 《阜阳双古堆西汉汝阴侯墓发掘简报》). *Cultural Relics* (*Wen Wu*, 《文物》), 1978, (8): 12.

Table 9.5 Different annotations in *Plain Questions* and *Miraculous Pivot* of the original literature

The original literature	Chapter 3 in *Miraculous Pivot*	Chapter 54 in *Plain Questions*
Whenever the needles are used, what is depleted of proper qi, it must be replenished. What is full of evil qi,[56] it must be drained. What has accumulated over an extended period of time, that is to be discarded. Where evil has been victorious, it is to be depleted… What is called repletion and depletion, these are conditions as if something were present, and as if something were absent. One observes the later and the earlier developments, as if there was something still there, as if something had been lost[57]…	When it says: "what is depleted, it must be replenished", that is: a depletion noticeable at the qi opening requires supplementation "What is full, it must be drained" is: an abundance noticeable at the qi opening requires drainage "What has accumulated over an extended period of time, that is to be discarded" is: blood is to be removed from the vessels "Where evil qi has been victorious, they are to be depleted" is to say: the evil qi is to be drained from all the conduits displaying an abundance… "What is called repletion and depletion, these are conditions as if something were present, and as if something were absent" is to say: in the case of repletion qi are present; in the case of depletion qi are absent "One observes the later and the earlier developments, as if there were something still there, as if something had been lost" is to say: this concerns the sequence of supplementation and draining in accordance with the presence of depletion and repletion. It is to be observed to what degree the qi has already been brought down and to which degree they are still present[58]…	As for "when piercing a depletion, then replenish it", that is, pierce until there is heat below the needle. When the qi is replete, then heat is present As for "when there is fullness, then discharge it", that is, pierce until there is cold below the needle. When the qi is depleted, then cold is present As for "what is densely compacted and old, eliminate it", that is, release the bad blood As for "in case evil dominates, deplete it", that is, remove the needle and do not press the hole pierced … As for "when one speaks of repletion and depletion", these are differences in the quantities of cold and warm qi As for "sometimes it is not present, sometimes it is present", that is, it is fast and cannot be recognized As for "investigate later and earlier", that is to get to know the earlier and later course of a disease[59]…

Note The Original literature is in Chapter 1 "The Nine Needles and the Twelve Origin Openings" of *Miraculous Pivot*

[56] "Evil qi" refers to "pathogenic qi" in Paul U. Unschuld's translation version of *Plain Questions* and *Miraculous Pivot*.

[57] Chapter 1 "The Nine Needles and the Twelve Origin Openings" of *Miraculous Pivot* of Paul U. Unschuld's translation version.

[58] Chapter 3 "Explanatory Remarks on the Small Needles" of *Miraculous Pivot* of Paul U. Unschuld's translation version.

[59] Chapter 54 "Explanations on the needles" of *Plain Questions* of Paul U. Unschuld's translation version.

applying modifications and supplements. In the latter case, the two works were likely completed at a later time, while incorporating some long-established knowledge. In other words, in these two works, some content was completed earlier, while some other later, the whole process reflects the development of medicine rather than the compilation process of the two works.

In addition, if we reject the notion that *Plain Questions* and *Miraculous Pivot* are the "18 volumes of *Huangdi's Internal Classic*", then the view that they are the representative works of the "school of Huang Di" would also be refuted. Next chapter on the study of Bian Que's works on pulse also demonstrates how works of the school of Bian Que were incorporated in *Plain Questions* and *Miraculous Pivot*, revealing that *Huangdi's Internal Classic* is probably based on various medical works recorded in "Catalog of Literary Works". This not only indicates the origin of the significantly expanded content within Huangdi's Internal Classic but also explains the inconsistencies found within it.

Overall, during the Warring States Period, due to the inconsistencies of the characters used by various states and the praise for the "arts of war" under turbulent conditions, it was relatively unlikely for medicine and other natural subjects to make progress. Medicine remained somewhat secretive or restricted, taught only within families or by mentors. Although the theory of meridians had been proposed, its practical application was still limited, comparable to the medical works from Mawangdui. Additionally, there was an evident parallelism between medicine and witchcraft. Until Emperor Wen (reigning from 179 to 157 BC), there were still proposals such as "setting physicians and witch doctors to treat diseases".[60] But during the reign of Emperor Xuan (73BC -49 BC), a shift occurred. In a letter to a general, the Emperor encouraged consulting physicians and cultivating a peaceful mind to extend one's lifespan naturally.[61] This indicates that in the Western Han Dynasty, medicine began to supersede witchcraft as clinical medicine made great strides. During this period of time, classical prescriptions, the medical instrument "nine needles", and the theories of meridian and Zang-Fu organs all basically started taking shape. These achievements not only form the foundation and crucial elements of *Plain Questions* and *Miraculous Pivot*, but also represent their significant academic value.

Afterthought:

As to the meaning of "Su Wen (素问, Plain Questions)", medical historians mostly quoted the annotation by Lin Yi of the Song Dynasty (AD 960–1279):

> Quan Yuanqi explained the name of "Su Wen" as follows: "Su (素) means the root, and wen (问) refers to the act of questioning by Huang Di." The work states the source and origin of the temperaments and five elements, hence this was named Su Wen.

Quan Yuanqi was the first who has annotated *Plain Questions* in history. However, Lin Yi was not entirely satisfied with this explanation and proposed another interpretation.

> Although there is such an interpretation, its meaning is not very clear. According to *A Road to the Heaven* (*Qian Zao Du*, 《乾凿度》), "All the visible matter comes from the

[60] Volume 49 in *Book of the Former Han*.

[61] Volume 59 in *Book of the Former Han*.

> invisible, and there is the concept of Great Change, Great Origin, Great Commencement, and Great Simplicity. Great Change refers to the absence of qi; Great Origin is where the qi originates; Great Commencement is where the physique commences; Great Simplicity is where the substance begins." When there is a presence of qi, physique, and substance, various diseases can arise. Therefore, Huang Di inquired about these diseases. The meaning of "Plain Questions" may stem from the concept of Great Simplicity, where the substance begins.[62]

In essence, the two arguments do not differ significantly. Lin Yi simply substituted the word "origin" of Quan Yuanqi with the "beginning" of the tangible substance. However, after recently reading Luo Yijun's *On the Modern Significance of Dong Zhongshu's View of Plain King* (*Lun Dong Zhong Shu Su Wang Guan De Xian Dai Yi Yi*, 《论董仲舒"素王"观的现代意义》),[63] I gained a newfound understanding of the meaning of "Su Wen". Dong Zhongshu was a Confucian, so he revered Confucius as the plain king and regarded the *Spring and Autumn Annals* (*Chun Qiu*, 《春秋》) as "the words of the plain king". The concept of "Su Wang" first appeared in *Zhuang Zi*, "Using this (inaction) in a subordinate position is the way of the sage-king without a crown".[64] According to this perspective, a "plain king" is a distinguished ruler without an official title. The *Plain Questions* was written in the format of Huang Di posing questions and his ministers providing answers. Could it be that Huang Di was inquiring about medicine in his role as the plain king? As described in the first chapter of the *Plain Questions*, "In former times there was Huang Di. When he came to life, he possessed the powers of a deity. Even in his infancy, he could speak. As a young child, he was exceptionally perceptive. In his maturity, he was sincere and skillful. And when he reached adulthood, he ascended to heaven". It's a pity that Huang Di did not hold any earthly official position, making it fitting to bestow the title of "Plain King" upon this revered Chinese ancestor.

---

[62] *Plain Questions*, Volume One. A secondary annotation by Wang Bing (the Tang Dynasty), proofread by Lin Yi (the Song Dynasty).

[63] China Research Association on the History of the Qin and Han Dynasties, *A Collection of Essays on History of the Qin and Han Dynasties*, Album 6, pp. 309–319.

[64] Volume 5 in *Commentaries on 'Zhuang Zi'* (*Zhuang Zi Zhu*, 《庄子注》) by Guo Xiang of the Jin Dynasty.

Chapter 10
Study on Bian Que's Sphygmology

Bian Que, an eminent ancient physician in China, is credited as the pioneer of sphygmology. But it remains uncertain whether the "sphygmology" refers to the theory of meridians or the method of pulse diagnosis (Fig. 10.1).

According to *Records of the Grand Historian* (*Shi Ji*, 《史记》), Chunyu Yi, an eminent physician in the Western Han Dynasty (202 BC–AD 8), acquired the pulse diagnosis books written by Huang Di and Bian Que and the method of observing five-colored facial complexions, from his master Gongsheng Yangqing.[1] According to the chapter "Catalog of Literary Works" in the *Book of the Former Han* (*Han Shu*,《汉书》), there used to be nine volumes of *Bian Que's Internal Classic* (*Bian Que Nei Jing*,《扁鹊内经》), 12 volumes of *Bian Que's External Classic* (*Bian Que Wai Jing*,《扁鹊外经》) and 23 volumes of *Prescriptions of Bian Que and Yu Fu in the Period of Yellow Emperor* (*Tai Shi Huang Di Bian Que Yu Fu Fang*,《泰始皇帝扁鹊俞拊方》). Unfortunately, all these texts have been lost. Moreover, many doctors and scholars of the later generations believe that, the *Classic of Difficulties* (*Nan Jing*,《难经》), written in the Eastern Han Dynasty (AD 25–220), is the work of Bian Que (Qin Yueren). This belief has been widely adopted in later historical works of TCM, but no one had ever paid any attention to the contents of "Bian Que's sphygmology" quoted in *Pulse Classic* (*Mai Jing*,《脉经》) written by Wang Shuhe, the Minister of Imperial Medical Bureau in the Western Jin Dynasty (AD 266–316). Initially, I also didn't pay much attention to this matter. It was only when I conducted a critical study of the nature and content analysis of the received version of *Huangdi's Internal Classic* (*Huang Di Nei Jing*,《黄帝内经》), that I came to understand how to apply these discourses on "Bian Que's sphygmology". This sheds clearer light on the dissemination of Bian Que's works and the essence of sphygmology. Moreover, a comprehensive study of significant figures and their academic achievements during the formative period of basic theories directly contributes to the reconstruction of the medical history of the Qin and Han Dynasties (221 BC–AD 220).

[1] The 45th chapter "The Biographies of Bian Que and Cang Gong" in *Records of the Grand Historian*, Volume 150.

Y. Liao, *Reconstructing the Image of Medicine in the Qin and Han Dynasties*, Understanding China, https://doi.org/10.1007/978-981-95-6801-7_10

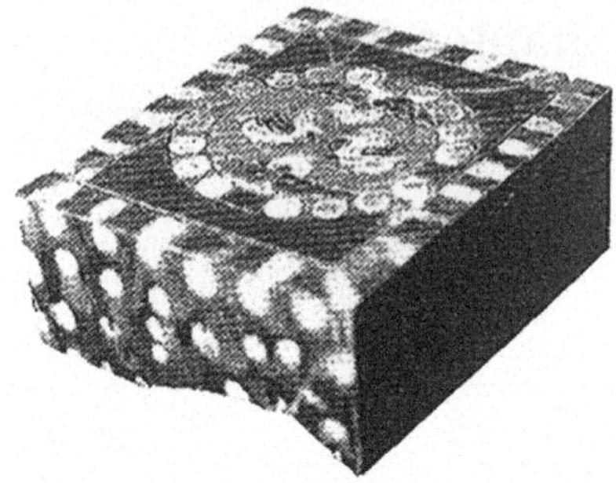

Fig. 10.1 Porcelain cushion for pulse diagnosis. Private collection in Xi'an, named "tri-colored glazed pulse diagnosis cushion of the Tang Dynasty with beaded pattern and a pair of mandarin ducks"

10.1 Textual Research on Bian Que's Sphygmology

First of all, is it really true that Bian Que's works mentioned in the *Records of the Grand Historian* and the *Book of the Former Han* cannot be traced?

In the chapter "The Biography of Bian Que" in *Records of the Grand Historian*, Sima Qian concludes by stating, "Up to now, all the pulse's theory and practice are based on Bian Que's sphygmology". This clearly illustrates that during the early Western Han Dynasty, sphygmology was closely associated with Bian Que by physicians and medical classics of that time. As mentioned earlier, Chunyu Yi received Bian Que's pulse books from Gongsheng Yangqing and shared this knowledge with others, providing strong support for this viewpoint. More specific evidence about the dissemination of "sphygmology" can be found in the chapter "The Biography of Cang Gong" in the *Records of the Grand Historian.* For example, it mentions, "The King of Jibei sent his imperial physicians Gao Qi and Wang Yu to study with me, and I taught them the meridians distributed in the upper and lower parts of the human body......"; "Du Xin, the Steward Assistant of the Feudal Prince Gao Yong, was very interested in pulse diagnosis and came to visit me. Then I taught him the distribution of the meridians in the upper and lower body as well as the method of observing five-colored facial complexion"; "Tang An from Zhaoli Linzi also came for learning and I taught him the method of observing five-colored facial complexion, the distribution of the meridians, etc.".[2] Taking the whole society into consideration, it is reasonable to assume that there were multiple cases of disseminating Bian Que's sphygmology during that time, as recorded in the *Records of the Grand Historian*: Gongsheng Yangqing → Chunyu Yi → Gao Qi, Wang Yu, Feng Xin, Du Xin, Tang An, etc. Certainly, there must be more similar dissemination cases between masters and apprentices. The long-term preservation of books and medical knowledge would have been greatly facilitated if they were passed on to imperial doctors or officials. Then, the origin of both *Bian Que's Internal Classic* and *Bian Que's External Classic* may indeed be traceable. These books could have been written by Bian Que (or sorted out by his descendants) as recorded in *Seven Categories of Writings* (*Qi Lve*,《七略》)

[2] The 45th chapter "The Biographies of Bian Que and Cang Gong" in *Records of the Grand Historian*, Volume 150.

compiled by Liu Xiang and his son Liu Xin. The content related to "sphygmology" in these books primarily originates from Bian Que.

In Chapter 9, I have elaborated on various pieces of evidence to prove that the received version of *Huangdi's Internal Classic* is not the same as those eighteen volumes recorded in the chapter "Catalog of Literary Works" in the *Book of the Former Han*. Then it can be inferred that the received version of *Huangdi's Internal Classic*, including *Plain Questions* (*Su Wen*,《素问》) and *Miraculous Pivot* (*Ling Shu*,《灵枢》), was actually "developed from" rather than "juxtaposed with" the seven kinds of "medical classics" recorded in the chapter "Catalog of Literary Works" in the *Book of the Former Han*. Therefore, it is likely that the works from later periods incorporated content from the earlier medical classics. Once evidence is found, it will further support the aforementioned conclusion. Theoretically speaking, it is also possible that some content from Bian Que's works was directly quoted or included in later medical books. Based on this inference, I realized the importance of studying and utilizing the descriptions of Bian Que's sphygmology. After conducting research on related literature, I did find some clues that align perfectly with my expectations.

Wang Shuhe (AD 210–280), the head of the Imperial Medical Bureau in the Western Jin Dynasty, mentioned in the preface of *Pulse Classic*: "Now I have collected some contents about pulse from Qi Bo (2697–2597 BC) and Hua Tuo (AD 145–208) and compiled them into ten volumes".[3] This statement is consistent with the objective facts. Volume 3, for example, contains references to "newly compiled" material, the "*Classic of Four Seasons* (*Si Shi Jing*,《四时经》)", "*Plain Questions*", "*Needling Classic* (*Zhen Jing*,《针经》)", and "Zhang Zhongjing" for theoretical guidance. Moreover, the titles of each chapter in volume 5 are as follows:

Zhang Zhongjing's Treatise on Pulse-taking

Bian Que's Yin-Yang Pulse-taking Method

Bian Que's Pulse-taking Method

Bian Que and Hua Tuo's Essentials for Auscultation and Observation

Bian Que's Essentials for Diagnosing Various Incongruous, Inconsistent and Death Pulse

This implies that Wang Shuhe had access to the works on sphygmology by Bian Que when he wrote the *Pulse Classic*. Alternatively, he could ascertain Bian Que's statements in other works. Upon comparison, similarities were found between "Bian Que's sphygmology" and statements in the received version of *Huangdi's Internal Classic* and *Classic of Difficulties*. The internal relationship among these documentary records is therefore worth exploring. It is unlikely that Wang Shuhe extracted and attributed some texts from *Plain Questions*, *Miraculous Pivot* and *Classic of Difficulties* to "Bian Que's sphygmology" without noting the source. Given Wang Shuhe's academic attitude and other contents recorded in the *Pulse Classic*, he would have acknowledged the source if he had made such a reference.

This means that Wang Shuhe did peruse the original materials of Bian Que's sphygmology, enabling him to specifically identify certain contents in *Plain Questions*, *Miraculous Pivot* and *Classic of Difficulties* as derived from Bian Que's work.

---

[3] Wang Shuhe (the Western Jin Dynasty), The chapter "Self-Preface" in *Pulse Classic*.

Additionally, since there are no mentions of Bian Que's name in these classics, it is plausible that the received version of *Huangdi's Internal Classic* incorporated elements of "Bian Que's sphygmology" without explicitly crediting him.

Table 10.1 illustrates the traces of quotations from "Bian Que's sphygmology" in *Plain Questions*, *Miraculous Pivot* and *Classic of Difficulties*. These traces are referred to as such because it is impossible to recreate the complete view of "Bian Que's Sphygmology". As a result, we cannot determine the presence of other content from "Bian Que's Sphygmology". Among these traces, the most typical one is the chapter "Discussion of Unusual Diseases" in *Plain Questions*, which is entirely included in "Bian Que's Essentials for Diagnosing Various Incongruous, Inconsistent and Death Pulse", the fifth chapter of the *Pulse Classic*. Furthermore, this chapter lacks conversations between Huang Di and Qi Bo, and their names are absent, which differs from the stylistic approach of *Plain Questions*. Doesn't it just serve as collateral evidence indicating that this chapter has another source?

Secondly, the question of whether the author of the *Classic of Difficulties* is Bian Que has always been a matter of contention, with both sides having a comparable number of proponents. Even today, some people still believe that this book was indeed written by Bian Que.[4] There is no record of this book in "Catalog of Literary Works" in the *Book of the Former Han*. This is unlikely to be an oversight; it is obvious that the book was written after Liu Xiang and his son Liu Xin compiled the catalog.[5] Moreover, its language is easy to understand, showing a difference from *Plain Questions* and *Miraculous Pivot*.[6] Especially in terms of theoretical structure, there are obvious differences compared with *Plain Questions* and *Miraculous Pivot*. For example, regarding the explanation of the "triple energizer" in the theory of zang-fu organs, both *Plain Questions* and *Miraculous Pivot* regard it as one of the six fu-organs with a specific form, which is consistent with the description of the six fu-organs in *Comprehensive Discussions in the White Tiger Hall* (*Bai Hu To Yi*,《白虎通义》) a book compiled in the early Eastern Han Dynasty. However, the *Classic of Difficulties* first proposed the "theory of the intangible triple energizer", which has triggered a longstanding debate that remains unresolved to this day. The purpose of the *Classic of Difficulties* in establishing the theories of "the intangible triple energizer" and "the two kidneys being respectively the 'kidney' and the 'life-gate'" was, of course, to transform the theory of "five zang-organs and six fu-organs", which

[4] For details of the views of the two academic schools, please refer to the two famous Japanese works published many times by People's Health Publishing House: *A Study of Medical Records before the Song Dynasty* (*Song Yi Qian Yi Ji Kao*,《宋以前医籍考》) and *A Study of Chinese Medical Records* (*Zhong Guo Yi Ji Kao*,《中国医籍考》). The present-day viewpoint is expressed in Cheng Hongru's "The causerie of Qin Yueren and the *Classic of Difficulties*", in *Chinese Journal of Medical History* (*Zhong Hua Yi Shi Za Zhi*,《中华医史杂志》), No. 3, 1982.

[5] The chapter "Catalog of Literary Works" in the *Book of The Former Han* was compiled on the basis of *Seven Categories of Writings*.

[6] TambaMototsugu (Japan) talked about the *Classic of Difficulties* in *A Study of Chinese Medical Records* (p.79), which said: "Compared with *Plain Questions* and *Miraculous Pivot*, the tone of the *Classic of Difficulties* is slightly weaker, which seems to be written by people living in the west of Luoyang".

Table 10.1 Comparison of Bian Que's sphygmology in the received version of *Huangdi's internal classic* and *classic of difficulties*

Bian Que's sphygmology	The received version of Huangdi's internal classic	Classic of difficulties
Lesser yang pulse is at times small, at times large, at times short, and at times long… Greater yang pulse is vast, big and extended. It floats over the sinews… Yang brightness pulse is at the surface and it is big and short	The chapter "Discourse on Phenomena Reflecting the Status of Qi in a Normal Person" in the *Plain Questions*	The chapter "The Seventh Difficulty" in the *Classic of Difficulties* (Notes: What the ellipsis omits are specific seasons and days that are governed by respective pulse qi. And these contents are included in *Classic of Difficulties* but excluded in *Plain Questions*)
Lesser yin pulse is tight and fine … Greater yin pulse is tight, fine and long … Reverting yin pulse is sunken, short and tight …		
A sick person with a yellow face and green–blue eyes will not die. But if the green–blue color is like that of the young grass, the patient will die A sick person with a yellow face and red eyes will not die. But if the red color is like that of rotten blood, the patient will die A sick person with a yellow face and white eyes will not die. But if the white color is like that of withered bones, the patient will die A sick person with a yellow face and black eyes will not die. But if the black color is like that of soot, the patient will die A sick person whose face and eyes are the same color will not die A sick person with a black face and green–blue eyes will not die A sick person with a green–blue face and white eyes will die A sick person with a black face and white eyes will not die A sick person with a red face and green–blue eyes will die in six days…	The chapter "The Generation and Completion of the Five Depots[7]" in *Plain Questions* (Notes: the contents are not exactly the same.)	

(continued)

[7] "Depots" refer to "Zang-organs" in Paul U. Unschuld's translation version of *Plain Questions*. He also uses "palaces" to explain "Fu-organs".

Table 10.1 (continued)

Bian Que's sphygmology	The received version of Huangdi's internal classic	Classic of difficulties
A green–blue complexion should resemble the gloss of greenish jade; it should not resemble indigo blue; A red complexion should resemble something of vermilion color wrapped up in white; it should not resemble ochre; A white complexion should resemble goose feathers; it should not resemble salt; A black complexion should resemble the color of multi-layered lacquer; it should not resemble the charcoal; A yellow complexion should resemble realgar wrapped up in gauze; it should not resemble clay	The chapter "Discourse on the Essentials of Vessels and the Subtleties of the Essence" in *Plain Questions*	
If a liver disease with white skin appears, the patient will die at geng and xin If a heart disease with black eyes appears, the patient will die at ren and gui If a spleen disease with green–blue lips appears, the patient will die at jia and yi If a lung disease with red cheeks and swollen eyes appears, the patient will die at bing and ding If a kidney disease with a swollen face and yellow lips appears, the patient will die at wu and ji	The chapter "Discourse on Phenomena Reflecting the Status of Qi in a Normal Person" in *Plain Questions* (Notes: some words are omitted.)	
When examining the scrofula with frequent alternate chills and fever, one sees red vessels in the eyes extending into the pupil from above and below, prognosis is as follows: If one sees one vessel, death follows within one year If one and a half vessels are visible, he will die within one and a half years If two vessels are visible, he will die within two years If two and a half vessels are visible, he will die within two and a half years If one sees three vessels, death follows within three years	The chapters "Cold and Heat Diseases" and "Discussing Illness; Examining the Foot-long Section" in *Miraculous Pivot* (Notes: There are no "scrofula, 瘰疬", in both chapters.)	

(continued)

Table 10.1 (continued)

Bian Que's sphygmology	The received version of Huangdi's internal classic	Classic of difficulties
It is stated in the classic that some diseases may end in death, some diseases may be cured by themselves without treatment, and some diseases may persist for months or years. Can death and survival, the presence and absence of a disease be diagnosed through pulse-taking? … If the patient has an enlarged abdomen and suffers from diarrhea, the pulse should be fine, faint, and unsmooth. If, on the contrary, the pulse is tight, large, and slippery, death is a certainty		The chapter "The Seventeenth Difficulty" in *Classic of Difficulties* (Notes: This paragraph constitutes the full text of "The Seventeenth Difficulty")
A person with a diseased physical condition but a healthy pulse manifestation will survive, while a person with a healthy physical condition but a diseased pulse manifestation will die		The chapter "The Twenty- Fifth Difficulty" in *Classic of Difficulties*
Fullness of evil qi in the liver, kidneys or lungs is repletion which gives rise to dropsy, …		
When the heart vessel is full and big, this goes along with convulsions and spasms; the sinews are cramped When the liver vessel is small and tense, this goes along with convulsions and spasms; the sinews are cramped …… When the heart vessel beats and is smooth and tense, this is a sign of heart elevation illness When the lung vessel is in the depth and beats, this is a sign of lung elevation illness… (32 articles in total)	The chapter "Discourse on Very Strange Diseases" in *Plain Questions* (Notes: This paragraph constitutes the full text of the chapter "Discourse on Very Strange Diseases".)	
When a person exhales once, the pulse beats twice and the qi proceeds three cun.[8] … Beating three times during one exhalation indicates the loss of essence. … In the first stage of an injured pulse, the pulse beats once when a person exhales once	The chapter "The 50-fold Circulation" in *Miraculous Pivot* (Notes: it is said in the chapter "The 50-fold Circulation" that, "When man exhales once, his vessels move twice, and the passage of his qi covers a distance of three cun., … the water in the clepsydra descends by 100 marks, … the course of 13.500 breathings". And this paragraph constitutes the full text of this chapter.)	The chapters "The First Difficulty" and "The Fourteenth Difficulty" in *Classic of Difficulties* (Notes: more description, generalization, and explanation on this part are contained in *Classic of Difficulties* than in *Miraculous Pivot*)

[8] In TCM, fingers are used as a measuring scale in determining acupoints because every individual's body size is unique. 1 cun is equal to the width of one's thumb, 2 cun is equal to the width of one's forefinger, middle finger and ring finger; 3 cun is equal to the width of one's forefinger, middle finger, ring finger, and little finger.

did not conform to the yin-yang and odd–even matching relationship, into the theory of "six zang-organs" (zang-organs belong to yin and should be an even number) and "five fu-organs" (fu-organs belong to yang and should be an odd number). This also indicates that there is a certain time gap between the *Classic of Difficulties* and all previous medical works that adhered to the theory of "five zang-organs and six fu-organs". Therefore, among the two views on whether this book was written by Bian Que, the latter view should be considered correct. Although the *Classic of Difficulties* was not written by Bian Que, it does not prevent it from incorporating the content of Bian Que's sphygmology. In other words, precisely because it was written later, it had the possibility of taking Bian Que's sphygmology as one of its research objects.

In addition, while it is reasonable to believe that the *Classic of Difficulties* was written after *Plain Questions* and *Miraculous Pivot*, it is not accurate to consider the former as an explanation of the latter. Many topics discussed in the *Classic of Difficulties* are not mentioned in *Plain Questions* and *Miraculous Pivot.* Only in the "The Sixteenth Difficulty" chapter of the *Classic of Difficulties* is there a discussion of sphygmology, which states that the movement in the vessels can be examined by considering the sixty informants. This concept is also briefly mentioned in the *Plain Questions*.[9] "The Sixteenth Difficulty" also records the theory that pulse diagnosis includes three body parts and nine pulse-takings, differentiating between yin pulses and yang pulses, and variations in pressure. It also considers the changes in the vessels based on the four seasons. These pulse diagnosis methods by past medical sages have endured for a long time. However, as time has passed, doctors have developed their own methods, raising the question of which method is the most suitable. It should be noted that the pulse diagnosis methods mentioned in the *Classic of Difficulties* were quoted from various medical classics. Moreover, the *Classic of Difficulties* totally negates such ideas and initially proposes the theory of exclusive examination of the wrist pulse and the method of diagnosing yin-qi and yang-qi respectively according to the Cun and Chi pulsation. The *Classic of Difficulties* does not merely interpret the *Huangdi's Internal Classic* but instead absorbs a wide range of medical works from that time. It adopts an "argumentation" form of self-questioning, integrating the clarification of views and the construction of its own theoretical system through the 81 questions posed (as shown in Table 10.1).

The research on Bian Que's works can provide valuable insights when examining the origins and development of the received version of *Huangdi's Internal Classic.* If *Plain Questions* and *Miraculous Pivot* were the eighteen volumes of *Huangdi's Internal Classic* recorded in the chapter "Catalog of Literary Works" in the *Book of the Former Han*, it becomes difficult to determine if these texts incorporated content from the seven mentioned "medical classics". This, in turn, hinders any comparison between these three medical classics and Bian Que's sphygmology, as clearly stated

[9] The chapter "Discourse on Comparing Abundance and Weakness" in *Plain Questions*: The features *The Strange and the Normal* are outlined under sixty headings. To diagnose the circumstances of the union with the subtle. To pursue the changes of yin and yang......To know this, then, is sufficient for carrying out a diagnosis. Wang Bing noted: "A text with the title '奇恒势' with sixty entries is no longer transmitted nowadays" (translation version: Paul U. Unschuld, *Huang Di Nei Jing Su Wen*).

in the *Pulse Classic*. Furthermore, it prevents the recognition of important yet easily overlooked clues that contribute to the reconstruction of the medical history of the Qin and Han Dynasties.

Thus far, two presupposed hypotheses have been confirmed. Firstly, it is evident that Bian Que's sphygmology is integrated into the received version of *Huangdi's Internal Classic*. Secondly, Bian Que's sphygmology is also referenced in other medical works, indicating its influence beyond the confines of the *Internal Classic*.

10.2 Content Analysis of Bian Que's Sphygmology

As Sima Qian said, "Even today, those who talk about pulse diagnosis in the world all follow Bian Que". This well reflects Bian Que's achievements and contributions. Based on the Bian Que's pulse-taking knowledge preserved in *Pulse Classic*, some important aspects in the basic theories of TCM, like the concept of pulse-qi circulation (perhaps related to blood circulation), the five-color diagnostic method, and the determination of life or death, originated from Bian Que.

Bian Que stated, during one exhalation, the pulse beats twice and the qi (energy) flows three cun (a unit of measurement); similarly, during one inhalation, the pulse again beats twice, and the qi flows another three cun. In one complete respiratory cycle (exhalation and inhalation), the pulse beats five times in total. ...... Regarding the water clock (clepsydra) that measures time, he observed there are 100 marks in a full day-night cycle. Each mark corresponds to 135 breaths, totaling 13,500 breaths over 100 marks. Two marks constitute one circuit, during which the qi completes a full circulation around the body. Thus, the qi makes fifty complete circuits in a day-night cycle. Regarding pulse patterns: Two consecutive beats are termed one arrival. If three arrivals occur during one exhalation, this indicates deviation from normal physiology; four arrivals signify essence depletion; five arrivals predict death. In the first stage of an injured pulse, the pulse beats once when the person completes one exhalation or beats twice when the person completes one breath. During this time the qi proceeds for three cun; In the second stage of an injured pulse, the pulse beats once when the person completes one breath; In the third stage of the injured pulse, the pulse beats once when the person finishes one breath plus one exhalation; In the fourth stage of the injured pulse, the pulse beats once when the person finishes two breaths; In the fifth stage of the injured pulse, the pulse beats once when the person finishes two breaths plus one exhalation. The subsequent stages of the injured pulse involve variations in the number of breaths and exhalations, corresponding to specific pulse patterns.[10]

In order to assist readers in understanding the main points, this condensed text provides the following summarization:

[10] Wang Shuhe (the Jin Dynasty), the 5th chapter "Diagnosis on Damaging Pulse and Bout Pulse" in *Pulse Classic*, Volume 4.

(1) Normal pulse rate and its measurements: the pulse beats twice when the person completes an inhalation or an exhalation. And in the interval between inhaling and exhaling, the pulse will beat once more. So, the pulse beats five times when the person finishes one complete respiratory movement. That is the normal pulse rate. Such measurement method has been used to date. Pulse diagnosis in Traditional Chinese Medicine (TCM) not only focuses on observing pulse manifestations but also involves calculating pulse rate, similar to Western medicine's approach. TCM practitioners typically use the breath rate as a benchmark for pulse diagnosis, eliminating the need for a watch during clinical diagnosis.

Some metrological experts once said that there was a lack of "standards" in ancient China, but this is not the case. Like "foot", a standard unit based on the length of the Queen's foot, there were also some standard measurement units based on the human body in ancient China. For example, "cun"—the width of the middle finger, "chi"—the length between the tips of the thumb and the middle finger when the palm is straightened, and "xun"—the length between the fingertips of the two middle fingers when the two arms are straightened. Unlike Western countries, which base standards on one person's foot, China used the measurer themselves as the reference, leading to variations in standards. These subtle differences, often overlooked, contributed to disparities in the East–West development of science and technology. Nonetheless, it remains reasonable for doctors to use their own breath rate as a standard in pulse diagnosis, as patients' respiratory rates often change in sync with pulse beats.[11]

(2) Significant concepts and abnormal pulse rate: Bian Que had clearly explained the meaning of some concepts used in disease diagnosis. For example, "one bout" means that the pulse beats twice, and "the first damage (the first stage of abnormally slow pulse rate)" means that the pulse beats only once when a person completes one exhalation or inhalation (as shown in Table 10.2).

The concept of "pulse rate" is the first applied in pulse diagnosis, and it was considered the primary standard for testing pulse conditions and identifying diseases in the early stages. Both traditional Chinese medicine (TCM) and Western medicine prioritize observing the speed of pulse beat. Later, the concepts to describe the abnormal "rhythm" of pulse beat, known as "intermittent pulse" (like "extrasystole") in western medicine and "irregular-rapid pulse", "irregular-intermittent pulse", "regular-intermittent pulse" in TCM, also attracted great attention. TCM has further developed and applied dozens of "pulse manifestations" when research on pulse in Western medicine remained focused on rate and rhythm. Objectively speaking, the research and utilization of "pulse" in TCM are indeed more in-depth compared to Western medicine.

(3) The movement and length of the "pulse" are also examined. It is stated that one breath consists of one exhalation and one inhalation, during which the qi in the pulse travels six cun. One circuit signifies that the qi in the pulse has completed its movement around the body. In a day and night, there are 50 circuits, which equates

[11] Wang Shuhe (the Western Jin Dynasty), the 5th chapter "Diagnosis on Damaging Pulse and Bout Pulse" in *Pulse Classic*, Volume 4.

Table 10.2 Abnormal pulse rate and diseases

Faster than normal			Slower than normal		
Concept	Pulse Rate	Disease	Concept	Pulse Rate	Disease
Three bouts	1 breath: 6 beats	Loss of blood	The first damage	1 breath: 2 beats	The qi is too short to circulate round the body; deficiency of qi and listlessness and slackness of the body
Four bouts	1 breath: 8 beats	Loss of essence	The second damage	1 breath: 1 beat	This is called a departure from the regular: since the blood does not go where it should, there is blood in both the stools and urine
Five bouts	1 breath: 10 beats	Death	The third damage	1.5 breath: 1 beat	This is called wrestling: while the qi circulates on, the blood lingers behind. They cannot coordinate and both become faint. If the qi is blocked and therefore becomes replete, there is fullness in the chest
			The fourth damage	2 breaths: 1 beat	This is called blood collapse: the body becomes emaciated and fatigued, with only skin and bones. Qi and blood are both exhausted
			The fifth damage	2.5 breaths: 1 beat	This is called expiry: qi urgency (i.e., rapid breathing), inability to rise from bed, cold breath from the mouth, and the expiration of all pulses

to 13,500 respirations. Therefore, the qi in the pulse moves a total distance of 81,000 cun (equivalent to 810 zhang) in a day and night, averaging 16.2 zhang per circuit.

Let's compare these contents with the received version of *Huangdi's Internal Classic*.

Firstly, the core content of the quoted passage constitutes the whole chapter "The 50-fold Circulation" in *Miraculous Pivot* and the first paragraph of the chapter "Discourse on the Essentials of Vessels and the Subtleties of the Essence" in *Plain Questions* in the form of a dialogue between Huang Di and Qi Bo. Even though the original word "pulse" was replaced with "nutrient qi" in the fifteenth chapter "The 50-fold Circulation" in *Miraculous Pivot*, the descriptions of their movement remain the same (i.e., "50 circuits in one day and night" and "810 zhang in one day and night").

In the sixteenth chapter "Nutrient-Qi", it is elaborated that the nutrient qi circulates within the twelve regular meridians (from the meridian of hand greater yin to that of foot reverting yin), as well as two extraordinary meridians, the Conception Vessel (CV) and the Governor Vessel (GV), which are located on the front and rear midline

Table 10.3 "The measurements of the vessels"

Name of vessels	Length
Six hand yang meridians	5 chi for each
Six hand yin meridians	3.5 chi for each
Six foot yang meridians	8 chi for each
Six foot yin meridians	6.5 chi for each
Two walker vessels[12]	7.5 chi for each
Conception vessel (CV)	4.5 chi
Governor vessel (GV)	4.5 chi
Sum:	16.2 zhang

of the human body, in a continuous cycle. Undoubtedly, it is an explanation of the chapter "The 50-fold Circulation".

Then, in the seventeenth chapter "The Measurements of the Vessels", the specific length of each meridian was introduced, which is just a further extension of the former chapters. Although named "The Measurements of the Vessels", this chapter only introduces the length of fourteen meridians and two "heel vessels" (as shown in Table 10.3). Other four of the "eight extra meridians" including thoroughfare vessel (TV), belt vessel (BV), yin link vessel (Yin Lyu), and yang link vessel (Yang Lyu) were all not mentioned, let alone their respective length which is not included in the total length either. According to Table 10.3, the length of the 28 meridians adds up to exactly 16.2 zhang (54 m). This clarifies that the length of each meridian is not based on actual measurement but rather precise calculation. Without a deep understanding of TCM theories, it is unfounded and presumptuous to assert that TCM developed solely from observation and experience.

So far, the interrelation among the three chapters of *Miraculous Pivot* and the origin and development of the basic theory of TCM, including the course of the meridians and the length of vessels, are all clarified. At that time, the morphology of "vessel" obviously referred to blood vessels and blood circulation, and the "pulse beat" proposed by Bian Que undoubtedly referred to pulsation. However, the definite mutual relationship between the theory of pulse-qi circulation and blood circulation cannot be asserted. Some articles claim that the only difference between the two theories lies in accuracy,[13] but this is an overestimation of the theoretical development during that period.

Modern physiological research indicates that the speed of blood circulation can only be measured by volume velocity, which is equal to a particular cross-sectional area of blood vessels. The cross-sectional area of capillaries is approximately 800

[12] Walker vessels can be divided into "yin walker vessel" and "yang walker vessel". Although it is said that "Walker vessels are seven chi and five cun long (equals to one zhang and five chi)", it does not say whether it is the combination of yin and yang or the combination of left and right. Modern commentators may not be correct in their interpretation of "the combination of left and right".

[13] Liu Xueli et al., "Human Understanding on Blood Circulation", *Discovery of Nature* (*Da Zi Ran Tan Suo*, 《大自然探索》), 1987, (1): 170.

times larger than that of the aorta, resulting in a significantly slower linear velocity in capillaries compared to the aorta. If the movement of vessel qi described by Bian Que refers to blood circulation, the statement "the qi proceeding three cun" at a constant speed does not precisely explain the objective phenomenon. Although Bian Que's theory of pulse-qi circulation is not as advanced as the comprehensive understanding of blood circulation revealed by W. Harvey (AD 1578–1657), it is still highly valuable. It surpasses the tidal theory advocated by the renowned Roman doctor C. Galen (AD 129–200), who believed that blood returns to the heart in a manner similar to rising and falling tides. Therefore, any comparison between Chinese medicine and Western medicine should be based on the knowledge system from the same attribute or the same period.

10.3 Five-Colored Facial Observation and Life-Death Determination

A person who is sick with a yellow face and green–blue eyes will not die. However, if the green–blue color resembles that of young grass, the patient will die.

A person who is sick with a yellow face and red eyes will not die. But if the red color resembles that of rotten blood, the patient will die.

A person who is sick with a yellow face and white eyes will not die. However, if the white color resembles that of withered bones, the patient will die.

A person who is sick with a yellow face and black eyes will not die. But if the black color resembles that of soot, the patient will die. If a person is sick with a black face and white eyes, they will not die. A person who is sick with a red face and green–blue eyes will die in six days.

A person who is sick with a red face and white eyes will die in ten days. Anxiety, anger, deep thought, and worry have internally depleted the vital energy of the heart. Even if the facial complexion improves, it is urgent to prepare a coffin.[14]

In addition to the pulse-taking method, the methods of determining life and death by observing facial complexions mentioned above were also developed by Bian Que. These succinct sayings are included in the received version of *Huangdi's Internal Classic*. The combination of pulse-taking and observing facial complexions forms a significant diagnostic method in *Plain Questions*. Specifically, it states that "The methods about the observation of complexion and pulse-taking were valued by the heavenly emperor and passed down by ancient physicians"[15] and "The inspection of the five-colored facial complexions and the changes in the movements in the vessels are rules for diagnosis".[16] The core meaning can be summarized into three points:

(1) Normal and abnormal facial complexion.

[14] Wang Shuhe, the Jin Dynasty, the 4th Chapter "Bian Que and Hua Tuo's Essentials for Auscultation and Observation" in *Pulse Classic*, Volume 5.

[15] The chapter "Discourse on Moving the Essence and Changing the Qi" in *Plain Questions*.

[16] The chapter "Jade Tablet Discourse on the Essentials" in *Plain Questions*.

(2) Life-death determination based on facial complexion.
(3) The meanings of various facial complexions in diagnosis. (e.g., "red eyes indicate that the disease is in the heart, white eyes indicate that the disease is in the lung". When inspecting the blood vessels, a redder appearance suggests more heat in the patient, while a greener or bluer appearance suggests more pain.)

The mutual relationship between facial complexions and pulse conditions has practical significance and aligns with clinical practice and the five-phase theory. However, whether the five-phase theory was introduced into Traditional Chinese Medicine (TCM) after its systematization or if it was systematized based on natural science is still a question. In *Plain Questions* and *Miraculous Pivot*, there are various correspondences between the five elements and other concepts, indicating that the development of the five-phase theory aimed to conform to objective laws and provide a more accurate explanation of nature. While some scholars believe that philosophy defines medicine (natural science),[17] further exploration of their mutual relationship and influence is necessary, considering the dissemination of Bian Que's pulse diagnosis discussed in the previous paper. This is because the matching relationship between the "five elements" and "five facial complexions" in Bian Que's sphygmology aligns with the standard classical modes.

10.4 Formation of the Meridian and Collateral Theory

In recent years, there has been a growing interest and support for exploring the extraordinary functions of the human body. This has led to a reevaluation of Bian Que's ability to "diagnose all the diseases of five Zang-organs", as documented by Sima Qian. Some contemporary academic publications even refer to Bian Que as an ancient physician with extraordinary powers.[18] However, it is impossible for us to ascertain how he applied these extraordinary powers to diagnose and treat diseases. Even if such powers did exist, it is crucial for doctors to focus on improving their medical skills rather than relying solely on special powers for diagnosis and treatment. Furthermore, from a historical research perspective, the following points are worth considering:

(1) The formation of the meridian and collateral theory followed a specific path that cannot be explained solely by the general knowledge system.
(2) As pointed out in the "First Section—General Introduction", from the findings of meridians literature and cultural relics unearthed in Hunan and Hubei provinces, the meridian and collateral theory in these places lacks the conception of "qi and blood circulation". Additionally, the length of meridians with the same attribute varies. These observations suggest differences from Bian Que's sphygmology.

---

[17] Che Li, "*Huangdi's Internal Classic* and the relationship between heaven and man", *Chinese Journal of Medical History*, 1987, (2): 79.

[18] Su Li, "A Brief Study of the Name and Facts of Bian Que", *Chinese Journal of Medical History*, 1987, (1): 50.

(3) The origin of vital energy is a question that has been pondered and explained in various traditional medicines worldwide. In Bian Que's sphygmology, it is stated that "When a person exhales once, the pulse beats twice, and the qi proceeds three cun, and when a person inhales once, the pulse beats twice more, and the qi proceeds another three cun". This clearly indicates that the act of breathing was considered the source of power for the movement of qi and blood. It also explains why the lungs were likened to bellows. The chapter "The Conduit Vessels" in the *Miraculous Pivot* further establishes that the starting point of the circulation circle of twelve meridians is the middle energizer, indicating that essence derived from food and water was regarded as the material basis and source of power for qi and blood. The *Classic of Difficulties* expands on this concept by conceiving the "life gate" as the fundamental motivation for life activities.

Summarizing Bian Que's medical achievements based on the above aspects provides a more detailed and comprehensive understanding than merely introducing the three stories recorded in the *Records of the Grand Historian*. Moreover, it is not explicitly stated in the book that the "six principles of refusing treatment", attributed to Bian Que, were actually put forward by Bian Que himself. Instead, this seems more like Sima Qian's personal reflection after recounting the story of "Bian Que's inspecting the facial complexion of Duke Huan of Qi". Careful examination of the original text can help verify this inference.

Chapter 11
Historical Features of Medical Theory in *Classic of Difficulties*

The Eastern Han Dynasty, also known as the Later Han Dynasty, which spanned from AD 25 to 220, holds significant importance in the development of science, technology, and culture in Chinese history. However, there are only a few surviving medical works from this period that have been passed down to the present day. Among them, the *Classic of Difficulties* (*Nan Jing*, 《难经》) can be regarded as a typical representative. Unfortunately, the existing literature on this book in the field of medical history is inadequate and limited. It is often dismissed as a work lacking innovation or containing a mixture of disadvantages and advantage,[1] without recognizing the profound influence it had on the evolution of Traditional Chinese Medicine (TCM) theory during the Eastern Han Dynasty's cultural development.[2] Consequently, not only is the significance of studying this book ignored, but also the changes in TCM theory from the Western Han Dynasty (202 BC-AD 8) to the Wei and Jin Dynasties (AD 220–420) and their close interrelation with the social and cultural context remain unclear. The following discussion aims to delve deeper into this aspect.

11.1 Researches on the Author and Historical Features

The authorship of the *Classic of Difficulties* remains a subject of debate, with a long-standing controversy between those who believe it was written by Qin Yueren (Bian Que) and those who deny this claim. There's no need to elaborate on this. However, modern studies of medical history mostly agree that the book was completed during the Eastern Han Dynasty, so it's naturally hard to attribute it to pre-Qin medical

[1] Jia Dedao, *A Brief History of Chinese Medicine* (*Zhong Guo Yi Xue Shi Lue*,《中国医学史略》), p. 88.
[2] Zhao Pushan, *Ancient Chinese Medicine* (*Zhong Guo Gu Dai Yi Xue*, 《中国古代医学》), p. 44.

Y. Liao, *Reconstructing the Image of Medicine in the Qin and Han Dynasties*, Understanding China, https://doi.org/10.1007/978-981-95-6801-7_11

scholars. Only Cheng Hongru still held this view in his "random talk"[3]; there are also those who record both views without making a definite assertion.[4] What needs a brief discussion is whether a compromise explanation put forward by some people based on the above two views can hold water. This explanation is that the book was written by the descendants or disciples of Bian Que. during the Eastern Han Dynasty.

In Yu Shenchu's *A Brief History of Medicine in China* (*Zhong Guo Yi Xue Jian Shi*, 《中国医学简史》), the *Classic of Difficulties* is not mentioned in the context of the medical history of the Western and Eastern Han Dynasty. However, when discussing the Spring and Autumn and Warring States period (770-221 BC), the author notes that "the medical techniques of Qin Yueren were inherited by his disciples and successors. In the Han Dynasty (220 BC-AD 220), some individuals summarized his medical theories and experiences, compiling them into the *Classic of Difficulties*".[5] Li Bocong expands upon this in his dissertation, *A Study of Bian Que and the Bian Que School* (*Bian Que He Bian Que Xue Pai Yan Jiu*, 《扁鹊和扁鹊学派研究》), stating:

> The legend that it is Qin Yueren who wrote the *Classic of Difficulties* must and ought to be interpreted as follows: In the Eastern Han Dynasty, ancient physicians of the Bian Que School wrote the *Classic of Difficulties*. This book inherited and developed the academic thoughts of the Bian Que School before the Western Han Dynasty. Thus, the real authors of the *Classic of Difficulties* are the physicians of the Bian Que School.

The author underscores the importance of this conclusion by using strong language and adding emphasis to the statement. Furthermore, the supporting arguments presented align with those advocating for Qin Yueren (Bian Que) as the author. Some of the statements in the *Classic of Difficulties* are consistent with Bian Que's words as quoted in Wang Shuhe's *Pulse Classic* (*Mai Jing*, 《脉经》) from the Western Jin Dynasty (AD 266–316) and Sun Simiao's *Essential Prescriptions Worth a Thousand Gold for Emergencies* (*Qian Jin Fang*, 《千金方》) from the Tang Dynasty (AD 618–907). Based on the analysis of these materials, it has been determined that the so-called "words of classics" in the *Classic of Difficulties* are, in fact, quotations from the medical works of the Bian Que. School.[6]

The quotations from Bian Que's medical statements in the *Classic of Difficulties* provide valuable insights into why later generations regarded it as Qin Yueren's masterpiece. However, it would be inappropriate to solely attribute this book to the Bian Que. School in the Eastern Han Dynasty based on these quotations alone. There is no evidence to suggest that the so-called "Bian Que's words" in later medical books were really quoted from the *Classic of Difficulties*, although they may share similar content. Besides Bian Que's works, the *Classic of Difficulties* also incorporates content from other ancient medical classics, making it a compilation rather than purely an inheritance of a specific medical school in the pre-Qin period. Thus,

---

[3] Cheng Hongru, "The causerie of the Qin Yueren and *Classics of Difficulties*", *Chinese Journal of Medical History* (*Zhong Hua Yi Shi Za Zhi*, 《中华医史杂志》), 1982, (3): 44.

[4] Zhao Pushan, *Ancient Chinese Medicine*, p. 44.

[5] Yu Shenchu, *A Brief History of Chinese Medicine*, p. 40.

[6] Li Bocong, *A Study of the Bian Que and the Bian Que School*, Xi'an: Shaanxi Science and Technology Press, 1990, p. 211.

the arguments proposing Bian Que. or the descendants of the Bian Que. school as the authors of the *Classic of Difficulties* are not rigorous or accurate enough. It is important to note that while many medical theories of Bian Que. are reflected in the *Classic of Difficulties* and even in other medical texts like the *Plain Questions* and *Miraculous Pivot*, there are still many remarks of Bian Que. that have not been quoted in the *Classic of Difficulties*.

Although the author of the *Classic of Difficulties* is unknown according to current information, it is evident that Confucian culture in the Eastern Han Dynasty strongly influenced its medical treatises. To understand this connection, it is necessary to explore the characteristics of Confucianism throughout different historical periods.

Confucianism was one of the most prominent schools of thought among "the nine schools of philosophy" during the pre-Qin period. In the Han Dynasty, Confucianism became the sole venerated school of thought, with its "six arts" serving as the standard, while other schools were rejected. Consequently, Confucianism continued to be influential despite the rise of Taoism and Buddhism. Moreover, the emergence of Neo-Confucianism in the Song Dynasty (AD 960–1279) further revitalized Confucianism. It should be noted that although Confucianism traces its origins back to the doctrines of Confucius and Mencius, it evolved with distinct characteristics in different historical periods. Confucianism in the pre-Qin period primarily preached about "the five constant virtues" mainly through the texts of "six Confucian classics". The *Book of Changes* (*Yi Jing*,《易经》), although touching on the concept of yin and yang, primarily served as a divination book. The "Confucianism School" emphasized the study of the six Confucian classics, the importance of benevolence and righteousness, adherence to the orthodox examples of Yao and Shun (legendary sage kings in ancient China), observance of the etiquette of King Wen and King Wu of the Zhou Dynasty (approximately 1046-256 BC), and respect for Confucius as a teacher. These were the fundamental characteristics of Confucianism.[7] When studying the relationship between Confucian culture and Traditional Chinese Medicine (TCM), many people focus on the positive aspects, such as the adherence to lofty medical ethics, the concern for practical matters over mythical elements, and the promotion of self-cultivation. However, it is important not to overlook any potential negative influences.[8]

Actually, to achieve the goal of "assisting the monarch, conforming to nature, and emphasizing enlightenment", Confucianism itself has been constantly improved with the development of society. In the Western Han Dynasty, philosopher Dong Zhongshu (179-104 BC) played a crucial role in advancing Confucianism by incorporating the philosophical speculation of the yin-yang theory and five-phase theory into the ideology of the "five constant virtues". This integration allowed the thoughts from other schools to be incorporated into Confucianism. It is said that Dong Zhongshu's

---

[7] The tenth chapter "Catalog of Literary Works" in *Book of the Former Han* (*Han Shu*,《汉书》), Volume 30.

[8] Yan Liang and Zhang Zhibin, "Confucianism and the development of Chinese Medicine." Chinese Journal of Medical History, 1987, (3): 154; Li Liangsong and Guo Hongtao, *Chinese Traditional Culture and Medicine* (*Zhong Guo Chuan Tong Wen Hua Yu Yi Xue*,《中国传统文化与医学》). Xiamen: Xiamen University Press, 1990, pp. 22–27.

contribution to the unification cause of the Western Han Dynasty lies in his integration of different doctrines since the Warring States period and various schools of Confucianism under the name of Confucius and the Gongyang School.[9] Since "Wang Mang was fond of auspicious omens and Emperor Guangwu of the Eastern Han Dynasty rose to power with the help of prophetic symbols", books such as *Pattern on the Legendary Horse* (*He Tu*, 《河图》), *Pattern on the Supernatural Tortoise* (*Luo Shu*, 《洛书》), and *Seven Classics on Divination* (*Qi Jing Wei*, 《七经纬》) "became popular in the world". Although criticized by some scholars, including Kong Anguo, Mao Gong, Jia Kui, these books were annotated by Confucian classics masters like Zheng Xuan. Scholars specializing in the internal history of Confucian classics focus on the disputes between the modern and ancient script versions of the classics, as well as the compatibility of Zheng Xuan's school of Confucian classics studies. Sociologists pay more attention to the struggles for rights and interests between social groups and the changes in their status through the historical study of Confucian classics. Scientific historians, on the other hand, should focus on the academic atmosphere and thinking methods that emerged in the Eastern Han Dynasty, which were more advanced than those in Dong Zhongshu's time. These combined concepts such as Tai Ji (Supreme Ultimate), Liang Yi (Two Modes), Si Xiang (Four Constitution), Ba Gua (Eight Trigrams) from the *Book of Changes* with the yin-yang theory and the five-phase theory, and supplemented with numerical deduction to explain the formation of heaven and earth, political and human relations, the transformation and generation of all things, and the principles of physics. However, orthodox Confucianism criticized these ideas as "heretical and disturbing the classics of the Doctrine of the Mean". After Emperor Yang of the Sui Dynasty (AD 581-618) ascended the throne, he "sent envoys everywhere to search for and burn all the books related to divination and prophecies in the world. Since then, this learning no longer existed".[10] Nevertheless, the rise of Neo Confucianism in the Song and Ming Dynasties (AD 960-1644) can actually be regarded as a repetition of this kind of change, and later there also emerged the criticism of Neo-Confucianism by the Textual Research School in the Qing Dynasty. Regardless of the textual research and identification of the authenticity of Confucian classics as well as their academic origin and literature value, the two "heresies" emerging in the Eastern Han Dynasty and the Song Dynasty both had the same progressive tendency to explore the ultimate truth, explain all things in nature and construct a development model for the universe. The main difference between them lies in the fact that the "heresy" in the Eastern Han Dynasty emphasized the five-phase theory, with the yin-yang theory interspersed, while the "heresy" in the Song Dynasty took the yin-yang theory as the framework and incorporated the five-phase theory. In fact, these so-called "heresies" are the concrete expressions of Confucianism's emphasis on rationality. Joseph Needham also noted that

---

[9] Fan Wenlan, *A Compendium of General History of China* (*Zhong Guo Tong Shi Jian Bian*,《中国通史简编》), Volume 2. Beijing: People's Publishing House, fourth Ed., 1964, pp. 111–112.

[10] All of the above can be found in the *History of the Sui Dynasty* (*Sui Shu*,《隋书》), Volume 32.

"it promoted the germination of science".[11] The *Classic of Difficulties*, which was written in the Eastern Han Dynasty and became popular in the Song and Ming Dynasties, helps to understand certain concepts, such as why the "stirring qi between the kidneys" is regarded as the "primordial qi", the root of the twelve meridians, and the basis of "human life"; why the five Zang-organs and six Fu-organs are changed to six Zang-organs and five Fu-organs; why the "five transport points" are particularly emphasized; why diseases are mostly divided into five categories; why the pulse-taking method is simplified into the theory of yin and yang, and why only the methods of reinforcement and reduction are discussed in the needling method. It also makes it easy to understand why the status of this book gradually rose after the Song Dynasty and why it was widely valued by most doctors.

The discussions on various medical issues in the *Classic of Difficulties* are actually theoretical and principled. Unlike the *Plain Questions* (*Su Wen*, 《素问》) and *Miraculous Pivot* (*Ling Shu*, 《灵枢》), there are no clear records in this book about specific treatment methods and experiences, such as how to select acupoints and perform acupuncture for a certain disease. This characteristic of rational thinking was not unique to Confucianism, but in the Eastern Han Dynasty, it was represented by Confucian culture. Among them, the scholars of ancient-script Confucian classics were particularly good at this. So the discussions of the masters of modern-script Confucian classics were summarized by the scholars of ancient-script Confucian classics into *Comprehensive Discussions in the White Tiger Hall* (*Bai Hu Tong Yi*, 《白虎通义》). Moreover, the characteristic of the modern-script Confucian classics, which is to expound one's own ideas through the classics, is also reflected in the *Classic of Difficulties*. Therefore, we must not mistakenly think that the *Classic of Difficulties* inherits the words of a certain school or is a work to explain the *Huangdi's Internal Classic* (*Huang Di Nei Jing*,《黄帝内经》) just because there are a few sentences like "The classics say so-and-so" in it. Similar to the inclusive nature of Zheng Xuan's Confucian scholarship, the *Classic of Difficulties* constructed a consistent theoretical system using various previous medical theories as materials and the philosophical thought of "primordial qi, yin-yang, and the five elements", which represented the highest level of that era. The degree of close combination and the scope of application of the yin-yang and five-element theories in it are both beyond those in the extant version of the *Huangdi's Internal Classic*. In the *Plain Questions* and *Miraculous Pivot*, in many cases, diseases are discussed individually, and treatment methods, needling techniques, and acupoint selection are simply stated. Moreover, yin-yang and the five elements often operate independently. In some chapters, the reasoning is based either solely on yin-yang or only on the five elements. For example, classifying people into five types of metal, wood, water, fire, and earth, or five types of greater yin, greater yang, lesser yin, lesser yang, and balanced yin-yang can be regarded as different theoretical inclinations of different schools. In the *Classic of Difficulties*, however, yin-yang and the five elements are integrated. For example, in the chapter titled "The Tenth Difficulty", the assertion that "one specific movement in

[11] Joseph Needham, *The History of Science and Technology in China*, Chinese translation version, Volume 2, p. 12.

the vessels may undergo ten variations" is based on the combined pairing results of the five evils (five elements) and the concepts of hardness and softness (yin-yang), that is, $5 \times 2 = 10$. Another example can be seen in the chapter titled "The Fifty-Fifth Difficulty", where the notion of "accumulations pertaining to yin and gathering pertaining to yang" is based on the yin-yang theory, while the further subdivision of "accumulations" into five categories in "The Fifty-Sixth Difficulty" is based on the five-phase theory. Besides, although the specific content of *Ten Changes* (*Shi Bian*, 《十变》) quoted multiple times in the *Classic of Difficulties* is not elaborated upon in detail, it can be inferred that the theoretical basis of this book also includes a combination of the yin-yang theory and the five-elements theory. The generation and restriction relationships among the five transport points were also discussed in the *Huangdi's Classic of the Bright Hall* (*Huang Di Ming Tang Jing*, 《黄帝明堂经》). Only through an in-depth study of the *Classic of Difficulties* can we fully comprehend the significant changes that occurred in medical theory during the Eastern Han Dynasty.

11.2 Changes in Medical Theory

11.2.1 Evolution of Pulse Diagnosis Methods

Many traditional pulse diagnosis methods, including "three positions and nine indicators pulse-taking", "carotid and wrist pulse-taking", and "twenty antique meridians pulse-taking", have been documented in ancient medical texts like *Plain Question*s and *Miraculous Pivot*. The wrist pulse-taking method, also known as Cunkou pulsation or radial artery pulsation, is specifically recommended in the *Classic of Difficulties*, and continues to be used in clinical diagnosis of Traditional Chinese Medicine (TCM) to this day. The prominence of the wrist pulse-taking method in pulse diagnosis can be traced back to the era of the *Classic of Difficulties*, marking a significant turning point in the evolution of pulse diagnosis in TCM.

In *Plain Question*s and *Miraculous Pivot*, there is no explicit mention of the division of Chi and Cun by the wrist pulse. The term "Chi-Cun" actually refers to "Chifu, the forearm skin" and "Cunkou, the wrist pulse". According to Yang Shangshan in the Sui Dynasty, the wrist pulse was initially divided into "Chi and inch" by Qin Yueren. During the era of Hua Tuo (around AD 145–208, the late Eastern Han Dynasty), it was further divided into "Cun (inch), Guan (bar), and Chi (cubit)".

> Qin Yueren stated that the length of "Cun and Chi" is constant, with the yang Cun being nine fen long and the yin Chi being one inch long. However, he did not mention Guan. Hua Tuo proposed that the total length of Cun, Guan, and Chi is three cun,[12] as they are all one cun long individually. The theoretical basis for this view is unknown. Wang Shuhe, Huangfu

---

[12] In TCM, fingers are used as a measuring scale in determining acupoints because every individual's body size is unique. 1 cun is equal to the width of one's thumb, 2 cun is equal to the width of one's forefinger, middle finger, and ring finger; 3 cun is equal to the width of one's forefinger, middle finger, ring finger, and little finger.

Mi, and others hold different views on this matter, but they all acknowledge the existence of Guan.[13]

It is certain that the concept of Guan did not appear in the era of *Classic of Difficulties*. However, it remains debatable whether Qin Yueren was the one who initially divided the wrist pulse into Cun and Chi, as there is no preserved evidence in Wang Shuhe's *Pulse Classic* (*Mai Jing,*《脉经》) according to the content of "Bianque's Sphygmology". Yang Shangshan may have put forward this point of view probably because he regarded Qin Yueren as the author of the *Classic of Difficulties*. Nonetheless, it cannot be denied that the division of the wrist pulse into Cun and Chi in the *Classic of Difficulties*, based on the yin-yang theory and the method of carotid and wrist pulse-taking, represents a transitional stage in pulse diagnosis.

The methods of "carotid and wrist pulse-taking" and "Chi and Cun pulse-taking" are all based on the comparison of yin and yang and the distinction between the upper and lower parts of the human body. Compared to the former method which compares the upper yang carotid with the lower yin wrist pulse, the latter one focusing merely on the wrist pulse is less complicated. By using the radial styloid (Guan) at the wrist as a boundary, the Cun on the distal side, representing yang, corresponds to the original carotid pulse, while the Chi on the proximal side, representing yin, corresponds to the original wrist pulse. In addition, some terms, including "visceral exhaustion pulse" and "Guan (closure)–Ge (rejection)" used in the "carotid and wrist pulse-taking method", are also applied to describe various pulse manifestations in the "Chi and Cun pulse-taking method" (see "The Third Difficulty" for details). To better understand the essence of the "wrist pulse-taking method" advocated in the *Classic of Difficulties*, it should be noted that at that time, "Guan" was not yet one of the three divisions of the wrist pulse but rather an anchor point and an indispensable premise for the application of yin-yang theory in this method.

11.2.2 Renewal of the Meridian System

The concept and constitution system of the "eight extra meridians" was originally proposed in the *Classic of Difficulties*. In *Plain Questions* and *Miraculous Pivot*, the names, courses, and indications of the eight meridians were also recorded, but in a scattered way without any specific systematic descriptions of their definitions or meanings. This is because different chapters in *Plain Questions* and *Miraculous Pivot* were written by different authors. So, the concept and system of the "eight extra meridians" didn't exist until the Eastern Han Dynasty. In other words, in the era of *Huangdi's Internal Classic*, the eight meridians were considered inconsequential and were not included in the system of the "twelve meridians". For example, in the chapter "To Pierce Lower Back Pain" in *Plain Questions*, it discusses how various meridians can cause lumbago, including foot greater yang, foot lesser yang, foot yang

[13] Yang Shangshan (the Sui Dynasty), *Great Simplicity of Huangdi's Internal Classic* (*Huang Di Nei Jing Tai Su,*《黄帝内经太素》), p. 33.

brightness, foot lesser yin, and foot reverting yin, and lists the different symptoms caused by the disorder of these meridians. It also enumerates other meridians that may cause lumbago, such as the branch of the foot greater yang meridian, the collateral of the foot lesser yang meridian, the yang link vessel, the lateral collateral of the greater yang meridian, the meridian passing the perineum, the collateral of the foot greater yang meridian, the branch of the foot lesser yin meridian, the lateral collateral of the foot greater yin, and the branch of the yang link vessel. However, the yang link vessel mentioned here does not have any special significance even though it belongs to the "eight extra meridians". In *Classic of Difficulties*, it is stated that the "eight extra meridians" can hold the excessive qi and blood from the "twelve regular meridians", similar to a lake that accumulates the overflowing water from rivers.

> This analogy can be compared to the planning and construction of ditches and reservoirs by the wise people of ancient times. When the ditches and reservoirs are full, their excess contents flow into deep lakes because even the wise people were unable to find other means to seize and ensure the continuation of the flow. Similarly, when the conduits and network vessels of man are filled to capacity, the excess contents enter the eight single-conduit vessels where they are no longer part of the circulation because the twelve main conduits cannot accommodate this surplus. When the single-conduit vessels receive stagnant evil influences, swellings and heat may occur, and in such cases, one would need to stimulate the respective vessel using a sharp stone.[14]

It can be observed that the establishment of the "eight extra meridians" is entirely based on the analogy between the human body and nature. Reading Dong Zhongshu's *The Luxuriant Dew of the Spring and Autumn Annals* (*Chun Qiu Fan Lu*, 《春秋繁露》) and other Confucian works in the Han Dynasty, as well as Xiao Ji's *Gist of Five Phases* (*Wu Xing Da Yi*, 《五行大义》) which inherited the theory of divination, one can immediately understand how the author of the *Classic of Difficulties* was influenced by the prevailing customs of that time.

The *Classic of Difficulties* provides a clear explanation of the origin, end, circulation, and indications of the "eight extra meridians", which has become a theoretical reference followed by later generations. For example, when Yang Shangshan annotated relevant contents in *Great Simplicity of Huangdi's Internal Classic*, he quoted from *Huangdi's Classic of Eighty-one Difficulties* (*Huang Di Ba Shi Yi Nan*, 《黄帝八十一难经》), another name for the *Classic of Difficulties.* In later meridian theory and acupuncture and moxibustion studies, the "eight extra meridians" hold an equally important position as the "twelve regular meridians". Unfortunately, not many people have deeply studied what the original appearance of these contents was in *Plain Questions* and *Miraculous Pivot*, what status they held, and what changes occurred to them later in the *Classic of Difficulties.* People generally think that the theories of traditional medicine all originated from *Huangdi's Internal Classic.*

---

[14] Hua Shou (the Yuan Dynasty), The first volume "Twenty-eight Difficulties" in the *Original Meaning of Classic of Difficulties* (*Nan Jing Ben Yi*,《难经本义》).

11.2.3 Reinterpretation of the Zang-Fu Theory

It is generally believed that "the theory of invisible triple energizer" originated from the *Classic of Difficulties* and had a great influence on later generations. However, the so-called "theory of invisible triple energizer" in the *Classic of Difficulties* does not imply that the triple energizer is literally invisible.

In the Zang-Fu theory, out of the 17 names used to describe the organs, including the five Zang-organs, six Fu-organs, and extraordinary Fu-organs, 16 of them are still in use today and correspond consistently with the names used in modern anatomy. The only exception is the "triple energizer", which is not commonly used nowadays. It is verified that the word "焦" (energizer) in "三焦" (triple energizer) was initially written as "膲" which refers to "shrunken meat" according to Gao You's annotation in the chapter "Discussion on Astronomy" in *Writings of Prince Huainan* (*Huai Nan Zi*, 《淮南子》). Similarly, in the chapter "Root and Connection" in *Miraculous Pivot*, there is saying of "brittle body hair and scorched skin", and its meaning is all under-developed flesh, which is the specific form and nature of the triple energizer. As one of the six Fu-organs, the triple energizer is located in the abdominal cavity and is responsible for regulating qi and water within the human body. At that time, the naming and functional description of the triple energizer were not based on biochemistry but rather on the limited understanding of how food essence is transported throughout the body and how water is transferred from the intestines to the bladder. It was believed that the water and food ingested would be transformed into waste and excreted from the body after passing through the visible and interconnected passages of the gastrointestinal tract. The "energizer" located outside these passages was believed to act as a medium to transport food essence throughout the body and facilitate the movement of water into the bladder to form urine.[15] After the publication of *Plain Questions* and *Miraculous Pivot*, *Comprehensive Discussions in the White Tiger Hall*, a Confucian classic written in AD 72 also discussed the five Zang-organs and six Fu-organs:

> The triple energizer, the house of the pericardium, encompasses the passageways of water and grain in the organism. It is also seen as the beginning and end of the influences that course through the body. The upper energizer is compared to an orifice, the middle energizer to woven fabric, and the lower energizer to a ditch.[16]

There are several expressions that basically coincided with the above-mentioned descriptions. For example, in the chapter "The Production and Convergence of Nutrient Qi and Defense Qi", it is stated that "The upper energizer is like fog. The central energizer is like a humidifier. The lower energizer is like a ditch". Additionally, in the chapter "The Separation of the Five (Protuberance-Illnesses) Jin and Ye Liquids" in *Miraculous Pivot*, it is said that "The triple energizer emits qi to warm the

[15] Ancient cognition of urine production can be found in the sixth chapter of the second part of this book.

[16] Ban Gu (the Han Dynasty), *"Emotions and Temperament" of the chapter "Discourse on Virtues* in Comprehensive Discussions in the White Tiger Hall.

muscles, and nourish the skin...".[17] It is worth noting that the *Classic of Difficulties* was completed around the same time as *Comprehensive Discussions in the White Tiger Hall*, so why is the "triple energizer" considered a theoretical rather than a physical organ in the *Classic of Difficulties*?

In the *Classic of Difficulties*, the concept of the "triple energizer" is mentioned six times, with the most detailed discussion found in the chapter "The Thirty-First Difficulty":

> The triple energizer encompasses the passageways of water and grain in the organism. It is seen as the beginning and end of the influences that course through the body. The upper section of the triple energizer extends from below the heart, passing through the diaphragm, and ending at the upper opening of the stomach. Its main function is intake rather than discharge, and any disorders in this section are regulated at CV 17 (Tanzhong). The central section of the triple energizer is located in the central duct of the stomach; it does not extend further upward or downward. It is responsible for the spoiling and processing of water and grains. Disorders in this section are regulated at the sides of the navel. The lower section of the triple energizer begins exactly at the upper opening of the bladder and extends downward. It is responsible for separating the clear from the turbid portions and is primarily involved in discharge rather than intake. It acts as a transmitter for these functions, and disorders in this section are regulated one Cun below the navel.

So far, the views on the triple energizer in the *Classic of Difficulties* are not essentially different from those found in *Huangdi's Internal Classic* and *Comprehensive Discussions in the White Tiger Hall*. Moreover, in the chapter "The Thirty-Ninth Difficulty", where the number of Fu-organs is stated as five instead of six, the triple energizer is still considered a Fu-organ that is not directly associated with the five Zang-organs. In the chapter "The Thirty-Eighth Difficulty", it is mentioned that the triple energizer has a name but no form and is associated with the hand minor yang conduit. It is referred to as an external palace. Similar statements can be found in the chapter "To Consider the Transportation Openings as the Foundation" in the *Miraculous Pivot*, highlighting the corresponding relations between the five Zang-organs with six Fu-organs. Both sources acknowledge the fact that the triple energizer is one of the six Fu-organs.

To gain a thorough understanding of the "theory of the invisible triple energizer" in the *Classic of Difficulties*, it is important to consider additional evidence that supports this concept. In the chapter "The Twenty-Fifth Difficulty", it is mentioned that the pericardium and the triple energizer represent the external and internal aspects, respectively, and both are named entities without physical form. The pericardium, in this context, refers to the tissue surrounding the heart, which is considered a "non-independent entity" because it does not exist as a separate organ. Similarly, the triple energizer, which wraps around other organs in the abdominal cavity, shares the same morphological characteristics as the pericardium. Yang Shangshan further clarifies that the pericardium has its own meridian called "hand reverting yin" but no independent entity, similar to how the triple energizer has its own qi and meridian called "hand lesser yang" but no independent entity. This is why they are connected

[17] Paul U. Unschuld, *Huang Di Nei Jing Ling Shu* (*The Ancient Classic on Needle Therapy*), The complete Chinese text with an annotated English translation, p. 383.

to each other in terms of interior and exterior relations. [18] Thus, the character "别 (bié)" mentioned twice actually refers to the lack of an independent form separated from other organs, rather than the absence of objective physical substance. In conclusion, although the triple energizer has physical presence, it is still referred to as an "external Fu-organ" because it is merely the entity attached to organs rather than an independent organ itself.

According to the *Classic of Difficulties*, Zang-organs belong to yin, so the number of them should be even and Fu-organs belong to yang, so the number of them should be odd. To align the yin-yang attributes of Zang-Fu organs with the odd-even number concept, the number of Fu-organs was reduced from six to five by excluding the triple energizer. Conversely, the number of Zang-organs was increased from five to six by separating the two kidneys into the "left kidney" and "right life gate". This is precisely why the tangible triple energizer with its functional role was transformed into a nominal Fu-organ in this text.

In fact, the authentic "theory of invisible triple energizer" was put forward on the basis of dividing the human body into upper, middle, and lower portions. Consequently, the understanding of the triple energizer underwent a fundamental change as well. For example, the *Pulse Classic* states that "Cun governs the upper energizer, including the head, skin, hair, and extending to the hands. The Guan rules the middle energizer, including the abdomen and the lumbar region. And the Chi rules the lower energizer, including the lower abdomen and extending to the feet".[19] In *Bright Hall Treatise on the Five Zang-organs* (*Ming Tang Wu Zang Lun*,《明堂五脏论》), an ancient medical classic unearthed in Mogao Caves at Dunhuang, it is also said that:

> The upper energizer is from the head to the heart, the middle energizer is from the heart to the umbilicus, and the lower energizer is from the umbilicus to the foot. So, the triple energizer is nominal and invisible.[20]

The theory of the invisible triple energizer, which is based on the division of the human body into upper, middle, and lower parts, holds true significance as it avoids the confusion between the concepts of "intangible" and "tangible but non-independent". This clarification clears any ambiguity regarding whether the triple energizer refers to "invisible qi" or the "ínvisible place where qi moves". Consequently, it can be inferred that the triple energizer does not possess actual functions. At this point, readers need to notice the following interrelated changes in medicine between the Han and Jin Dynasties:

1. In the Han Dynasty, the triple energizer was recorded as a tangible organ with its function according to *Plain Questions*, *Miraculous Pivot* and the *Classic of Difficulties*. But in the Jin Dynasty (AD 265-420), it became an abstract concept referring to the upper, middle and lower parts of the human body.

---

[18] Yang Shangshan (the Sui Dynasty), *Great Simplicity of Huangdi's Internal Classic*, p. 108.

[19] Wang Shuhe (the Jin Dynasty), the third chapter "The Domination of Three Passes" in *Pulse Classic*, Volume 1.

[20] Ma Jixing, *Textual Research and Interpretation of Dunhuang Ancient Medical Books* (*Dun Huang Kao Yi Ji Kao Shi*,《敦煌古医籍考释》), p. 12.

2. Similar changes can be observed in pulse diagnosis methods. In the Eastern Han Dynasty, the wrist pulse was divided into two parts including Cun and Chi to coordinate with yin and yang based on the *Classic of Difficulties*. Conversely, in the Western Jin Dynasty, the wrist pulse was divided into three parts including Cun, Guan, and Chi, which respectively corresponded to the upper, middle, and lower parts of the human body, as described in the *Pulse Classic*.
3. The division of the human body into upper, middle, and lower parts in the Jin Dynasty bears resemblance to the division found in traditional Indian medicine. Whether this similarity is a result of foreign cultural influence remains uncertain. However, it is evident that during that time, the triple energizer solely represented the three divided parts nominally, as its original meaning had been silently and completely replaced. Regarding the proposed concepts of the "thoracic triple energizer" (proposed by Yv Tuan and Zhang Jiebin in the Ming Dynasty), the "stomach triple energizer" (proposed by Luo Mei in the Qing Dynasty), the "membrane triple energizer" (proposed by Tang Zonghai in the Republic of China, AD 1912-1949),[21] they are all variations of this "invisible theory" that reflect diverse academic viewpoints in the process of exploring and investigating the original meaning of classical texts.

11.2.4 Signs, Symptoms, and Acupoints

In *Classic of Difficulties*, the yin-yang theory is the guiding principle for pulse diagnosis while the five-phase theory forms the ideological basis for discussing signs, symptoms, and acupoints. For example, in the chapter "The Forty-Ninth DIfficulty" and "The Fiftieth Difficulty" are used to define the mutual interference among the "pathogenic qi" of the five Zang-organs and they are pathogen from mother-organ (the mother element in the generating relationship of the five elements), pathogen from child-organ (the child element in the generating relationship of the five elements), thief pathogen (the element being un-restricted in the restricting relationship of five elements), mild pathogen (the element being restricted in the restricting relationship of five elements), and regular pathogen (from within the element itself). The chapters "The Fifty-Third Difficulty" and "The Fifty-Fourth Difficulty" describe the transmission and transformation of diseases among the five Zang-organs and six Fu-organs using the concepts of mutual generation and mutual restricting. Moreover, in the chapters "The Fifty-Sixth Difficulty" and "The Fifty-Eighth Difficulty", the diseases including accumulation, diarrhea, cold damage, are categorized into five categories based on the number "5" (Fig. 11.1).

The most typical example is the chapter "The Fifty-Sixth Difficulty", in which the accumulation of five Zang-organs is related to five directions in the five-phase theory. Taking the stomach as the reference center, the accumulation in the liver is located on the left; the accumulation in the heart is located above; the accumulation

---

[21] For details, see Ren Yingqiu, *The Doctrine of the Schools of Traditional Chinese Medicine* (*Zhong Yi Ge Jia Xue Shuo*,《中医各家学说》), pp. 186–190.

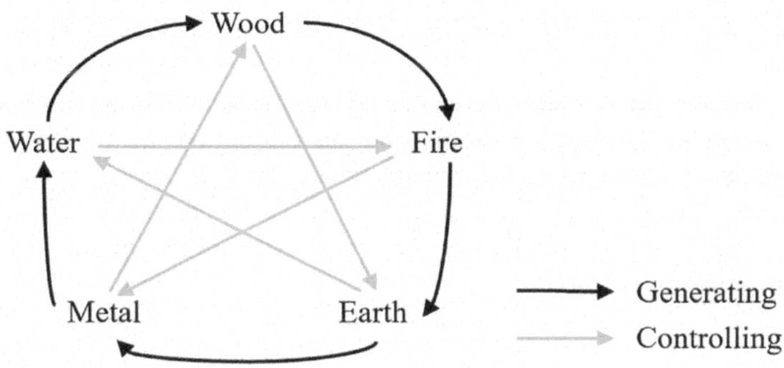

Fig. 11.1 Generation and restriction relationship of five phases

in the spleen is located in the middle; the accumulation in the lung is located on the right; the accumulation in the kidney is located below. Moreover, the cause of this is that when pathogenic factors spread according to the law of mutual restraint in the five-phase theory, and when they encounter the organ they can restrain during its "dominant" period, the pathogenic qi cannot continue to spread downward and stagnates, forming an "accumulation".

In the discussion of the acupoints, the *Classic of Difficulties* primarily focuses on the five transport points. Although six chapters in *Plain Questions* and *Miraculous Pivot* also mention the five transport points, the comprehensive relationship among "four seasons, five elements, and five transport points" had not been established at that time. The *Classic of Difficulties* systematically discusses the mutual coordinating relationship among "five transport points, five elements and five seasons".[22] It specifies the disease correspondences of the five transport acupoints according to the relationship among the five elements. Besides, on one hand, the relationship of the five acupoints from the "well point" to the "sea point" on each meridian aligns with the generating relationship of the five elements. The relationship between the same-named acupoints on yang meridian and yin meridian is in line with the restricting relationship of the five elements (for example, the well points on yang meridians correspond to metal and the well points on the yin meridians correspond to wood). This corresponding relationship appears to have existed prior to the completion of *Classic of Difficulties*, as evidenced by related content from *Ten Changes*. Meanwhile, the importance of the five transport acupoints in *Classic of Difficulties* precisely reflects the author's emphasis on the five elements and the theoretical methods of acupoint selection. It was not until the Eastern Han Dynasty that acupuncture and moxibustion therapy based on the meridian and collateral theory, integrated with the five-phase theory and underwent significant changes. However, this transformation

[22] When introducing these acupuncture points in the chapter "Discussion on Acupoints" in *Miraculous Pivot*, the first point of the Yin meridian is "Jing, Mu" and the first point of the Yang meridian is "Jing, Jin", but in the *Essentials of Huangdi's Internal Classic*, the word "Jing" is not followed by the word "Mu" or "Jin". So, it is clear that the word "wood" or "gold" was added by later generations.

was not superficial, but deeply rooted in the theory of five elements. Thus, acupoint selection theories that were not based solely on the meridian and collateral theory emerged. Examples include statements like "Through the wells one can master fullness below the heart. Through the brooks one can master body heat". This theory has been passed down through *Classic of Difficulties* and *A-B Classic of Acupuncture and Moxibustion* (*Zhen Jiu Jia Yi Jing*,《针灸甲乙经》) accomplished in the Western Jin Dynasty and it has consistently been regarded as one of the important guidelines for acupoint selection methods.

11.3 Contents Added by Later Generations

Some of the contents found in the received version of the *Classic of Difficulties* were added by later generations. For example, in the chapter titled "The Thirty-First Difficulty", the latter part of the sentence which reads "Disorders in this section are regulated at CV 17 (Tanzhong) located one cun and six fen[23] below CV 18 (Yutang), exactly in the fold between the two breasts" and the last sentence which states "Another copy of this text says 'throughway' of influences" are all the later added annotations. Besides, there are also some added contents involving the epochal characteristics of the medical theory.

11.3.1 Explanations of "Three Sections" in Five-Phase Theory

The explanation of "three sections" in the chapter "The Eighteenth Difficulty" was added by later generations. The following is the original text.

> The eighteenth difficulty: The movement in the vessels appears in three sections. Each section consists of four conduits, with the great-yin and the yang-brilliance conduits of the hands, and the great-yang and the minor-yin conduits of the feet, constructing the upper and the lower sections. What does that mean? Here is the explanation: The hand-great-yin and the hand-yang-brilliance conduits are associated with the phase of metal; the foot-minor-yin and the foot-great-yang conduits are associated with the phase of water. Metal generates water. As water flows downward, it is unable to ascend. Therefore, the foot-minor-yin and the foot-great-yang conduits can be felt at the section below the gate. The foot-ceasing yin and the foot minor-yang conduits are associated with the phase of wood; it generates the phase of fire, which is associated with the hand-great-yang and the hand-minor-yin conduits. The flames of fire ascend; they are unable to move downward. Hence, the hand-great-yang and the hand-minor-yin conduits can be felt at the section above the gate. The hand-heart-master and the hand minor-yang conduits are associated with the phase of fire; it generates the phase of soil, which is associated with the foot-great-yin and with the foot-yang-brilliance conduits. The soil phase rules the central palace and is, therefore, situated in the central section. All

[23] 1 fen is equal to 0.1 cun.

> of this corresponds to the mutual generation and nourishment of the Five Elements, akin to a child-mother relationship.
> The movement in the vessels appears in three sections and spans nine indicator levels. Which illnesses are governed by the movements in these sections and levels? Here is the explanation: The three sections concerned are the Cun section, the gate section, and the foot section. The nine indicator levels refer to the surface, center, and depth of each of the three sections. The upper section is patterned after heaven; it governs illnesses located from the chest upward to the head. The central section is patterned after man; it governs illnesses located below the diaphragm to the navel. The lower section is patterned after earth; it governs illnesses located below the navel to the feet. For treatment, one should first carefully examine which section displays which movement in the vessels before applying the appropriate treatment using needles.

In the chapter "The Eighteenth Difficulty", the mentioned "three sections" in the first paragraph refer to neither the "upper, middle, and lower" in the "three sections and nine indicators", nor the "Cun, Guan, and Chi" of the wrist pulse. Instead, the author uses this concept to integrate the twelve meridians into the five-phase theory. The difference between the "four meridians" in each section and the three pulses in each section of the "three sections and the nine indicators" lies not only in the number but also in their respective connotations (as shown in Table 11.1).

As mentioned earlier, the wrist pulse in the *Classic of Difficulties* is divided into Chi and Cun sections, which correspond to Yin and Yang and are used to explain the pathological changes in different parts of the human body and Zang-Fu organs. The inclusion of the Guan section would disrupt the self-consistency of the pulse and pathology system based on the "Chi-Cun and Yin-Yang theory". Therefore, it can be inferred that the sentence "The three sections are Cun, Guan, and Chi; the nine indicators are superficial, middle, and deep" was written by annotators who did not understand the meaning of "These are all the relationships of mutual generation and nourishment between the five elements and their offspring". Otherwise, why would Yang Shangshan of the Sui Dynasty say that Hua Tuo's method of positioning the three-section wrist pulse "has no known basis for what is said"? Cheng Hongru's argument about pre-Qin sphygmology based on this sentence in his previously cited "Casual Discussion" article is undoubtedly untenable, and it is also a misjudgment even if it is applied to the early Eastern Han Dynasty.

11.3.2 Measurement of the Five Zang-Organs and Six Fu-Organs

In the "Forty-Second Difficulty" chapter of the *Classic of Difficulties*, the author poses questions about the dimensions and capacities of the intestines and stomach. The subsequent paragraph then provides specific measurements of the stomach, small intestine, ileum, sigmoid colon, and rectum, concluding with a self-questioning statement: These are the figures of the dimensions of the intestines and of the stomach and of the amounts of water and grains they hold. In the succeeding paragraph, the weight, size, and function of the five Zang-organs are stated directly without raising

Table 11.1 Comparison between the "Three Sections and Four Meridians" and "Three Positions and Nine indicators"

Classic of Difficulties			*Plain Questions* "three positions and nine indicators" pulse diagnosis method
Three sections	Five elements (there are two fire elements)	Twelve meridians	
Upper section	Wood fire	Foot reverting yin (wood) Foot lesser yang (wood) Hand greater yang (fire) Hand lesser yin (fire)	Two arteries in the forehead Two arteries in the cheeks Two arteries before the ears
Middle section	Fire Earth	Hand reverting yin (fire) Hand lesser yang (fire) Foot greater yin(earth) Foot yang brightness (earth)	Hand greater yin Hand lesser yin Hand yang brightness
Lower section	Metal Water	Hand greater yin (metal) Hand yang brightness(metal) Foot greater yang(water) Foot lesser yin (water)	Foot reverting yin Foot lesser yin Foot greater yin

any questions. And the weight, shape, length, and volume of the six Fu-organs are introduced again but in different terms compared with the first paragraph.

For instance, in the chapter "Intestines and Stomach" and "A Healthy Person Ends the Ingestion of Grain" in *Miraculous Pivot*, the appellations of the small intestine, ileum, sigmoid colon, and rectum are the same as that in the first paragraph of the chapter "The Forty-Second Difficulty", but in the second paragraph, terms including the small intestine, large intestine and anus are applied, among which the word "anus" is not found in both *Plain Questions* and *Miraculous Pivot*. While in the chapter "The Forty-Fourth Difficulty", the "seven through-gates" are mentioned and the anus is called "corporeal soul gate", which is in line with the description in the chapter "Different Discussion on the Five Zang-Organs" in *Plain Questions*. The term "anus" (肛gang or 肛门) has no recorded history in the *Explanation and Study of Principles of Composition of Characters* (*Shuo Wen Jie Zi*, 《说文解字》). So, it can be estimated that the history of the word "肛gang" or "肛门 (anus)" is not so long,[24] suggesting it is a relatively recent term. The repetition of gastrointestinal tract records may be due to subsequent generations supplementing the characters based on other documents. Sentences such as "The liver weighs four jin and four liang (about 946.6 g).[25] It has three lobes on its left and four lobes on its right side" and "The heart has seven holes and three hairs (毛)"[26] are exactly the same as the description in *Illuminated Hall Treatise on the Five Viscera* in the Six Dynasties (AD 222-589). Although it cannot be asserted that these words in *Classic of Difficulties* originated from *Illuminated Hall Treatise on the Five Viscera*, it is obvious that they are not the original words of *Classic of Difficulties*.

11.3.3 The Traces of Taoism Infiltration

The chapter "The Eighth Difficulty" in the *Classic of Difficulties* mentions that "the moving qi between the kidneys" is referred to as "the spirit guarding against the evil". Some modern physicians, purely relying on the literal meaning and falling into the old trap of "applying scientific labels", interpret it as "the function of defending against the invasion of external pathogens".[27] However, it should be noted that although this term appears in the *Classic of Difficulties*, it is not an inherent term in the medical system, but rather similar expressions can be found in Taoist scriptures. For example, in the annotation of the sentence "The inner chambers and the purple palace are the spiritual gateways" in Chapter 17 "The Spirit Platform" of *A Regimen Classic of Taoism* (*Huang Ting Nei Jing Jing*, 《黄庭内景经》), *A Representative Classic of Taoism in the Six Dynasties* (*Da Dong Jing*, 《大洞经》) states that "Three-fen inside from the point directly above the space between the eyebrows is the 'Shou Cun'". Furthermore, Chapter 8 "The Spirit of the Heart" says that "The spirit of the heart, named Danyuan, is also called 'Shou Ling'".[28] Not only are the word-using and tone similar, but the idea that each part of the human internal organs has a main deity is also a typical manifestation of Taoist culture. (Fig. 11.2)

Situated 15 kilometers southwest of Guan County, Sichuan Province, this mountain is renowned for its magnificent overlapping peaks, towering ancient trees, as

---

24 *Huangdi's Classic of the Bright Hall* has mentioned "anal prolapse".

25 In this period, 1 jin is equal to 222.73 g and 1 liang is equal to 13.92 g. In modern day, the weight of liver in adult is about 1200–1400 g.

26 The "seven holes" is still easy to understand, but "three hairs" is quite difficult to understand, and the interpretation of each note is not satisfactory. Later, it is said that "the accumulation of bile", should be in terms of the pericardial fluid, so it is suspected that this is the process of reading and writing the word "membrane" written as "hair". In addition, the chapter "Forty-First Difficulty", which states "Why does the liver have two lobes", is different from the "seven lobes" of the "left three lobes and right four lobes in the chapter "Forty-Second Difficulty".

27 Nanjing College of Traditional Chinese Medicine, *Analysis on Classics of Difficulties* (*Nan Jing Jiao Shi*, 《难经校释》), Beijing: People's Medical Publishing House, 1979, p. 18.

28 *A Regimen Classic of Taoism*, Shanghai: Shanghai Chinese Classics Publishing House, 1990, p. 18.

Fig. 11.2 Wu Dong Tian archway in Qing Cheng Mountain

well as the notable landmarks of Shangqing Palace and Tianshi Cave. Legend has it that Zhang Daoling, a renowned Taoist master of the Eastern Han Dynasty, once immersed himself in Taoist practices on this sacred mountain.

The Preface of *Classic of Difficulties* written by Wang Bo, a great scholar in the Tang Dynasty, said that the book was initially passed down to Qin Yueren by Qi Bo, Huang Di, Yi Yin, Tang, Tai Gong, Wen Wang, Yi He, and so on. It then continued its journey to Hua Tuo through nine physicians, followed by six physicians who transmitted it to Huang Gong. Finally, Cao Fuzi received the book from Huang Gong. Cao Fuzi, also known as Yuan or Zhendao, remains an enigmatic figure, shrouded in mystery and unknown to most.[29] Based on this account, it may be inferred that Cao Fuzi was associated with Taoism. While it is uncertain whether the later part of this "chain of inheritance" exclusively involved Taoists, expressions like "the spirit guarding against evil" do indicate the influence of Taoism, which cannot be solely attributed to medical theory.

[29] Li Fang et al. (the Song Dynasty), The chapter "the Preface of Classic of Difficulties" in *Finest Blossoms in the Garden of Literature* (*Wen Yuan Ying Hua*, 《文苑英华》), Volume 735.

Chapter 12
Before and After the Establishment of Materia Medica

The term "materia medica" was commonly used in ancient China to refer to the pharmacology of traditional Chinese medicine (TCM). It encompassed the names, properties, functions, indications, producing areas of various drugs, as well as knowledge about their collection, preservation, and processing. Essentially, materia medica in TCM covers all aspects of drug-related knowledge.

The emergence of materia medica does not exactly represent the origin and the use of drug. Objectively, the term "materia medica" marks the establishment of TCM pharmacology, which originated later than the time when various drugs were widely applied with a certain rational understanding. Since medical knowledge within different schools hadn't been integrated and systematically organized by the end of the Eastern Han Dynasty (AD 25–220), we must examine the works written by Tao Hongjing (AD 456–536) during the Southern Dynasty (AD 420–589) to fully explore the subject of materia medica.

12.1 The Emergence of the Term "Materia Medica"

Undoubtedly, in the early stages of TCM development, there were extensive discussions on materia medica by practitioners. For example, *Discussion on Materia Medica* (*Yao Lun*, 《药论》), a book obtained by the renowned doctor Chunyu Yi (about 215–140 BC) in the Western Han Dynasty (202 BC-AD 8) from his master Gongsheng Yangqing (251–176 BC). It is also said that Bian Que's disciple Zi Yi once composed *Zi Yi's Materia Medica* (*Zi Yi Ben Cao Jing*, 《子义本草经》)[1] which

[1] It is stated in "Physician" of the chapter "Government Officials" in *The Rites of Zhou* (*Zhou Li*, 《周礼》) annotated by Zheng Xuan: "The combination formula of five kinds of materia medica, namely herbs, woods, insects, stones and grain were adopted by Shen Nong and Zi Yi". It's also annotated by Jia Gongyan in the *Classics and Books of Western Jin* (*Zhong Jing Bu*, 《中经簿》) that "*Zi Yi's Materia Medica* was composed by Zi Yi (子仪 or 子义)".

Y. Liao, *Reconstructing the Image of Medicine in the Qin and Han Dynasties*, Understanding China, https://doi.org/10.1007/978-981-95-6801-7_12

consisted of only one volume. During the Three Kingdoms Period (AD 220–280), theories of eight predecessors were quoted in Wu Pu's works on materia medica, including Shen Nong, Huang Di, Bian Que, Yi He, Qi Bo, Tong Jun, Lei Gong and Family Li.[2] However, when Liu Xiang (77–6 BC) and Liu Xin (50 BC-AD 23), father and son, sorted out ancient books and compiled the catalogue *Seven Categories of Writings* (*Qi Lve*, 《七略》) at the end of the Western Han Dynasty, they did not mention any pharmacological works. Does this indicate that books related to materia medica were not created until the Eastern Han Dynasty? Here is a brief discussion on this issue.

According to *Seven Categories of Writings* in the chapter "Catalog of Literary Works" in *Book of the Former Han* (*Han Shu*, 《汉书》), there are four categories of works related to prescriptions and techniques in TCM. The commentary on the latter two categories, "health preservation" and "sexual healthcare", does not mention drugs. However, in the commentary on the first category "medical classics", it states:

Medical classics explore human physiological characteristics, including blood vessels, meridians and collaterals, bone marrow, yin and yang, exterior and interior. Further, they elucidate the root causes of various diseases, distinguish the boundaries between death and life, and analyze features of such treatment methods as acupuncture, healing stone, decoction and moxibustion to concoct various drugs for different prescriptions to address clinical needs.

The current edition of *Huangdi's Internal Classic* (*Huang Di Nei Jing*, 《黄帝内经》) does not specifically describe "the concoction of various drugs for different prescriptions to address clinical needs". This not only supports the argument that "the current edition of this book is not the one recorded in the "Catalog of Literary Works" in the *Book of the Former Han*", but also illustrates that the original "medical classics" included knowledge related to materia medica. Moreover, it is mentioned in the commentary following the second category, "classical formula":

Classical formulas examine the cold and warm properties of herbs and minerals, measure the severity of diseases, utilize the nourishment of drug flavors, adapt to the appropriate qi sensations, distinguish the five bitter and six pungent flavors, and prepare prescriptions with the properties of water and fire to unblock blockages, resolve stagnation and restore the body to balance.

Obviously, the description "distinguish different flavors of materia medica targeting Five Zang- and Six Fu-Organs" is related to the knowledge of drug properties and flavors, which falls under the discipline of "materia medica". This shows that the early medical works are different from those of later generations. The former not only recorded formulas and prescriptions for specific diseases, but also encompassed pharmacological knowledge, which is the content of materia medica. In this regard, it is possible that the early theories from Shen Nong, Huang Di, Qi Bo, Yi He,

---

[2] It is recorded in chapter "Catalog of Classics and Books" in *History of the Sui Dynasty* (*Sui Shu*, 《隋书》): "Wu Pu, a disciple of Hua Tuo (around AD 145-208) in the Liang Dynasty (AD 502-557), has written six volumes of materia medica". His volumes were lost, but it is often quoted in *Readings of the Taiping Era* (*Tai Ping Yu Lan*, 《太平御览》) and other books, and now there is Shang Zhijun's edition of *Wu Pu's Materia Medica* (*Wu Pu Ben Cao*, 《吴普本草》) (Beijing: People's Medical Publishing House, 1987).

and Bian Que, quoted by *Wu Pu's Materia Medica* were based on ancient medical books recorded in the chapter "Catalog of Literary Works" in the *Book of the Former Han.* However, when the *Seven Categories of Writings* was completed, most of the pharmacological knowledge had not yet been compiled into separate monographs and could be found in medical classics and classical formulas attributed to figures like Huang Di, Bian Que, Shen Nong, etc. Hence, it can be said that by the time the book *Seven Categories of Writings* was accomplished, pharmacology had not been recognized as an independent branch of medicine in the medical field or even in the whole academic world.

The term "materia medica" was first mentioned in the "Records of Outskirts Sacrifices" in the *Book of the Former Han* which records that in the second year of Emperor Cheng's reign in the Western Han Dynasty (31 BC), "More than 70 people waiting for immortals were asked to return home, including prescription masters, messengers, assistants, and those who needed to wait for imperial edict with prescriptions and materia medica".[3] Another record in the fifth year of the Emperor Ping's reign (AD 5) mentioned "recruited people across China who mastered scattered Confucian classics, books recording ancient events, astronomy, calendar and arithmetic, phonology, philology, *Literacy Textbook for Children* (*Shi Zhou Pian*, 《史籀篇》), medicine and astrology, materia medica, and those who taught five great Confucian classics, *The Analects of Confucius* (*Lun Yu*, 《论语》), *The Classic of Filial Piety* (*Xiao Jing*, 《孝经》) and *The Dictionary* (*Er Ya*, 《尔雅》)……".[4] Moreover, there is a record in the "Records of the Errants" of the *Book of the Former Han*, which states that "a man named Lou Hu used to practice TCM with his father in the city Chang'an when he was a child and thus came into contact with the households of noblemen. Lou Hu read hundreds of thousands of words in books related to medical classics, materia medica and prescriptions".[5]

In the three aforementioned records, the phrase "those who needed to wait for imperial edict with prescriptions and materia medica" in the first record refers to an official title. Yan Shigu's annotation suggests that it refers to individuals who awaited an imperial edict involving prescriptions and materia medica. In the second year of Emperor Cheng's reign, hundreds of sacrificial sites that had been set up by the Emperor Gaozu, Liu Bang, were abolished. As a result, it is mentioned that more than 70 people waiting for immortals were asked to return home, including prescription masters, messengers, assistants, and those who needed to wait for imperial edict with prescriptions and materia medica. However, there is no concrete evidence regarding the origin of this official title. "Materia medica" in the second one can be regarded as either an abstract noun or a book name similar to *Literacy Textbook for Children*, five great Confucian classics, *The Analects of Confucius*, *The Classic of Filial Piety* and

[3] "Records of Outskirts Sacrifices" in *Book of the Former Han* (*Han Shu*, 《汉书》), Volume 25, 5th.

[4] "Annals of Emperor Ping" in *Book of the Former Han* (*Han Shu*), Volume 12, 12th.

[5] "Records of the Errants" in *Book of the Former Han* (*Han Shu*), Volume 92, 62nd.

The Dictionary.[6] However, the "materia medica" mentioned along with the medical classics and prescriptions in the third record undoubtedly is an abstract noun.

In the late Western Han Dynasty, Wang Mang's (45 BC-AD 23) political activities are worth noting. Although he officially ascended to the throne in AD 9, he had already been active on the political stage before that, just as that stated in "Annals of Emperor Ping" in the *Book of the Former Han*, "During Emperor Ping's reign, all political affairs were decided by Wang Mang". "Catalog of Literary Works" in the *Book of the Former Han* was compiled on the basis of Liu Xin's *Seven Categories of Writings*. It was compiled during the reign of Emperor Ai of the Han Dynasty, so it does not reflect the cultural and literary activities of the following 20 years. In the fourth year of Emperor Ping's reign (AD 4), Wang Mang initiated a nationwide recruitment of individuals with special talents, resulting in the gathering of thousands of people. These individuals were allowed to express their opinions, which were recorded at court with the intention of rectifying absurdities and aligning various viewpoints.[7] Immediately afterwards, there came the record of recruiting talents who were proficient in "materia medica" nationwide in the year of AD 5. It can be speculated that the works related to "materia medica" must have been mentioned in the accounts of these people. Lou Hu, who was able to "recite hundreds of thousands of words about medical classics, materia medica and prescriptions", lived during the same period as Wang Mang. Consequently, it can be concluded that the fundamental meaning and the pharmaceutical works under the name of "materia medica" were primarily formed in the first decade after AD, between the reign of Emperor Ping in the late Western Han Dynasty and the time when Wang Mang declared himself emperor. From then on, TCM pharmacology became an independent system. Whether in the eyes of doctors or other scholars, "materia medica" was seen as a distinct branch of TCM. As a result, *Plain Questions* (*Su Wen*, 《素问》), *Miraculous Pivot* (*Ling Shu*, 《灵枢》), and even *Classic of Difficulties* (*Nan Jing*, 《难经》) held the same status as the works categorized in the "Medical Classics" of the *Seven Categories of Writings*. However, newly written "medical classics" rarely included specific content about pharmacology.

12.2 Early Works of Materia Medica

Given the historical context of the establishment of materia medica, "the early works of materia medica" could not be traced back to the pre-Qin Period (before 221 BC) like other branches of TCM. The early works of materia medica did not appear before *Seven Categories of Writings* of Liu Xin in the Western Han Dynasty. However, there are mentions of "Shen Nong and Zi Yi's technology" when annotating *The Rites of Zhou* and Jia Gongyan annotated "Zi Yi's Materia Medica" in the *Classics and Books*

[6] In the collation edition of *Book of the Former Han* published by Zhonghua Book Company, all these mentioned books are added with book title marks.

[7] "Annals of Wang Mang" in *Book of the Former Han*, Volumes 99, 69th.

of Western Jin. During the Three Kingdoms Period, Wu Pu introduced the theories of eight great masters, including Bian Que without any reference to the theory of Zi Yi. Thus, the so-called *Zi Yi's Materia Medica* is probably only a small part of the works of the school which is characterized by teacher-apprentice inheritance and represented by his master, Bian Que. Whereas, this small part is probably an integral part of the "Bian Que's internal and external classics" recorded in *Seven Categories of Writings*. At the time of its composition, the medical and academic communities did not view materia medica as a separate branch, so books related to this field were not categorized separately. In a cultural context that emphasized "learning from the master" and adherence to domestic disciplines, even if Zi Yi had put forward relative theories of materia medica, it is reasonable to assume that Zi Yi's theories were incorporated into those of his master, Bian Que. Additionally, considering that most of the contents of the so-called "Bian Que's internal and external classics" were likely written by his disciples and successors, later generations might have attributed this pharmacological knowledge to "Bian Que's saying". Alternatively, because materia medica had already become an independent branch, it's natural to call it as "*Zi Yi's Materia Medica*".

After the establishment of "materia medica", it experienced unprecedentedly rapid development. During the compilation of ancient book catalogues in official history for the second time, the quantity and types of books about materia medica were significant (Fig. 12.1). For example, in the chapter "Catalog of Classics and Books" in the *History of the Sui Dynasty*, the commentary following the eight volumes of *Shennong's Classic of Materia Medica* (*Shen Nong Ben Cao Jing*, 《神农本草经》) states "In the Liang Dynasty, there are five volumes of *Shennong's Classic of Materia Medica*, two volumes of works related to *Shennong's Classic of Materia Medica*, seven volumes of *Cai Yong's Materia Medica*, six volumes of *Wu Pu's Materia Medica*, ten volumes of *Tao Hongjing's Materia Medica*, nine volumes of *Sui Fei's Materia Medica*, six volumes of *Qin Chengzu's Materia Medica*, three volumes of *Wang Jipu's Materia Medica*, one volume of *Li Dangzhi's Materia Medica*, one volume of *Tan Daoshu's Materia Medica*, five volumes of *Essentials of Materia Medica for Pathogenesis* by general Xu Shuxiang from the Liu Song Dynasty (AD 420–479), ten volumes of *Essentials of Internal Medicine and Miscellaneous Diseases* by Xu Shuxiang and other three masters, two volumes of *Wang Mo's Collected Writings of Materia Medica for the Pediatric*, nine volumes of Gan Junzhi's *Collected Writings of Materia Medica for Carbuncle, Ears and Eyes*, seven volumes of *Tao Hongjing's Collected Annotations on the Classic of Materia Medica*, one volume of *Zhao Zan's Materia Medica*, one volume of *Simplified Version of Materia Medica* and one volume of *The Application of Materia Medica*, that's all". The commentary following three volumes of *Tong Jun's Records of Materia Medica* (*Tong Jun Yao Lu*, 《桐君药录》) states that in the Liang Dynasty, there are four volumes of *Newly Collected Records of Materia Medica* by Xu Tao, who has been dubbed as Yun Hui General, six volumes of *Li Dangzhi's Records of Materia Medica*, forty two volumes of *Methods of Materia Medica*, three volumes of *Precepts for Materia Medica*, two volumes of *Properties of Materia Medica*, two volumes of *Medicinal Combination*, three volumes of *The Compendium of Materia Medica*, two volumes of

Shennong's Classic on Collection of Materia Medica, one volume of *Contraindications of Materia Medica*. Moreover, in the chapter "Catalog of Classics and Books" in the *History of the Sui Dynasty*, the following works on materia medica are contained:

1. *Shennong's Materia Medica* (*Shen Nong Ben Cao*, 《神农本草》), eight volumes; 2. *Shennong's Materia Medica*, four volumes, annotated by Lei Gong; 3. *Family Zhen's Materia Medica* (*Zhen Shi Ben Cao*, 《甄氏本草》), three volumes; 4. *Tong Jun's Records of Materia Medica*, three volumes; 5. *A Summary of Tai Qing Materia Medica* (*Tai Qing Cao Mu Ji Yao*, 《太清草木集要》) by Tao Hongjing, two volumes; 6. *Shennong's Classic of Materia Medica*, three volumes; 7. *Classic of Materia Medica* (*Ben Cao Jing*, 《本草经》) by Cai Ying, four volumes; 8. *The Compendium of Materia Medica* (*Yao Mu Yao Yong*, 《药目要用》), two volumes; 9. *Synopsis of Classic of Materia Medica* (*Ben Cao Jing Lve*, 《本草经略》), one volume; 10. *Materia Medica* (*Ben Cao*, 《本草》) by Xu Taishan, two volumes; 11. *Essentials of Classics of Materia Medica* (*Ben Cao Jing Lei Yong*, 《本草经类用》), three volumes; 12. *Pronunciation and Meaning of Materia Medica* (*Ben Cao Yin Yi*, 《本草音义》) by Yao Zui, three volumes; 13. *Pronunciation and Meaning of Materia Medica* (*Ben Cao Yin Yi*, 《本草音义》) by Zhen Liyan, seven volumes; 14. *Collection of Materia Medica* (*Ben Cao Ji Lu*, 《本草集录》), two volumes; 15. *Writings of Materia Medica* (*Ben Cao Chao*, 《本草钞》), four volumes; 16. *Miscellaneous and Essential Teachings of Materia Medica* (*Ben Cao Za Yao Jue*, 《本草杂要诀》), one volume; 17. *Prescriptions of Materia Medica* (*Ben Cao Yao Fang*, 《本草要方》), three volumes; 18. *Properties of Materia Medica* (*Yi Ben Cao Lu Yao Xing*, 《依本草录药性》), three volumes; 19. *Beautiful Illustration of Materia Medica* (*Ling Xiu Ben Cao Tu*, 《灵秀本草图》) by Yuan Pingzohng, six volumes; 20. *Illustration of Ganoderma Lucidum* (*Zhi Cao Tu*, 《芝草图》), one volume; 21. *Method of Collecting Materia Medica in Forest* (*Ru Lin Cai Yao Fa*, 《入林采药法》), two volumes; 22. *Collection of Materia Medica in Taichang* (*Tai Chang Cai Yao Shi Yue*, 《太常采药时月》), one volume; 23. *Collection of Materia Medica in Four Seasons and the Catalogue* (*Si Shi Cai Yao Ji He Mu Lu*, 《四时采药及合目录》), four volumes; 24. *Records of Materia Medica* (*Yao Lu*, 《药录》) by Li Mi, two volumes; 25. *Different Names of Various Materia Medica* (*Zhu Yao Yi Ming*, 《诸药异名》) by Shamen Xingju, eight volumes; 26. *Properties of Various Materia Medica* (*Zhu Yao Yao Xing*, 《诸药要性》), two volumes; 27. *Methods of Planting Materia Medica* (*Zhong Zhi Yao Fa*, 《种植药法》), one volume; 28. *Planting Ganoderma Lucidum* (*Zhong Shen Zhi*, 《种神芝》), one volume.

From these bibliographies alone, it can be seen that there were already some monographs on specific aspects of materia medica at that time. For instance, there were works on the explanations of pronunciation and meanings (refer to 12 and 13), the textual research of drug aliases (refer to 25), the appropriate time for collecting herbs (refer to 21, 22, and 23), artificial cultivation (refer to 27 and 28), and herb diagrams depicting the specific forms of drugs (refer to 19 and 20). Most of these early works on materia medica have been lost. However, to a large extent, this kind of loss belongs to the situation of "the name is lost but the essence remains". It is true that many ancient books and documents were destroyed due to natural and human-made factors. However, it is important to note the updating and progress of knowledge in newly created books, which demonstrate desirable innovation and development based on the experiences of previous generations. Therefore, this type of loss is a process of abandonment and inheritance in the long course of historical development, accompanied by scientific and technological progress. Viewing the development history of materia medica from this perspective, we can observe that the extent and speed of discarding and losing old books are directly proportional to

Fig. 12.1 Stone inscription in the Southern Dynasties: Exorcising evils. After the Eastern Jin Dynasty (AD 317–420), the north and the south of China were AQ2 211 split up. The Southern Dynasties exerted an enormous influence on the development 212 of Chinese culture

the advancement of pharmacology. Nonetheless, this elimination is unfortunate for historians who strive to uncover all the details of its development process. Thus, they often have to rely on speculation to connect scattered historical materials.

Among the above-mentioned works on materia medica, *Shennong's Classic of Materia Medica* and some works written by Tao Hongjing in the Liang Dynasty have been well preserved. Besides, the general contents of *Wu Pu's Materia Medica* can also be seen from books such as *Readings of the Taiping Era*. To some extent, these historical materials can reflect the development process of the materia medica system.

12.3 *Wu Pu's Materia Medica* and the Theories of Eight Masters

Wu Pu, a disciple of Hua Tuo, is mentioned in the biographies of Hua Tuo in the *Book of the Later Han* (*Hou Han Shu*, 《后汉书》) and "*History of the Wei Dynasty*" in the *Records of the Three Kingdoms* (*San Guo Zhi*, 《三国志》). According to these records, Wu Pu learned TCM from Hua Tuo and was taught "five animal exercises". Through regular practice, Wu Pu was able to maintain good health well into his nineties, with clear senses and strong teeth. In the chapter "Maxim" in *Keeping Innocence* (*Bao Pu Zi*, 《抱朴子》) by Ge Hong, in the Jin Dynasty (AD 265–420), it states that Wu Pu lived for more than 100 years. Hence, it is generally estimated that Wu Pu was born in the middle of the second century and died in the middle of the third century.

The chapter "Catalog of Classics and Books" in the *History of Sui Dynasty* states that in the Liang Dynasty, Wu Pu, a disciple of Hua Tuo, wrote six volumes of materia medica, but the book has been lost to history, and only quotations preserved in *Readings of the Taiping Era* and other books remain for research. It is worth noting that this book records the theories of eight great masters, including Shen Nong, Huang Di, Qi Bo, Bian Que, Yi He, Tong Jun, Lei Gong and Li Dangzhi. Therefore, although this book was completed between the time of *Shennong's Classic of Materia Medica* and that of *Variorum of the Classic of Materia Medica* (*Ben Cao Jing Ji Zhu*, 《本草经集注》) of Tao Hongjing in the Liang Dynasty, its contents are not all based on the former one. The available textual materials on the theories of the eight masters cited in the book offer limited information about alias, place of origin, properties and flavors, indications, and processing of materia medica in each master's theory. We can only infer the main inclusions of *Wu Pu's Materia Medica* and the theories of the eight masters based on the quoted information, leaving some uncertainties about what is not involved. The existing historical materials provide some general insights:

(1) The early masters reached some consensus regarding the toxicity of materia medica (see Table 12.1). This might be regarded as the first step taken by the ancients in understanding these drugs. When considering a substance as either food or medicine, whether it is toxic and edible takes precedence over its taste, indications, and effects.

(2) Drug flavor: The flavors of materia medica (pungent, sweet, sour, bitter, salty, etc.) can be regarded as an objective record of the TCM experts' taste sensation.

Table 12.1 A comparison of drug toxicity among the theories of eight matters cited in *Wu Pu's Materia Medica*

	Shen Nong	Huang Di	Qi Bo	Tong Jun	Yi He	Bian Que	Lei Gong
Huang Di	▲▲▲▲▲▲▲ ▲▲▲▲▲▲▲ ▲▲▲▲▲▲▲ ▲▲▲						
Qi Bo	▲▲▲▲▲▲▲ ▲▲▲▲▲▲▲ ▲▲▲▲ ××	▲▲▲▲▲▲▲ ▲▲▲▲▲▲▲ ▲▲ ×					
Tong Jun	▲▲▲▲▲▲▲ ▲▲▲▲▲▲▲	▲▲▲▲▲▲▲ ▲▲▲▲▲▲▲ ▲	▲▲▲▲▲▲▲ ▲▲▲▲▲▲▲ ▲				
Yi He	▲ ×	×	▲▲				
Bian Que	▲▲▲▲▲▲▲ ▲▲▲▲▲ ×××	▲▲▲▲▲▲▲ ▲▲▲▲▲▲▲ ××××	▲▲▲▲▲▲▲ ▲▲▲ ××	▲▲▲▲▲▲▲ ▲▲▲▲▲ ××	▲▲		
Lei Gong	▲▲▲▲▲▲▲ ▲▲▲▲▲▲▲ ▲▲▲▲▲▲▲ ▲▲▲▲▲▲▲ ▲▲▲▲▲ ×	▲▲▲▲▲▲▲ ▲▲▲▲▲▲▲ ▲▲▲▲▲▲ ××	▲▲▲▲▲▲▲ ▲▲▲▲▲▲▲ ▲▲▲▲▲ ×	▲▲▲▲▲▲▲ ▲▲▲▲▲▲▲ ▲▲▲▲ ×	▲ ×	▲▲▲▲▲▲▲ ▲▲▲▲▲▲▲ ××	
Li Shi	▲	▲	▲▲				▲▲

Table 12.2 A comparison of drug flavor between the theories of eight matters cited in *Wu Pu's Materia Medica*

	Shen Nong	Huang Di	Qi Bo	Tong Jun	Yi He	Bian Que	Lei Gong
Huang Di	▲▲▲▲▲▲▲ ▲▲▲▲▲▲▲ ▲▲▲▲▲▲▲ ▲▲▲▲▲▲▲ ▲▲ ××××××× ××××××× ××						
Qi Bo	▲▲▲▲▲▲▲ ▲▲▲▲▲▲▲ ▲▲▲▲▲▲▲ ▲▲▲▲▲▲ ××××××× ××××××× ××××××	▲▲▲▲▲▲▲ ▲▲▲▲▲▲▲ ▲▲ ××××××× ××					
Tong Jun	▲▲▲▲▲▲▲ ▲▲▲▲▲▲▲ ▲▲▲▲ ××××××× ××××××× ××	▲▲▲▲▲▲▲ ▲▲▲▲▲▲▲ ▲ ×××××	▲▲▲▲▲▲▲ ▲ ××××××				
Yi He	▲ ×××	▲ ××	▲ ×	▲▲			
Bian Que	▲▲▲▲▲▲▲ ▲▲▲▲▲▲▲ ▲▲ ××××××× ××××××× ××××	▲▲▲▲▲▲▲ ▲▲▲▲▲▲▲ ××××	▲▲▲▲▲▲▲ ▲▲▲▲▲ ××××××× ××	▲▲▲▲▲▲▲ ▲▲▲▲▲▲ ×××	▲▲▲ ×		
Lei Gong	▲▲▲▲▲▲▲ ▲▲▲▲▲▲▲ ▲▲▲▲▲▲▲ ▲▲▲▲▲▲▲ ▲▲▲▲▲▲▲ ▲▲▲▲▲▲ ××××××× ××××××× ××××××× ××××	▲▲▲▲▲▲▲ ▲▲▲▲▲▲▲ ▲▲▲▲ ××××××× ×××	▲▲▲▲▲▲▲ ▲▲▲▲▲▲▲ ▲▲▲▲▲▲▲ ▲ ××××××	▲▲▲▲▲▲▲ ▲▲▲▲▲▲▲ ▲ ×××××	××	▲▲▲▲▲▲▲ ▲▲▲▲▲ ××××××× ×××××	
Li Shi	▲ ××	▲	▲ ×	▲			▲ ××

Note: Same: ▲; Different: ×.

However, in reality, there is a little consistency among different experts, with very few cases where all of them have the same view. This indicates that in the records of drug flavors, objective sensations are only one aspect. On the other hand, it is very likely that some experts have incorporated the theory of "pungent and sweet flavors are dispersing and belong to yang, while sour and bitter flavors are purging and belong to yin”,[8] and even the theoretical associations of the five qi, five flavors, and the aversions and desires of the five zang-organs in the five-phase theory into their records of drug flavors. In addition, the inconsistent records of flavors clearly demonstrate that these eight masters were independent of each other. Through comparison, it can be seen that there is no pair of their records that can be considered relatively consistent (see Table 12.2).

(3) Drug property: The concept of “property” (cold, heat, warm, cool, and neutral) is a further understanding following the toxicity and flavor. This understanding is actually not a primary description, but an abstract conception based on abstract natures of diseases (cold, heat, excess and deficiency). The records of drug properties mainly rely on the principles of treating febrile diseases with cold drugs and cold diseases with heat drugs, as well as the manifestations after taking herbs. However, without abstract attributes of diseases and symptoms, there would be no corresponding heat or cold property of materia medica. In the theories of the eight masters, the historical materials about the drug property are far less abundant than those about toxicity and flavor, which is certainly not entirely due to the absence of citation. Instead, it is largely attributed to the relatively late development of this further theoretical understanding. For example, in the existing historical materials of *Wu Pu's Materia Medica*, there are 108 references to the flavor in the quotation from Shen Nong, while only 22 to the property. Whereas in later times, both property and flavor of each materia medica have been mentioned.

Wu Pu's Materia Medica extensively quotes from the theories of the eight masters, showcasing a sense of “inheriting the past”. Its contribution in “paving the way for the future” is manifested in *Materia Medica* compiled by Tao Hongjing in the Liang Dynasty. Besides taking the 365 kinds of materia medica in *Shennong's Classic of Materia Medica* as the main body, a considerable amount of the additional 365 kinds of materia medica and the supplementary materials for the 365 kinds in *Shennong's Classic of Materia Medica* are taken from *Wu Pu's Materia Medica.*

12.4 *Shennong's Classic of Materia Medica*

12.4.1 The Era

Regardless of the view that the book was written in remote antiquity, some scholars still argue that it predates the burning of books by the First Emperor of the Qin Dynasty (259–210 BC). For example, Tao Hongjing in the Liang Dynasty, collated the book, considering its authenticity. He proclaimed “Previous theories all called this book *Shennong's Classic of Materia Medica.* I think it is credible”. Of course, he

[8] The chapter “Comprehensive Discourse on Phenomena Corresponding to Yin and Yang” in *Plain Questions* of Paul U. Unschuld's translation version.

also knew that the book was not likely to have been originated from the Shen Nong Period. So, he added "But before the Huang Di Period, characters have not been invented yet… The book's knowledge of drug properties would have been passed down orally. Or how can it be learned otherwise? They compare it to the *Plain Questions*, indicating that it may have undergone revisions by later generations. It is mentioned that Qin Shi Huang did not destroy books related to medicine or divination, which supports the book's authenticity". As for the question that the names of origin places of the materia medica in the book are mostly from the Eastern Han Dynasty, he pointed out that the names of provinces and counties where the materia medica were produced are used in the Eastern Han Dynasty, which may be recorded by Zhang Zhongjing (around AD 150–219) and Yuan Hua.[9] Yan Zhitui of the Northern Qi (AD 550–577) also commented on the loss of ancient books due to past events, "The rulers of the Qin Dynasty wiped out scholarship and Dong Zhuo (AD ?-192) burned books, which led to the destruction of ancient books. More than that, the names of such provinces and counties where produce materia medica as Yu Zhang, Zhu Ya, Zhao Guo, Chang Shan, Feng Gao, Zhen Ding, Lin Zi, Feng Yi in this book and the theory of Shen Nong were not the original text but added by later generations".[10] Ma Jixing, a contemporary expert in medical history literature, believes that the book summarizes the rich experience of drug use during the Warring States Period (475–221 BC) and was continuously transcribed and supplemented by TCM doctors in the Qin and Han Dynasties (221 BC-AD 220).[11] However, it is too early to assume that the book was completed during the Warring States Period just because it summarizes the drug-using experience of that time. The materia medica recorded in this book are not as primitive and peculiar as those in *Classic of Mountains and Seas* (*Shan Hai Jing*, 《山海经》). Even the medical books unearthed in Mawangdui and other relevant documents of medical history in the early Western Han Dynasty show that medicine during that time had not reached the same level.

When considering the textual research of the bibliography, since the book was not recorded until the Liang Dynasty, it can be inferred that it was composed much later. Liang Qichao (AD 1873–1929), a scholar in modern times, suggests that the book probably appeared in the Eastern Han Dynasty and the Three Kingdoms Period, but underwent further development during the Liu Song and Southern Qi Dynasties (AD 420–502). The exact author and era cannot be determined, but it is likely to be between the late Eastern Han Dynasty and the Liu Song and Southern Qi Dynasties.

---

[9] Tao Hongjing (the Liang Dynasty), "Preface" of *Collected Annotations on Materia Medica* (*Ben Cao Ji Zhu*, 《本草集注》). Quoted from *Texual Research and Interpretation of Dunhuang Ancient Medical Books* (*Dun Huang Gu Yi Ji Kao Shi*, 《敦煌古医籍考释》), p. 337.

[10] Yan Zhitui (the Northern Qi Dynasty), the 17th entry of the chapter "Textual Research" in *Instructions of Family Yan* (*Yan Shi Jia Xun*, 《颜氏家训》). Note: the book Complete Library in the *Catalog of Four Branches of Literature* (*Si Ku Quan Shu Zong Mu*, 《四库全书总目》) was verified as written by "Yan Zhitui in the Sui Dynasty".

[11] Editorial Board of Chinese Medical Encyclopedia, "Medical History" of *China Encyclopedia of Medicine* (*Zhong Guo Yi Xue Bai Ke Quan Shu*, 《中国医学百科全书》), p. 172.

Chen Bangxian, a medical historian in modern times, proclaimed that this opinion is "relatively reasonable".[12]

It should be noted that there can be a time gap between when books were written and the actual content they cover. In AD 4, Wang Mang summoned individuals with specialized knowledge to record their expertise, followed by a search for talents proficient in "materia medica" in the year of AD 5. The works on materia medica written by them actually did not represent the latest medical development as the contemporary scientific papers do, but were intended to elaborate on what they know about the ancient medicine. As a result, what they wrote about was only their overall knowledge of materia medica, thus leading to the time gap between "writing" and "content". In *Shennong's Classic of Materia Medica*, a significant portion of the records of pharmaceutical effects contain the theory of immortality. However, it would be incorrect to claim that the book was written "when the theory of immortality during Emperor Wu of Han's time hadn't declined significantly".[13] It can be reasonably speculated that the book reflects the scope and features of medication used by alchemists during that era. Tao Hongjing noted that since the Wei and Jin Dynasties (AD 220–420), Wu Pu, Li Dangzhi, and others had made further additions and deletions to the book. Different versions of the book had varying numbers of materia medica entries, including 595, 431, and 319, with descriptions that were sometimes vague in terms of grade (top, medium, and low), property (heat natured and cold natured), category (herbs, stones, insects, and animals), and indications.[14] Based on this, it can be concluded that the original text must have been written before the Three Kingdoms Period. Considering that "the names of the prefectures and counties mentioned mostly conform to the administrative divisions of the Later Han Dynasty. Except for the names of Yongzhou and Liangzhou which have a longstanding history, most other names date back to before Yangjia (the reign title of Emperor Shun, AD 132–135)",[15] the most likely time of its compilation was in the early Eastern Han Dynasty. However, a comparison between the "Shen Nong's Theory" cited by Wu Pu and the current version of *Shennong's Classic of Materia Medica* shows that the book in Wu Pu's time was not as complete as the present one. Evidently, certain contents were added or deleted after Wu Pu's era.

To illustrate this issue, it is important to determine whether the "Shen Nong's Theory" cited by Wu Pu is in the same as *Shennong's Classic of Materia Medica*. By comparison, the similarities and differences in terms of property and flavor between the two are as follows:

---

[12] Chen Bangxian: *The History of Chinese Medicine* (*Zhong Guo Yi Xue Shi*, 《中国医学史》), p. 108.

[13] Fan Xingzhun, *A Brief History of Traditional Chinese Medicine* (*Zhong Guo Yi Xue Shi Lve*, 《中国医学史略》), p. 30.

[14] Tao Hongjing (the Liang Dynasty), "Preface" of *Collected Annotations on Materia Medica*. Quoted from *Textual Research and Interpretation of Dunhuang Ancient Medical Books* edited by Ma Jixing, p. 337.

[15] Li Ding, An Interpretation of the Origin of Materia Medica in the *Classic of Materia Medica*, *Journal of Medical History* (*Yi Shi Za Zhi*, 《医史杂志》), 1952, (4): 167.

Drug properties { same 15; different 4 }

Drug flavor { same 83; different 17 }

The above analysis shows that the two sources are consistent in many aspects. The difference lies in that there are 52 references to "toxic" or "non-toxic" in *Shen Nong's Theory* cited by Wu Pu, but few such records are found in the contents of *Shennong's Classic of Materia Medica* in *Classified Materia Medica* (*Zheng Lei Ben Cao*, 《证类本草》) written in the Song Dynasty (AD 960–1279). The "Shen Nong's Theory" cited by Wu Pu also includes herbs without records of medicinal properties or flavors. Furthermore, there are more records of drug flavors compared to drug properties in it (108: 22), whereas the current version of *Shennong's Classic of Materia Medica* includes both drug properties and flavors.

When discussing the time of completion of this book, scholars have differing opinions due to the unclear definition of the "era of completion". Therefore, it is difficult to determine who is right or wrong because these opinions reflect the scholars' respective concerns.

12.4.2 The Author

The book is entitled with the name of "Shen Nong", which is generally believed to be influenced by two factors. First, the legend of "Shen Nong tasting a hundred of herbs" associates the field of materia medica with Shen Nong. Second, during that time, there was a tendency to revere ancient figures and disregard the present. Therefore, authors would entitle their works with the names of ancient sages such as Shennong and Huang Di in order to gain acceptance and recognition during turbulent times. As a result, these works were perceived as if they were written by the ancient sages themselves, and were highly esteemed.[16] It is widely acknowledged that Shennong, being a legendary figure, did not actually write the *Classic of Materia Medica*, and even those who revere the ancient are aware of this fact. Therefore, it is generally believed that *Shennong's Classic of Materia Medica* is a compilation of medicinal experiences before the Han Dynasty and is titled as such for the aforementioned reasons.[17] However, this common belief may not be entirely accurate. While

[16] Gao You (the Han Dynasty) annotated: *An Overall Interpretation of Philosophical Work of Master Liu An* (*Huai Nan Hong Lie Jie*, 《淮南鸿烈解》), Volume 19, Discussion on Virtues.

[17] Fan Xingzhun's *A Brief History of Traditional Chinese Medicine* (*Zhong Guo Yi Xue Shi Lve*), "It is a pharmaceutical manual summarizing the related knowledge before the late Western Han Dynasty" (p. 30); Yu Shenchu's *A Brief History of Traditional Chinese Medicine* (*Zhong Guo Yi

it is true that the book is titled with the name of Shennong, regarding it as a post-compilation pseudonym ignores the existence of various academic schools in the development of pharmacology, just like in other branches of medicine. The theories of the eight masters cited in *Wu Pu's Materia Medica* demonstrate differences in the understanding of materia medica among different systems. "Shen Nong" represents only one academic school, and nearly half of its records on drug flavors are different from those of other schools. Hence, the following two points should be considered.

First, the books of materia medica attributed to Shen Nong only represent one school of pharmacology at that time. Thus, *Shennong's Classic of Materia Medica* cannot be considered a comprehensive summary of pharmacology prior to the Western Han Dynasty. Particularly, the extensive descriptions of immortal classics and magical techniques like "longevity" and "immortality" may not reflect the pharmacological knowledge system of clinical doctors who prioritize treatment.

Second, *Shennong's Classic of Materia Medica* is not the first book to use Shen Nong's name. In the "Catalog of Literary Works" in the *Book of the Former Han*, there are seven volumes of *Dietary Prohibitions of Shennong and Huangdi* (*Shen Nong Huang Di Shi Jin*, 《神农黄帝食禁》). Additionally, Jia Gongyan's annotation to "Physician" of the chapter "Government Officials" in *The Rites of Zhou*, mentions the "Catalog of Literary Works" in the *Book of the Former Han*, which states "Seven volumes of *Medicated Diet of Shennong and Huangdi* (*Shen Nong Huang Di Shi Yao*, 《神农黄帝食药》)". Whether it is "dietary prohibition" or "medicated diet", these examples indicate that medical works have long used the name of Shennong. Hence, *Shennong's Classic of Materia Medica* can be regarded as a great work from one academic school that took its inspiration from Shen Nong. Although this school lacks medical theory books like the medical classics, it made early progress in the field of materia medica and eventually played a leading role in this area.

12.4.3 Content Structure

The original book of *Shennong's Classic of Materia Medica* has long been lost over time. The current version is compiled on the basis of officially revised materia medica after the Ming Dynasty (AD 1368–1644).[18] It can be seen from the book titles in the above-mentioned "Catalog of Classics and Books" of the *History of the Sui Dynasty* that there were at least three versions of "*Shennong's Materia Medica*":

Xue Jian Shi), "It was written by a number of TCM physicians who summarized the knowledge of materia medica before the Han Dynasty and entitled with the name of "Shen Nong" (p. 52); Xue Yu's *Historical Materials of Chinese Pharmacology* (*Zhong Guo Yao Xue Shi Liao*, 《中国药学史料》), "The book *Classic of Materia Medica* is added with a title and then called *Shennong's Classic of Materia Medica*)" (p. 88).

[18] The compilation of *Shennong's Classic of Materia Medica* started with three volumes of *Orthodox Classic of Materia Medica* (*Ben Cao Zheng Jing*, 《本草正经》) of Wang Yan in the Southern Song Dynasty (around AD 1188), which has been lost, followed by those of Lu Fu in the Ming Dynasty and Sun Xingyan in the Qing Dynasty.

an eight-volume version, a three-volume version, and a four-volume version with Lei Gong's Collected Annotations. The eight-volume version also mentions a five-volume version from the Liang Dynasty, but it is unclear how similar the contents are among these versions. Tao Hongjing in the Liang Dynasty compiled a three-volume edition of Materia Medica based on the four-volume version. This book mainly focuses on the three grades of materia medica mentioned in *Shennong's Classic of Materia Medica*, which consist of 365 kinds of medicinal substances. In addition to these, the book includes another 365 kinds of materia medica used by famous physicians, resulting in a total of 730 kinds. The 730 kinds of materia medica are distinguished by red and black annotations. The red annotations represent the originals from *Shennong's Classic of Materia Medica*, while the black annotations represent those used by famous physicians. Subsequently, during the Tang (AD 618–907) and Song Dynasties (AD 960–1279), many significant works on materia medica were created based on this book. These works have since become important sources for the investigation and study of *Shennong's Classic of Materia Medica* today (Figs. 12.2, 12.3 and 12.4).

According to these records, *Shennong's Classic of Materia Medica* contains 365 kinds of materia medica categorized into three grades: top, medium, and low grades. It is believed that long-term use of top-grade materia medica can prolong one's life,

Fig. 12.2 Top grade materia medica: Ren Shen (Ginseng)

Fig. 12.3 Medium grade materia medica: Wu Wei Zi (Schisandra Chinensis)

Fig. 12.4 Low grade materia medica: Ba Dou (Croton) (From left to right)

while medium and low-grade substances are suitable for treating diseases but should not be taken for extended periods. Moreover, the records of each item generally include its name, property, flavor, indications, origin, and alias.

12.5 Tao Hongjing's Works on Materia Medica

The knowledge in *Shennong's Classic of Materia Medica* has been passed down thanks to *Variorum of the Classic of Materia Medica* compiled by Tao Hongjing in the Southern Dynasty. And the supplements by Tao are mostly from the comments of the predecessors on classics. In order to have a comprehensive knowledge about history in this period, we need to introduce Tao Hongjing's compilation and revision of materia medica.

Tao Hongjing, also known as Tongming, was born in Moling, Danyang (now Jurong, Jiangsu province) in 456 AD during the Xiaojian reign of the Song Dynasty and died in 536 AD during the Datong reign of the Liang Dynasty, living to be 81 years old.[19]

Tao Hongjing has been clever and studious since childhood. He was said to have read thousands of books, excelled in playing musical instruments and chess, and was good at writing cursive and official scripts. In the late Liu Song Dynasty, he served as a reading attendant for the princes and was highly appreciated. When he was ten years old, he got Ge Hong's *Biographies of Divine Immortals* (*Shen Xian Zhuan*, 《神仙传》) and loved it so much that he became deeply interested in health-preservation techniques. In the 10th year of Emperor Yongming's reign in

[19] According to *Southern History* Volume 76, "Seclusion" of "Tao Hong Jing" in "Biography, 66th". While *History of the Liang Dynasty* (*Liang Shu*, 《梁书》) volume 51, "Tao Hong Jing" of "Hermit" in "Biography, 45th" states, "He died in the second year of Datong at the age of 85".

the Qi Dynasty (AD 492), he hung his court-robe at the Shenwu Gate, submitted a letter to resign from his official position, and then retreated to live in seclusion in Ju Qu Mountain in Jurong. He built a house in the mountain and named himself Hua Yangtao. From then on, he dedicated himself to writing and alchemy. Tao Hongjing inherently cherished beautiful sceneries like mountains and rivers. Tao Hongjing loved mountains and rivers and cherished the scenery. When he was an official, "even though he was in a prominent position, he kept a low profile and didn't interact with the outside world, only focusing on reading". After resigning from office, he traveled all over famous mountains to search for medicinal herbs. He had in-depth research on various subjects, including yin and yang, the five elements, divination and astrology, geography, local products, traditional Chinese medicine techniques, and materia medica. Tao Hongjing made significant contributions in various fields throughout his life. He created an astronomical instrument called "Hun Tian Xiang" and authored numerous books, including 100 volumes of *Academic Garden* (*Xue Yuan*, 《学苑》), *Collected Annotations on the Classic of Filial Piety* (*Xiao Jing Ji Zhu*, 《孝经集注》), *Collected Annotations on the Analects of Confucius* (*Lun Yu Ji Zhu*, 《论语集注》), *Almanac of Emperors' Reign* (*Di Dai Nian Li*, 《帝代年历》), *Collected Annotations on the Classic of Materia Medica*, *Effective Formulas* (*Xiao Yan Fang*, 《效验方》), *Formulas to Keep Up in One's Sleeve* (*Zhou Hou Bai Yi Fang*, 《肘后百一方》), *Records of Ancient and Modern Provinces and Counties* (*Gu Jin Zhou Jun Ji*, 《古今州郡记》), and *Collection of Illustrations* (*Tu Xiang Ji Yao*,《图像集要》).[20]

According to the chapter "Catalog of Classics and Books" in the *History of the Sui Dynasty*, Tao Hongjing's works on materia medica include the following:

(1) In the eight volumes of *Shennong's Materia Medica*, it is annotated, "In the Liang Dynasty … the hermit Tao (namely Tao Hongjing) compiled ten volumes of *Materia Medica* … Tao Hongjing's variorum of materia medica consist of seven volumes".
(2) Two volumes of *A Summary of Tai Qing Materia Medica,* written by Tao Hongjing during his seclusion.

There is a belief among later generations that Tao Hongjing also wrote the three volumes of *Miscellaneous Records of Famous Physicians* (*Ming Yi Bie Lu*, 《名医别录》) mentioned in the *History of the Sui Dynasty*, which is considered a reference to supplement *Shennong's Classic of Materia Medica* and expand its content. However, further verification is needed to confirm which works on materia medica were written by Tao Hongjing and whether he wrote *Miscellaneous Records of Famous Physicians*.

[20] The introduction of Tao Hongjing is based on *Southern History*, Volume 76, "Seclusion" of "Tao Hong Jing" in "Biography, 66th".

12.5.1 *Volumes of Tao's* Materia Medica

The chapter "Catalog of Classics and Books" in the *History of the Sui Dynasty* was the first to mention that "During the Liang Dynasty, … Tao Hongjing's *Collected Commentaries on 'Shennong's Classic of the Materia Medica* consisted of seven volumes. Although the records in the chapter "Catalog of Classics and Books" in the *Old History of the Tang Dynasty* (*Jiu Tang Shu*, 《旧唐书》) and "Catalog of Literary Works" in the *New History of the Tang Dynasty* (*Xin Tang Shu*, 《新唐书》) have slight textual differences, it's evident that they all refer to this book. However, Tao Hongjing's own description of his materia-medica work doesn't match these records. In the first paragraph of the fragment of the *Preface to Collected Annotations on the Classic of Materia Medica* [21] (hereinafter referred to as the "Preface") from a Kaiyuan-era manuscript unearthed in Dunhuang, the so-called "Autobiographical Preface by Tao Hongjing" (hereinafter referred to as the "Autobiographical Preface"), it is clearly stated that "including this preface, there are three volumes in total". In the following text, it is mentioned again that "I have now compiled these three volumes". It can be inferred that the original book must have consisted of three volumes rather than seven. As collateral evidence, we can cite what Tao Hongjing said in the preface of *Supplement to Formulas to Keep Up in One's Sleeve* (*Bu Que Zhou Hou Bai Yi Fang*, 《补阙肘后百一方》): "All the above-mentioned methods have been fully recorded in the upper volume of the materia medica I compiled". If the original book had been divided into seven volumes, it would not have been possible to refer to it as "the upper volume". As for why there was a seven-volume edition of *Collected Commentaries on the Classic of Materia Medica*, the details will be described later.

12.5.2 *The Title of Tao's* Materia Medica

The beginning of the "Preface and Records" section is originally incomplete. So, it's unknown whether the original had a title. The note at the end of this volume reads "Preface and Records of the First Volume of Collected Annotations on Materia Medica", followed by the six-character text "written by Hua Yang Tao Yinju" (Tao Hongjing). This is consistent with the record in the "Biography of Tao Hongjing" in the *Southern History* (*Nan Shi*, 《南史》) that one of his works was "*Collected Annotations on Materia Medica*". However, in the "Catalog of Classics and Books" in the *History of the Sui Dynasty*, it is named "*Collected Annotations on the Classic of Materia Medica*". Although the two names differ only by the character "Jing" (classic), their implications are different. Naming it "*Collected Annotations on the*

[21] This remnant book is stored in RyuKoku University, Kyoto, Japan. There is a photocopy of *Ji Shi An Series* (*Ji Shi An Cong Shu*, 《吉石庵丛书》) in China. In the photocopy of Shanghai Qunlian Publishing House in 1955, it is with a title of *Variorum of the Classic of Materia Medica*. *Textual Research on Dunhuang Ancient Medical Books* edited by Ma Jixing also contains this book, which can be used for reference.

Classic of Materia Medica" clearly regards the nature of this book as an "annotated compilation" of the *Shennong's Classic of Materia Medica*, which contradicts the objective fact that only 50% of the contents in the original book are from the *Shennong's Classic of Materia Medica.* The name "Collected Annotations on Materia Medica" does not carry such a meaning. The reason for the change of the book title might be due to a transcription error, but the underlying cause is that the idea of the supreme status of the *Shennong's Classic of Materia Medica* had gradually formed in people's minds. People had not only long forgotten about *Wu Pu's Materia Medica*, Li Dangzhi's *Records of Materia Medica*, and the theories of Qi Bo, Bian Que, and other experts quoted by Wu Pu, but also ignored the principle of compiling the book stated in Tao Hongjing's "Self-Preface", which said "All details and general aspects are included without omission... It is actually the work of one person". They regarded the 365 newly added "medicines by famous doctors" (the same number as the medicines in the *Shennong's Classic of Materia Medica*) as annotations to *Shennong's Classic of Materia Medica.* From the perspective of the history of medical thought, it can be said that it was not until then that *Shennong's Classic of Materia Medica* had gained a supreme position in the pharmaceutical system. However, *Collected Annotations on Materia Medica* is not the original title of Tao's materia medica work. Right after Tao's "Self-Preface" in the "Preface and Records" section comes the table of contents of the book, namely:

> The first volume of *The Classic of Materia Medica*: origin of drug properties, names of the diseases and their diagnoses, inscriptions and catalogue, and the detailed application.
>
> The second volume of *The Classic of Materia Medica*: jade, herbs and woods, totaling 356 kinds.
>
> The third volume of *The Classic of Materia Medica*: there are 374 types of materia medica in total, that is, 195 kinds of insects and beasts, fruits and vegetables, staple foods, and 179 kinds of materia medica with only names but no physical object.

These three volumes collectively feature 730 forms of materia medica, with a meticulous catalog accompanying each volume. Notably, the catalog includes annotations in both red and black ink. It is worth noting that the entire book is divided into seven volumes.

The catalog mentioned above is presented in white characters on a black background in the *Classified Materia Medica*, which implies that it is associated with the original text of *Shennong's Classic of Materia Medica.* Ma Jixing's annotation also states that "the above three items are all from the original catalogue of *Shennong's Classic of Materia Medica*".[22] However, this view is not tenable. First of all, when referring to *Shennong's Classic of Materia Medica* in "Tao's Preface", it is clearly stated that "what exists now are these four volumes, which are its original text". It can be seen that despite the existence of multiple transmitted versions at that time, Tao Hongjing relied on the four-volume version. Moreover, both the small-character original annotations and the subsequent large-character text in it are sufficient to show that this catalog is for the 730 types of medicinal herbs, so it can only be

[22] Edited by Ma Jixing: *Textual Research on Dunhuang Ancient Medical Books*, p. 338.

the catalog of Tao's *Materia Medica*, rather than the catalog of the 365 types of medicinal herbs in *Shennong's Classic of Materia Medic*a. In addition, it exactly coincides with the number of volumes of the Tao's materia medica works that he mentioned several times. As for why this catalog evolved into the text of *Shennong's Classic of Materia Medica* in "red characters" or "white characters on a black background", it should be attributed to the copying errors during the four-hundred-year period when "there is no original copy for the red and black characters to match".[23] The internal cause of this error is still the influence of the subconscious belief in people's minds that *Shennong's Classic of Materia Medica* is supreme, thinking that whenever the "Classic of Materia Medica" is mentioned, it must refer to *Shennong's Classic of Materia Medica.* However, it was not the case during the Six Dynasties (AD 222–589). For example, the chapter "Catalog of Classics and Books" in the *History of the Sui Dynasty* records that Wang Jipu's *Classic of Materia Medica*, Li Dangzhi's *Classic of Materia Medica*, and the *Classic of Materia Medica* compiled by Cai Ying. From this, we can infer that the original title of Tao's book was not "*Collected Annotations*", but "*Materia Medica*" (as stated in the "Preface" of the *Supplement to Formulas to Keep Up in One's Sleeve*, and "Ten volumes of Tao Yinju's Materia Medica" in the chapter "Catalog of Classics and Books" in *History of the Sui Dynasty*) or *The Classic of Materia Medica* (as stated in this catalogue). This is consistent with the ideological awareness and the purpose of writing the book shown in his "Preface", where he did not subordinate his work to *Shennong's Classic of Materia Medica.* Instead, he "combined various classics, studied and summarized details", "selected both the essence and the general, leaving nothing out" to create a "self-contained work", aiming to "leave it to those who understand the principles" after his death. Furthermore, considering that Wang Shuhe (AD 210–280) and Huangfu Mi (AD 215–282) in the Jin Dynasty integrated the achievements of their predecessors to compile works like the *Pulse Classic* (*Mai Jing*, 《脉经》) and *A-B Classic of Acupuncture and Moxibustion* (*Jia Yi Jing*, 《甲乙经》), and also taking into account that "Catalog of Classics and Books" in the *History of the Sui Dynasty* records many works written by people of the Six Dynasties that were self-titled as "classics", it can be inferred that the characteristic of medical development in that era was not to emphasize venerating ancient classics and relying on ancient authorities. The research nature in literature summarization far outweighed textual research and annotation. The purpose and the actual effect were to construct independent and all-encompassing systems that strived to be complete and incorporate knowledge from both ancient and modern times in various branches of medicine.

[23] Tang Shenwei (the Song Dynasty), *Classified Materia Medica*, Volume 1, "Preface" of "Re-edited in Kaibao".

12.5.3 The Source of Seven Volumes Version of Collected Annotations

The small-character annotation in the catalogue of Tao's *Materia Medica* reads, "The main text is divided into seven volumes". Judging from the context before and after, it should refer to the middle and lower volumes that contain 730 kinds of herbs. There are two possibilities for its formation: One is that in Tao's research on materia medica, he first made annotations and textual research on specific herbs. He wrote seven volumes according to the so-called "natural attribute classification method" such as minerals, herbs, and trees, and then summarized them into the middle and lower volumes and incorporated them into his *Materia Medica.* Another possibility is that later generations separated this part and made it into seven volumes. However, regardless of which possibility it is, the content of the seven-volume version of the *Collected Annotations on the Classic of Materia Medica* should only contain the content of the middle and lower volumes of Tao's *Materia Medica.* In this way, adding one volume of the "Preface" and two volumes of *A Summary of Tai Qing Materia Medica*, as mentioned in the sentence, completes the ten-volume compilation of Materia Medica by Tao Yinju during the Liang Dynasty. This statement is not just based on the coincidence of the number of volumes. We should also note that the book *A Summary of Tai Qing Materia Medica* is indeed classified as a work on materia medica works in the chapter "Catalog of Classics and Books" in the *History of the Sui Dynasty.*

In addition, in the above-cited catalogue, the sentence "and are annotated in red and black" should also be understood as referring to the contents of the middle and lower volumes. It does not mean that the "Preface" in the first volume also has red and black characters and annotations.

12.5.4 The Preface of Tao's Materia Medica

This is the first volume of Tao's *Materia Medica.* It is an overview of materia medica, and it should be noted that there are no annotations (Zi Zhu, "子注") in this volume. The term "Zi Zhu" has two aspects. First of all, "Zi" is an ancient honorific title for men and teachers, not a self-proclaimed one. In the middle and lower volumes of Tao's *Materia Medica*, besides his own annotations, the theories of Huangfu Mi, Xu Yi, Guo Jingchun, Li Dangzhi, Tong Jun, and other scholars, as well as those from immortal classics, Taoism, prescription masters, alchemists, incense makers, pharmacists, and ordinary people, are cited as "Zi Zhu". Whereas "Preface" only involves Tao's own theory, which distinguishes it from the other two volumes. Although there are quotations from the "original text", there are no "Zi Zhu" annotations. Secondly, as small characters beneath the main text, "Zi Zhu" is used to provide explanations for issues where deletion would result in a loss of meaning, yet recording everything would be verbose. Under the "Table of Seven Emotions" in the "Preface", the small characters are used for all content except for the names of materia medica. Removing these small characters would result in missing content. Therefore, these small characters do not belong to Zi Zhu but are merely a unique form of writing.

There are other two types of small characters in the text. One type is used to mark the pronunciation of the characters using fanqie (a traditional Chinese method of indicating pronunciation), and the other type indicates that the classification method of materia medica is no longer commonly used. Some scholars argue that these small characters were added by later generations and do not require further analysis. The remaining small characters provide instructions for the contents of each volume in the "catalogue". It is uncertain whether these small characters were written by Tao or later generations. However, since the following large characters are also a further explanation of the contents of the three volumes, these small characters are more likely to be added by later generations.

As for the preface, under the first ten paragraphs of quotations in it, there is a note saying "This is what the original text says". In the *Classified Materia Medica*, these ten paragraphs are written in white on a black background to indicate that they are the original text of *Shennong's Classic of Materia Medica*. However, since Tao marked "This is from the original text" under each paragraph, it is unnecessary to use color to distinguish the quotations from the rest of the text. The section on "Commonly used materia medica for all diseases" in the *Classified Materia Medica* is also written in black and white to distinguish the materia medica in *Shennong's Classic of Materia Medica* from those used by famous physicians. However, in Tao's original text, it is mentioned that red dots indicate herbs of heat property, black dots indicate herbs of cold property, and those without dots are of neutral property. It is inferred that the names of materia medica are not written in two colors. Therefore, the use of two colors in the *Classified Materia Medica* does not represent the features of the preface accurately.

What is more important is to examine whether these ten paragraphs of "original statements" are truly the original texts of *Shennong's Classic of Materia Medica*. So far, no one has analyzed the character "Shuo" here. The *Origin of Chinese Characters* (*Shuo Wen Jie Zi*, 《说文解字》)states that the character "Shuo" means explanation. The "Catalog of Literary Works" in the *Book of the Former Han Dynasty* describes the annotations of classics as follows: "Explanations of five-character texts can reach 20,000 to 30,000 words". Although the "Shuo" in medical classics is not as excessive as that, the text of "Shuo" is still more extensive than that of the "classics". It is easy to understand, and there may be inconsistent interpretations among different "Shuo". For example, regarding the explanation of the four-character phrase "Replenish the deficiency" in the chapter "Nine Needles and the Twelve Origin Openings" in *Miraculous Pivot*, the chapter "Explanatory Remarks on the Small Needles" in *Miraculous Pivot* interprets it as "What is meant by replenishing the deficiency is to tonify when the qi at the qi-kou (wrist pulse) is deficient". Whereas in the chapter "Explanations on the Needles in *Plain Questions*", it states that "When needling a deficiency to replenish it, there should be a feeling of heat under the needle. When qi is abundant, there will be heat". The understandings of the two are different.

The first "original statement" quoted in Tao's "Preface" reads, "The 120 kinds of superior herbs serve as the sovereign drugs. They are mainly used for nourishing life to correspond with the heavens, are nontoxic, and do no harm to people even if taken in large quantities for a long time. These who want to improve physical fitness, boost qi, stay young and extend their lifespan should refer to the upper section of the classic". And there are corresponding explanations about "medium-grade herbs" and "inferior herbs". As long as one notices that these words are "Shuo" (explanation) rather than the "Jing" (original classic), it's quite easy to think that the sentences like "Generally, superior herbs nourish life, medium-grade herbs nourish one's nature, and inferior herbs treat diseases" quoted from *Shennong's Classic of Materia Medica* in the *Readings of the Taiping Era* and *Classified Excerpts from Ancient Writers* (*Yi Wen Lei Ju*, 《艺文类聚》) seem more like the original text. If one wonders why Tao quoted the "explanations" instead of the "original classic" in the era when both "the classics" and "the explanations" were accessible, we should consider that the content of the "explanations" is richer, easier to understand, more specific and targeted, and it holds as important a classic status as the "original classic". For example, in the current version of *Huangdi's Internal Classic*, all sentences starting with the phrase "so-called" can be regarded as "explanations", but their academic significance has never been doubted in the past.

Here are two more examples. The content of the "original text" in Paragraph 7 is: "When using toxic herbs to treat diseases, start with a dose as small as a millet grain. Stop the treatment once the illness is cured. If it's not cured, double the dose. If it still doesn't work, increase the dose ten-fold. The key is to stop when the illness is cured". In Tao's subsequent explanation, he then quotes the "original classic" that matches this content: "According to the 'original classic': 'For one toxic herb, take one pill as small as a fine hemp seed. For two toxic herbs, take two pills as large as a common hemp seed…' Up to ten herbs, the pills should be as large as Chinese holly seeds, with the number of pills corresponding to the number of herbs". This can be regarded as an example of not quoting the "original classic". The next example more clearly demonstrates the relationship between the "explanation" and the "original classic".

> …The seasons and years have their own early or late timings, so there's no need to completely follow the original text. The "explanation" says that "drying in the yin" means drying the herbs during the yin days of the six jia..[24] According to the method of divination, on the day of jiazi, the corresponding yin is guiyou, and the herbs should be placed in the location corresponding to "you". I don't think it must be so. It simply means drying the herbs in a shaded place, not exposed to the sun or wind.

The phrase "dry in Yin" is derived from the "original classic", and is also mentioned by Tao in "this text". However, the description of "dry in Yin" in the "explanation" lacks solid grounding and does not convince Tao. This can be taken as an example of "not quoting the explanation".

[24] "Six Jia" refers to the six combinations beginning with the first Heavenly Stem in the sixty combinations of the ten Heavenly Stem and the twelve Earthly Branches.

The tenth "original text" discusses "main diseases" and provides a list of over 40 types of diseases. Generally speaking, the names of these diseases are basically two-character names, which hardly conforms to the characteristics of early disease names. For example, before Zhang Zhongjing, the disease was mostly referred to as "water disease" (or "suffering from water"), rather than being collectively called "edema". It was not until in the book *Treatise on the Pathogenesis and Manifestations of All Diseases* (*Zhu Bing Yuan Hou Lun*, 《诸病源候论》) in the Sui Dynasty that it was listed as a special category (there is an "edema syndrome" in volume 21). Another example is that disease names such as "pi–shi" (abdominal mass due to improper diet), "liu–yin" (retained fluid), and "inability to pass urine and feces" did not exist in ancient times, but all appear in the *Treatise on the Pathogenesis and Manifestations of All Diseases.* This fully shows that the formation time of this section of the "original text" could not be very early.

To summarize, besides the specific documentation of 365 types of materia medica, *Shennong's Classic of Materia Medica* also includes legends about Shen Nong. But Tao did not record this portion as it held no medical value. These legends are found in the "lost texts of *Shennong's Classic of Materia Medica*" in various literary works and books. Furthermore, the book contains some "explanations" that hold theoretical value and reflect a wealth of knowledge regarding medication. Tao selectively quoted what he considered accurate from this section and developed them into the "Preface".

12.5.5 Miscellaneous Records of Famous Physicians

The relationship between Tao Hongjing's work on materia medica and *Miscellaneous Records of Famous Physicians* has been extensively studied, with Shang Zhijun being the most notable scholar in this regard. In addition to compilation and collation of *Miscellaneous Records of Famous Physicians*, Shang also discusses the origin, writing process, basic contents, characteristics and values of this book in detail. Basically, he believes that before Tao Hongjing wrote *Collected Annotations*, the *Miscellaneous Records of Famous Physicians* referred to the additional information provided by renowned doctors in *Shennong's Classic of Materia Medica.* Upon the completion of *Collected Annotations*, Tao Hongjing incorporated these additional records into the *Miscellaneous Records of Famous Physicians.*[25] However, this statement lacks solid support due to the following reasons:

(1) *Miscellaneous Records of Famous Physicians* was written prior to the time of Tao. Tao's annotations under the terms "Xuanshi (magnetitum)" and "Nvwei (stem of Clematis apiifolia DC.)" in *Shennong's Classic of Materia Medica* mention that magnetitum has an alias as Xuanshi, whereas in *Miscellaneous Records of Famous Physicians*, these terms are listed separately. Additionally, while *Shennong's Classic of Materia Medica* mentions Nvwei but not

[25] Edited by Shang Zhijun: *Miscellaneous Records of Famous Physicians*, Beijing: People's Medical Publishing House, 1986, p. 318.

Weirui (Rhizoma Polygonati Odorati), the *Miscellaneous Records of Famous Physicians* includes Weirui but not Nvwei.

(2) The *Miscellaneous Records of Famous Physicians* is not a book of materia medica, but a book of prescriptions. It is categorized among prescription books in the chapter "Catalog of Classics and Books" in the *History of the Sui Dynasty* [between *A Collection of Empirical Formulas from Famous Physicians* (*Ming Yi Ji Yan Fang*, 《名医集验方》) and *Concise Medical Formulas* (*Shan Fan Fang*, 《删繁方》)], which is far from the catalog of books of materia medica. Scholars who research *Miscellaneous Records of Famous Physicians* always take the black characters in the *Classified Materia Medica* as the reference. Shang Zhijun took "moxa" as an example and stated that the information on materia medica in the *Miscellaneous Records of Famous Physicians* includes its name, property and flavor, toxicity or non-toxicity, main indications, alias, place of origin, time for collection, etc.[26] However, the record of "moxa" in the *Miscellaneous Records of Famous Physicians* merely states that it is cold in property when raw and hot in property when cooked. It is indicated for hematochezia, bleeding from five sense organs or subcutaneous tissues and purulent dysentery. It can be boiled or made into pills and powder for use (see contents of *Miscellaneous Records of Famous Physicians* cited by Tang's edition of *Classified Materia Medica*). Diverse sources citing the *Miscellaneous Records of Famous Physicians* mainly focus on the indications of materia medica, sometimes accompanied by relevant formulas. To this end, it is categorized as a book of prescriptions in the chapter "Catalog of Classics and Books" in the *History of the Sui Dynasty*.

(3) The *Miscellaneous Records of Famous Physicians* has no connection to Tao Hongjing's book on materia medica, and its author remains difficult to identify. As Tao's works in the chapter "Catalog of Classics and Books" in the *History of the Sui Dynasty* are explicitly attributed to Tao Hongjing or Tao Yinju, it can be reasonably concluded that Tao Hongjing did not write the *Miscellaneous Records of Famous Physicians*.

(4) The confusion between "Supplementary Herbs of Famous Physicians" and the *Miscellaneous Records of Famous Physicians*. In his works, Tao Hongjing only used the term "supplementary herbs" to generalize the drugs outside of *Shennong's Classic of Materia Medica*. Besides the statement in the "Preface" that there are also 365 supplementary herbs of famous physicians", it can also be found in the annotation under the entry of "Shi Jue Ming (Concha Haliotidis)": "This kind of herb was originally also attached to the entry of Jueming. Since its shell is of a different kind, it is now regarded as a supplementary herb". Similarly, under the entry of "Wen Ge" (Meretricis Concha), Tao's annotation also says: "This is of a different kind but in the same entry. If distinguishing it, the number would be large. Now it is regarded as an attached entry and within the scope of supplementary herbs. There are four such herbs in total". At that

---

[26] Edited by Shang Zhijun: *Miscellaneous Records of Famous Physicians*, Beijing: People's Medical Publishing House, 1986, p. 319.

time, there was no issue of Tao referring to the "supplementary herbs" as of the herbs in the *Miscellaneous Records of Famous Physicians*, nor referring to the revised texts of famous physicians like Wu Pu and Li Dangzhi as the texts in the Records of Famous Physicians.

Up to the Tang Dynasty, when the emperor inquired about the changes in ancient materia medica, Yu Zhining replied, "In the past, Tao Hongjing annotated *Shennong's Classic* (*Shen Nong Jing*, 《神农经》) together with the separate records of various scholars…". The emperor then asked, "Why are the *Materia Medica* and *Miscellaneous Records* two different things?" Yu Zhining replied, "… *Miscellaneous Records* were written by Wu Pu and Li Dangzhi during the Wei and Jin Dynasties. They describe the shapes and colors of flowers and leaves, and the compatibility of herbs, and provide explanations based on the classics. So Tao Hongjing combined and recorded them".[27] It was at this time that the term "miscellaneous records" first appeared. Yu Zhining used "miscellaneous records" instead of "Supplementary Herbs of Famous Physicians" perhaps because the latter can only summarize the medicines outside *Shennong's Classic of Materia Medica* and cannot include the supplementary materials added by famous doctors under the medicines in *Shennong's Classic of Materia Medica*. However, we must note that "miscellaneous records" is not a book title here. Later generations, on the ground that Yu Zhining participated in the compilation of the *Newly Revised Materia Medica* (*Xin Xiu Ben Cao*, 《新修本草》) in the Tang Dynasty and had seen the *Miscellaneous Records of Famous Physicians*, cited his words as evidence. However, they all mistakenly used the two characters "miscellaneous records" as the title of the book, thus leading to the following two wrong reasoning processes.

In reality, the term "Miscellaneous records" mentioned by Yu Zhining is equivalent to *Miscellaneous Records*, *Miscellaneous Records of Famous Physicians* and the content written in black in Tao's *Materia Medica*.

The *Miscellaneous Records* was cited when compiling the *Materia Medica* in the Tang Dynasty. Therefore, it was natural to use the *Miscellaneous Record* to proofread the content written in black in Tao's *Materia Medica*.[28]

12.5.6 Classification of Materia Medica

Tao Hongjing's innovative reform in the drug classification method is the thing most highly praised in the past research on Tao's herbal medicine. In short, Tao changed the classification method of the three grades (superior, medium, and inferior) into classifying drugs into various sections such as jade and stones, herbs, trees, insects and beasts, fruits, vegetables, grains, and "having names but no real substances" according to the natural sources (or natural attributes) of the drugs. This statement

---

27 *New History of the Tang Dynasty*, volume 140, "Yu Zhining" in Biography, 29th.

28 Edited by Shang Zhijun: *Miscellaneous Records of Famous Physicians*, Beijing: People's Medical Publishing House, 1986, p. 318.

is inaccurate. As Shang Zhijun pointed out, "it can be inferred that the prototype of this classification method already existed in *Wu Pu's Materia Medica*. Or perhaps the classification method established in *Wu Pu's Materia Medica* was the pioneer of the classification method based on the natural sources of drugs".[29]

Throughout history, people have classified materia medica based on their natural attributes, which can be seen in various aspects:

(1) Ancient people naturally divided objects according to their natural attributes, as demonstrated by the radicals of Chinese characters and their functions.
(2) A systemic manifestation of this way of thinking can be exemplified by *Erya* (*The Classic of Etymology*). In it, Sects. 13 to 19 are respectively dedicated to "Explanation of Grasses", "Explanation of Trees", "Explanation of Insects", "Explanation of Fish", "Explanation of Birds", "Explanation of Beasts", and "Explanation of Livestock". Moreover, even Zhang Hua's *Naturalis Historia* (*Bo Wu Zhi*, 《博物志》), which is similar to a miscellany, records things according to the classification method of "Unusual Beasts, Unusual Birds, Unusual Insects, Unusual Fish, and Unusual Plants". Thus, it shows that it is the most basic and common classification method.
(3) In the medical field, *The Rites of Zhou* first records that "five types of medicinal herbs are used to treat diseases". Zheng Xuan annotated, "The five types of medicinal herbs refer to herbs, trees, insects, stones and grains". This is undoubtedly a classification method based on natural attributes. Later, *Plain Questions* and *Miraculous Pivot* mention the concepts of five grains, five fruits, five livestock, and five vegetables. Though influenced by the theory of the five elements, the classification system is consistent.
(4) *Shennong's Classic of Materia Medica* is indeed a crucial book of materia medica. However, regarding it as a comprehensive summary overlooks the issue of different academic schools in the development of medicine during the Qin and Han Dynasties, leading to the regret of "blind men feeling an elephant". It should be noted that in the history of the development of medicine and pharmacy during the Qin and Han Dynasties, *Shennong's Classic of Materia Medica* is merely a masterpiece that synthesizes the theories of one school. Whether it is the numerous statements about immortality in the records of medicinal effects or its three-grade classification method, neither is sufficient to reflect the full picture of the development of pharmacology before the Western Han Dynasty. Undoubtedly, pharmacy that prioritizes clinical treatment cannot follow the three-grade classification method. The classification method of Tao Hongjing is actually a combination of these two classification methods. He categorizes materia medica into jade, herbs, trees, insects and beasts, fruits, vegetables, grains and "materia medica with only a name but no physical substance". Under each of these categories, there is a further classification of upper, middle, and lower grades. This classification system reflects Tao Hongjing's dual identity as a medical scholar and a Taoist. On one hand, he didn't abandon the inheritance and pursuit of the

[29] Edited by Shang Zhijun, etc.: *Wu Pu's Materia Medica*, Beijing: People's Medical Publishing House, 1987, p. 91.

path to immortality, and on the other hand, he had the consciousness to adhere to scientific classification methods.

A detailed analysis of Tao Hongjing's works on materia medica demonstrates several key points. First, in the history of ancient medical development, *Shennong's Classic of Materia Medica* did not hold the supreme position. Thus, Tao compiled a more comprehensive *Materia Medica* by referencing a wide range of sources. It is worth noting that modern people's reverence for ancient texts has become even stronger compared to the past. Second, the "milestone" for comprehensively summarizing previous knowledge of medicinal substances did not occur during the Han Dynasty. Despite the high reputation of *Shennong's Classic of Materia Medica*, it is not actually a representative work of a comprehensive summary of pharmacology. Lastly, before the Northern and Southern Dynasties, pharmacology can be said to have always developed through multiple channels. It was Tao Hongjing's writing of the *Materia Medica* that achieved the first systematic and comprehensive summary.

Chapter 13
The Old Theory of Cold Damage and the New School of Warm Diseases

Zhang Zhongjing and his masterpiece *Treatise on Cold Damage and Miscellaneous Diseases* (*Shang Han Za Bing Lun*, 《伤寒杂病论》) hold significant importance in the study of medical history during the Qin (221–207 BC) and Han (202 BC-AD 220) Dynasties. From another perspective, "cold damage", a general term encompassing all diseases caused by the "six exopathogenic factors,[1]" accounts for half of the exogenous diseases in ancient Chinese medicine, making it an extremely important topic. To truly grasp the reasons behind Zhang Zhongjing's revered status as a "medical sage" and the transformation of *Treatise on Cold Damage and Miscellaneous Diseases* from a mere "remedy book" to a medical classic, it is essential to examine its theory on cold damage. This theory was first separated into a specialized text during the Jin Dynasty (AD 266–420), and then further refined and clarified in the Song Dynasty (AD 960–1279). As a result, the "six-meridian pattern identification" became a widely accepted standard. In addition, during the Ming (AD 1368–1644) and the Qing (AD 1636–1912) Dynasties, with the growth of population and the spread of pestilence, a series of questions emerged. What are the similarities and differences between "pestilence" and "warm disease"? Can the treatment methods for cold damage be effectively applied to warm diseases?

[1] "Six exopathogenic factors" refers to the six external climatic factors that can contribute to illness when excessive or untimely. These factors include wind, cold, summer-heat, dampness, dryness, and fire, also known as the six climatic pathogenic factors.

Y. Liao, *Reconstructing the Image of Medicine in the Qin and Han Dynasties*, Understanding China, https://doi.org/10.1007/978-981-95-6801-7_13

13.1 Historical Development of Cold Damage and Warm Diseases

13.1.1 Cold Damage and Cold Damage School

Cold damage in TCM is different from typhoid caused by salmonella typhi in modern medicine. It can be interpreted in both broad and narrow senses. As stated in the chapter "The Fifty-Eighth Difficult Issue" in *Classic of Difficulties* (*Nan Jing*, 《难经》), "There are five kinds of cold damage: being hit by wind, being harmed by cold, wetness, warmth, heat illness, and warmth illness". In this context, "cold damage" (shang han) serves as a general term encompassing various externally contracted heat diseases. Its origin can be traced back to the explanations in the chapter "Discourse on Heat" in *Plain Questions* (*Su Wen*, 《素问》), which states, "Now, as for heat diseases, they are all categorized as 'harm caused by cold'", and "When a person is harmed by cold, then the disease that develops is heat". However, according to *Plain Questions*, "shang han", meaning "damaged by cold", was merely described as a cause of disease, and the specific diseases resulting from "shang han" were referred to as heat diseases. In a narrower sense, "shang han" represents one type of cold damage, alongside "zhong feng" (wind attack), "shi wen" (dampness-warmth), "re bing" (heat disease), and "wen bing" (warm disease). However, its interpretation and denotation have varied over time and under different circumstances. Though the *Classic of Difficulties* mentions that "patients with different types of cold damage have different symptoms", it only records five pulse manifestations for syndrome differentiation without providing any explanations of these symptoms. The therapeutic principles recorded in the book are also limited to the broader concept of cold damage, emphasizing only the sweating and purgation methods in the chapter "Discourse on Heat" in *Plain Questions*. By the late Eastern Han Dynasty (AD 25–220), when Zhang Zhongjing compiled *Treatise on Cold Damage and Miscellaneous Diseases*, specific definitions for "zhong feng", "shang han", and "wen bing" were established:

> Greater yang disease, characterized by fever, sweating, aversion to wind, and a moderate pulse, is referred to as "zhong feng" (wind attack).
>
> Greater yang disease, characterized by either already having fever or not yet having fever, along with aversion to cold, body pain, nausea, and a tight pulse in both yin and yang (Chi and Cun pulsations), is referred to as "shang han" (cold damage).
>
> Greater yang disease, characterized by fever, thirst, and no aversion to cold, is referred to as "wen bing" (warm disease).

However, it is important to note that the three definitions provided earlier are designed to distinguish the initial stage of greater yang disease (Stage 1 of externally contracted febrile disease). There are additional stages in this disease progression, including yang brightness disease (Stage 2), lesser yang disease (Stage 3), lesser yin disease (Stage 4), greater yin disease (Stage 5), and reverting yin disease (Stage 6). Each of these stages exhibits different clinical manifestations and disease nature compared to the initial stage. Meanwhile, these six stages are all included in *Treatise on Cold Damage and Miscellaneous Diseases.* As such, it is impossible to fully define the concept of "cold damage", or to comprehensively understand the book based solely on the three definitions mentioned earlier. These definitions primarily aim to clarify the application of two therapeutic principles for the early stage of externally contracted febrile disease: Ma Huang Decoction (to promote sweating and release the exterior) and Gui Zhi Decoction (to release the muscles). Furthermore, they help distinguish between "shang han" (cold damage) in a narrow sense and warm disease, based on whether the patient experiences symptoms of aversion to wind and cold or not).

In addition to the climate-based law (cold in winter, warm in spring, hot in summer), another theory of that era based on etiology provided a different perspective on externally contracted febrile disease. According to this theory, if a person becomes ill in winter due to cold invasion, it is called cold damage; if a person is invaded by cold in winter but does not fall ill immediately, and the cold pathogen remains dormant under the skin until spring when it emerges, it is called warm disease. If the cold pathogen remains dormant until summer before causing illness, it is called summer-heat disease. In this context, cold damage, warm disease, and summer-heat disease all result from the invasion of cold in winter, while other types of externally contracted febrile diseases caused by sudden climate changes (such as cold in spring or cool in summer) are referred to as "shi qi" (epidemic qi).[2] However, this theory limited the cause of cold damage to the invasion of cold specifically in winter, assuming that cold is the primary influence during that season. It supported its theoretical framework with the "latent pathogen qi theory", which posits that pathogenic cold remains hidden under the skin and later develops into warm disease in spring and summer-heat disease in summer. However, the clinical relevance of this theory is limited. The *Treatise on Cold Damage and Miscellaneous Diseases* predominantly uses medications with a pungent-warm nature. As a result, the distinctions between cold damage and warm disease received little attention until the emergence of therapeutic principles emphasizing the release of the exterior with pungent-cool treatments. By assessing the ratio of cold and heat syndromes of clinical patients, it appears that cold damage refers to externally contracted diseases where cold syndromes predominate over heat syndromes, indicating therapeutic principles of releasing the exterior with pungent-warm treatments and warming the middle to restore yang. On the other hand, warm disease refers to diseases with more heat syndromes than cold syndromes, indicating therapeutic principles of releasing the exterior with pungent-cool treatments and extinguishing wind to settle fright (Fig. 13.1).

In order to fully grasp the concept of "cold damage" and its derivatives, it is important to consider the following three points:

---

[2] This statement comes from the "Examples of Cold Damage" in *Treatise on Cold Damage Diseases* (*Shang Han Lun,* 《伤寒论》) and *Arcane Essentials from the Imperial Library* (*Wai Tai Mi Yao,* 《外台秘要》) by Wang Tao in the Tang Dynasty. While both classics attribute this statement to the *Great Statement of Yin and Yang* (*Yin YangDa Lun,* 《阴阳大论》), but this book has been lost for centuries.

Fig. 13.1 The Memorial temple to the Medical Sage. Situated in Nanyang city, Henan province, the hometown of Zhang Zhongjing

(1) The broad sense of cold damage did not disappear with the emergence of the narrow sense. Although the study of warm disease flourished during the Ming and the Qing dynasties, resulting in numerous renowned medical practitioners and works, the medical classification still only recognized "cold damage" and not "warm diseases" as a separate category.
(2) In the Han Dynasty, the term "shang han" was already understood as the opposite of warm disease, but it never specifically referred to external disease caused by deficiency cold. For example, "yang brightness disease" recorded in *Treatise on Cold Damage and Miscellaneous Diseases* is characterized by symptoms such as high fever, extreme thirst, profuse sweating, a large and surging pulse, and hard stools, which are very similar to those of warm disease. Furthermore, its therapeutic principles—clearing and purgation—and the representative formulas such as Bai Hu Decoction (White Tiger Decoction) and Cheng Qi Decoctions (Purgative Decoctions)[3] were also adopted by the warm disease school in later generations. Therefore, some experts believed that the principles of cold damage could also be applied to warm disease, suggesting that there was no fundamental difference between the two. The reasons behind the emergence of the warm disease school and the subsequent interpretation of the *Treatise on Cold Damage Diseases* (*Shang Han Lun*, 《伤寒论》) as specifically addressing externally contracted diseases caused by deficiency cold will be discussed further in subsequent sections.

[3] Cheng Qi Tang refers to various purgative decoctions, such as Da Cheng Qi Tang (Major Purgative Decoction), Xiao Cheng Qi Tang (Minor Purgative Decoction), Tiao Wei Cheng Qi Tang (Stomach-Regulating Purgative Decoction), etc.

(3) *Treatise on Cold Damage Diseases* was published by the Office of Revising Medical Books during the Northern Song Dynasty (AD 960–1127). This important medical work laid the foundation for the emergence of the "cold damage school" and subsequently attracted a significant following. In the late Qing Dynasty and the early Republic of China era (AD 1912–1949), the "classical formulas school" emerged. The "Cold damage school" can indeed be divided into two branches. In a broad sense, it refers to those who highly regard Zhang Zhongjing's *Treatise on Cold Damage Diseases* and its system of six-meridian pattern identification. They utilize these principles as general guidelines for externally contracted diseases and miscellaneous illnesses. In a narrow sense, it refers to various schools that study and annotate *Treatise on Cold Damage Diseases* from different perspectives. These include the school of three principles, the school of maintaining the old theory, the school of syndrome differentiation based on formulas, and the school of syndrome differentiation based on therapeutic principles.[4] Additionally, the "classical formulas school" refers to practitioners who view Zhang Zhongjing's formulas as classic formulas and strictly adhere to the original formulas, including dosages and administration methods, in their clinical practice.

13.1.2 Warm Disease and Warm Disease School

The term "warm disease" originated in the *Plain Questions*. The term was mentioned twice in this book. The first is "If someone falls ill due to exposure to the cold during winter, they may develop a warm disease in the spring".[5] The second is "Therefore, if there is a proper storing of essence, there will be no occurrence of warm diseases in the spring".[6] The former can be considered the origin of the theory of "latent pathogen qi", and the latter highlights an important pathology: the underlying cause of warm disease is not necessarily external contraction, but rather yin deficiency with internal heat (inability to store essence). Symptoms of warm disease mentioned in *Plain Questions* include "the pulse is rapid", fever, "when sweat leaves the body, the heat returns", and "the patient utters crazy words and cannot eat",[7] indicating heat syndromes. However, compared with those in *Treatise on Cold Damage Diseases*, the symptoms of warm disease recorded in *Plain Questions* show little variation.

[4] Editorial Board of China Encyclopedia of Medicine: "History of Medicine" volume in *China Encyclopedia of Medicine* (*Zhong Guo Yi Xue Bai Ke Quan Shu*,《中国医学百科全书》), p78.

[5] Chapter "Discourse on how the Generative Qi Communicates with Heaven" in *Plain Question*.

[6] Chapter "Discourse on the True Words in the Golden Chest" in *Plain Question*.

[7] Chapter "Discourse on Phenomena Reflecting the Status of Qi in a Normal Person" in *Plain Question*, "When man exhales once and his vessels exhibit three movements and when he inhales once and his vessels exhibit three movements and are racing, with heat appearing in the foot-long section, that is called 'to suffer from warmth'". Chapter "Discourse Deliberating upon Heat Disease" in *Plain Question*, "Someone suffers from warmth. When sweat has left his body, the heat returns always and the movement in the vessels races. The disease is not weakened by the sweating. The patient utters crazy words and cannot eat".

During that time, the term "warm disease" was used, but it was actually classified under the category of cold damage. Since the Song Dynasty (AD 960–1279), diseases contracted externally were treated based on different principles depending on whether they were categorized as cold or heat natured (i.e., "to treat cold with heat" and "to treat heat with cold"). According to Zhu Gong, in the late Northern Song Dynasty, there were two distinct groups of medical practitioners. One group favored medications with cold nature and hesitated to use medications with heat nature, such as Fu Zi (Radix Aconiti Lateralis Preparata) and Liu Huang (Sulfur). They advised patients to consume cold beverages and prescribed cold natured medications like San Huang Wan (Three Yellows Pill) even in freezing winter. On the other hand, the second group preferred medications with heat nature and avoided using medications with cold nature, such as Da Huang (Radix et Rhizoma Rhei) and Mang Xiao (Natrii Sulfas). They instructed patients to take medications processed with the method of Zhi (stir-bake with adjuvant) and Duan (calcine), as well as Jin Ye Dan (Golden Liquid Pills) in sultry summer. In this context, "occasionally, there were patients who had read some remedy books and possessed some knowledge of TCM. They believed they had a certain understanding of their own conditions. Hence, if they perceived their diseases to be of a heat nature, they would seek out doctors known for their expertise in treating diseases of heat nature, and vice versa".[8] Apparently, this phenomenon marks a noticeable separation between cold and heat branches of the cold damage school. However, it also sheds light on an interesting observation: during that period, the greatest advantages and basics of TCM, namely distinguishing yin and yang, understanding cold and heat, and regulating deficiency and excess, were mastered by patients instead of medical practitioners. In certain cases, some patients took it upon themselves to practice syndrome differentiation, so as to enable them to locate the most suitable doctors for their specific needs. This situation is regrettable, as it highlights a disparity in knowledge and expertise between patients and medical practitioners. However, it is important to acknowledge that such a phenomenon persists even in modern times.

Liu Wansu (AD 1110–1200), one of the four distinguished medical practitioners during the Jin (AD 1115–1234) and Yuan (AD 1271–1368) period, is widely recognized for his significant contributions to the development history of warm disease school. However, due to the prevailing theory of Zhang Zhongjing, he made few innovations in the study of warm disease. His emphasis on treating heat syndromes with medications of cold nature contributed to the theoretical and therapeutic principles of warm disease school. Until then, the warm disease school had not yet developed into an independent branch at that time. It was during the Ming Dynasty that a significant shift occurred. In 1642, Wu Youxing compiled *Treatise on Pestilence* (*Wen Yi Lun*, 《瘟疫论》), and introduced a groundbreaking perspective. Wu Youxing highlighted the distinct differences in etiology, pathogenesis, and treatment between pestilence and cold damage disorders, despite their superficial similarities. He proposed that pestilence was caused by the "pestilential qi" between heaven and earth, and was

---

[8] Zhu Gong (the Song Dynasty), "Preface" in *Book to Safeguard Life Arranged by Categorized Patterns* (*Lei Zheng Huo Ren Shu*, 《类证活人书》), Beijing: The Commercial Press, 1955.

highly contagious. "Whoever (the old, the young, the strong and the weak) could be affected by the pestilential qi, which entered the body through the mouth and nose, they get infected". This expanded understanding surpassed the limitations of the *Treatise on Cold Damage Diseases*, and further advanced the theoretical framework of externally contracted febrile diseases of TCM. From that point onward, the warm disease school gradually developed into an independent school, accompanied by the emergence of the contemporary formula school, which posed a formidable challenge to the established school of classical formulas.

Zhang Zhongjing, revered as a "medical sage", is considered the founding figure of cold damage school. His renowned masterpiece, the *Treatise on Cold Damage Diseases*, is regarded as the classic work in this discipline. However, in contrast to the cold damage school, the warm disease school lacks a singular influential founder or a widely recognized masterpiece. This absence has allowed for greater freedom and diverse development within the warm disease school. For example, Ye Tianshi, one of the "four famous medical practitioners of the warm disease school" in the Qing Dynasty, proposed the system of defense, qi, nutrient and blood pattern identification, which expanded beyond the traditional six-meridian pattern identification system. Additionally, Wu Tang, another esteemed member of the "four famous medical practitioners", established his own unique system of triple energizer pattern identification. However, there is ongoing debate regarding the equivalence of the terms "wen (温, warm disease)" and "wen (瘟, pestilence)". Opinions differ among experts on this subject. For example, Xue Shengbai and Wu Youxing argued that "wen (温, warm disease)" and "wen (瘟, pestilence)"[9] are indeed equivalent. However, Lei Feng, the author of *Discourse on Seasonal Diseases* (*Shi Bing Lun*, 《时病论》), held the perspective that wen (温) pertains to warm heat disease, while wen (瘟) refers specifically to pestilence. Although the two terms share the same pronunciation, they refer to different diseases. The former is caused by "constant qi", while the latter stems from the influence of "pestilential qi". He further argued that "medical books on warm disease cannot be used to guide the treatment of pestilence, and those on pestilence cannot be used to guide the treatment of warm disease".[10]

So, what is warm disease? This probably poses a significant challenge for experts in the field. There were numerous monographs on warm disease or pestilence in the Ming and the Qing Dynasties. However, a consensus on the exact nature of "warm disease" remained elusive. For example, some held that "warm disease" referred to "pestilence". According to this view, the difference between cold damage and warm disease lay in the mode of pathogen invasion. Cold damage was thought to be externally contracted through the skin and body hair, while warm disease (pestilence) resulted from the invasion of pathogen via the mouth and nose. However,

---

[9] Xue Shengbai (the Qing Dynasty), "Random Notes on Daily Lectures". This article was recorded in *Wu's Compiled Medical Discourses* (*Wu Yi Hui Jiang*, 《吴医汇讲》) compiled by Tang Lishan in the Qing Dynasty, p14.

[10] Lei Feng (the Qing Dynasty), *Treatise on Seasonal Disease* (*Shi Bing Lun*, 《时病论》), Beijing: People's Medical Publishing House, 1956, Volume 8, p142.

other experts contradicted this interpretation, arguing that "the so-called pestilential qi is nothing but stagnant heat",[11] which was contrary to the view that the pathogen invades people via the mouth and nose. Another viewpoint, exemplified by Wu Tang, believed that "pestilence" was merely a manifestation of warm disease. He classified nine subtypes of "warm disease", including wind-warmth, warm heat, pestilence, warm toxin, summerheat-warmth, dampness-warmth, autumn-dryness, winter-warmth, and dampness malaria.[12] In his opinion, the difference between cold damage and warm disease was that "cold damage disease starts from foot greater yang (bladder) meridian, and progresses from the exterior to the interior", while "warm disease starts from hand greater yin (lung) meridian, and develops from the upper to the lower". In a word, there is a common sense in the school of warm disease that febrile diseases can be categorized into two distinct types: cold damage and warm disease, based on their different natures. They believed that the etiologies, pathogenesis and treatment principles of these two categories are also markedly different. By challenging the prevailing theory from *Plain Questions*, which posits that "all febrile diseases pertain to cold damage", the warm disease school sought to expand upon existing knowledge and refine the definitions of febrile illnesses more precisely.

Warm disease, at its core, is fundamentally caused by dampness and heat, which distinguishes it from cold damage. This internal dampness and heat constitute the essence of the so-called "fu xie" (latent pathogenic qi). Due to the presence of internal dampness and heat, the disease will inevitably manifest, regardless of whether there is an external trigger such as wind-cold or pestilential qi in the environment. Therefore, the clinical manifestations of warm disease may or may not include symptoms of external contraction. Whether the disease occurs sporadically or as an epidemic, the treatment principle remains the same. This is because TCM focuses on addressing the internal cause of disease rather than the external factors.

[11] Zhou Sizhe (the Qing Dynasty), "Statement on Warm Epidemics". This article was recorded in *Wu's Compiled Medical Discourses* compiled by Tang Lishan in the Qing Dynasty, p38.

[12] Wu Tang (the Qing Dynasty), *Systematic Differentiation of Warm Diseases* (*Wen Bing Tiao Bian*, 《温病条辨》), Volume 1.

Fig. 13.2 Zhang Zhongjing's Tomb. Located in now Nanyang city, Henan province

13.2 Zhang Zhongjing and *Treatise on Cold Damage and Miscellaneous Diseases*

13.2.1 A Brief Introduction of Zhang Zhongjing

Zhang Zhongjing, also known as Zhang Ji,[13] widely recognized as the founder of cold damage school, was born in Nanyang prefecture (now Nanyang city, Henan province)[14] in the late Eastern Han Dynasty. Although there are no official historical records about him, various historical references indicate that he approximately lived from the middle of the second century to the early third century.[15] According to *Biography of Famous Physicians* (*Ming Yi Zhuan*, 《名医传》) by Gan Bozong of the Tang Dynasty (AD 618–907), Zhang Zhongjing "was appointed as the governor of Changsha county through the selection process known as 'ju xiao lian' (upholding filial piety and honesty, 举孝廉)[16]" (Fig. 13.2).

[13] *List of Famous Physicians (Ming Yi Lu,* 《名医录》): A native of Nanyang, his given name is "Ji" and he styled himself as "Zhongjing".

[14] Various historical records provide conflicting information regarding Zhang Zhongjing's ancestral native place. *Introduction of Xiangyang* (*Xiang Yang Fu Zhi*, 《襄阳府志》) states that he was from Zaoyang, Nanyang prefecture. But Zaoyang hasn't been established during Zhang Zhongjing's time. On the other hand, *Introduction of He'nan* (*He Nan Tong Zhi*, 《河南通志》) records that he was from Nieyang, Nanyang prefecture, or Nieyang, Nanyang region. But there is no reliable record that can verify this statement.

[15] The exact dates of Zhang Zhongjing's birth and death remains unknown. This book speculates on these dates by drawing upon from related individuals and records found in various medical books.

[16] The "Preface" of the *Treatise on Cold Damage Diseases* (*Shang Han Lun*, 《伤寒论》) attributes authorship to "Zhang Ji, a native of Nanyang who was the governor of Changsha prefecture in the Han Dynasty". But historians in later generations cast doubt on this statement due to the absence of any record of a Changsha prefecture governor named Zhang Ji.

Zhang Zhongjing initially learned medicine under the guidance of Zhang Bozu in the same prefecture. "Zhang Bozu was a native of Nanyang. He was modest and dedicated to TCM. He was meticulous and considerate when treating patients, and his treatments were highly effective, earning him great respect at that time. Zhang Zhongjing, from the same prefecture, admired him greatly and became his apprentice". As Zhang Zhongjing's skills developed, he gained prominence as a renowned physician. It is said that after assuming the position of the governor of Changsha prefecture, Zhang Zhongjing "practiced medicine in the capital city, where he gained significant recognition for his exceptional medical expertise. People at that time regarded him as being on par with ancient medical luminaries like Bian Que. and Cang Gong[17]". Zhang Zhongjing's medical achievements are encapsulated in his work, *Treatise on Cold Damage and Miscellaneous Diseases*. In the preface of the book, it is revealed that Zhang Zhongjing's family was once large, but "in less than ten years since the first year of Jian'an (AD 196), two thirds of his family tragically passed away, with roughly 70 percent of these deaths attributed to cold damage". Driven by his sorrow, Zhang Zhongjing "diligently learned ancient medical theories and extensively referenced various medical formulas". He compiled 16 volumes in his masterpiece, *Treatise on Cold Damage and Miscellaneous Diseases*, by referencing ancient TCM classics, including the *Plain Questions*, *Nine Volumes* (*Jiu Juan*, 《九卷》), *Yellow Emperor's Canon on Eighty-one Difficult Issues* (*Huang Di Ba Shi Yi Nan Jing*, 《黄帝八十一难》), *Comprehensive Discourse on Yin and Yang* (*Yin Yang Da Lun*, 《阴阳大论》), and the *Tai Lu Medication List* (*Tai Lu Yao Lu*, 《胎胪药录》).

13.2.2 *Inheritance of* Treatise on Cold Damage and Miscellaneous Diseases

The contents related to "cold damage" in the *Treatise on Cold Damage and Miscellaneous Diseases* were preserved through the efforts of Wang Shuhe, the Director of the Imperial Medical Bureau during the Western Jin Dynasty (AD 265–317). However, before the Song Dynasty, the titles, volume numbers, contents, and classification of Zhang Zhongjing's works underwent frequent changes as they were privately copied. For example, during the Liang Dynasty (AD 502–557), there were ten volumes of *Zhang Zhongjing's Treatise on Cold Damage* (*Zhang Zhong Jing Bian Shang Han*, 《张仲景辨伤寒》), and one volume of *Zhang Zhongjing's Discourse and Critical Prescriptions for Diseases* (*Zhang Zhong Jing Ping Bing Yao Fang*, 《张仲景评病要方》). In the "Records of Classics and Books" chapter of the *History of the Sui Dynasty* (*Sui Shu*, 《隋书》), there were 15 volumes of *Zhang Zhongjing's Prescriptions* (*Zhang Zhong Jing Fang*, 《张仲景方》), and two volumes of *Zhang Zhongjing's Prescriptions for Female Diseases* (*Zhang Zhong Jing Liao Fu Ren Fang*, 《张仲景

[17] Zhang Gao (the Song Dynasty), *A Record of the Life by Various Physicians* (*Yi Shuo*, 《医说》), Volume 1.

疗妇人方》). In the Northern Song Dynasty, the Office of Revising Medical Books conducted a thorough examination and concluded that "the *Treatise on Cold Damage Diseases* consists of ten volumes, 22 chapters, 397 therapeutic methods, and 112 prescriptions (excluding repeated formulas)".[18] Subsequently, the Office undertook the proofreading of the *Canon of the Golden Coffer and Jade Sheath* (*Jin Gui Yu Han Jing*, 《金匮玉函经》) which shared the same content as the *Treatise on Cold Damage Diseases*. This revised version, staying true to the original content, is composed of eight volumes, 29 chapters, and 115 prescriptions.[19] It is important to note that these two versions in the Song Dynasty elaborate on cold damage only, and leave other miscellaneous diseases unmentioned. Later, Wang Zhu, a scholar of the Imperial Academy in the Northern Song Dynasty, discovered three volumes of *Synopsis and Prescriptions of the Golden Chamber and Jade Letter* (*Jin Gui Yu Han Yao Lve Fang*, 《金匮玉函要略方》) in old worm-eaten books in library. The first volume deals with cold damage, the second volume with miscellaneous diseases, and the third volume with prescriptions for miscellaneous diseases, including female diseases. Bureau for Revising Medical Books of the Northern Song Dynasty made revisions by deleting the content on cold damage and attaching relevant prescriptions into the corresponding chapters on miscellaneous diseases. This resulted in the creation of a new book known as the *Synopsis of the Golden Chamber* (*Jin Gui Yao Lve*,《金匮要略》) as it stands today. Therefore, there are two kinds of Zhang Zhongjing's works. One is *Treatise on Cold Damage Diseases*, which concentrates on cold damage, while the other is the *Synopsis of the Golden Chamber*, which covers miscellaneous diseases. Unfortunately, the original version of Zhang Zhongjing's works remains unknown. It is believed that *Synopsis and Prescriptions of the Golden Chamber and Jade Letter* discovered by Wang Zhu possibly represents an original version of *Treatise on Cold Damage and Miscellaneous Diseases*.[20] This opinion holds valid as it is evident that Wang Shuhe (Director of the Imperial Medical Bureau of the Western Jin Dynasty) consciously selected the contents on cold damage and collected them into *Treatise on Cold Damage Diseases*.[21] This leads to debates surrounding Wang Shuhe's merits and demerits. Some held that Wang Shuhe's deeds could be compared to "cutting a beautiful brocade into pieces, and decorating the fragmented brocade with rags. The later generations would have no access to the original works of Zhang Zhongjing…and because of his choices, the inheritance of Qi Huang (Qi Bo and

[18] Lin Yi et al., (the Song Dynasty), "Preface" of *Annotations on 'Treatise on Cold Damage Diseases'* (*Shang Han Lun Zhu Shi*, 《伤寒论注释》).

[19] Lin Yi et al., (the Song Dynasty), Commentary on 'The Jade Case Classic of the Golden Cabinet'. Quoted from *Verification on Pre-Song Medical Works* (*Song Yi Qian Yi Ji Kao*, 《宋以前医籍考》) by Okanisi Tameto, p577.

[20] Jia Dedao, *A Brief History of Chinese Medicine* (*Zhong Guo Yi Xue Shi Lve*, 《中国医学史略》), p96.

[21] Huangfu Mi (the Jin Dyansty), "Preface" of *A-B Classic of Acupuncture and Moxibustion* (*Zhen Jiu Jia Yi Jing*, 《针灸甲乙经》).

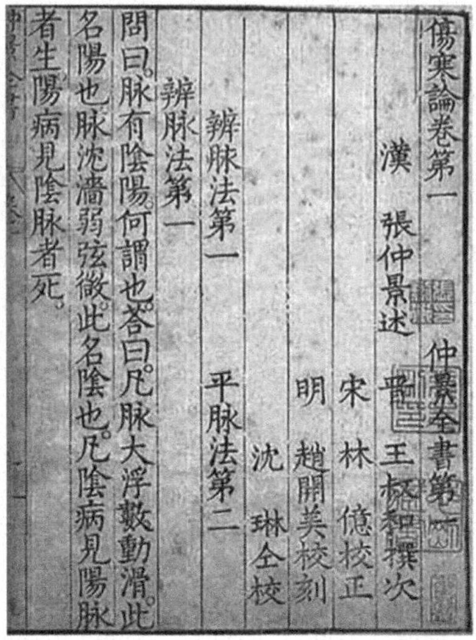

Fig. 13.3 One version of the *Treatise on Cold Damage Diseases* in the Ming Dynasty (Provided by Huying Chong)

Huang Di, referring to TCM) is doomed to vanish".[22] There are differing opinions regarding the historical significance of Zhang Zhongjing's works, with some suggesting that his contributions might have been lost to history had it not been for the efforts of Wang Shuhe in collecting and organizing them. Moreover, ongoing debates persist regarding the authorship of certain chapters within the *Treatise on Cold Damage Diseases*, with disagreements arising as to whether they are original works by Zhang Zhongjing or additions by Wang Shuhe (Fig. 13.3).

13.2.3 System of Six-Meridian Pattern Identification

Based on the theory of six-meridian pattern identification, the course of heat disease is divided into six stages according to clinical symptoms. That is, three yang (greater yang, yang brightness, lesser yang) and three yin (greater yin, lesser yin, reverting yin). The term "six-meridian" is employed because of the parallelism between the stages of heat disease and the yang-yin meridians in TCM theory. But what is the relationship between the three yang and three yin of heat disease and those of meridians?

The categorization of heat disease into three yang and three yin stages was firstly introduced in the chapter "Discourse on Heat" in *Plain Question.* Therefore, it is widely believed that *Treatise on Cold Damage Diseases* adopted the six-meridian system described in the chapter "Discourse on Heat" of the *Plain Questions* to establish its theory of pattern identification and treatment. However, a comparison of the

[22] Yu Chang (the Qing Dynasty), first chapter of *Tracing Back to the 'Treatise on Cold Damage'* (*Shang Lun Pian*, 《尚论篇》).

symptoms of the six-meridian disease in these two books reveals that the three yang aspects are roughly consistent, while the three yin aspects are different significantly (see Table 13.1).

It can be seen from Table 13.1 that the symptoms of six-meridian diseases align to some extent with those of meridian and collateral diseases. However, the symptoms of six-meridian diseases in *Treatise on Cold Damage Diseases* have gone beyond the realm of meridian and collateral theory and showed distinct clinical characteristics. It can be concluded that the system of six-meridian pattern identification of heat disease is originated from the meridian and collateral theory in *Plain Questions*. But by the time the *Treatise on Cold Damage Diseases* was compiled, this system had already diverged from the meridian and collateral theory, serving as the source of different stage names and pattern names of heat disease.

The system of six-meridian pattern identification in the chapter "Discourse on Heat" from *Plain Questions* was not particularly important in the establishment of therapeutic principles, given the relative simplicity of those principles at that time. Diseases in the three yang stages were treated with diaphoresis, while diseases in the three yin stages were treated with purgation. However, in the *Treatise on Cold Damage Diseases*, the therapeutic methods for six-meridian diseases were greatly

Table 13.1 Symptoms of six-meridian disease recorded in the chapter "Discourse on Heat" in *Plain Questions* and *Treatise on Cold Damage Diseases*

	Chapter "Discourse on Heat" in *Plain Questions*	*Treatise on Cold Damage Diseases*
Greater yang disease	Headache, neck pain, and lower back and spine stiffness.	Floating pulse, stiffness and pain of the head and nape, and aversion to cold.
Yang brightness disease	Generalized fever, pain in eyes, dry nose, and hard to lie down.	Stomach excess (high fever, great thirst, profuse sweating and hard bound stool, etc.)
Lesser yang disease	Pain in chest and flanks, and deafness.	Bitterness in the mouth, dryness of the throat and dizziness
Greater yin disease	Abdominal fullness and dry throat.	Abdominal fullness, vomiting, inability to eat, severe diarrhea, and frequent abdominal pain. If purgation is used to treat it, it will cause hardness and lump below the chest
Lesser yin disease	Dry mouth and tongue, and thirst.	Faint and feeble pulse and sleepiness
Reverting yin disease	Vexation and fullness, and the scrotum shrinks	Wasting-thirst, counterflow of qi against the heart, pain and heat in the chest, hunger without desire to eat, vomiting or vomiting of worms after taking food

enriched, including the first six of the eight methods that emerged in later generations: promoting sweating, promoting vomiting, purgation, harmonizing, warming, and clearing.

With the systematical summarization and reorganization by medical practitioners throughout history, the basic content of six-meridian pattern identification can be roughly summarized as follows:

(1) Greater yang disease—the first stage of cold damage. The pathogen resides in the exterior and is characterized by a floating pulse, stiffness and pain in the head and nape, fever, and aversion to cold. If there is sweating due to exterior deficiency, Gui Zhi Decoction (Cinnamon Twig Decoction) is recommended. If there is no sweating due to exterior excess, Ma Huang Decoction (Ephedra Decoction) is recommended to promote sweating. If the pathogenic heat invades viscera and bowels (bladder and small intestine) via the related meridian, it will lead to symptoms like water-retention, urination-retention, difficult urination, blood amassment, etc.
(2) Lesser yang disease—the second stage of cold damage. The pathogen lies between the exterior and interior, leading to alternating chills and fever. In addition, since the lesser yang meridian is connected to the gallbladder, patients may experience symptoms such as bitter taste in the mouth, dry throat, dizziness, vexation, and vomiting. Xiao Chai Hu Decoction (Minor Bupleurum Decoction) is most commonly used for lesser yang disease, and depending on different clinical manifestations, other variations of this formula may be employed, such as Da Chai Hu Tang (Major Bupleurum Decoction), Chai Hu Gui Zhi Tang (Bupleurum Cinnamon Twig Decoction), and Chai Hu Gui Zhi Gan Jiang Tang (Bupleurum, Cinnamon Twig and Dried Ginger Decoction), collectively known as “Chai Hu Bei” (Bupleurum class) formulas.
(3) Yang brightness disease—the third stage of cold damage. Pathogenic heat invades the stomach, causing symptoms such as a floating, surging and long pulse, fever without chills, profuse sweating, severe thirst, and hard, dry stools. For heat patterns without excess, Bai Hu Tang (White Tiger Decoction) is used to clear the heat. For heat pattern with excess (indicating accumulation), Cheng Qi Tang (Purgative Decoction) is used for purgation.
(4) Greater yin disease—the fourth stage of cold damage. Symptoms include a sunken pulse, abdominal fullness, vomiting, inability to eat, intermittent abdominal pain, and no thirst. The pathogen invades the spleen and stomach, and as the saying goes, “excess patterns indicate yang brightness disease, and deficiency patterns indicate greater yin disease”, the treatment primarily involves warming with formulations from the Li Zhong lineage, such as Li Zhong Wan (Middle-Regulating Pill) and its related variations.
(5) Lesser yin disease—the fifth stage of cold damage. It is similar to septic shock in modern medicine, and it is characterized by a faint and fine pulse and a tendency to sleepiness. Medications commonly used in lesser yin disease include Fu Zi (Radix Aconiti Lateralis Preparata) and Ren Shen (Radix Ginseng).

(6) Reverting yin disease—the sixth stage of cold damage, characterized by a critical and severe condition. In this stage, symptoms include a sunken and fine pulse, cold limbs, bluish nails, and overall body chills. The main therapeutic principle is to "restore yang and rescue from collapse".

However, the system of six-meridian pattern identification in Zhang Zhongjing's *Treatise on Cold Damage and Miscellaneous Diseases* is not extensively elaborated. It coexists with various miscellaneous diseases, food taboos, and prescriptions for miscellaneous ailments (i.e., the contents of *Synopsis of the Golden Chamber*). Recognizing the significance of the system of six-meridian pattern identification, Wang Shuhe collected the relevant contents on treating cold damage from *Treatise on Cold Damage and Miscellaneous Diseases* and organized them into a separate book. Therefore, our comment on Wang Shuhe shall not be limited to the praise and criticism of predecessors (Fig. 13.4).

This book, written by Wang Tao in the Tang Dynasty, consists of 40 volumes and incorporates dozens of medical works before the Tang Dynasty. The first chapter in it focuses on "Cold damage", reflecting the prevailing belief at the time that diseases are caused by "six exopathogenic factors".

During the span of eight centuries, from Wang Shuhe's compilation of *Treatise on Cold Damage Diseases* to the early Northern Song Dynasty, this system of six-meridian pattern identification did not receive significant attention from medical practitioners. For Sun Simiao of the Tang Dynasty, *Treatise on Cold Damage and*

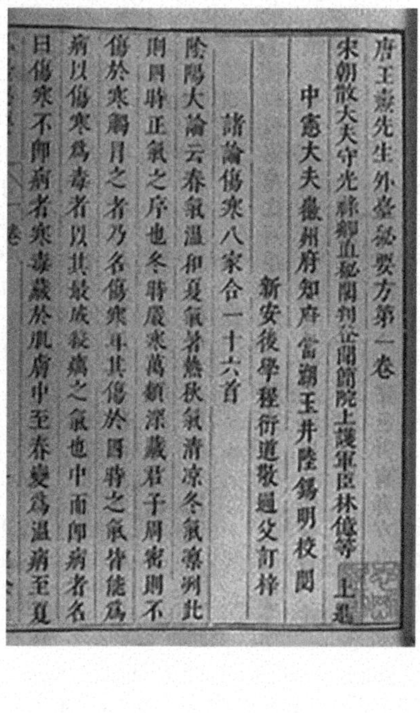
唐王燾先生外臺秘要方第一卷
宋朝散大夫守光祿卿直秘閣判登聞檢院上護軍臣林億等上進
中憲大夫徽州府知府當湖玉井陸錫明校閱
新安後學程衍道敬通父訂梓
諸論傷寒八家合一十六首
陰陽大論云春氣溫和夏氣暑熱秋氣清涼冬氣凜冽此
則四時正氣之序也冬時嚴寒萬類深藏君子周密則不
傷於寒觸冒之者乃名傷寒耳其傷於四時之氣皆能為
病以傷寒為毒者以其最成殺厲之氣也中而即病者名
曰傷寒不即病者寒毒藏於肌膚中至春變為溫病至夏

Fig. 13.4 One version of the *Arcane Essentials* from the Imperial Library in the Ming Dynasty

Miscellaneous Diseases was just a mere prescription book.[23] Besides, there was no record of Zhang Zhongjing's theory in volume I ("Discourse on Eight Great Physicians of Cold Damage School") of *Arcane Essentials from the Imperial Library* (*Wai Tai Mi Yao*, 《外台秘要》) compiled by Wang Tao. Only in the section containing prescriptions from different schools was Zhang Zhongjing and his *Treatise on Cold Damage Diseases* mentioned. Hence, at that time, the importance of *Treatise on Cold Damage Diseases* was on par with other popular prescription books, such as *Prescriptions by Fan Wang* (*Fan Wang Fang*, 《范汪方》), *Prescriptions by Shen Shi* (*Shen Shi Fang*, 《深师方》), *Formulary of Trifles* (*Xiao Pin Fang*, 《小品方》), *Records of Proved Prescription* (*Gu Jin Lu Yan Fang*, 《古今录验方》), *Pocket Prescriptions for Emergency* (*Sui Shen Bei Ji Fang*, 《随身备急方》), *Prescriptions by Cui Shi* (*Cui Shi Fang*, 《崔氏方》), *Essential Prescriptions Worth a Thousand Gold for Emergencies* (*Bei Ji Qian Jin Yao Fang*, 《备急千金要方》), *Supplement to the Essential Prescriptions Worth a Thousand Gold* (*Qian Jin Yi Fang*, 《千金翼方》), and *Handbook of Prescriptions for Emergencies* (*Zhou Hou Bei Ji Fang*, 《肘后备急方》). Therefore, when Lin Yi and his colleagues of Office of Revising Medical Books in the Northern Song Dynasty revised *Treatise on Cold Damage Diseases* (1065), they commented that

> His words are characterized by refinement and profound wisdom, and his therapeutic methods are both simple and effective, offering detailed insights that surpass the comprehension of those with limited vision and shallow understanding. Although more than 800 years have passed since the composition of his book, only Wang Shuhe truly grasped its essence. Throughout this period, there were talented and renowned medical practitioners like Ge Hong, Tao Jing, Hu Qia, Xu Zhicai and Sun Simiao, but they all focused on their respective schools and ultimately failed to fully comprehend the true essence of Zhang Zhongjing's work.[24]

Till the Song Dynasty, *Treatise on Cold Damage Diseases* gained recognition as a medical book on therapeutic principles and methods. Medical practitioners extensively studied the book from various perspectives, leading to a comprehensive understanding of its content. The compilation and annotation of the articles, along with detailed elucidation and discussion on basic concepts such as yin and yang, exterior and interior, cold and heat, deficiency and excess, three yin and three yang, etc., played a crucial role in revealing the essence of the six-meridian pattern identification system. As a result, the impression of the book gradually evolved from a prescription book to a medical classic. Ever since, Zhang Zhongjing was initially honored as "the lesser sage", and ultimately came to be revered as "the medical sage".

From the Song Dynasty to the Jin and the Yuan Dynasties, at least 80 scholars dedicated themselves to researching *Treatise on Cold Damage Diseases*, resulting in a wealth of systematic theories and insights. Up to now, about 400–500 scholars have delved into the principles, methods, formulas, and medicinal substances recorded in the book, giving rise to almost a thousand monographs, thus forming a distinct

[23] Sun Simiao (the Tang Dynasty), *Supplement to the Essential Prescriptions Worth a Thousand Gold Pieces* (*Qian Jin Yi Fang*, 《千金翼方》), Volume 9.

[24] Lin Yi et al., (the Song Dynasty), Preface of *Annotations on 'Treatise on Cold Damage Diseases'*.

and prominent school of cold damage disease in the academic history of traditional Chinese medicine.[25] Although these scholars have employed various research methods and perspectives, the system of six-meridian pattern identification remains the general outline of *Treatise on Cold Damage Diseases.*

13.3 Warm Disease School

13.3.1 The Emergence of Warm Disease School

The warm disease school focuses on the study and understanding of febrile diseases. They explore the etiology, pathogenesis, syndrome identification, and treatment methods for these conditions. In support of their perspective, they assert that warm diseases and pestilences are distinct from cold damage. This assertion can be seen as making two key points. Firstly, not all fever is caused by external wind-cold. Their perspective challenges the etiological explanation of heat diseases recorded in the "Discourse on Heat" chapter of the *Plain Questions*, which states "Now, as for heat diseases, they all are of the type 'cold damage'". Secondly, the pathogenesis, syndrome identifications, and treatments of warm diseases and pestilences do not align with the theory of the *Treatise on Cold Damage Diseases.* Hence, they proposed a new approach to address these diseases. It is generally believed that the warm disease school flourished during the Ming and Qing Dynasties as a new school, representing a significant advancement in TCM theory and treatment for externally contracted diseases and infectious diseases. As for the origin of the warm disease school, it is more commonly attributed to Liu Wansu, one of the Four Great Medical Practitioners during the Jin and Yuan periods, and the founder of the Cold and Cool School. According to his theory, "excess of any of the six qi transform into fire and heat, causing various diseases", so medical practitioners of the Cold and Cool School typically use medications with cold and cool properties to treat diseases. However, this understanding is quite superficial. In the chapter "Discourse on Heat", the six-meridian diseases are described, primarily exhibiting symptoms of a heat-related nature. Specifically, the symptoms of the three yin diseases are markedly different from those described in the *Treatise on Cold Damage Diseases.* Additionally, the same chapter mentions the phenomenon known as "liang gan yu han" (pathogenic cold invading both the exterior and interior related meridians simultaneously). This phenomenon is sequentially recorded as follows: on the first day of the disease, the greater yang meridian and lesser yin meridian are affected simultaneously; on the second day, the yang brightness meridian and greater yin meridian are affected simultaneously; and on the third day, the lesser yang meridian and reverting yin meridian are affected simultaneously (refer to Table 13.2). These characteristics align more closely with the description of warm heat disease as described by the warm disease

[25] Editorial Board of China Encyclopedia of Medicine: "Medical History" volume in *China Encyclopedia of Medicine*, p78.

school. For example, the early stages of the disease, the presence of "dry mouth" clearly indicates yin deficiency and fluid depletion. Subsequently, the appearance of "delirious speech" reflects the "reverse transmission to the pericardium", as described by Ye Tianshi, a renowned scholar of warm diseases. In terms of treatment, Chen Yanzhi, in his *Formulary of Trifles* from the Northern and Southern Dynasties (420–589 AD), already emphasized the need to differentiate between pestilence and cold damage. He emphasized that since ancient times, cold damage has been considered a difficult ailment to treat, while "tian xing" (epidemic infectious diseases) and pestilence are believed to be caused by toxic pathogenic qi. However, some physicians fail to distinguish between these conditions and treat them with the same therapeutic methods, as they mistakenly consider them different names for the same disease. The term "cold damage" is used by scholars, while "tian xing" or pestilence are used by the general population. In fact, these three diseases require "different prescriptions and therapeutic principles, as evidenced by the medical classics".[26] Although the *Treatise on Cold Damage Diseases* was published at the end of the Eastern Han Dynasty (25–220 AD), its principles and methods were not specifically applied to cold damage in subsequent dynasties. From the Tang Dynasty's *Arcane Essentials from the Imperial Library*, we can see a multitude of prescriptions primarily focused on clearing heat and detoxifying, widely used in the treatment of cold damage. Below are a few examples for illustration[27]:

Shi Gao Tang (Gypsum Decoction) from *Prescriptions by Shen Shi*: Shi Gao (Gypsum Fibrosum), Huang Lian (Rhizoma Coptidis), Huang Bai (Cortex Phellodendri), Huang Qin (Radix Scutellariae), Dan Dou Chi (Semen Sojae Preparatum), Zhi Zi (Fructus Gardeniae), Ma Huang (Herba Ephedrae).

Ge Gen Tang (Pueraria Decoction) from *Formulary of Trifles*: Ge Gen (Radix Puerariae), Sheng Jiang (Fresh Ginger), Long Dan (Radix Gentianae), Da Qing Ye (Folium Isatidis), peeled Rou Gui (Cortex Cinnamomi), Gan Cao (Radix Glycyrrhizae), Ma Huang (Herba Ephedrae), Yu Zhu (Rhizoma Polygonati Odorati), Chi Shao (Radix Paeoniae Rubra), Huang Qin (Radix Scutellariae), Shi Gao (Gypsum Fibrosum), Sheng Ma (Rhizoma Cimicifugae).

Collection of Proven Prescriptions (*Ji Yan*, 《集验》): Da Qing Ye (Folium Isatidis), Gan Cao (Radix Glycyrrhizae), E Jiao (Ass Hide Glue), Dan Dou Chi (Semen Sojae Preparatum).

Essential Prescriptions Worth a Thousand Gold for Emergencies: Chai Hu (Radix Bupleuri), Chi Shao (Radix Paeoniae Rubra) or Bai Shao (Radix Paeoniae Alba), Zhi Zi (Fructus Gardeniae), Zhi Mu (Rhizoma Anemarrhenae), Dan Dou Chi (Semen Sojae Preparatum), Shi Gao (Gypsum Fibrosum), Huang Qin (Radix Scutellariae), Da Qing Ye (Folium Isatidis), Sheng Ma (Rhizoma Cimicifugae), Ku Xing Ren (Semen Armeniacae Amarum).

Huang Lian Jie Du Tang (Coptis Detoxification Decoction) from *Prescriptions by Cui Shi*: Huang Lian (Rhizoma Coptidis), Huang Qin (Radix Scutellariae), Huang Bai (Cortex Phellodendri), Zhi Zi (Fructus Gardeniae).

[26] Wang Tao (the Tang Dynasty), *Arcane Essentials from the Imperial Library* (*Wai Tai Mi Yao Fang*, 《外台秘要方》), Volume 1.

[27] The following formulas are all recorded in volume 1 and 2 of *Arcane Essentials from the Imperial Library* by Wang Tao in the Tang Dynasty.

Table 13.2 Symptoms of "liang gan yu han" (pathogenic cold invading both the exterior-interiorly related meridians at the same time) recorded in chapter "Discourse on Heat" in *Plain Questions*

Exterior-interiorly related meridians	Symptoms
Greater yang and lesser yin	Headache, dry mouth, vexation, and fullness
Yang brightness and greater yin	Abdominal fullness, generalized fever, anorexia, and delirious speech
Lesser yang and reverting yin	Deafness, retracted testicles, fainting, being unable to eat or drink, or recognize people

Shao Yao Di Huang Tang (Radix Paeoniae Alba and Radix Rehmanniae Recens Decoction) from *Formulary of Trifles*: Chi Shao (Radix Paeoniae Rubra) or Bai Shao (Radix Paeoniae Alba), Sheng Di Huang (Radix Rehmanniae Recens), Mu Dan Pi (Cortex Moutan Radicis), small pieces of Xi Jiao (Rhinoceros Horn).

For another example, Sun Simiao, a prominent medical expert in the Tang Dynasty, pitied that the doctors of his time preferred to treat cold damage with medications with cold and cool nature.[28] Similarly, Zhu Gong, a TCM expert in the late Northern Song Dynasty, noted that externally contracted diseases were treated separately (i.e., to treat cold with heat, and to treat heat with cold) in the Song Dynasty. Hence, it can be concluded that the method of treating externally contracted heat diseases with medications with cold and cool nature has persisted throughout different dynasties. Hence, it raises the question: why did the warm disease school emerge after the Song Dynasty? Fundamentally, with the rise of cold damage school in the Song Dynasty, *Treatise on Cold Damage Diseases* was honored as a medical classic and its author, Zhang Zhongjing was revered as a medical sage, which inevitably resulted in a pervasive veneration of *Treatise on Cold Damage Diseases* in the treatment of externally contracted heat diseases.[29]

Another important reason is that traditional Chinese medicine was deeply influenced by the rise of Neo Confucianism during the Song and Ming Dynasties. The scholarly method of the Tang Dynasty, which focused on compiling and referencing various medical texts, was replaced by a scholarly pursuit of "qiong li" (tracing the very essence of things) and "ge wu" ((investigating the principles of all things). This intellectual atmosphere was a major factor in the emergence of the cold damage school and also served as the ideological background for the rise of the warm disease school. In other words, although various methods of treating warm diseases with cold and cool medications existed before the Song Dynasty, there was a lack of systematic and theoretical integration of these practices. The warm disease school played a pivotal role in summarizing and advancing the treatment of warm diseases by building upon the experiences of previous practitioners. Through the accumulation of treatment experience, the refinement of methods, and the expansion of medication options, TCM therapeutics were further developed.

[28] Sun Simiao (the Tang Dynasty), *Supplement to the Essential Prescriptions Worth a Thousand Gold Pieces*, Volume 9.

[29] Wu Youxing (the Qing Dynasty) "Preface" of *Treatise on Pestilence* (*Wen Yi Lun*, 《瘟疫论》).

13.3.2 Wu Youxing and *Treatise on Pestilence*

Wu Youxing, also known as Wu Youke, was a medical expert in the late Ming Dynasty, from Wuxian county, Jiangsu province. In 1642, he compiled the *Treatise on Pestilence* (*Wen Yi Lun*, 《瘟疫论》), which became the most influential medical classic of the warm disease school for two reasons. Firstly, Wu Youxing was one of the earliest TCM experts in the warm disease school during the Ming and Qing Dynasties. Secondly, "since ancient times, there was no specialized book solely focusing on pestilence, but *Treatise on Pestilence* was the pioneering and definitive work on this subject, establishing it as the first of its kind".[30] The academic thoughts of Wu Youxing and his masterpiece *Treatise on Pestilence* can be summarized as follows:

(1) Warm disease was pestilence. In the past, the concepts of warm disease, "tian xing", pestilence, and similar terms were quite ambiguous. Wu Youxing aimed to clarify this conclusion by asserting that warm disease was essentially pestilence. This perspective was closely related to his extensive clinical experience in treating numerous epidemics and infectious diseases. However, it is important to note that the scope of warm disease is broader than that of pestilence, making it inappropriate to equate the two terms directly.

(2) The special cause of pestilence: Wu Youxing believed that pestilence was primarily caused by the invasion of pathogenic factors through the mouth and nose, independent of the external climate. This was an accurate observation, as it correctly identified one mode of transmission for epidemics. However, from the perspective of etiology, this assumption has its limitations. Pathogenic microorganisms, the source of infection, exist consistently regardless of time and place. Yet, why do some individuals fall ill while others remain unaffected under the same environmental conditions? As an ancient adage goes, "Where pathogen invades, there is qi deficiency", "When an individual possesses an abundance of healthy qi internally, the pathogen is unable to cause illness". Modern medicine attributes this difference to the immune system's role in resistance. However, these explanations only provide a partial understanding of the broader factors involved. Wu Youxing noticed an important phenomenon: "Among individuals severely affected by a pestilence, a visible symptom is the presence of a thick, powdery tongue fur, covering the entire tongue with no gaps. In milder cases, the tongue coating is thinner and white, indicating less internal heat".[31] However, he mistakenly attributed the thickness of the tongue fur solely to the severity of the pathogen invasion, overlooking its significance as an indicator of the disease's cause.

(3) There are nine distinct ways in which pestilence can transmit and evolve. These include transmission solely via the exterior, transmission via the exterior

---

[30] *History of the Qing Dynasty* (*Qing Shi Gao*, 《清史稿》), Beijing: Zhonghua Book Company, 1977, Volume 520, p13867.

[31] Wu Youxing (the Qing Dynasty), Chapter "Beginning of Pestilence" in the first volume of *Treatise on Pestilence*.

> followed by another round of exterior transmission, transmission solely via the interior, transmission via the interior followed by another round of interior transmission via both the exterior and interior separately, transmission via both the exterior and interior separately followed by another round, transmission where the exterior is more prominent or the interior is more prominent, transmission initially via the exterior and then the interior, and vice versa. However, a closer examination of these "nine transmissions" reveals that they can be summarized in two key concepts: exterior and interior. The therapeutic principles are also straightforward. That is, the pathogen in the exterior shall be cleared, and the pathogen in the interior shall be purged.

Wu Youxing, a physician known for his speculative thinking, was greatly influenced by the concepts of "qiong li" and "ge wu" from the Neo Confucianism of the Song and Ming Dynasties. He meticulously investigated the root causes of the plague, delving into the pathogenic factors, examining the routes of infection, scrutinizing the affected areas of the body, and thoroughly understanding the modes of transmission.[32] He noted that goats exhibit immunity to the pathogenic factor affecting cows, chickens resist the pathogenic factor impacting ducks, and animals generally display resilience against the pathogenic factor affecting human beings. Finally, different creatures are susceptible to different pathogenic factors due to variations in their qi.[33] If Wu Youxing had access to a microscope during his time, he might have found out the essence of the pathogenic qi under its different forms along this path. Furthermore, he postulated that by identifying the pathogenic qi and its corresponding medication, the disease could be treated effectively with the specific medication, making complex formulas unnecessary.[34] To some extent, this assumption became a reality with the introduction of antibiotics. Generally, Wu Youxing's way of thinking aligns surprisingly well with modern Western medicine. However, his logical thinking was not fully developed, leading to his misunderstanding of the causal relationship between the observed phenomenon of the white-flour-like tongue fur and his assumption that the pathogen invades people via the mouth and nose. Thus, he incorrectly concluded that "the pestilence is caused by miscellaneous qi, and no specific solution has been found yet. In this case, physicians have no choice but to utilize therapeutic methods including promoting sweating, vomiting, and purgation".[35] This statement fundamentally deviates from the essence of traditional Chinese medicine. If Wu Youxing had access to various antibiotics, he would have understood that the approach of using one specific medication for each type of pestilence would be ineffective if the underlying cause, as indicated by white-flour-like tongue fur, is not addressed. Consequently, he would have recognized that his assertion that physicians must resort to methods such as diaphoresis, emesis, and purgation in treating pestilence is flawed.

---

[32] Wu Youxing (the Qing Dynasty), *Treatise on Pestilence*, Preface.

[33] Wu Youxing (the Qing Dynasty), *Treatise on Pestilence*, Volume 2.

[34] Wu Youxing (the Qing Dynasty), *Treatise on Pestilence*, Volume 2.

[35] Wu Youxing (the Qing Dynasty), *Treatise on Pestilence*, Volume 2.

13.3.3 Ye Gui and the System of Defense, Qi, Nutrient, and Blood Pattern Identification

Ye Gui, styled Ye Tianshi and named Xiangyan, was a native of Wuxian county, Jiangsu province. According to research, he lived during the reigns of Kangxi and Qianlong of the Qing Dynasty (AD 1667–1746).[36] Ye came from a family of TCM practitioners, with his father and grandfather both being TCM doctors. Upon the death of his father, he studied under Zhu, who was a disciple of his father. Whenever he encountered an accomplished physician, he would seek to learn from them. In total, he had 17 teachers, and his enthusiasm for learning TCM and his exceptional medical expertise gave rise to numerous folk stories. Ye devoted his entire life to practicing medicine. His academic ideas are preserved in the book *Discourse on the Treatment of Warm Disease* (*Wen Zheng Lun Zhi*, 《温证论治》), which was compiled by his follower Gu Jingwen and further refined by Tang Dalie.[37] It serves as the primary source for studying Ye Gui's academic thought and contributions in the field of TCM.

Ye Gui formulated a four-stage classification system for the progression of warm disease, "wei" (defense), qi, "ying" (nutrient), and "xue" (blood). Building upon these four stages, he developed his own system of pattern identification. "Generally, the progression of warm disease moves successively from the "wei" (defense) stage to the "xue" (blood) stage. In the "wei" (defense) stage, treatment involves inducing sweating; at the qi stage, the focus is on clearing the qi. At the "ying" (nutrient) stage, the pathogen may still be drawn back to the qi aspect, and treatment may include medications such as Xi Jiao (Rhinoceros Horn), Xuan Shen (Scrophularia ningpoensis Hemsl), and Ling Yang Jiao (Cornu Saigae Tataricae). At the "xue" (blood) stage, the treatment aims to cool the blood and disperse blood stasis to prevent pathological blood consumption or bleeding. Ye Gui further explained that "when the warm disease pathogen invades the body, the lung is the first to be affected", then comes the "reverse transmission" to the pericardium. He recognized that the lung governs qi and pertains to the "wei" (defense) aspect; while the heart controls blood and belongs to "ying" (nutrient) aspect. The pathogenesis of warm disease shares similarities with cold damage, but their therapeutic methods and principles are quite different".

During Ye Gui's time, his medical insights were not widely recognized due to two main reasons. Firstly, Ye believed that pattern identification and treatment at each stage of warm disease were more important than theoretical discussions. Secondly, Ye's masterpiece *Discourse on the Treatment of Warm Disease* first appeared in *Collection of Papers of Physicians from Wu Areas* (*Wu Yi Hui Jiang*), a journal established by Tang Dalie in 1792 AD. Therefore, not all of Ye's contemporaries were familiar with his medical thoughts. Actually, it was in later generations that his medical insights gained recognition as a fundamental framework for pattern

---

[36] Guo Aichun, *Chronology of Chinese Medical History* (*Zhong Guo Yi Shi Nian Biao*, 《中国医史年表》), Heilongjiang People's Publishing House, 1984, p165, p185.

[37] Tang Lishan (the Qing Dynasty), *Wu's Compiled Medical Discourse.*

identification in warm disease. As a clinical therapist, Ye Gui often prescribed sets of medications accordingly without providing specific prescription names. Another renowned TCM expert, Wu Tang, developed numerous famous formulas based on Ye's medical records, such as Sang Ju Yin (Mulberry Leaf and Chrysanthemum Decoction), which was derived from Ye's treatment of Qin's wind-warmth case. Similarly, Qing Gong Tang (Palace-Clearing Decoction) was based on Ye's treatment of Ma's warm heat disease, and Lian Mei Tang[38] (Coptis teeta, Fructus Mume Decoction) was based on Ye's treatment pf Gu's summer-heat disease.

13.3.4 Wu Tang and the System of Triple Energizer Pattern Identification

Wu Tang, also known as Wu Jutong, was from Huaiyin county in Jiangsu province. Born in 1758 (the 23rd year of Emperor Qianlong's reign), he passed away in 1836[39] (the 16th year of Emperor Daoguang's reign). Wu Tang assimilated a diverse range of academic thoughts spanning from ancient times to his era, drawing insights from sources like *Plain Questions*, Zhang Zhongjing, Wu Youxing, and Ye Tianshi. He asserted that a true understanding of cold damage reveals the effectiveness of Zhang Zhongjing's therapeutic methods, emphasizing that understanding warm diseases exclude using pungent-warm treatments.[40]

Wu Tang categorized the progression of warm disease into three distinct stages: upper, middle, and lower stages. This conceptual framework forms the theoretical foundation of his triple energizer pattern identification system.

> Warm diseases enter the body through the mouth and nose, with nasal breathing connecting to the lung and oral breath connecting to the stomach. Therefore, they initially affect the lung. If the lung impairment progresses reversely, it affects the pericardium. The warm disease primarily targets the upper energizer. Without proper treatment, it progresses to involve the middle energizer (the stomach and spleen). Failure to cure the warm disease at the middle energizer stage results in its spread to the lower energizer (the liver and spleen). In summary, the warm disease sequentially impacts the upper energizer first, with the lower energizer being affected last.[41]

Building upon this theoretical framework, he discussed the progression and pathological changes of 11 warm diseases, categorizing them into upper, middle, and lower stages, including wind-warmth, warm heat, pestilence, warm toxin, summer-heat-warmth, dampness-warmth, autumn-dryness, winter-warmth, warm malaria,

[38] Ren Yingqiu, *The Doctrine of the Schools of Traditional Chinese Medicine* (*Zhong Yi Ge Jia Xue Shuo*, 《中医各家学说》), p142.

[39] Editorial Board of China Encyclopedia of Medicine: "Medical History" volume in *China Encyclopedia of Medicine*, p146.

[40] Wu Tang (the Qing Dynasty), *Systematic Differentiation of Warm Disease* (Wen Bing Tiao Bian, 《温病条辨》), Volume 1.

[41] Wu Tang (the Qing Dynasty), Systematic Differentiation of Warm Disease, Volume 2.

latent summer-heat, and cold-dampness. Over time, subsequent generations of physicians primarily used two systems for diagnosing warm diseases: Ye Gui's system of defense, qi, nutrient, and blood pattern identification and Wu Tang's system of triple energizer pattern identification. These two systems form the theoretical foundation for pattern identification and treatment in the warm disease school.

13.4 Similarities and Differences of Historical Conditions between Cold Damage School and Warm Disease School

The Cold Damage school, which emerged in the late Eastern Han Dynasty, is primarily based on Zhang Zhongjing's *Treatise on Cold Damage Diseases*. The Warm Disease school was formed during the Ming and Qing Dynasties, with physicians such as Wu Youxing and Ye Tianshi at its core. Notably, the two schools are separated by a significant time gap of approximately 1500 years. Examining the similarities and differences in the objective historical conditions that led to their emergence can greatly deepen our understanding of their respective treatment systems.

13.4.1 Both Schools Are Developing with the Times

Pestilence has been documented throughout the various dynasties of ancient China. Its prevalence was particularly notable during the late Eastern Han Dynasty to the Three Kingdoms period (second to third centuries), and the late Ming Dynasty to the Qing Dynasty (sixteenth to eighteenth centuries). As shown in the figure, the frequency of pestilence exhibits an upward trend from ancient to modern times, which can be attributed to the greater availability of historical materials in more recent periods. However, it is important to note that the outbreak frequency in the third century ranked highest among the centuries before the eleventh century, indicating that the curve trend is influenced by the passage of time and the abundance of records, but is not solely determined by them. The cold damage school and warm disease school are two major treatment systems for externally contracted diseases (or heat diseases, including epidemics). The two systems appeared during the respective peak periods illustrated in the aforementioned curve, indicating the close relationship between objective needs and the evolution of medical practices (Fig. 13.5).

For example, the Biography of Wang Xiu of the Wei Dynasty in *Records of the Three Kingdoms* presents that Wang Xiu, who was twenty years old, traveled to Nanyang and stayed with Zhang Feng. Zhang Feng's family was suffering from a serious illness, with no one to care for them. Wang Xiu showed great empathy and personally nursed them back to health, after which he departed. This suggests that the epidemic had affected the Zhang family of Nanyang. Wang Xiu's visit to

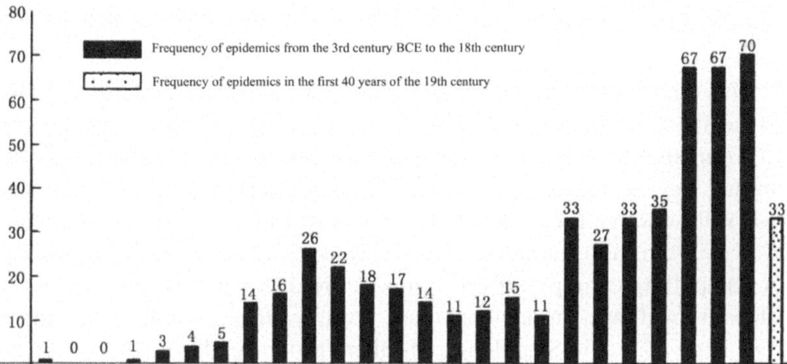

	B.C.							A.D.																		
Century	7	6	5	4	3	2	1	1	2	3	4	5	6	7	8	9	10	11	12	13	14	15	16	17	18	19
Number of epidemics	1	0	0	1	3	4	5	14	16	26	22	18	17	14	11	12	15	11	33	27	33	35	67	67	70	33

Fig. 13.5 Comparison of frequency of pestilence in different centuries

Nanyang was likely in the third year of Jianning (170). In his preface to *Treatise on Cold Damage and Miscellaneous Diseases*, Zhang Zhongjing expressed deep sorrow for the mass deaths caused by previous pandemic in his hometown. Similarly, Wu Youxing's preface to *Treatise on Pestilence* described the major epidemic of the Xin Si year in Chongzhen reign (1641), which affected both northern and southern Shandong and Zhejiang provinces, particularly in May and June, causing widespread infections among households. These accounts illuminate the relationship between cold damage, warm disease, and past epidemics.

13.4.2 Climate Change and Pestilence

What factors contributed to the two peaks of pestilence in the late Eastern Han Dynasty and the Ming and Qing Dynasties? While it is commonly believed that the pestilence is more likely to occur after periods of war, statistical analysis reveals that post-war outbreaks only accounted for approximately a quarter of the total cases during those times.[42] Moreover, during the Kangxi, Yongzheng, and Qianlong reigns (1662–1795) of the Qing Dynasty, which were considered "peaceful and prosperous", the incidence of epidemics remained high. This suggests that the two outbreaks cannot be explained solely by war-related factors. However, by comparing the historical shifts in climate between cold and warm periods in ancient China with the curve depicted in the figure, valuable insights can be gained. To shed light on this, let's

[42] Zhang Zhibin, A Study on Various Factors of Epidemics in Ancient Times, *Chinese Journal of Medical History* (*Zhong Hua Yi Shi Za Zhi*, 《中华医史杂志》), 1990, (1): 28.

consider Mr. Zhu Kezhen's overview of "climate change over the past five thousand years".[43]

(1) During the period of Yangshao Culture (5000–3000 BC) and the era of the Ruins of the Shang Dynasty (1319–1046 BC), the climate in northern China, especially in Xi'an and Anyang areas, was notably mild. Abundant subtropical plants and animal species flourished in these regions. Remarkably, around 3000 years ago, the Yellow River Basin, located in northern China, experienced a climate that was as warm and humid as today's Yangtze River Basin in southern China. A compelling example of this temperature difference is the migration pattern of barn swallows. In ancient times, barn swallows would fly as far north as Tancheng county in Shandong, a northern province, during the Spring Equinox, while now they only reach Shanghai. This shift in their migration indicates a noticeable change in temperature. (Note: The mild climate during this era served as a critical natural foundation for the emergence of the Yellow River culture in the Central Plains region. Because humans have the distinction of having the least amount of body hair among mammal species. If the climate was cold during that period, and people did not have warm clothes, survival in the low temperatures would have been challenging, let alone the thriving and development of the Yellow River culture.)
(2) The tenth century BC marked a cold period in China's climate history.
(3) Following that, a warm period ensued, including the Spring and Autumn and the Warring States Period (770–221 BC), the Qin Dynasty, and the Western Han Dynasty.
(4) However, a noticeable shift occurred from the beginning of the A.D. era, as the climate in China gradually became colder. By the time of Cao Cao (155–220 AD) (Cao Cao was a renowned figure in the Three Kingdoms period), the climate had become colder compared to the reign of Emperor Wu (141–87 BC) of the Han Dynasty. In 225 AD, the freezing of the Huaihe River was recorded for the first time, marking a significant indicator of the cooling climate. This cold period reached its peak in the first half of the fourth century, with the average annual temperature being about 2–4 °C lower than that of today.
(5) However, a shift back to warmer climate conditions occurred during the Sui and the Tang Dynasties (581–907 AD). By the middle of the seventh century, China experienced a mild climate, where plum trees and orange trees in Xi'an flourished and producd abundant fruits.
(6) As the eleventh century began, another cooling trend emerged, leading to a significant drop in temperature by the early twelfth century. In 1111 AD, the Taihu Lake was frozen over with such substantial and hardened ice that vehicles could traverse its surface (Taihu Lake spanned Jiangsu and Zhejiang provinces, and covered an area of 2250 square kilometers.). Meanwhile, in Fuzhou (capital of Fujian province), litchi trees suffered fatal damage from freezing twice during the twelfth century.

[43] Zhu Kezhen, "A Study on Climate Change in China in the Past Five Thousand Years" in *The Works of Zhu Kezhen* (*Zhu Kezhen Wen Ji*, 《竺可桢文集》), Beijing: Science Press, 1979, p.475.

(7) The climate was generally warm from the early to the middle of the thirteenth century, after which it transitioned back to colder conditions.
(8) From the fifteenth century to the twentieth century, the coldest climate was observed in the seventeenth century, especially from 1650 to 1700 AD. During this period, Jiangxi province's orange trees suffered severe damage from two cold waves, resulting in complete destruction. Notably, Taihu Lake, the Hanshui River, and the Huaihe River each froze four times, while Dongting Lake froze three times. Even the large and southern Poyang Lake froze. Even in the tropical regions of southern China, ice and snow became frequent during this half-century period.

As depicted in the figure, two significant peaks of pestilence can be observed in the history of China. The first one occurred during the second and third centuries, coinciding with the cold period that started in the latter half of the second century. The second peak took place during the sixteenth to eighteenth centuries, coinciding with the cold period centered around the seventeenth century. In the first period, the cold damage school emerged in response to the prevailing conditions, as documented in the *Treatise on Cold Damage and Miscellaneous Diseases*. Conversely, during the second period, the Ming and Qing Dynasties witnessed the emergence of the warm disease school, which focused on diseases associated with warmer climates.

Now, let's examine the research report on the climate of the middle and lower reaches of the Yellow River during the Qin and the Han Dynasties.[44] Firstly, it is notable that the Han Dynasty recorded significantly more years of cold climate in the middle and lower reaches of the Yellow River Basin. There were 29 traceable years and 6 untraceable years of cold climate during this period. In contrast, there were only 13 traceable years and 1 untraceable year of hot climate. Hot weather was relatively more common in the second century B.C., but the frequency of hot weather declined significantly in Chang'an (today's Xi'an) thereafter. This deviation from the general trend, where records typically become simpler further back in time, is an interesting observation. Typically, low temperatures are more prevalent in later periods compared to earlier ones. However, in the first century, cold climate was more predominant in the middle reaches of the Yellow River and Chang'an (Xi'an), whereas in the second century, the lower reaches of the Yellow River and Luoyang experienced more frequent cold climate. When it comes to the extent of damage, hot climate was generally less severe. Only one record, dating to 107 B.C., mentions "many people died of heatstroke". In contrast, low temperature had a far more devastating impact, as evidenced by records of events such as "premature frost", "heavy snow reaching a depth of over 11 feet", and "snow piling up to a depth of more than 11 feet". These extreme cold conditions resulted in damage to crops, reduced grain production, the death of mulberry trees leading to a decline in silk farming, the death of wildlife, and even the occurence of famines.

---

[44] *Wen Huanran, Study on the climate of the middle and lower reaches of the Yellow River in the Qin and Han Dynasties* (*Qin Han Shi Dai Huang He Zhong Xia You Qi Hou Yan Jiu*, 《秦汉时代黄河中下游气候研究》), Beijing: The Commercial Press, 1959, pp. 40–54.

By examining the severe cold climate and the resulting damages such as famines caused by abnormal cold waves during the late Eastern Han Dynasty, it is easier to understand the sudden shift—from heat to cold, and from excess to deficiency—in the treatment system for externally contracted diseases during that period. Given the primary pathogenic factors of that era, "famine" and "cold" were likely the most significant objective factors. Therefore, according to *Treatise on Cold Damage and Miscellaneous Diseases*, medications with pungent-warm properties are used to release the exterior and dissipate cold. Additionally, the text recommends taking hot porridge after consuming the decoction "to assist the medicinal efficacy" and includes the addition of maltose to warm the middle and dissipate cold as auxiliary methods These methods were particularly relevant and effective for treating famine-stricken patients. Otherwise, they might not have been as necessary or effective. Regarding the description in *Treatise on Cold Damage and Miscellaneous Diseases*: "lesser yin disease is characterized by faint and feeble pulse and sleepiness". However, nowadays, these symptoms are typically observed in the elderly and the weak. Therefore, the six-meridian patterns described in the text are merely common characteristics among patients suffering from famine and cold. During the Ming and the Qing Dynasties, social conditions were quite different, which led physicians of the warm disease school to encounter distinct disease characteristics.

In addition, population density is another important factor contributing to the differences between the cold damage school and the warm disease school. From the Western Han Dynasty to the early Ming Dynasty in ancient China, the population remained relatively stable at about 50 million, which undoubtedly facilitated the control of various infectious diseases. However, the situation changed by the middle of the Qing Dynasty when China's population exceeded 100 million, and it reached 300 million by the end of the Qing Dynasty.[45] This rapid growth created a different scenario, as the larger population provided necessary "carriers" for the spread and continuation of numerous diseases. The more economically developed regions in southern China, especially in Jiangsu and Zhejiang provinces, attracted a significant influx of migrants from the northern regions, resulting in a drastic increase in population density.[46] Therefore, Jiangsu province ranked first in the frequency of pestilence outbreaks with 72 incidents, followed by Zhejiang province with 68 incidents. The frequency of pestilences outbreaks in these two provinces was 2–3 times higher than that of Jiangxi (37 incidents), Fujian (24 incidents), Anhui (23 incidents), and other provinces in the southern region of the Yangtze River. Experts of the warm disease school primarily practiced medicine in the densely populated areas of Jiangsu and Zhejiang, where pestilence was commonly observed. Taking all of these factors into consideration, the differences between the cold damage school and the warm disease school become more apparent.

---

[45] Liang Fangzhong, *Statistics of Population, Farmland, and Farm Tax in China's Past Dynasties* (*Zhong Guo Li Dai Hu Kou, Tian Di, Tian Fu Tong Ji*, 《中国历代户口、田地、天赋统计》), Shanghai: Shanghai People's Publishing House, 1980, pp.4–11.

[46] Liang Fangzhong, *Statistics of Population, Farmland, and Farm Tax in China's Past Dynasties*, Shanghai: Shanghai People's Publishing House, 1980, pp.273–279.

Chapter 14
Relationship Between the Establishment of Basic Theories and Morphological Cognition (Anatomy)

The exploration of life's mysteries can be approached from two perspectives: the study of its form and function. One is to research and map the physical structures of animals or plants, while the other is to figure out how these structures work through experiments.[1]

Even at an early stage, the scientific study of life has relied on two key elements: observation and investigation with precise measurement and equipment. In contrast, TCM takes a different approach. While "observation" is also a crucial aspect, TCM's understanding of function is primarily based on imagination rather than experiments.

In the early TCM classics, there are records and descriptions of some viscera, bones, and muscles of the human body. For example, the chapter "Intestines and Stomach" in *Miraculous Pivot* (*Ling Shu*, 《灵枢》) records that the length of the human esophagus is 1.6 chi (a unit of length, 1 chi $\approx$ 3.33 m), while the total length of large and small intestines is 56.8 chi. Although the measurement system used in ancient times was different from that of today, the ratio between these two parts (1:35.5) aligns reasonably well with the ratio found in modern anatomical studies (1:37), implying that the objectivity of these ancient records. Therefore, in the 1950s, scholars highlighted that ancient Chinese people possessed some knowledge of human anatomy.[2] However, it seems that there has been a lack of recognition regarding the relationship between this ancient anatomical knowledge and the establishment of the theoretical system of TCM. As we all know, anatomy serves as a fundamental pillar in modern Western medicine. Yet, TCM lacks anatomy as its

[1] Whitfield (UK), translated by Fan Yizu, *Illustrated History of Science and Technology in the World* (*Cai Tu Shi Jie Ke Ji Shi*, 《彩图世界科技史》), pp. 364.

[2] Huang Shengbai, "Human Anatomy of Chinese People 2000 years Ago", *Journal of Traditional Chinese Medicine* (*Zhong Yi Za Zhi*, 《中医杂志》), 1955, (4): 42–43; Liu Guangzhou, "Understanding of Anatomy and Physiology in Chinese Medicine", *Journal of Traditional Chinese Medicine*, 1955, (1): pp. 49–51; Hou Baozhang, "History of Anatomy in China", *Medical History and Health Care Organization* (*Yi Xue Shi Yu Bao Jian Zu Zhi*, 《医学史与保健组织》), 1957, (2): 64–73.

Y. Liao, *Reconstructing the Image of Medicine in the Qin and Han Dynasties*, Understanding China, https://doi.org/10.1007/978-981-95-6801-7_14

explicit theoretical foundation, rendering it distinctive in this regard. It reveals that the previous studies of anatomy in ancient China were primarily focused on exploring the accomplishments themselves,[3] neglecting the deeper connection between ancient anatomical knowledge and TCM theories.

By examining the relevant contents of classical medical books, it becomes evident that, although the theoretical system of TCM differs significantly from modern Western medicine, its key theories are still connected to the anatomical knowledge of that time. This perspective not only helps us understand the development process of TCM theories but also provides insights into resolving several longstanding contentious issues.

14.1 Functions of the Gallbladder and the "Extraordinary Organs"

The theory of Zang-Fu organs is an important component of TCM basic theory. Although the concepts of Zang-Fu organs in TCM do not directly correspond to the organs in modern medicine, their basic concepts are undoubtedly rooted in the understanding of visceral entities. Before the Western Han Dynasty (202 BC-AD 8), there was no unified concept regarding the concepts of viscera and bowels. For example, in the *Zhuang Zi* (《庄子》), there are theories of five Zang-organs[4] and six Zang-organs.[5] In the *Plain Questions* (*Su Wen*), there are theories of nine Zang-organs, eleven Zang-organs, and twelve Zang-organs.[6] Additionally, different perspectives considered various organs as Zang-organs, Fu-organs, or both. For example, some considered the brain and marrow as Zang-organs; while others considered the intestines and stomach as Zang-organs; and even as Fu-organs.[7] The chapter "Further Discourse on the Five Depots[8]" in *Plain Questions* first introduces criteria to classify the viscera and bowels. It states that as for the so-called five Zang-organs, they store the essence qi and do not drain it. Hence, even if they are full, they cannot be

[3] Ma Jixing, "Anatomic Diagram of Human Body in the Song Dynasty", *Medical History and Health Care Organization*, 1957, (2): pp. 125–128. Wei Yizong: *History of Chinese Orthopedic Techniques* (*Zhong Guo Gu Ke Ji Shu Shi*, 《中国骨科技术史》), Shanghai: Shanghai Scientific and Technological Literature Press, 1983, 18–23.

[4] Guo Xiang (the Jin Dynasty) annotator, the Outer Chapter "The Eighth Section of Extra Discussion", "The Eleventh Section of Just Being There" in *Commentaries on "Zhuang Zi"* (*Zhuang Zi Zhu*, 《庄子注》), Volume 4.

[5] Guo Xiang (the Jin Dynasty) annotator, the chapter "Second Section of Discussion on Equality of All Things" in *Commentaries on "Zhuang Zi"*, Volume 1.

[6] It can be seen in the chapter "Discourse on the Six Terms of a Year and on Phenomena Associated with the Condition of the Depots" and "Discourse on the Hidden Canons in the Numinous Orchid Chambers" in *Plain Questions*.

[7] The chapter "Further Discourse on the Five Depots" in *Plain Questions*.

[8] "Depots" refer to "Zang-organs" in Paul U. Unschuld's translation version of *Plain Questions*. He also uses "palaces" to explain "Fu-organs".

Table 14.1 Classification of the Zang-Fu organs

Classification	Organs	Criteria
Zang-organs	Heart, Liver, Spleen, Lung, Kidney	They store the essence qi and do not drain it. Hence, even if they are full, they cannot be replete
Fu-organs	Stomach, Large Intestine, Small Intestine, Bladder, Triple Energizer, Gallbladder	They transmit and transform things, but do not store them. Hence, they may be replete, but they cannot be full
Extraordinary Organs	Brain, Marrow, Bone, Blood Vessel, Gallbladder, Uterus	They store and do not drain

replete. These include the heart, liver, spleen, lungs, and kidneys. And it also records that as for the six Fu-organs, they transmit and transform things but do not store them. Hence, they may be replete, but they cannot be full, including the stomach, large intestine, small intestine, bladder, triple energizer, and gallbladder. Besides, those so-called "extraordinary organs" include the brain, marrow, bones, vessels, gallbladder, and uterus. They "store and do not drain". (As shown in Table 14.1).

We can see that the theory of Zang-Fu organs is based on morphology. However, understanding how the explanation of function goes beyond the specific organ itself is a different matter. It is worth noting that according to the above classifications, the gallbladder appears in both categories of Zang-organs and the extraordinary organs, which requires an explanation considering both its morphological and functional characteristics. The classification standards in *Plain Questions* primarily rely on the functions of the organs, and the understanding of these functions is rooted in morphological observations. For example, regarding the Fu-organs, "When water and grain enter the mouth, then the stomach is replete and the intestines are empty. When the food moves down, then the intestines are replete and the stomach is empty. Hence the text states: replete but not full".[9]As for the description of the stomach, it is more anatomically oriented. In the chapter "On Swelling" in *Miraculous Pivot*, the stomach is depicted as having "five orifices". In the Ming Dynasty (AD 1368–1644), Zhang Jingyue interpreted the five orifices as "pharynx, cardia and pylorus, ileocecal junction and outlet for discharging dregs (i.e., the anus)".[10] However, the same chapter mentions that the Zang-Fu organs have their own boundaries. If the five orifices extend from the throat to the anus, they would clearly exceed the anatomical boundaries of the stomach. Therefore, Zhang's interpretation appears to be inaccurate in this context. To better understand the "five orifices" of the stomach, it is important to consider that the duodenum was not recognized in ancient anatomical classifications. Therefore, the five orifices likely refer to the upper and lower openings of the stomach, as well as the openings of the liver, gallbladder, and pancreas into the duodenal

[9] The chapter "Further Discourse on the Five Depots" in *Plain Questions.*

[10] Zhang Jiebin (the Ming Dynasty), the chapter "Distension of the Zang-Fu Organs" in *Classified Classic* (*Lei Jing*, 《类经》), Volume 16.

region. Obviously, this knowledge was derived from ancient anatomical practices, which directly contributed to the establishment of medical theories.

At that time, although people could observe through dissection practices that the liver, gallbladder, and pancreas were interlinked to the stomach and intestines via ducts, but they lacked the understanding of the specific digestive roles played by bile and the pancreas. For instance, although they were aware of a condition similar to diabetes called "consumptive thirst" and had various treatments for it, they did not understand its connection to the pancreas or know what insulin was. Due to their subjective reasoning, ancient medical physicians mistakenly regarded those ducts as ways of absorbing food essence. They thought that "the qi and blood produced by the stomach flowed through conduit-channels[11]". It was believed that the gallbladder should contain abundant "essential liquid"[12] which should be stored and not drained. If the gallbladder was flabby and the essential liquid was insufficient, it would be referred to as the "cowardly male.[13]" "If the gall liquid flows off, then one has a bitter flavor in his mouth".[14] Thus, in ancient China, there was a lack of understanding that the bile is produced by the liver and aids in the digestion of lipids in the small intestine. Instead, they thought it was produced by the stomach and intestines. Even in the Qing Dynasty (AD 1636–1912), Wang Qingren, a renowned anatomist, adhered to this misconception. In his Zang-Fu diagram, he depicted the opening of the common bile duct and named it the "fluid gate". It shows that although ancient anatomical knowledge was rudimentary, it laid the foundation for the establishment of TCM theories.

Similarly, the distinction of the gallbladder from other Zang-Fu organs is noticeable in *Plain Questions*. While the other five of the six Fu-organs are named "containers", they are described as "able to transform waste and facilitate the process of taste and digestion, allowing substances to enter and exit".[15] Because the gallbladder has the general characteristics of the Fu-organs in morphology, and what it contains is the essence "that could store and not drain", which is different from the other five Fu-organs, it is also called "extraordinary organs" along with the brain, marrow, bones, vessel, and uterus with similar properties.

Thus, it is evident that the theory of Zang-Fu organs is built upon the understanding of their forms and functions. The knowledge of their morphology stems

---

[11] The chapter "The Jade-Tablets" in *Miraculous Pivot* (translation version: Paul U. Unschuld, *Huang Di Nei Jing Ling Shu*, The Ancient Classic on Needle Therapy. The complete Chinese text with an annotated English translation).

[12] The chapter "The Forty-second Difficulty" in *Classic of Difficulties* (*Nan Jing*, 《难经》).

[13] The chapter "On Courage" in *Miraculous Pivot* (translation version: Paul U. Unschuld, *Huang Di Nei Jing Ling Shu*, The Ancient Classic on Needle Therapy. The complete Chinese text with an annotated English translation).

[14] The chapter "The Four Seasonal Qi" in *Miraculous Pivot* (translation version: Paul U. Unschuld, *Huang Di Nei Jing Ling Shu*, The Ancient Classic on Needle Therapy. The complete Chinese text with an annotated English translation).

[15] The chapter "Discourse on the Six Terms of a Year and on Phenomena Associated with the Condition of the Depots" in *Plain Questions* (translation version: Paul U. Unschuld, *Huang Di Nei Jing Su Wen*. An annotated translation of *Huang Di's Inner Classic-Basic Questions*).

from anatomy, while the understanding of functions relies on imaginative thinking. During that era, the visual observation of morphology and anatomy, coupled with the relatively straightforward transport function of the gastrointestinal tract, enabled people to gain a solid understanding and reliable data. However, it was impossible for them to reach conclusions that align with modern medicine when it came to aspects that require explanations based on modern biochemistry, glandular secretion, and other advanced knowledge. In essence, without a preliminary grasp of morphology and anatomy in ancient China, the concept of "extraordinary organs" would not have emerged in TCM. Moreover, without examining the differences from a historical perspective, we shall never truly comprehend how the understanding of organ function differs between ancient and contemporary people.

14.2 Relationship Between the "Triple Energizer" and Anatomy

The term "triple energizer" first appeared in the received version of *Huangdi's Internal Classic* (*Huang Di Nei Jing*, 《黄帝内经》), which explicitly identifies the triple energizer as one of the human organs, grouping it together with the gallbladder, stomach, large intestine, small intestine, and bladder as part of the "six Fu-organs". While due to the lack of clear descriptions of its shape in the text, there have been debates in the medical field over the past two thousand years regarding its shape and nomenclature. Various explanations emerged, such as the theory of invisible triple energizer,[16] the theory of cavity-based triple energizer, the theory of stomach-related triple energizer, the theory of an omentum-related triple energizer, and the three-section theory of the triple energizer.[17] By the Qing Dynasty, the concept of the triple energizer became even more ambiguous when the school of warm diseases used it to delineate different stages of heat diseases.

Ren Yingqiu, a modern scholar, has raised concerns about the illogicality of the theory regarding the invisible nature of the triple energizer. The theory has recognized that the triple energizer is a Fu-organ with functions related to qi movement and water dredging, so it is inaccurate to claim that the triple energizer is intangible… in comparison, the cavity theory seems more reasonable. Ren further noted that "Tang Zonghai (1862–1918) may have initially been influenced by the cavity theory,

[16] The theory of "invisible triple energizer" is derived from the *Classic of Difficulties*. For example, in the chapter "The Twenty-fifth Difficulty", it says that the heart-master and the triple energizer represent the outside and inside. Both have a name but no form. In the chapter "The Thirty-eighth Difficulty", it says that it is like this. The existence of the triple energizer is to be named as the reason for the fact that there are six Fu-organs. The triple energizer represents an additional source of original influences; it governs all the influences circulating in the body. It has a name but no form. It is associated with the hand-lesser-yang conduit. This is an external Fu-organ. Since then, many people in later generations have followed this theory.

[17] For details, please refer to *The Doctrine of the Schools of Traditional Chinese Medicine* (*Zhong Yi Ge Jia Xue Shuo*, 《中医各家学说》) by Ren Yingqiu, pp. 186–190.

as he regarded the interconnected omentum with the coelomic cavity as the triple energizer". Tang's ideas were also influenced by anatomy and physiology of western medicine. Due to the limitations of western medicine at that time and Tang's lack of scientific knowledge, many of his arguments may appear somewhat specious......In short, it is difficult to achieve a seamless integration through textual analysis alone.[18]

Ren's observations regarding the flaws in the invisible theory resonate with many researchers. But the challenge lies in clarifying the shape and quality of the triple energizer. That is why Ren did not arrive at a reasonable conclusion after analyzing various theories put forth by different scholars. Moreover, using modern medical theory to study ancient concepts presents significant difficulties when attempting to uncover the truth. It is important to consider the limitations and immaturity of nineteenth-century Western medicine as well as the scientific knowledge available to Tang Zonghai. How could anatomy and physiology have matured thousands of years ago? If we just simply apply modern medical theories without caution to interpreting the ancients, it would be unavoidable to make mistakes and oversights. If we were to imagine what ancient physicians might discover while examining the structures of the pleuroperitoneal cavity, it becomes evident that in addition to the heart and lungs, the pericardium could also be observed in the pleural cavity. This observation might have led to the identification of a vessel known as the vessel of the pericardium (Hand Heart Governor), and which was believed to bear pathogens for protecting the heart. In the abdominal cavity, a large amount of "Jiao (energizer)" could also be observed, in addition to the stomach, large intestine, small intestine, liver, spleen, kidney, bladder, and uterus.

The term "Jiao" in ancient texts refers to empty or hollow flesh. In Gao You's interpretation in the chapter "Discussion on Astronomy" in *Writings of Prince Huainan* (*Huai Nan Zi*, 《淮南子》), "Jiao" is interpreted as hollow flesh. Similarly, the chapter "Root and Connection" in *Miraculous Pivot* mentioned that "The body hair is broken short, and the skin structures are scorched", which also alludes to the hollow flesh. So, it can be understood that the triple energizer is characterized by this specific shape and quality. However, some people mistakenly interpreted the term "Jiao" as fire, leading to the misconception that "triple energizer refers to lesser yang and minister fire". But it is important to note that "the San Jiao (triple energizer, 三焦)" was written as "Jiao (膲)" in ancient times in various texts, including *Miraculous Pivot*, *Pulse Classic* (*Mai Jing*, 《脉经》), *A-B Classic of Acupuncture and Moxibustion* (*Zhen Jiu Jia Yi Jing*, 《针灸甲乙经》), etc.

In the chapter "Discourse on the Hidden Canons in the Numinous Orchid Chambers" in *Plain Questions*, it states that "The triple energizer is the official functioning as the opener of channels. The paths of water originate in it". The chapter "To Consider the Transportation Openings as the Foundation" in *Miraculous Pivot* records that "The triple burner[19] is the central conduit functioning as a moat. The individual waterways leave from it. It is linked to the urinary bladder. It is a solitary

[18] Ren Yingqiu, *The Doctrine of the Schools of Traditional Chinese Medicine*, pp. 188–189.

[19] "Triple Burner" refers to "triple energizer" in Paul U. Unschuld's translation version of *Miraculous Pivot*.

repository. These are the links of the six repositories". Moat was written as "Du" in ancient texts. In *Origin of Chinese Characters* (*Shuo Wen Jie Zi*, 《说文解字》), it interprets "Du" as ditches; and Duan's annotation is that all the passages that water passes through, regardless of size, are called "Du". In the chapter "Root and Connection" in *Miraculous Pivot*, there is also an annotation describing "Du" as meaning that the skin and flesh are shriveled and weak. We use "Jiao" to explain "Du" and Fu-organ in charge of dredging to explain triple "Jiao". By considering the mutual interpretations of these two characters, it becomes apparent that the bowel organ of the triple energizer actually refers to the different parts of extra-organ tissues which are surrounded by the visceral peritoneum in the abdominal cavity. It naturally covers the abdominal cavity in the theory of cavity triple energizer and the mesentery of the small intestine in the theory of omentum triple energizer. The visceral peritoneum not only encapsulates many ducts, lymph nodes, etc., but also forms various hollow "empty flesh" due to its abundance of fat. Furthermore, it covers the surface of organs as a whole. Therefore, there is a statement in *Miraculous Pivot*: "These are the links of the six short-term repositories".

As for the precise location of the triple energizer, the descriptions provided in the chapter "The Conduit Vessels" in *Miraculous Pivot* offer clarity: "The hand minor yang[20] conduit vessels of the triple burner: They originate from the tip of the finger next to the little fingers... They enter the suprasternal fossa, spread in the chest center, and dissipate in the heart enclosure. They descend through the diaphragm and follow a connection with the triple burner". Similarly, the pericardium meridian, which is internally and externally related to the triple energizer, is explained in the same chapter, it says that "The hand ceasing yin qi conduit vessels of the heart ruler: They originate in the chest. Where they appear and connect with the heart-enclosing network. They descend through the diaphragm and connect with the triple burner". In terms of collocation, shu (pertaining to) is used for the root meridian, while luo (collateral) is used for the meridian being internally and externally related to it, which indicates that the location of the triple energizer is below the diaphragm and does not include the parts above it. Many of the previously mentioned theories failed to clearly articulate this point. It is important to note that the *Miraculous Pivot* only mentions "xun shu" and "li luo" for these two meridians among twelve meridians discussed. For the other ten meridians, only "shu" is mentioned in relation to the root Zang-organs (or Fu-organs), and only "luo" is mentioned in terms of their internally and externally related organs. This indicates that the scope of the triple energizer is much broader than that of the general Zang-Fu organs, which aligns with the actual distribution of the abundant "Jiao" in the abdominal cavity.

The records regarding the triple energizer imply that the ancients provided such vivid descriptions primarily based on their anatomical observations. But what are the functions of the triple energizer?

---

[20] "Minor yang" refers to "lesser yang" in Paul U. Unschuld's translation version of *Miraculous Pivot*. He also uses "major yang/yin" to explain "greater yang/yin".

14.3 Metabolic Theory

Digestion, absorption, and excretion are significant topics in the field of life sciences. Besides, they are also important questions for the ancient Chinese medical theory to address. However, given the limited understanding and cognitive abilities of ancient peoples, they could not fully grasp the fundamental processes of assimilation and dissimilation, which encompass the entire metabolic absorption process within organisms. TCM theory extensively uses the concept of qi, referring to the entire metabolic absorption process as "qi transformation". While some contents of TCM's foundational theories may have been influenced by personal experiences of qi movement during meditation, a careful analysis of the theory of metabolism and absorption may suggest that this is not the case at all. Rather, it is a comprehensive theory based on morphological observations and human perception.

The entire process, from the entry of food into the mouth to its eventual discharge, is completed within a continuous cavity that extends from the mouth to the anus. This accurate description was made possible by the ancient scholars' foundational knowledge of morphology. However, there remain significant questions: How is the essence effectively absorbed? Furthermore, without a direct connection between the gastrointestinal tract and the bladder, does the liquid matter, once in the stomach, achieve the separation of the pure and turbid to reach the bladder? These are the significant challenges encountered in the metabolic absorption theory. Although modern medicine has clearly established that nutrient absorption primarily occurs in the small intestine, ancient medical texts suggested that this process primarily took place in the stomach.

> When grains first enter the stomach, their finest essence is first emitted by the two burners of the stomach to nourish the five Zang-organs. They separate into two pathways, following the routes of the nutritive and defensive qi.[21]

There are some explanations for the functions of the upper and middle energizers of the stomach in TCM.

> The upper burner opens and disperses the essence of the five grains to nourish the skin, fill the body, and moisturize the hair, much like the way mist and dew provide moisture. This is what is called "qi". The middle burner receives the essence, extracts its juices, transforms them, and turns them red, which is what is called "blood".

The middle burner also emerges from the stomach, following the upper burner. The qi receives is separated into dregs, steamed to produce body fluids, and transformed into fine essence. This essence then flows upward into the lung vessels, where it is

[21] The chapter "On Flavors" in *Miraculous Pivot* (translation version: Paul U. Unschuld, *Huang Di Nei Jing Ling Shu*, The Ancient Classic on Needle Therapy. The complete Chinese text with an annotated English translation); "Camp qi" and "guard qi" refer to "nutrient qi" and "defensive qi" in Paul U. Unschuld's translation version of *Miraculous Pivot*.

further converted into blood, which is then distributed throughout the body to sustain life. There is nothing more precious than this.[22]

Therefore, as mentioned in the chapter "The Conduit Vessels" in the *Miraculous Pivot*, "diseases of the stomach are generated by the blood". Similar texts reinforce this notion, indicating that absorption took place in the stomach and was facilitated by "the two energizers of the stomach". Why did such an absorption theory appear at that time? This is because, in the abdominal cavity of humans or animals, the visceral peritoneum, which is closely connected to the stomach, forms the greater and lesser omentum, especially in the liver and stomach ligament formed in the left part of the lesser omenta. Specifically, the hepatogastric ligament, formed in the left part of the lesser omentum, encompasses the left and right gastric arteries, veins, gastric superior lymph nodes, and nerves. Similarly, the hepatoduodenal ligament, formed in the right part, envelops three important structures, namely common bile duct, proper hepatic artery, and portal vein as well as lymph and nerves. Both of these structures possess the nature of "Jiao" and "Du". Based on the statements in the *Plain Questions*, that "the qi and blood leaving the stomach follow the conduit-channels" and that "the qi of food enters the stomach. The stomach spreads essence to the liver. Excessive qi flows into the sinews",[23] it can be deduced that the ancient people considered these paths crucial for absorption. This incorrect understanding also led to the gallbladder being distinguished from the six Fu-organs and classified as an extraordinary organ due to its storage function rather than drainage.

In addition, there is another path for absorption: "all the four limbs are supplied with qi by the stomach, but the stomach qi is unable to reach the conduits directly. It is only because of the spleen that the four limbs receive their supplies". This is because "the spleen and the stomach are connected through a membrane, and the spleen can move the body fluids on behalf of the stomach".[24] In the human body, the spleen and the stomach are both intraperitoneal organs, enveloped by the visceral peritoneum and connected as a whole. Just as what has been mentioned above, they are connected through a membrane. This indicates that the qi transformation theory established by physicians at that time was based on their anatomical observations. They integrated objective observations with subjective reasoning to explain the functions of the human body and establish the theoretical system.

In the basic theory of TCM, the metabolism of water and humor is irrelevant to the kidney in ancient times due to a lack of knowledge about the physiology and biochemistry of renal circulation and urine production. Sometimes when mentioning

[22] The chapter "Camp Qi and Guard Qi - Generation and Meeting" in *Miraculous Pivot* (translation version: Paul U. Unschuld, *Huang Di Nei Jing Ling Shu*, The Ancient Classic on Needle Therapy. The complete Chinese text with an annotated English translation).

[23] The chapter "Further Discourse on the Conduit Vessels" in *Plain Questions* (translation version: Paul U. Unschuld, *Huang Di Nei Jing Su Wen*. An annotated translation of *Huang Di's Inner Classic-Basic Questions*).

[24] The chapter "Discourse on the Major Yin and on the Yang Brilliance Conduits" in *Plain Questions* (translation version: Paul U. Unschuld, *Huang Di Nei Jing Su Wen*. An annotated translation of *Huang Di's Inner Classic-Basic Questions*).

that water disease is associated with the kidney, it stems from its role in the five-phase theory, where "the kidney governs water". Until the nineteenth century, Wang Qingren stated that "there are two tubes in the concavity of the two kidneys, leading to the general conduit. The kidneys themselves are solid with no orifices, and therefore, they cannot store essence".[25] He also believed that "the bladder has a lower orifice but no upper orifice". This explains why, at that time, both Chinese and Western medicine considered that water would infiltrate into the bladder. Let's examine the statements in the *Miraculous Pivot*:

> The foot major yang conduits, externally, link up with the clear waters of the Qing river; internally, they are connected to the urinary bladder. They communicate with all water pathways. The hand major yang conduits, externally, link up with the waters of the Huai river; internally, they are connected to the small intestine, from which the water pathways originate. The (bladder) meridian of the foot greater yang connects to the water passages; the (small intestine) of the hand greater yang……guides water passages.[26]

> The lower burner directs the flow into the ileum and then into the urinary bladder, where fluids seep in. Water and grain are present together in the stomach. There they are transformed into dregs and then descend together into the large intestine, constituting the lower burner, where the fluids seep in. A separate juice is secreted along the lower burner and seeps into the urinary bladder.[27]

From these records, water metabolism occurs through the small and large intestines before reaching the bladder. This transportation from the large and small intestines to the bladder is facilitated through "Jiao", which are distinct areas formed in the lower part of the abdomen associated with the peritoneum. When water and grain enter the intestines and the stomach, some parts separate into the curved intestine, and some remain in the lower burner. If these fluids cannot seep into the urinary bladder, the lower burner swells. The water then spills over, leading to a condition known as water swelling, or[28] ascites, characterized by the accumulation of a significant amount of fluid in the peritoneal cavity.

In addition to deriving qi and blood from food, the human body also needs to absorb fluid and liquid, which refer to the beneficial liquids present in the body. Fluid refers to liquid with lower concentration, while liquid contains higher levels of protein and sugar, such as cerebrospinal fluid and fluid in the articular cavity. However, there are inconsistencies regarding the pathway of fluid and liquid absorption as described in the *Miraculous Pivot* and *Plain Questions*. In the *Miraculous Pivot*, it states that "the urinary bladder acts as the official regional rectifier. It stores the body fluids,

---

[25] Wang Qingren (the Qing Dynasty), the chapter "Revised Figure of Zang-Fu Organs" in *Correction of Errors in Medical Classics* (*Yi Lin Gai Cuo*, 《医林改错》), Volume 1, Shanghai: Shanghai Science and Technology Press, 1966, pp. 7.

[26] The chapter "The Conduit Waters" in *Miraculous Pivot* (translation version: Paul U. Unschuld, *Huang Di Nei Jing Ling Shu*, The Ancient Classic on Needle Therapy. The complete Chinese text with an annotated English translation).

[27] The chapter "Camp Qi and Guard Qi - Generation and Meeting" in *Miraculous Pivot* (translation version: Paul U. Unschuld, *Huang Di Nei Jing Ling Shu*, The Ancient Classic on Needle Therapy. The complete Chinese text with an annotated English translation).

[28] The chapter "The Separation of the Five (Protuberance-Illnesses) Jin and Ye Liquids" in *Miraculous Pivot*.

and when qi is transformed, urine can be produced from there".[29] It means that the fluid, humor and water enter the bladder through the water passage of the triple energizer, and then the bladder extracts the essence. If the bladder fails to transform qi properly, then the essence will be lost with the urine, resulting in a condition called unctuous strangury. In the *Plain Questions*, it describes the process as follows: "all water and grain enter the mouth. There are five distinct flavors, each of which blends into its designated sea. The jin and ye liquids each move along their specific paths. The triple burner emits qi to warm the muscles and the flesh and to fill the skin. These are the jin liquids. Those that flow into the body but do not move further are the ye liquids". It implies that before infiltrating into the bladder, the liquid has been extracted from the triple energizer, "internally seeping into the hollow spaces in the bones and supplementing the brain with bone marrow".[30] This perspective arose from the observation that the entire mesentery covering the small intestine is rooted in the inner part of the vertebrae, leading to the belief that the essence of food directly penetrates into the "bone cavity" (bone marrow cavity) to "nourish the brain marrow". This concept persisted until the Qing Dynasty, when Wang Qingren, an anatomy expert, wrote *Corrections of Errors in Medical Classics* based on this morphological understanding, the chapter "The Conduit Vessels" in *Miraculous Pivot* defines the main disease of the large intestine as "diseases generated by the jin and ye liquids" and the main disease of the small intestine as "diseases generated by the ye liquids". This represents an advancement from the *Plain Questions*, which attributed the absorption of body fluids to the qi transformation of the bladder. This progress is due to the development of morphology. Classical medical texts provide a comprehensive description of the process, from the intake of food and drink, through nutrient absorption and water metabolism, to the excretion of waste from the anus – The po-gate, too, is engaged by the five Zang-organs.[31]

Historically, it has been generally believed that ancient Chinese medicine paid little attention to the internal form, structure, and functioning of the human body. However, with the evolution of the Chinese academic thought, this tendency became increasingly popular. This phenomenon primarily emerged after the Eastern Han Dynasty (AD 25–220), with the advent of the medical era represented by the *Classic of Difficulties*. Therefore, the theories of the life gate and the source qi gained prominence, replacing the previous theory of the triple energizer. As medical theories continued to develop, new theories such as midnight-noon ebb-flow, the five movements and six qi correspondences, and the application of the *Book of Changes* (*Zhou Yi*, 《周易》) emerged. These theories contributed to the perception that TCM was detached from practical reality, appearing mysterious and abstract. Therefore, the interpretation of classical TCM texts often became divergent and misleading.

---

[29] The chapter "Discourse on the Hidden Canons in the Numinous Orchid Chambers" in *Plain Questions* (translation version: Paul U. Unschuld, *Huang Di Nei Jing Su Wen*. An annotated translation of *Huang Di's Inner Classic-Basic Questions*).

[30] The chapter "The Separation of the Five (Protuberance-Illnesses) Jin and Ye Liquids" in *Miraculous Pivot* (translation version: Paul U. Unschuld, *Huang Di Nei Jing Ling Shu*. The Ancient Classic on Needle Therapy. The complete Chinese text with an annotated English translation).

[31] The chapter "Further Discourse on the Five Depots" in *Plain Questions*.

In the study of metabolism and absorption in TCM, a commonly referenced statement can be found in the chapter "Further Discourse on the Conduit Vessels" in *Plain Questions:* "Beverages enter the stomach, where they overflow with essence qi, which is transported upward to the spleen. The spleen qi disperses this essence, which rises to the lung. The lung regulates the waterways and transports the water downward to the urinary bladder". However, the interpretation of this passage often diverges from its original meaning. For example, in modern times, the pathway of water metabolism is typically explained as liquid entering the stomach, then passing through the spleen to the lung, and finally to the bladder. This interpretation not only veers into the realm of the mysterious and abstract but also leaves the concept unclear and difficult to understand. It should be noted that the "liquid" referred to here is not limited to H_2O, but also includes substances such as wine, medications, and liquid foods. Among these, substances such as alcohol, volatile oils, and aromatic hydrocarbons can be directly absorbed by the gastric wall and react quickly with the body, representing the original meaning of essential qi being distributed around.[32] Undoubtedly, this statement stems from body sensation, and individuals can personally verify it. However, it should be noted that the H_2O from the liquid doesn't attribute to the lung upward but rather transports downward to the bladder through the water passage of the triple energizer. This statement in the chapter "Further Discourse on the Conduit Vessels" in *Plain Questions* is consistent with the pathway of water metabolism aforementioned.

14.4 The Concept of Vessel

The theory of meridians and vessels is a significant component of the theoretical system of TCM. It not only serves as a foundational principle for acupuncture and moxibustion practice but also provides insights into human physiology and pathology, condition analysis, and disease diagnosis. While the exact nature of meridians and vessels remains unclear, the theory essentially reflects the physiological functions of living organisms. Both scientists at home and abroad have provided evidence to support the existence of this unique conduction system, which differs from blood vessels and nerves in humans and animals.

This conduction system was initially described as "vessels" in ancient Chinese medicine. Over time, during the periods corresponding to the *Plain Questions* and

[32] There is such a paragraph in the chapter "Camp Qi and Guard Qi – Generation and Meeting" in *Miraculous Pivot*, which could confirm that people at that time had already noted the difference between wine and water:

Huang Di: "When someone drinks wine, the wine enters the stomach. The grain consumed earlier has not been digested yet, and the urine is discharged separately first. How is that?" Qi Bo: "Wine is a liquid made of fermented grain. Its qi is wild and clear. Hence, even if it is ingested only after an earlier ingestion of grain, it will precede the grain and leave the stomach as a liquid first". Huang Di: "Good! I have been informed: The upper burner is like fog. The central burner is like a humidifier. The lower burner is like a ditch. That is what is meant here".

Miraculous Pivot, a more systematic categorization emerged, distinguishing straight-running channels as meridian vessels, transverse running ones as collateral vessels and tiny branches as tertiary collateral vessels, thus the word "meridian and collateral" being used. Therefore, the morphological basis of meridian and collateral is analogous to that of vessel.

When studying the meridians or collaterals as described in Chinese medicine, researchers have found that the small blood vessels and capillaries within the arteriovenous system align with certain collaterals and tertiary collateral vessels in the meridian and collateral theory. However, large blood vessels do not correspond to the twelve regular meridians. To explain this discrepancy, it is necessary to consider the influence of ancient anatomical practices on the concept of the pulse.

As is known to all, meridians are not visually observable. However, ancient medical physicians did not possess the same conception. According to the chapter "The Conduit Vessels" in *Miraculous Pivot*: The twelve conduit vessels run hidden in the partings of the flesh; they are in the depth and cannot be seen. Those vessels that are always visible at the surface are all network vessels. When the color of the vessels is greenish blue, then cold is present, and there is also pain. When they are red, then they contain heat. Similar descriptions about meridians and vessels can also be found in other chapters of this book, such as "The conduit vessels decay and leak".[33] "When the blood vessels are filled to abundance, when they run hard across and are of red color, and ascend and descend without a regular location, with the small ones being thin as a needle, and the big ones resembling a sinew".[34] "When using the needles, it is imperative to first examine whether in the conduits and network vessels there is a condition of repletion or depletion. One presses the finger into the vessels and follows their course. Once squeezes the skin and pulls it up. One checks which movements correspond, and then removes the evil qi by moving them downward".[35]It shows that the "vessel" is visible in the theory of Chinese medicine. Most of what is mentioned in the sections above refers to blood vessels.

In addition, in the chapter "Cold and Heat Sensations" in *Miraculous Pivot*, it mentions that "patients experience alternating cold and heat sensations, and have a rope-of-pearls-illness at the neck and in the armpits......Those at the surface in the vessels that have not yet entered the interior to attach themselves to the muscles and the flesh, and that has generated pus in the exterior, they are removed easily". In this context, the "vessels" specifically refers to the lymphatic vessels.

In a word, the understanding of "vessels" at that time was mainly based on the visible vascular system of the human body, so it states in TCM classics: "That which

[33] The chapter "Obstruction- and Impediment-Illnesses" in *Miraculous Pivot* (translation version: Paul U. Unschuld, *Huang Di Nei Jing Ling Shu*. The Ancient Classic on Needle Therapy. The complete Chinese text with an annotated English translation).

[34] The chapter "The Blood Network Vessels" in *Miraculous Pivot* (translation version: Paul U. Unschuld, *Huang Di Nei Jing Ling Shu*. The Ancient Classic on Needle Therapy. The complete Chinese text with an annotated English translation).

[35] The chapter "Piercing to Regulate True and Evil Qi" in *Miraculous Pivot* (translation version: Paul U. Unschuld, *Huang Di Nei Jing Ling Shu*. The Ancient Classic on Needle Therapy. The complete Chinese text with an annotated English translation).

contains the camp qi lest they disappear, that is what is called 'vessels'". The chapter "The Biographies of Bian Que and Cang Gong" in *Records of the Grand Historian* (*Shi Ji*, 《史记》) records "incising skin and muscles, unblocking meridians and binding collaterals"; the chapter "Biography of Wang Mang" in *the Book of the Former Han* (*Han Shu*, 《汉书》) records that "Wang Mang asks imperial doctors, Shangfang (an officer in charge of household utensils for the emperor) and skilled butcher to dissect the body together, measures the length among five Zang-organs and uses thin bamboo branches to unblock the meridians in order to know the start and endpoints of it". These records are both circumstantial evidence for the understanding of this opinion. However, at that time, there was no distinct classification of arteries, veins, and lymph, leading to the lymph also being considered as vessels; they may explain "blood spurting" (puncturing the arteriole) as "abundant blood qi and excessive Yin qi"; and explain "scant, black and turbid blood" (puncturing venules) as "yang qi has accumulated and remain at one place for an extended period of time without being drained".[36] Meanwhile, because the anatomical technology at that time had not reached the level of vascular stripping, it was impossible for people to comprehensively understand the human vascular system (vascular system and lymphatic system). At the preliminary stage of understanding the anatomy and cognition of these tracts, there was a mistaken belief that the branches and connections of these tracts correlated with the deep and invisible "twelve regular meridians". This confusion arose from a mix of observations and subjective descriptions. Hence, it is stated that "The twelve conduit vessels extend concealed within the recesses of the body; they exist within the depths and remain unseen". This does not imply their non-existence, but rather their hidden nature.

This misconception had persisted for thousands of years until the introduction of western medicine to China and modern anatomy stripped the complete vascular system of the human body, people came to realize that blood vessels were not the same as the meridians. The inherent definitions of TCM concepts such as vessel, meridian, and vessel and collaterals have been redefined. However, these new definitions represent contemporary Chinese medicine rather than ancient Chinese medicine.

The primary reason for this confusion can be attributed to the limitations in morphological understanding at that time, which hindered a correct comprehension of the basic structure of the human body. It is important to approach the study of TCM development from a historical perspective rather than attempting to explain it solely through modern medicine. By doing so, we can avoid the assumption that ancients could distinguish vascular visible vessels and invisible meridian systems or that "meridians and vessels" in Chinese medicine are unrelated to the blood vessels, lymph and nerves in western medicine. Otherwise, it would be impossible to explain why there are a large number of written records regarding blood vessels as meridians in ancient medical literature, despite the fact that the meridian system is not the same as the vascular system.

---

[36] The chapter "The Blood Network Vessels" in *Miraculous Pivot* (translation version: Paul U. Unschuld, *Huang Di Nei Jing Ling Shu*. The Ancient Classic on Needle Therapy. The complete Chinese text with an annotated English translation).

To sum up, the development of traditional Chinese medicine during its crucial historical period when its theoretical foundation was laid not completely based on philosophical thinking and the summary of clinical experience. Many important fundamental theories, including explanations for physiology and pathology, were directly derived from morphological observations. Physicians of that time made efforts to establish a basis in morphology, especially in the description of the pulse. Because the theory of meridians does not depend on the morphology, its theoretical development was unique. Subsequently, Wang Mang could ask the imperial doctor and skilled butcher to dissect the human body and measure the length of the pulse channels. However, the scientific experiments of using bamboo branches to unblock the meridians failed to achieve the goal of understanding the starting and ending points of the pulse. If the technological capabilities at that time had allowed for a comprehensive understanding of the vascular system, all theoretical modifications would have likely occurred.

When considering the differences between Chinese and foreign medicine systems, it is important to focus on the development of anatomy rather than simply whether anatomy is included. Western anatomy has been continuously and extensively developed as an independent discipline, while ancient Chinese anatomy has been closely linked with medical theories from its inception. Therefore, many important theories concerning the functions of viscera and bowel organs, metabolism and absorption theory, and the descriptions of meridians and vessels incorporate morphological knowledge to varying degrees. However, due to the combined influence of feudal ethics and other reasons, anatomy research in ancient China did not reach significant depths; the economic structure and cultural background of the society were not conducive to cultivating physicians who could specialize in research separate from clinical treatment. Therefore, theories centered around the invisible concepts of "principles" and "qi" gradually became dominant, leading to a decline in and eventual abandonment of anatomical studies. During the Qing Dynasty, Wang Qingren personally examined the viscera and bowels at execution sites, examined and wrote *Corrections of Errors in Medical Classics*, which was regarded as being influenced by western learning, but in fact, it was not. However, Wang Qingren's actual contributions did not surpass the anatomical level of the Han Dynasty (202 BC–AD 220). Instead, he merely repeated the processes of that era. His studies on function and morphology were almost identical to those of ancient times. Due to a neglect of ancient investigations, Wang Qingren was mistakenly seen as establishing a new medical theory. To cite just one example, he criticized the "three-section triple energizer theory" (as shown in Fig. 14.1a), which had appeared after the Jin Dynasty (AD 266–420). However, the theory revised after his personal observations was exactly the triple energizer theory of "empty flesh" in the period of *Inner Classic*-Water Outlet (as shown in Fig. 14.1b).[37]

[37] Wang Qingren (the Qing Dynasty), *Correction of Errors in Medical Classics*, Shanghai: Shanghai Science and Technology Press, 1966, Edition One, pp. 5,7.

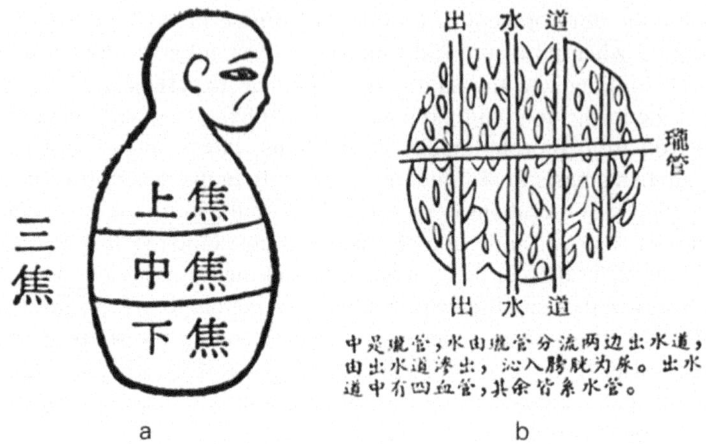

Fig. 14.1 "Water Outlet" (a) Refers to the "Ancient Diagram of Zang-Fu Organs"; (b) Refers to Wang Qingren's "Revised Diagram of Zang-Fu Organs"

Chapter 15
Physiology of Circulation and Respiration

"Qi" and "blood" are fundamental concepts in TCM theory. No matter how abstract and philosophical the meanings of "qi" and "blood" are, their original meanings refer to the "qi" derived from nature and the physical existence of "blood" within the vessels.

The circulation of "qi" and "blood", accompanied by life activities, is the foundation of life. However, how does the circulation of "qi" and "blood" work? Is it possible that TCM and modern medicine share a common understanding of these concepts based on different knowledge systems?

15.1 Introduction

Researchers in the field of science history seem to have discovered substantial evidence in TCM classics to show that the ancient people had possessed knowledge of the mechanism of respiration and blood circulation. As a result, these discoveries have garnered significant praise.

For example, when medical historians evaluated the value of *Huangdi's Internal Classic* (*Huang Di Nei Jing*, 《黄帝内经》), Jia Dedao claims that "*Huangdi's Internal Classic* has already demonstrated a clear and definite understanding of blood circulation" and that "the respiratory function of the lungs and their relations to blood circulation seems to have been basically understood".[1] Fan Xingzhun believes that "the theory of blood circulation in *Huangdi's Internal Classic* shows

[1] Jia Dedao, *A Brief History of Chinese Medicine* (*Zhong Guo Yi Xue Shi Lve*, 《中国医学史略》), p. 53.

Y. Liao, *Reconstructing the Image of Medicine in the Qin and Han Dynasties*, Understanding China, https://doi.org/10.1007/978-981-95-6801-7_15

similarities to modern physiology".[2] Yu Shenchu says: "It aligns with modern physiological theory".[3] Yu Ying'ao also believes that *Huangdi's Internal Classic* contains "a generally correct overview of the systemic circulation and pulmonary circulation".[4] Foreign scholars describe it "the first indication of the circulatory system" and view it as "an astonishing prophecy".[5] There are of course those who are silent and do not comment on this matter,[6] but there is a lack of negative evaluation or comprehensive analytical discussion in general.

In contemporary society, it is often assumed that the ancients already possessed an understanding of respiration and circulation, as these are now considered basic physiological knowledge. However, before delving into detailed discussions, it is important to note that *Huangdi's Internal Classic*, along with other TCM classics notably has never mentioned the heartbeat, which is a fundamental physiological phenomenon and vital sign. Recognizing this fact is essential for an objective assessment of the ancients' understanding of respiration and circulation. To understand why the theory of qi and blood circulation in TCM is an "ideological theoretical framework", which does not align even remotely with modern Western medicine, it is necessary to review how Western medicine, based on empirical science, gradually developed its understanding of the physiological mechanisms of respiration and circulation. This comparison with traditional Chinese medicine is essential for clarity.

(1) Cognition on blood circulation: Before the discovery of blood circulation, Galen (AD 129–200) proposed the tidal theory, suggesting that blood returns to the heart by the same path, rising and falling like tides. Later this theory had been dominant for a long time. Although Galen's explanation of the movement of heart blood is later found to be incorrect, "his experiment on living animals to investigate the function of the heart", "is one of the two most noteworthy experiments in antiquity".[7] Later, Leonardo Da Vinci and Vesalius, through autopsy, doubted and denied the existence of small holes in the ventricular septum; the pulmonary circulation was described by M. Servetus (AD 1509–1553) and M. R. Colunbus (AD 1516–1559); and the systemic circulation was speculated by A. Cesalpino (AD 1519–1603). In AD 1603, Fabricius ab Aquapendente (AD 1537–1619) discovered venous valves. All of these scientific achievements laid

[2] Fan Xingzhun, *A Brief History of Chinese Medicine* (*Zhong Guo Yi Xue Shi Lve*, 《中国医学史略》), p. 28.

[3] Yu Shenchu, *A Brief History of Medicine in China* (*Zhong Guo Yi Xue Jian Shi*, 《中国医学简史》), p. 35.

[4] Yu Ying'ao, "The Early Achievements of Chinese Medicine from Two TCM Classics", in *Scientific and Technological Achievements in Ancient China* (*Zhong Guo Gu Dai Ke Ji Cheng Jiu*, 《中国古代科技成就》), edited by the Institute for the History of Natural Sciences, Beijing: China Youth Publishing Group, 1978, p. 428.

[5] Svensmeier (Germany), translated by Ma Boying, *5000 Thousand Years of Medicine* (*Shi Jie Yi Xue Wu Qian Nian*, 《世界医学五千年》), pp. 16–17.

[6] For example, *Ancient Chinese Medicine* (*Zhong Guo Gu Dai Yi Xue*, 《中国古代医学》), written by Zhao Pushan.

[7] Dampier (UK), *History of Science* (*Ke Xue Shi*, 《科学史》), pp. 103–104.

the foundation for the discovery of blood circulation by W. Harvey (AD 1578–1657). In 1628, Harvey's *Movement of the Heart and Blood* was published, providing a correct description of blood circulation.[8] According to Harvey's own words, this founding was not based on critical thinking or a priori reasoning, but on "repeated vivisection".[9]

(2) Understanding of respiration: In physiology, respiration is closely intertwined with blood circulation, and the physiological significance of pulmonary circulation can only be revealed by clarifying the mechanism of air exchange.

In 1617, Fludd conducted an experiment where glassware was inverted over water, objects were burned inside it, and as a result, the volume of air inside decreased, leading to the extinguishing of the flames. Studies by R. Hooke (AD 1635–1703) and others have shown an active component in the air, later identified as oxygen, which is vital for combustion and respiration. Hooke further demonstrated that a continuous stream of air blown into the lungs eliminated the need for chest wall motion to sustain life. R. Lower (AD 1631–1691) used Hooke's artificial respiration experiment to figure out that the change in blood color was entirely due to the interaction between air in the lungs and the absorption of some of it by the blood. J. Mayow clarified in 1669 that "the air contains a vital constituent necessary for life, which enters the blood during breathing". However, "all the valuable research results were subsequently forgotten and remained unrecognized until Lavoisier rediscovered them a century later".[10]

From the brief recapitulation above, it is easy to know that the elucidation of the physiological mechanisms of respiration and circulation has predominantly relied on empirical research methods. The role of the "intermediate steps" is particularly important, as they not only form the necessary basis for reaching a complete and correct understanding but also the divergence between Eastern medicine ("traditional medicine") and Western medicine ("modern medicine") has grown with each of these intermediate steps.

Now, let's take a look at the understanding of respiration and circulation in ancient Chinese medicine which has not conducted such experimental research.

15.2 The Concept of Heartbeat Doesn't Exist in Ancient Chinese Medicine

In TCM classics, there are numerous discussions about heart function and its pathological changes. However, it is interesting to note that the physiological phenomenon of "heartbeat" is not explicitly mentioned. While this may initially seem surprising,

---

[8] Editorial Board of China Encyclopedia of Medicine, "History of Medicine Volume" in *China Encyclopedia of Medicine* (*Zhong Guo Yi Xue Bai Ke Quan Shu*, 《中国医学百科全书》), pp. 252–253.

[9] Dampier (UK), *History of Science*, p. 183.

[10] Dampier (UK), *History of Science*, pp. 185–186.

a closer examination of ancient classics without the influence of the "physiological knowledge" in modern medicine reveals the true historical context.

In the received version of *Huangdi's Internal Classic*, the typical discourse on heart function centers around statements such as: "The heart is the official functioning as ruler. Spirit brilliance originates in it"; "The heart is the basis of life; it is responsible for changes of the spirit". Although there are mentions of "the heart rules the body's blood and vessels", this is only a classification method based on the five-phase theory. Similar classifications can be found for other organs, such as "the lung rules the body's skin and body hair, the liver rules the body's sinews and membranes, the spleen rules the body's muscles and flesh, and the kidneys rule the body's bones and marrow". These classifications do not carry the same physiological significance as understood in modern medicine.

The descriptions of the disorder of heart and other internal organs are likewise classified according to the five elements. As a substantial organ, one prominent pathology concerning the heart is "heart pain". In the chapter "The Receding Qi Diseases" in *Miraculous Pivot* (*Ling Shu*, 《灵枢》), it is stated that "If the pain in the heart is severe and breaks out during the daytime, the patient will die in the evening. If the pain breaks out in the evening, he will die at dawn".[11] This is basically consistent with modern myocardial infarction. Diseases of the various internal organs are described in the chapter "To Consider the Long-term Depots as Foundations" in *Miraculous Pivot*:

> When the heart is small, then it is at peace, and evil qi cannot harm it. However, it is easily harmed by sadness; When the heart is big, then sadness cannot harm it, but it is easily harmed by evil qi. When the heart is elevated, then it fills the lungs, causing vexation and forgetfulness, and difficulty in expressing oneself. When the heart has sunken into the depth, then the long-term depot[12] lose their qi to the outside and are easily harmed by cold. They are easily made to fear someone else's words. When the heart is firm, then the long-term depots are in peace and serve as protector and stabilizer. When the heart is brittle, then one tends to suffer from melting dan-illness, and heat in the body center. When the heart stands upright, then it is harmonized, flexible and difficult to be harmed. When the heart is inclined, then it cannot fulfill its tasks and provides neither protection nor control.

In summary, upon analyzing the descriptions of heart function and disorder, there is no mention of specific terms such as "heartbeat", "rhythm" or "pulse rate".

This is because ancient people did not possess knowledge regarding the basic physiological phenomenon and the vital sign of heartbeat. Therefore, in situations requiring emergency resuscitation, their observations were primarily focused on "breathing" and body temperature. For example, in the pre-Qin period, the renowned physician Bian Que diagnosed the Crown Prince of the Guo State with "deathlike reversal". Bian Que believed that the Crown Prince was suffering from fainting and said: "If you doubt my words, please go in and examine the prince. You will hear ringing in his ears and observe his nostrils opening and closing. If you run your

[11] The chapter "The Receding Qi Diseases" in *Miraculous Pivot*.

[12] "Depot" refers to "Zang-organ" in Paul U. Unschuld's translation version of *Plain Questions* and *Miraculous Pivot*.

hand down his legs, you could still feel warm.[13]" Similarly, in the *Treatise on Cold Damage and Miscellaneous Diseases* (*Shang Han Za Bing Lun*, 《伤寒杂病论》), written in the late Eastern Han Dynasty (AD 25–220), the treatment of people who hanged themselves also focused on similar vital signs, "One man put his hand on the patient's chest and pressed repeatedly …… for a while, and then the patient's breathing returned and his eyes opened". Even the most famous ancient forensic medical monograph, *Collected Cases of Injustice Rectified* (*Xi Yuan Ji Lu*, 《洗冤集录》) completed in the Song Dynasty (AD 960–1279), still relied on vital signs of respiration and body temperature in its "life-saving formula": "If there is still warm under the heart, even after a day, the patient can be saved"; "If the mouth is closed and there is still a slight breath, …"; "If there is still warm in the heart, even after a day, the patient can be saved"; or "If the muscles have not yet become cold, …", etc.

The ancients' failure to recognize the concept of "heartbeat" can be attributed to two possible reasons. Firstly, although there are records of human autopsies conducted for medical research in ancient China,[14] it is impossible to directly observe the heartbeat during these procedures. Secondly, due to the prevalent understanding that the heart is the "official" responsible for thinking, it is hard to conceive that this "official" would continuously beat without cessation. The symptomatic experience of "heartbeat" will only manifest in times of imminent danger. For example, it is stated in the chapter "The State of Zhao" in *Strategies of the Warring States* (*Zhan Guo Ce*, 《战国策》): "Zhao Xiangzi went to the toilet, and suddenly his heart started pounding inexplicably. Then he inquired about the identity of the toilet handyman, and found it was Yu Rang (who wanted to take this opportunity to assassinate Zhao)". In the chapter "Biography of Gaozu" in *Records of the Grand Historian*, it is recorded: "On his way to Dong Yuan city, Gaozu passed by Bairen county, where Guan Gao, the Prime Minister of the state of Zhao, tried to kill Gaozu with his man, and Gaozu couldn't shake the feeling of imminent danger, causing his heart to race with unease. He promptly left the area, thus escaping from the assassination". In these instances, the "heart racing" is clearly not indicative of the rhythmic movement of the heart. Likewise, TCM practitioners also hold the belief that "When someone is mournful and grievous, then his heart will be excited. When the heart is excited, it affects the functioning of the five long-term Zang-organs and six short-term Fu-organs". In the chapter "Fearful Throbbing" (Zheng Chong) in *Yunlin Medical Book* (*Yun Lin Shen Gou*, 《云林神彀》) written by Gong Tingxian in the Ming Dynasty (AD 1368–1644), there is even mention of "the heart beats and stops".

However, we are left to ponder, were the ancients really been unaware of the heartbeat that resonates on the left side of the chest, which could be felt, touched, and even seen with the naked eye? Indeed, the ancient people were aware of the phenomenon of beating in the precordial region but failed to associate it with the beating of the heart:

---

[13] The 45th chapter "Biographies of Bian Que and Cang Gong" in *Records of the Grand Historian* (*Shi Ji*, 《史记》), Volume 150.

[14] The 69th chapter of "Biography of Wang Mang" in the *Book of the Former Han* (*Han Shu*, 《汉书》), Volume 99.

> The large network vessel of the stomach is called xu li. It penetrates the diaphragm and connects with the lung. It surfaces below the left breast. The movement in this vessel is reflected by movements of the garment. This is caused by the basic qi in the vessels. If it is abundant, panting, frequent, and interrupted, then the disease is in the center; In case of transverse knottings, the qi has accumulations; If it is interrupted and does not arrive at all, that is called "fatal". As for the movement reflected by the garments below the breast, if it is interrupted and does not arrive, in this case the basic qi leaks.[15]

The "stop of heart beating" was considered a sign of impending death and referred to as the "basic qi of the pulse". Although the ancient people recognized its importance, their theoretical conception framework attributed the essence of life to the stomach. They believed that the stomach generated the vital energy through its digestion and absorption functions. This perspective led them to interpret the apical beat, which is located "below the left breast" and whose movement "is reflected by the movement of the garments", as the beat of the "large network vessel of the stomach". As the ancient people were unaware of the autonomic nervous system's role in regulating the orderly beating of the heart, their theoretical explanation appeared to a natural and even "reasonable" conclusion.

15.3 The Stomach-Centered Circulation System

In the theoretical system of ancient Chinese medicine, although there was no recognition of the blood circulation system with the heart as the source of power, this doesn't imply a lack of understanding of qi and blood circulation. Otherwise, there would not have led to so many comments and theories about the blood circulation in ancient China. Therefore, we should focus on understanding how the ancient people explained the movement of qi and blood within the framework of TCM.

In ancient Chinese culture and medicine, the description of the heart as the "official of the sovereign" can only be used to explain the spiritual activities of human beings. The primary location for physiological functions such as the generation and movement of qi and blood is attributed to the "stomach":

> All five Zang-organs are supplied with qi by the stomach. Hence the stomach is the basis of the five Zang-organs; man requires water and grain as his basis. Hence, once a man is cut off from water and grain, he will die. Once the vessels contain no stomach qi, death is imminent too.[16]

> That from which man receives his qi, that is the grain. Where the grain flows, is the stomach. The stomach is the sea where water and grain, the qi and the blood gather. The qi and the blood leaving the stomach, follow the conduit channels.[17]

---

[15] The chapter "Discourse on Phenomena Reflecting the Status of Qi in a Normal Person" in *Plain Questions* (*Su Wen*, 《素问》).

[16] The chapter "Discourse on the Jade Mechanism and the True Qi of the Depots" in *Plain Questions*.

[17] The chapter "Jade Tablet Discourse on the Essentials" in *Miraculous Pivot*. The "conduit channel" refers to "meridian" in Paul U. Unschuld's translation version of *Plain Questions* and *Miraculous Pivot*.

> When the grains have just entered the stomach, their finest essence first is emitted by the stomach's two burners to moisten the five long-term Zang-organs. They move in two separate streams, along the paths of the nutrient and defense qi.[18]

Based on this physiological understanding of the stomach as the source of qi and blood, the meridian circulation system, centered around the stomach, is constructed. The movement of qi and blood begins at the lung meridian of hand greater yin. And the lung meridian of hand greater yin does not begin at the lung, but at the "middle energizer" where the stomach and intestines are located. From there, the qi and blood pass through the circle formed by twelve meridians, as well as the five Zang-organs and six Fu-organs, before returning to the lungs.[19] In the earliest extant medical books, i.e., the silk manuscripts *Moxibustion Classic of the Eleven Yin-Yang Meridians* (*Yin Yang Shi Yi Mai Jiu Jing*, 《阴阳十一脉灸》) and *Moxibustion Classic of the Eleven Foot-Arm Meridians* (*Zu Bi Shi Yi Mai Jiu Jing*, 《足臂十一脉灸经》) unearthed from the Mawangdui Han Dynasty Tombs, Hunan province in 1973, the movements of the twelve meridians were initially considered independent and unrelated to each other. In the chapter "The Conduit Vessels" in *Miraculous Pivot*, more detailed and accurate descriptions of the meridians were provided compared to the silk manuscripts from Mawangdui. However, the main streams of the meridians remained independent, and their interconnections were completed by the "branch vessels". For example, the liver meridian of foot reverting yin is traced from the foot to the head, and its connection with the lung meridian is described as this: "Further branches originate from the liver again, separately penetrate the diaphragm and having further ascended pour into the lung".

It is not our primary focus to delve into the existence of the "meridian and collateral" in the objective world. However, even if one acknowledges that some contents of the meridian and collateral theory provide objective descriptions of the life phenomenon that has not yet to be fully understood, it becomes apparent that the inclusion of "branch vessels", which link the various internal organs, was likely an attempt to construct a circular circle capable of explaining the movement of qi and blood in the human body.

There is a saying in TCM: "The meridians are incessantly circulating, and the circumference is endless". However, it is important to note that this should not be confused with the concept of blood circulation in modern medicine. Some individuals may mistakenly assume this due to a lack of in-depth understanding between the two medical systems, leading to their blind inference that, "In general anatomy, it has been understood that the heart and the vessels are connected, and the blood is incessantly

[18] The chapter "The Five Flavors" in *Miraculous Pivot*.

[19] The description of the twelve meridians in the chapter "The Conduit Vessels" in *Miraculous Pivot* is as follows: the first meridian is "the hand major yin vessels (also called hand greater yin meridian) of the lung: They start from the triple burner (also called triple energizer), reach downward, and connect with the large intestine. They wind around the stomach opening, extend upward to the diaphragm, and connect with the lung. From the lung they connect diagonally with, and appear below, the armpits"; the last meridian is the liver meridian of foot reverting yin and it terminates at "further branches originate from the liver again, separately penetrate the diaphragm and having further ascended pour into the lungs".

circulating and governed by the heart".[20] This does not take into account the fact that, as far as "general anatomy" is concerned, it can be said that the entire human body is interconnected with the vessels, not just with the heart. Therefore, it also makes it impossible for people to realize that the circulation of blood is governed by the heart. This is clearly illustrated in the chapter "Transports" in *Miraculous Pivot*, which explains why the arteries at the back of the feet beat: "Huang Di: The foot yang brilliance conduit, for what reason does it move? Qi Bo: The qi of the stomach ascends and flow into the lung. ......That is the path of the stomach qi that is separate from the yang brilliance conduit". According to the five elements theory in TCM, the heart dominates blood and vessels. However, the close connection between the vessels and the stomach can be observed in the context of pathology "When the stomach is not replete, then all vessels are depleted. When the stomach is empty, then the stem vessels are depleted".

This circulation, centered around the stomach, is summarized in TCM as the movement of the two types of qi, nutrient qi and defensive qi, that is:

> Man receives his qi from the grain. The grain enters the stomach, and from there its qi is transmitted to the lung. This way, all the five long-term depots and six short-term repositories receive qi. Their clear parts become the camp qi. The turbid parts become the guard qi. The camp qi is in the vessels. The guard qi is outside the vessels. They circulate without a stop. After 50 circulations a grand meeting happens.[21]

The explanation regarding the movement of nutrient qi is quite clear. It begins in the stomach, then passes to the lungs, and circulates throughout the body along the twelve meridians of hands, feet, and the mutual distribution of yin and yang. However, once the nutrient qi reaches the last meridian, the liver meridian of foot reverting yin, it does not connect with the lungs through the "branch vessel" as described in the chapter "Conduit Vessels" in *Miraculous Pivot* but forms another circulation pattern. That is, a total of fourteen meridians are involved in this circulation, including the conception vessel (CV) and governor vessel (GV) (Fig. 15.1).

Although the defensive qi also originates from the stomach, it belongs to "the violent qi of water and grain. This qi is fast and unrestrained and cannot enter the vessels. Hence, it moves inside the skin and in the partings of the flesh. It steams against the huang-membrane and it spreads in the chest and abdomen".[22] The descriptions of its circulation are confusing, but in a nutshell, it can be summarized that "the movement of the guard qi completes fifty circulations during one day and one night. During day time they move through the yang realm and complete 25 circulations. During the night they move through the yin realm and complete 25 circulations. The circulation includes the five long-term depots".[23] Since it has little to do with blood circulation, thus can be ignored.

---

[20] Wei Yizong, *History of Chinese Orthopedic Techniques* (*Zhong Guo Gu Ke Ji Shu Shi*, 《中国骨科技术史》), Shanghai: Shanghai Science and Technology Press, 1983, p. 23.

[21] The chapter "Camp Qi and Guard Qi-Generation and Meeting" in *Miraculous Pivot*. The "camp qi" refers to "nutrient qi" and "guard qi" refers to "defense qi".

[22] The chapter "Discourse on Blocks" in *Plain Questions*.

[23] The chapter "Movements of the Guard Qi" in *Miraculous Pivot*.

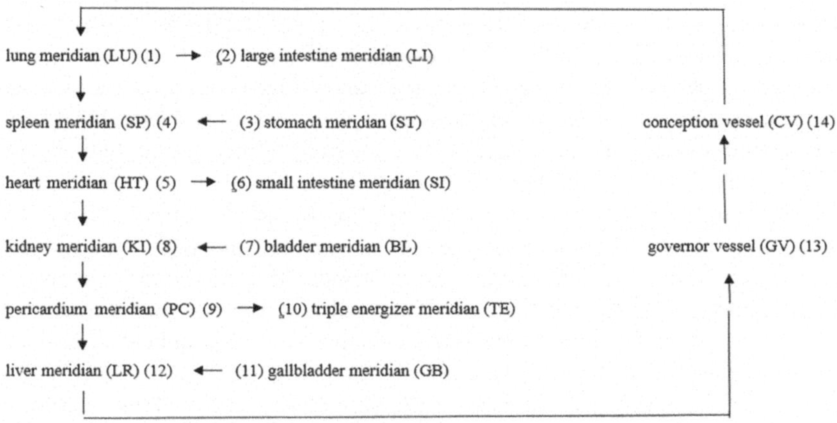

Fig. 15.1 Schematic diagram of the nutrient qi circulation

15.4 Stirred Pulse & Pulse Stirring

In TCM classics, the beating pulse is referred to as the "stirred pulse". There is a statement in the chapter "Discourse on the Three Sections and Nine Indicators" in *Plain Questions*: "The heaven indicator of the upper section: the moving vessels on the two sides of the forehead. The earth indicator of the upper section: the moving vessels on the two sides of the cheeks. The main indicator of the upper section: the moving vessels in front of the ears". Additionally, the first chapter in *Classic of Difficulties* (*Nan Jing*, 《难经》) says: "All the twelve conduits have sections where the movement in these vessels can be felt". At first glance, this may appear to align with the term "artery" used in modern medicine, but in fact, the two terms have distinct differences in their concepts and definitions.

First of all, in TCM, the concept of "pulse" is rather broad, encompassing both visible blood vessels and invisible meridians and collaterals. This confusion arises from the lack of comprehensive knowledge about the entire vascular system in ancient Chinese medicine.[24] Thus, the term "stirred pulse" does not have a corresponding conceptual meaning to "vein". For example, when the network vessels are pierced, "blood comes out and shoots" (piercing the small arteries) is interpreted as "both blood and yin qi are abundant"; "blood is black and turbid" (piercing the small veins) is interpreted as "yang qi accumulates and is not released".[25]

Without recognizing that the beating of the heart causes the beating of the pulse, the "stirred pulse" subordinate to the meridians is considered to be independent. This concept is exemplified in the pulse diagnosis method known as the "three positions and nine indicators".

---

[24] Refer to the discussion of ancient anatomical knowledge and theoretical systems in Chapter 14 of this book.

[25] The chapter "The Blood Network Vessels" in *Miraculous Pivot*.

The "three positions and nine indicators" is a pulse diagnosis method used in traditional Chinese medicine (TCM) to examine the stirred pulse in three specific locations on a person's head, hands and feet, and to infer the location of disease by comparing the pulse changes in these nine places (as shown in Table 15.1). If the pulses in these nine parts are generally consistent, it means that "yin and yang are evenly balanced" and the body is healthy. Therefore, there is a statement: "If the pulses at the nine indicators are identical, this is called 'normal person'".[26] If there is an imbalance of yin and yang, the pulse changes may occur, indicating the presence of a disease, so "Investigate the nine indicators. If the pulse at any one of them is hasty, this indicates disease. If the pulse at any one of them slow, this indicates disease. If the skin at any one of them is hot, this indicates disease. If the pulse at any one of them is slow, this indicates disease. If the skin at any one of them is hot, this indicates disease. If the skin at any one of them is cold, this indicates disease. If any one of them has sunk down, this indicates disease".[27] The terms "hasty" or "slow" mean that the pulse in one place beats faster or slower than in other places. The "three positions and nine indicators" pulse diagnosis method has a strong numerical component,[28] and does not necessarily align with the theory of meridians and collaterals. For example, in order to meet the numerical requirement of "three times three equals nine", an additional indicator called "qi in the chest" has to be added to the "middle" of the heart and lungs to make up the "three". However, the diagnostic criteria on which this pulse diagnosis method is based, can be traced back to the *Moxibustion Classic of the Eleven Foot-Arm Meridians*[29] unearthed from the Mawangdui Han Dynasty Tombs. These are some diagnostic criteria: "When brought together but cannot be made to fit, this indicates disease. When all the pulses at the nine indicators in the three sections do not conform with each other, this indicates death. When the pulses above and below, on the left and on the right correspond to each other like the pounding in a mortar, this indicates that the disease is severe".[30] An interpretation by Cai Jingfeng suggests that the phrase "pulse beating like the pounding in a mortar" refers to the occurrence of "third heart sound gallop", which is associated with a serious heart attack, which means that each heartbeat splits into three sounds. As Cai's article discusses, the reason is that "such sounds can only be perceived by applying the ear". In nineteenth century, Western medicine reported this symptom "in conjunction with examination by stethoscopes and modern scientific instruments". However, Cai added: "the ancients could listen to the heartbeat by pressing their ears directly against the chest, as there were no stethoscopes available

[26] The chapter "Discourse on Regulating the Conduits" in *Plain Questions*.

[27] The chapter "Discourse on the Three Sections and Nine Indicators" in *Plain Questions.*

[28] Refer to the discussion on the evolution of pulse diagnosis in Chapter 17 of this book.

[29] *Moxibustion Classic of the Eleven Foot-Arm Meridians* says, "if the patient's pulse beats like the pounding in a mortar, he cannot live more than three days". See Editing Group of Silk Book in Mawangdui Han Dynasty Tombs, *Silk Manuscripts from Mawangdui Han Dynasty Tombs [Part Four]* (*Ma Wang Dui Han Mu Bo Shu [Si]*,《马王堆汉墓帛书》〔肆〕), "Notes on Interpretation", p. 5.

[30] The chapter "Discourse on the Three Sections and Nine Indicators" in *Plain Questions.*

Table 15.1 Three positions and nine indicators pulse diagnosis method

Three positions	Nine indicators	Pulse position	Indications
Upper	Heaven Earth Human	The moving vessels on the two sides of the forehead The moving vessels on the two sides of the cheeks; The moving vessels in front of the ears	The qi at the corners of the head The qi of mouth and teeth The qi of ears and eyes
Middle	Heaven Earth Human	Hand greater yin Hand yang brightness Hand lesser yin	Lungs Qi in the chest Heart
Lower	Heaven Earth Human	Foot reverting yin Foot lesser yin Foot greater yin	Liver Kidney Spleen and stomach

at the time".[31] From a technical perspective, this may appear less relevant from the common method of pulse-taking. From a psychological perspective, it seems too "scientific", falling back on the fictitious premise that the ancient people had knew the relationship between heartbeat and pulse.

Historically, ancient Chinese medicine did not establish a direct relationship between the heart and the pulse beat, which led to a lack of awareness regarding the overall consistency of the pulse throughout the whole body. Therefore, changes in pulse rate in certain areas were often taken as diagnostic indicators, such as: "The right pulse is deep and tight, while the left pulse is superficial and slow".[32] Similarly, ancient medical texts describe pulse diagnosis methods as: "When both sides are examined together, the pulse beats can be felt by the doctor's hand. If the pulse rates on the left and right are not synchronized, this is a sign of impending death". This tendency to view pulse beats in isolation eventually evolved during the Ming Dynasty (AD 1368–1644), particularly under the influence of Li Shizhen, to the point where the Cunkou pulse (the radial artery of the hand) was divided into three sections—inch, guan, and chi—with each section differentiated by pulse rates of slow or fast.[33]

In Li Shizhen's *Binhu's Sphygmology* (*Bin Hu Mai Xue*, 《濒湖脉学》), it is mentioned that "A slow-beating Cun pulse indicates cold in the upper energizer; a slow-beating Guan pulse suggests cold and pain in the middle energizer; a slow-beating Chi pulse points to kidney deficiency, often accompanied by heaviness in waist and feet". Furthermore, "A fast-beating Cun pulse indicates sores in the throat, mouth and tongue; a fast-beating Guan pulse reflects heat in the stomach and liver; a fast-beating Chi pulse suggests internal heat due to yin deficiency". Li Zhongzi's

[31] Cai Jingfeng, *World Records in the History of Chinese Medicine—Records of Abnormal Pulse Rhythms* (*Zhong Guo Yi Xue Shi Shang De Shi Jie Ji Lu Yi Chang Mai Xiang De Ji Zai*, 《中国医学史上的世界记录·异常脉律的记载》), Hunan Science and Technology Press, 1983, pp. 10–20.

[32] The chapter "Discourse on Disease Manifestations" in *Plain Questions*.

[33] Tamba Motoyasu (Japan), Volume "Diagnosis of the Left and Right" in *Yi Sheng* (『醫賸』).

The Techniques of Diagnostician (*Zhen Jia Zheng Yan*, 《诊家正眼》) inherited Li Shizhen's opinions and stated that: "A slow-beating Cun pulse signifies cold in the upper body; a slow-beating Guan pulse indicates cold in the middle energizer; a slow-beating Chi pulse suggests yang deficiency". Additionally, "A fast-beating Cun pulse is associated with coughing; a fast-beating Guan pulse indicates heat in the stomach; a fast-beating Chi pulse reflects ministerial fire"; "a fast-beating pulse of right hand indicates heat exuberance, left hand indicates yin deficiency". The same distinction is also made for the pulse rate changes of the skipping, bound and intermittent. Other notable works, such as Li Jun's *Introduction to Medicine* (*Yi Xue Ru Men*, 《医学入门》), Xu Chunfu's *Ancient and Modern Medical Unification* (*Gu Jin Yi Tong*, 《古今医统》), and Zhang Jiebin's *Complete Works of Jingyue* (*Jing Yue Quan Shu*, 《景岳全书》) all include such discussions.

The above shows that TCM perspective on the "stirred pulse" focuses on the "pulsation" itself, considering it separate from the heartbeat. Thus, since *Moxibustion Classic on the Eleven Yin and Yang Channels*, a Silk Book unearthed from Mawangdui Han Dynasty Tombs, describing the direction of each meridian, there are statements like "...... beating indicates a certain disease". And *Rules of the Meridians* (*Mai Fa*, 《脉法》) unearthed in Mawangdui Han Dynasty Tombs and the *Book on Meridians* (*Mai Shu*, 《脉书》) unearthed in the Han Tombs of Zhangjia Mountain in Jiangling region, both say: "A single pulse beating among tranquil pulses manifests diseases. There are regular beating pulses such as lesser yin meridian in the shinbone and reverting yin and lesser yin meridians in arms. The abnormal fast beating of these pulses indicates diseases. That is the explanation of over-beating pulse". Or as *Huangdi's Internal Classic* states, "If now the vessels are suddenly excited, this is always so because evil qi have settled in them..."[34]; "The yang brilliance qi is in permanent movement. The great yang qi and the minor yang qi do not move. While usually they do not move, in the present case, they move with great speed. This is the manifestation of that disease".[35] In the transition from "inspecting the meridians and taking for the pulse" (i.e., separately examining the pulsations on the parts of the meridians) to "taking the Cun pulse only" (only examining the radial artery), the relationship between pulse rate and rhythm and the heartbeat was not mentioned. Instead, it was associated with the pathology of the five Zang-organs based on mathematical concepts. Such as "If there are 50 movements without a single intermittence, all the five long-term depots are supplied with qi. If within 40 movements there is one intermittence, one long-term depot is without qi. If within 30 movements there is one intermittence, two long-term depots are without qi. If within 20 movements there is one intermittence, three long-term depots are without qi. If within 10 movements there is one intermittence, four long-term depots are without qi. If there is one intermittence within less than ten movements, all five long-term depots are without qi".[36] This is the explanation for the abnormal pulse rhythm. In addition, *Classic of Difficulties* and *Pulse Classic* both have detailed discussions on the pulse rate,

[34] The chapter "The Conduit Vessels" in *Miraculous Pivot*.

[35] The chapter "Discourse on Disease Manifestations" in *Plain Questions*.

[36] The chapter "Root and Connection" in *Miraculous Pivot*.

ranging from "three times in one breath" to "six times in one breath", and from "one time in one breath" to "one time in four breaths". A pulse rate below normal is referred to as a "damaging pulse". The relationship between the damaging pulse and the five Zang-organs is explained as follows: "The first stage of damage (abnormally slow pulse rate) affects the skin and hair (lungs). The second stage of damage affects the blood vessels (heart). The third stage of damage affects the muscles (spleen). The fourth stage of damage affects the tendons (liver). The fifth stage of damage affects the bones (kidneys)".[37]

15.5 Understanding of Respiration

The difference between respiration and heartbeat lies in the fact that respiration is so clearly associated with the body's surface that it is unmistakably perceived by every normal person, and it is easily understood that there is an inherent connection between breathing and life. However, to correctly explain respiratory physiology, one must have knowledge of pulmonary circulation and gas exchange; otherwise, it is impossible to conceive that the respiratory process is confined solely to the lungs. If air is merely inhaled into the lungs and then exhaled, what is the significance of breathing? In traditional Chinese medicine, there was a lack of basic knowledge of respiration, so how did they explain the phenomenon of respiration that coexists with life?

In the chapter "The Five Flavors" of the *Miraculous Pivot*, it is stated, "All the five zang-organs and six fu-organs derive their qi by the stomach.......When the vital energy is abundant but fails to circulate, it accumulates in the chest, called the 'sea of qi.' The qi emitted by the lung travels through the throat. Therefore, it is exhaled when one breathes out and inhaled when one breathes in". Seemingly, its understanding of respiration is that air is exhaled from the lungs, but the circulation path appears to be:

Qi → enter the stomach → "a mutual beating of massive qi and their failure to move on" → accumulate in the chest
↳ (By meridian vessels) to the five viscera and six bowels →Out of the lungs →Out through the throat

There is also a passage in the chapter "Further Discourse on the Conduit Vessels" in the *Plain Questions* that is worth studying:

> The qi from food enters the stomach, which disperses its essence to the liver, and the liver spreads its qi to the sinews; The qi from food also enters the stomach, And the turbid qi is directed to the heart, where it infuses its essence into the vessels. The qi in the vessels circulates through the conduits, eventually returning to the lung. The lung coordinates with the one hundred vessels, distributing essence to the skin and body hair. The hair vessels

[37] The chapter "The Fourteenth Difficulty" in *Classic of Difficulties.*

> unite the essence and they move qi to the palaces.[38] If the essence of the palaces and the spirit brilliance remain in the four depots, the qi turns to a balance. Because this balance is maintained, at the qi-opening, a section of one inch is established to decide about a patient's death or survival.

Almost all the commentaries view this passage as describing the process of digestion and absorption of "food" into the stomach. However, throughout the *Huangdi's Internal Classic*, descriptions of digestion and absorption typically involve the phrase "grain enters the stomach".[39] And usually include the generation of blood. However, in the specific passage under discussion, there is no mention of blood generation, but the word "qi" is mentioned for nine times. Obviously, it is inappropriate to interpret the phrase "Shi qi" as "eating food". The word "Shi" is used as a verb to mean "to eat" or "to receive food". In fact, the inhalation of air is clearly referred to as "eating qi" in *Ten Questions* (*Shi Wen*, 《十问》), a Silk Book unearthed from Mawangdui Han Dynasty Tombs. And *Writings of Prince Huainan* (*Huai Nan Zi*, 《淮南子》) and *On Balance* (*Lun Heng*, 《论衡》) both refer to the technique of breathing and inhalation as "eating qi". If this explanation is valid, it can be seen as an explanation of the respiration physiology at that time, and thus we can gain insight into many texts, such as the one quoted above, "The stomach is the sea where water and grain, the qi and the blood gather. The qi and the blood leaving the stomach follow the conduit-channels".[40] Furthermore, it can aid in understanding the methods of inserting and withdrawing the needle in accordance with the breath to supplement and drain qi in the body. That is, the qi of breathing runs along the meridians and spreads throughout the body, therefore, "Wait until the patient exhales to pull the needle out. When the exhalation is completed, the needle is removed, allowing a large amount of qi to be expelled. Hence, this is called 'draining.'" On the other hand, "Wait for inhalation and pull the needle. The qi must not leave. At each place where a needle was inserted, push the hole and close the door and thereby letting the spirit qi be preserved. Large quantities of qi stay where they are. Hence this is called 'supplementation'".

Upon the completion of the *Classic of Difficulties*, a new understanding of how qi circulates in the human body was put forth: "That which is exhaled originates from the heart and lungs; that which is inhaled enters the kidneys and liver. During the transition between exhalation and inhalation, the spleen receives the taste influences of the grains". For individuals with contemporary scientific knowledge, comprehending the precise meaning of these ancient perspectives may pose a challenge, but the annotations provided by the previous scholars prove invaluable in unraveling the thought processes of the ancients.

---

38 The "palace" refers to "Fu-organ" and the "inch" refers to "cun", a traditional Chinese measurement which is equal to the width of one's thumb.

39 As the chapter "The Conduit Vessels" in *Miraculous Pivot* says: "The grain enters the stomach. As soon as the vessel paths are passable the blood and the qi will move"; the Chapter "Oral Inquiry" says: "The grains enter the stomach. The stomach qi ascends and pours into the lung"; the chapter "Camp Qi and Guard Qi - Generation and Meeting" says: "Man receives his qi from the grain. The grain enters the stomach....... Their clear parts become the camp qi. The turbid parts become the guard qi". etc.

40 The chapter "The Jade Tablets" in *Miraculous Pivot*.

Lyu Guang of the Period of the Three Kingdoms (AD 220–280) commented:

> The heart and lungs, located above the diaphragm, are both yang Zang-organ governing the internal, thus can control exhalation; the kidneys and liver, situated under the diaphragm, are both yin Zang-organ, thus they control inhalation.

Ding De of the Song Dynasty commented:

> The kidneys and liver are located below the diaphragm and govern the internal. During exhalation, the qi is expelled out through heart and lungs, hence people consider the heart and lungs to dominate exhalation. Additionally, the heart and lungs are situated above the diaphragm and govern the external, so during inhalation, the qi, associated with yin, enters the stomach and the liver.

"The Eleventh Difficulty" in *Classic of Difficulties* repeats the same point again in explaining "The pulse that stops before reaching 50 beats means that one zang-organ lacks qi":

> When inhaled qi cannot reach the kidneys and instead upon reaching the liver, it is known that if one Zang-organ lacks qi, the kidney is the first to be affected.

The *Classic of Difficulties* also conceives of an "origin of life qi," which is the central point in the three-dimensional space of the human body. It states:

> The twelve meridians are all connected to the origin of life qi. The term "origin of life qi" refers to the foundation of the twelve meridians and is also known as the "stirring qi in the region of the kidneys". This represents the root of five Zang-organs, six Fu-organs, and the twelve meridians, as well as the gate for breathing.

Whether it is "the origin of life qi" or "the stirring qi of the kidney region", these concepts are abstract and enigmatic to modern readers, leaving TCM theories seemingly difficult to comprehend. However, its origin and real meaning may be much simpler and more intuitive, rooted in our perception of the abdominal aorta beating. The reason is that in the chapter "Sixteen Difficulties" in *Classic of Difficulties*, when discussing the "internal syndrome" of the five Zang-organs, it also talks about "stirring qi" around the periumbilicus, and states that "pressing it firmly causes pain". In this way, the *Classic of Difficulties* identifies a driving force for breathing and the circulation of the qi and blood of the twelve meridians. And this force is separate from the digestive and absorptive functions of the stomach and the bellow mechanism of the lungs. But it is closely connected to the innate essence (umbilicus).

Another theoretical conception related to respiration can be found in the *Treatise on the Pathogenesis and Manifestations of All Diseases* (*Zhu Bing Yuan Hou Lun*, 《诸病源候论》), written during the Sui (AD 581–618) and Tang (AD 618–907) Dynasties. Although this book also upholds the basic idea that "qi circulates through the organs and meridians", it proposes that the qi circulation begins in the lungs: "All five Zang-organs are endowed with qi from the lungs".[41] If the qi, after completing

[41] This is different from the statement "The five Zang-organs are all endowed with qi in the stomach", which was quoted above from the chapter "Discourse on the Jade Mechanism and the True Qi of the Zang-organs" in *Plain Questions*. However, attention should be paid to the following: the qi that goes to the stomach should be the essence qi of food and drink; the qi that goes to the lungs should be the air that comes from the natural world.

its circulation through the Zang-organs and Fu-organs, reverts back to the lungs, it can lead to a series of lung-related diseases:

> Long-term cough syndrome: the lungs are extremely weak. Since the lungs are extremely weak, qi still passes through them.
>
> Syndrome of coughing with lung qi ascending counterflow: the lungs are deficient, coughing while qi still gathers in the lungs.
>
> Syndrome of long-term coughing with lung qi ascending counterflow: when the lungs are deficient, pathogens can invade, causing qi to rise in a counterflow.
>
> Syndrome of coughing with lung qi ascending counterflow and vomiting: the five Zang-organs are all endowed with qi from the lungs; the cold pathogen fights against the vital qi, causing it to gather back in the lungs.
>
> Symptoms of coughing and shortness of breath: when coughing, the qi returns to the lungs, causing the lungs to swell.

These explanations of the etiology and pathology of the lungs highlight the tangible nature of the "lungs" as real organs; the "deficiency" refers to a real space that can contain qi; and the "qi" is air. Neither the lungs nor qi has abstract functions; nor is deficiency an invisible, intangible decline in function. Although the book does not specify where the qi that should not return to the lungs should be exhaled from, it is clear that it is not exhaled from the lungs.

In fact, the ancients long regarded the lungs as the "gates" (bellows)—the driving force for the flow of qi and blood. The chapter "The 50-fold Circulation" in *Miraculous Pivot* states: "When one exhales, the vessels move twice, and the qi travels 3 cun; When one inhales, the vessels move twice again, and the qi travels 3 cun. Over 13,500 breaths in a day, the qi circulates 50 times through the body……The total distance traveled by the qi is 810 zhang".[42] This clearly interprets respiration as the source of power for pulse beating and the circulation of qi and blood. The *Treatise on Cold Damage Diseases* (*Shang Han Lun*, 《伤寒论》) also reinforces this perspective by stating, "Breath is the beginning of the pulse". Chen Yan of the Song Dynasty (AD 960–1279) emphasized, "Before taking a pulse, one must first understand the terms 'pulse' and 'breath'. The pulse refers to blood, and the breath refers to qi. The pulse does not move on its own; it's the qi that drives it". Shen Kuo, a renowned figure of the Song Dynasty, provided a clear and precise explanation of the lung's "bellows" function and the circulation of air within the five Zang-organs in his work *Dream Pool Essays* (*Meng Xi Bi Tan*, 《梦溪笔谈》): "The throat connects to the five Zang-organs, serving as the passageway for the intake and outflow of air. The five Zang-organs govern respiration, much like the bellows used in metal forging".[43]

---

[42] Zhang is a traditional Chinese measurement, and 1 zhang is equal to 3.33 m.

[43] Shen Kuo (the Song Dynasty), the chapter "Medicine Discussion" in *Dream Pool Essays* (*Meng Xi Bi Tan*, 《梦溪笔谈》), Volume 26.

15.6 Discussion

To establish a comprehensive understanding of respiratory physiology and blood circulation, it is crucial to conduct proper experimental studies. In ancient China, numerous circulation theories thrived, including the "cycle of five virtues" derived from the theory of the five elements, and the "five circuits and six qi" that govern the interactions between heaven and earth. If we do not overlook these theories, we should clearly recognize that the concept of "circulation" in TCM, which believes that qi and blood flowing ceaselessly along the meridians, does not equate with the modern physiological understanding of blood circulation.

With a limited understanding of physiology and biochemistry in ancient times, how can ancient people explain the "principle of life" and construct a theoretical system? In analyzing "the flow of qi and blood", we see how the ancient people employed the concept of "intention" to develop theories such as "the circulation system with the stomach as the driving force", "the circulation system with the lungs as the driving force" and "the midpoint of the body as the driving force". In fact, there are many more examples, and the approach can be applied to every concept, terminology, and theory of TCM. It is essential to engage in "historical analysis" that sheds light on the real meanings of these concepts, terms and theories across different periods; thus, we can gain a deeper understanding of the entirety of TCM, encompassing these concepts, terms, theories, and healing techniques.

In conclusion, the previous discussion highlights that many medical terms, which may appear mysterious and incomprehensible to modern people (such as qi, blood, the origin of life qi, the stirring qi of the kidney region, etc.) were originally based on the most direct and realistic objective existence. These intuitive understandings were used to construct an explanation system for the "principle of life"—the basic theory of traditional medicine—which reflects the process of TCM practitioners using "intention". Over time, certain concepts that were once precise and well-defined have acquired new connotations that are challenging to elucidate in contemporary terms (for example, the concept of "deficiency"), and this is also a process of TCM practitioners using "intention" to develop TCM. After the introduction of Western modern medicine, many concepts and theories in traditional medicine underwent explanations by modern medicine, and they became more abstract. The purpose is to make them more scientific and to "resolve" the contradictions between traditional medicine and modern medicine. This, we can say, is also a process of modern and even contemporary TCM practitioners using "intention" to advance the development of TCM.

Chapter 16
Yin-Yang Theory

The Yin-yang theory represents the most profound philosophical theory conceived by ancient Chinese philosophers. Joseph Needham, a famous historian of science, believed that this theory "was the most ultimate principle of which the ancient Chinese could conceive".[1] Scholars specializing in ancient culture, ideology, philosophy, and history of science widely acknowledge the crucial role of the yin-yang theory in fostering the development of traditional Chinese medicine. Also, this theory was extensively applied and developed in ancient Chinese medicine. However, it is worth considering whether there exists any difference between the "yin-yang theory" developed by the yin-yang school of the pre-Qin period (before 221 BC) and the "yin-yang theory" in traditional Chinese medicine. In a concise assessment, after the introduction of yin-yang theory into traditional Chinese medicine, there was a change in medical theory from adhering to the main idea of yin-yang theory that explains the health and illness with waxing-waning and circulation of yin and yang, to emphasizing the opposition and balance between yin and yang. It is in this process that yin-yang theory experienced its own evolution and development.

16.1 Yin-Yang School and Its Main Theories

The combination of yin-yang and five-phase, yin-yang and qi, as well as other related theories, constitutes the philosophical system of ancient China and forms the basic elements of traditional Chinese scientific thought. The origins, development, and specific applications of these concepts have always been focal points of research for scholars studying the history of ancient Chinese philosophy, intellectual history, and the history of science. Yin-yang and five-phase have their own separate theories in the early stage. Therefore, they should be discussed separately when studying the early

[1] J. Needham, *Science and Civilization in China* (*Zhong Guo Ke Xue Ji Shu Shi*, 《中国科学技术史》), the Chinese version, Volume. 2, p. 254.

Y. Liao, *Reconstructing the Image of Medicine in the Qin and Han Dynasties*, Understanding China, https://doi.org/10.1007/978-981-95-6801-7_16

cultural thought. Yin-yang theory had an earlier influence on medical development compared to the five-phase theory. Medical historical materials of the pre-Qin period contain few records on five-phase theory, while yin-yang theory had been used to explain the generation of diseases and human physiology.

The term "yin-yang" has a long history.[2] In the Han Dynasty (202 BC–AD 220), the origin of academic thoughts in the pre-Qin period was known as "yin-yang school":

> The Yin-yang school probably derived from the Xihe (astronomical officers). Those officers reverently complied with the natural laws and were skilled in observing the movements of the sun, moon and stars, enabling them to create a calendar and guide the people on conducting ritual sacrifices and agricultural work according to the seasons. However, when obstinate individuals adopted this theory, they became constrained by various taboos and trivial numerology, abandoning human efforts and relying blindly on ghosts and gods.[3]

It seems that the studies of the "yin-yang school" primarily focus on astronomy and calendar, without an explicit philosophical dimension. This is because the philosophical implications of yin-yang theory were not developed overnight; rather, they were often implicitly embedded in discussions regarding the relationship between nature and humanity at the early stage. Although astronomy and ephemeris were included within the scope of yin-yang theory, these elements were merely superficial. The true purpose was not limited to observing astronomical phenomena, creating or issuing calendars, but rather to explain the laws governing the interactions of yin and yang, and to emphasize the importance of aligning with these patterns: "The four seasons, the eight cardinal directions, the 12 divisions and the 24 solar terms each have their own set of rules and taboos. Those who follow them will flourish, while those who defy them will either perish or suffer great loss". Originally, the concept of yin and yang referred to the natural changes perceptible to human senses, such as the presence or absence of sunlight. Therefore, the change of the four seasons was particularly emphasized: "The order of the four seasons cannot be abandoned". The sequence of the four seasons is sprouting in spring, flourishing in summer, harvesting in autumn, and storing in winter. The Yin-yang school emphasized that "This sequence serves as a major principle of natural law and it is impossible to develop the universal order and rule if we fail to follow it".[4] This principle was not only applied to the natural changes of "generation, growth, transformation, harvest, and storage" of all living things, as well as artificial farming and animal husbandry, but also extended to politics and human affairs. This is the basic content of the theory of yin-yang school. It should be noted that the theory of the yin-yang school, which advocates conforming to the changing laws of the four seasons and yin and yang, focuses on the alternation and cyclical nature of yin and yang, rather than their opposition. The essence of the

---

[2] For example, the chapter "Fangshu (divination)" in the *Book of the Later Han* (*Hou Han Shu*, 《后汉书》) (Vol. 112, No. 72) mentioned that "The knowledge of yin-yang, calendar and celestial phenomena are often recorded in ancient classics and literature".

[3] The chapter "Catalog of Literary Works" in the *Book of the Former Han* (*Han Shu*, 《汉书》), Volume 30, No. 10.

[4] All can be found in the *Records of the Grand Historian* (*Shi Ji*, 《史记》), Volume 130, No. 70.

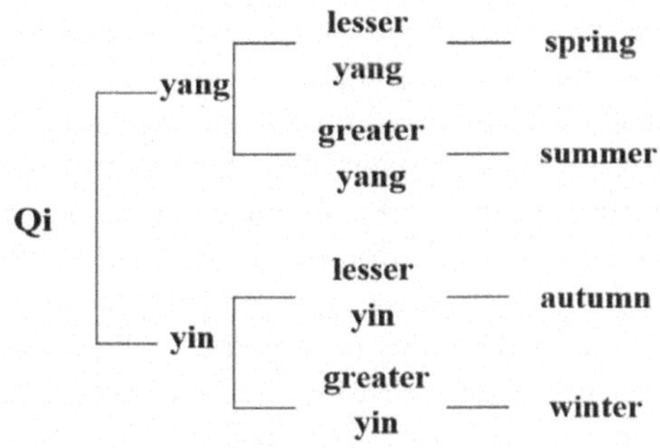

Fig. 16.1 Illustration of greater, lesser, yin and yang. At this time, the core of yin-yang theory is not "opposition" but the "circulation" of qi (the order of the four seasons)

waning and waxing of yin and yang is not a simultaneous process where one diminishes as the other grows. Instead, they each undergo independent phases of growth and decline. Specifically, spring and summer are the stages where yang emerges and reaches its peak, while autumn and winter are the stages where yin emerges and reaches its peak. In these two stages, yin and yang each experience the process from emergence to peak and then to decline: "when yang reaches its peak, yin begins to emerge; while when yin reaches its peak, yang begins to emerge". And "This follows the constant rule of yin and yang and the natural order of the universe" (Fig. 16.1).[5]

In the *Master Lyu's Spring and Autumn Annals* (*Lyu Shi Chun Qiu*, 《吕氏春秋》), there is an article entitled "The Round Principles", which illustrates the cycle law that "existing in an endless cycle of repetition, the universe is somewhat similar to a cartwheel" by listing the phenomena such as the daily cycle of day and night; the moon's passage through the twenty-eight constellations; the continuous movement of vital energy throughout the four seasons; the stages of life of "germination-generation-growth-maturity-withering-declining-concealing"; perpetual westward movement of clouds; the constant flow of water to the east day and night; and the never-ending sources of clouds and oceans. These phenomena are to demonstrate the rule of cycle, that is, "the motion of the heaven and earth continuously repeats itself, and when it reaches its peak, it will reverse its course". Its thinking mode aligns with the yin-yang school's emphasis on conforming to the order of four seasons and the law of yin and yang. Therefore, in the theoretical system that "qi" is divided into yin qi and yang qi, and yin and yang are further divided into greater yin, greater yang, lesser yin and lesser yang which correspond to the four seasons, and the key point of the theoretical system does not focus on the opposition of yin and yang.

If the cyclic factors of the four seasons and the day-night cycle are disregarded, the opposite qualities of yin and yang become naturally evident. For instance, the south side of a mountain and the north side of a river are yang; while the north side of a mountain and the south side of a river are yin. Similarly, the sun is yang and the moon is yin; the daytime is yang while the nighttime is yin; the heaven is yang

[5] Wei Zhao (the Wu Kingdom), the annotation for the *Discourses of the States* (*Guo Yu*, 《国语》), Part B, Volume 21.

and the earth is yin. In this way, various pairs, such as male and female, fire and water, wind and rain, and even male and female animals, are viewed as opposite sides. This understanding forms the foundation of the yin-yang theory, which carries philosophical connotations. In the early stages, the concept of yin-yang opposition did not greatly influence medical theory, as the yin-yang school focused more on the concept of conforming to the order of the four seasons. However, over time, the concept of yin-yang opposition became the mainstream and played a more significant role in medical theory than the cyclic concept of the four seasons. Therefore, there was an intriguing evolution of yin-yang theory and five-phase theory during the pre-Qin period and the Han Dynasty. The early emphasis on the circulation concept of yin-yang school gradually lost attention from researchers as the opposite attribute of yin and yang gained more prominence and elucidation. While the five-phase theory, initially a theory of categorization, was regarded as the typical theory related to circulation due to its prevalent theory of inter-promotion and inter-restraint.

The development of yin and yang is an interesting topic to explore. In the early stage, yin and yang specifically referred to yin qi and yang qi, rather than being the abstraction of the nature of all things. The term "yang qi" had already been established during the reign of the King Xuan of the Western Zhou Dynasty (about 827 BC). By the end of the Western Zhou Dynasty (about 1046–771 BC), Bo Yangfu explained the causes of earthquakes by the movement of yang qi and yin qi, which were well known to all. This can be seen in "The Discourses of Zhou" (Part A) of *the Discourses of the States* (*Guo Yu*, 《国语》). The chapter "The Old Tunes" of *Master Lyu's Spring and Autumn Annals* states that "In ancient times, during the reign of the Zhu Xiang tribe. It was excessively windy and there was too much yang qi accumulated. All things were blown away and scattered around, resulting in the inability of fruits to ripen. To address this issue, a liege called Shi Da invented a five-string zither to attract yin qi so as to tranquilize all living things". Ancient people believed that music could connect the yin qi and yang qi between the heaven and the earth, which is different from the modern understanding on music. There are extensive discussions about music and its effects in the chapter of the "Discourses of Zhou" of the *Discourses of the States*. Similarly, as "Cast the metal into bells, grind the stones into chimes, combine strings and wood into plucked instruments, and drill gourd and bamboo into instrument Sheng and Xiao (flute). Play drums to respond to the winds from eight directions so as to adjust the weather. Therefore, the yin qi won't be stagnant and the yang qi won't be diffuse; yin and yang are in order; the wind blows and the rain falls in a timely manner. Heaven endows happiness frequently and people gain well-being from it. All things are attuned to compose harmonious music and the heaven above and people below enjoy it. This is called just intonation". It is thought that the conduction of sound is the direct communication with the yin qi and yang qi between the heaven and the earth and would have some impacts. The abnormal exuberance of tangible yin qi and yang qi is also one of the causes of diseases. In the chapter "The First-year Reign of Zhaogong" of *The Commentary of Zuo* (*Zuo Zhuan*, 《左传》), there is a record of a medical practitioner named Yi He, who said: "There are six kinds of weather (normal qi) in the sky. As they intersect and descend, they generate five flavors, manifest as five colors, and produce five sounds. Once their limits are exceeded, six illnesses will arise. The six qi are yin, yang, wind, rain, night, and day. They are divided into four seasons and five elements, namely spring, summer, long summer, autumn, and winter. If any of them becomes excessive, it will bring disaster. Excessive yin qi causes cold diseases; excessive yang qi produces heat diseases; excessive wind harms the limbs; excessive rain causes abdominal illnesses; excessive nighttime sexual activities lead to mental confusion; excessive daytime exertion results in mental exhaustion". That is, the abnormal excess of yin qi or yang qi can cause diseases. However, as time progressed, yin and yang evolved into abstract concepts, with less focus on their physical manifestations.

16.2 Different Manifestations of the Circulation and Opposition of Yin and Yang in Medicine

The application of "the order of four seasons" to governance led to the establishment of various related systems recorded in the chapter "Proceedings of Government in the Different Months" of *The Book of Rites* (*Li Ji*, 《礼记》). This is a typical representation of the yin-yang school's teachings. Any deviation from the order of the four seasons is bound to lead to diseases. There are many related records in works similar to "Proceedings of Government in the Different Months", for example:

> If the autumn order is issued in the first month of spring (lunar January), there will be a great plague;
>
> if the spring order is implemented in the last month of summer (lunar June), citizens will suffer from wind cough;
>
> if the spring order is implemented in the second month of winter (lunar November), the people will be more likely to suffer malignant sores.

In reality, it seems to describe and understand the relationship between abnormal climate and the onset of diseases, but in fact, the so-called "orders" refer to government decrees and human affairs. Therefore, while both the yin-yang school and pre-Qin Taoist thought contain many philosophical elements, their essence is the social philosophy of political theory. It was only during the Qin and Han Dynasties (221 BC–AD 220) that these elements gradually evolved into a philosophy of cosmology.[6] In this process, some knowledge systems that modern people now consider part of natural science were not merely passive recipients of prevailing thoughts and theories of the time, but actively contributed to the development of cosmological philosophy, thereby driving the progress of social culture and ideology. In the field of medicine, we can clearly observe an evolution from conforming to the concept of "the order of four seasons" proposed by the yin-yang school to placing greater emphasis on the opposition and balance of yin and yang.

---

[6] Li Zehou, *On Traditional Chinese Thoughts* (*Zhong Guo Gu Dai Si Xiang Shi Lun*, 《中国古代思想史论》), Beijing: People's Publishing House, 1985, p. 314.

In the current version of *Huangdi's Internal Classic* (*Huang Di Nei Jing*,《黄帝内经》), there are many records illustrating the importance of conforming to the order of the four seasons:

> Now, the yin and yang qi of the four seasons are the fundamental roots of the all things. Hence, sages nourish yang in spring and summer, and yin in autumn and winter, thus following their roots. In this way, they coexist with all things at the gate of growth and decline. To oppose one's root is to attack one's foundation and corrupt one's true qi. Hence, yin and yang and the four seasons are the beginning and end of all things, they are the basis of death and life. Opposing them results in catastrophe and harms life. If one follows them, severe disease will not emerge. This is called "to achieve the Way".[7]

> In spring, the qi of heaven begins to disperse and become open; the qi of the earth begins to flow out. Frozen ground thaws, ice melts, water flows and the conduits are passable. Hence, the human qi is in the vessels. In summer, the conduits are full and the qi overflows. It enters the tertiary network vessels which thereby receive blood, making the skin full and robust. In late summer, both the conduits and the network vessels are abundant, and their contents internally overflow into the muscles. In autumn, the qi of heaven begins to collect, the pores and interstices close, and the skin becomes tight and drawn. In winter, all things are stored and conserved. Blood and qi are in the center, attaching to the bones and marrow, penetrating the five depots.[8] Hence, evil qi[9] regularly follows the flow of proper qi and blood of the four seasons when it enters the body and settles in it. However, when it comes to changes and transformations, they cannot be quantified in terms of regular time periods.[10]

Influenced by the order of four seasons and the concept of yin-yang circulation, there were "pulse diagnosis of four seasons" and the acupuncture therapy based on the theory of "four seasons". Understanding the principle of these diagnostic and therapeutic methods becomes challenging without the knowledge of the theory of yin-yang school in the early stage. For example, in the theory of sphygmology of later generations, the floating pulse (a superficially located pulse that can be felt by light touch and grows faint on hard pressure) was regarded as the sign of exterior pattern, while in *Huangdi's Internal Classic*, it is recognized as a normal pulse condition in spring; in acupuncture and moxibustion of later generations, the depth of acupuncture depends on the symptoms, while at that time, it depended on seasons.[11] In subsequent generations, traditional Chinese medicine largely abandoned the content related to four seasons, and solely inherited the ideas of sprouting in spring, flourishing in summer, harvesting in autumn, and storing in winter in health preservation.

As for the limitations of the yin-yang school that "being constrained by various taboos and trivial numerology" is also reflected in the medical field. For example:

---

[7] The chapter "Comprehensive Discourse on Regulating the Spirit in Accordance with the Qi of the Four Seasons" in *Plain Questions* of Paul U. Unschuld's translation version.

[8] "Depots" refer to "Zang-organs" in Paul U. Unschuld's translation version of *Plain Questions* and *Miraculous Pivot*. He also uses "palaces" to explain "Fu-organs".

[9] "Evil qi" refers to "pathogenic qi" in Paul U. Unschuld's translation version of *Plain Questions* and *Miraculous Pivot*.

[10] The chapter "Discourse on Opposition and Compliance in Piercing in the Course of the Four Seasons" in *Plain Questions* of Paul U. Unschuld's translation version.

[11] For more details, refer to the discussion on "The development of the methods of pulse diagnosis" in the "second part" of this book.

> In the first month and in the second month, the qi of heaven begins to spread; the qi of the earth begins to be effused; the qi of man is in the liver. In the third month and in the fourth month, the qi of heaven spreads properly; the qi of the earth is effused firmly; the qi of man is in the spleen. In the fifth month and in the sixth month, the qi of heaven is abundant; the qi of the earth has moved upward; the qi of man is in the head. In the seventh month and in the eighth month, the yin qi begins to kill; the qi of man is in the lung. In the ninth month and in the tenth month, the yin qi begins to freeze; the qi of the earth begins to be closed in; the qi of man is in the heart. In the eleventh month and in the twelfth month, the freezing is repetitive; the qi of the earth is enclosed; the qi of man is in the kidneys.

Accordingly, there are laws on needling methods. In spring, summer, autumn, and winter, there are advisable needling techniques for needling proper regions, respectively.

> Spring, summer, autumn, and winter, each season has its specific location to be pierced. Take as pattern the location where it is situated. If in spring one pierces the summer section, the movement in the vessels loses its order and the qi is feeble. The evil qi invades the bones and their marrow and the disease that was to be healed cannot heal. This lets a person wish not to eat. In addition, one is short of qi. If in spring one pierces the autumn section, the sinews will be cramped. Qi moving contrary to its regular course cause coughing within one circulation period. The disease that was to be healed does not heal. This lets a person frequently show signs of fright. In addition, he weeps. If in spring one pierces the winter section, the evil qi attached itself to the depots. This lets a person show signs of distension. The disease that was to be healed does not heal. In addition, that person wishes to talk.[12]

There are details regarding the needling techniques in summer, autumn, and winter, which won't be covered here. Actually, these rules, being of an imaginary nature, do not hold practical significance in acupuncture and moxibustion therapy. However, the reviewers who didn't know the theory of the yin-yang school offered inaccurate commentary, asserting that "The rules stated in this section violate the treatment principles. For example, the needling in spring is applied to the part for summer needling; needling in summer is applied to the part for autumn needling. Needling the wrong part will result in deterioration instead of recovery. Therefore, accurately targeting the affected parts is the most important in clinical practice".[13] In fact, based on the discourse on "the location of Renqi" in different months, it becomes apparent that the essence of these rules is "being constrained by various taboos".

In the current version of *Huangdi's Internal Classic*, the yin-yang theory has gradually developed into an abstract concept emphasizing opposition and balance, which is substantially different from the earlier theory developed by the yin-yang school. The concept of yin and yang includes not only the integration of yin qi and yang qi that form the noumenon of all things in the universe, but also pays attention to the different attributes of yin and yang, which is related to the theory of "image". In medical theory, the former aspect of yin and yang can explain the formation of life, the quality of endowment, and the characteristics of emotion and body, while the

---

[12] The chapter "Discourse on The Essentials of Diagnosis and On Exhaustion in the Conduits" in *Plain Questions* of Paul U. Unschuld's translation version.

[13] Teaching-Research Team of Nanjing College of Medical Classics. *Translation and Annotation to Plain Questions of Huangdi's Internal Classic* (*Huang Di Nei Jing Su Wen Yi Shi*, 《黄帝内经素问译释》). Shanghai: Shanghai Science and Technology Press, 1959, p. 113.

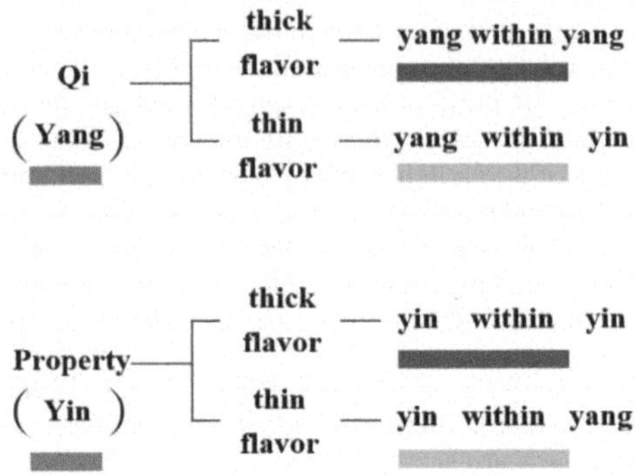

Fig. 16.2 The yin and yang properties of qi and flavor

latter can guide the process of diagnosis and treatment, as the identification of yin and yang is pivotal of pattern identification and treatment. These have been thoroughly discoursed in *Huangdi's Internal Classic*, and the highest degree of generalization is that:

> As for yin and yang, they are the Way of heaven and earth, the fundamental principles governing the myriad beings, father and mother to all changed and transformations, the basis and beginning of generating life and killing, the palace of spirit brilliance. ...Yin is tranquility, yang is agitation. Yang gives life, yin stimulates growth. Yang kills, yin stores. Yang transforms qi, yin completes physical appearance. Cold at its maximum generates heat; heat at its maximum generates cold. Cold qi generates turbidity; heat qi generated clarity.
>
> Water is yin; fire is yang. Yang is qi; yin is flavor. ...That which is of strong flavor is yin; that with weak flavor is yang of yin. That which is of strong qi is yang; that with weak qi is yin of yang. ...Acrid flavor and sweet flavor are effused and disperse and are classified as yang; sour flavor and bitter flavor cause gushing up and outflow and are yin.[14]

In this context, the yin-yang theory no longer emphasizes the concept of circulation but focuses on the opposing attributes of yin and yang. The dichotomous classification is no longer solely based on the concept of "the sequence of four seasons (a temporal concept). Instead, the method of "determining nature based on quantity" has been applied to the theories of qi and flavor, which are integral to drug theory (Fig. 16.2). In essence, yin and yang represent the highest level of abstraction and generalization. Their applicability extends across various aspects, following the principle that "from one can know ten, and from ten, know a hundred". Therefore, yin-yang constitutes the most vital component of the foundational theories in Chinese medicine.

[14] The chapter "Comprehensive Discourse on Phenomena Corresponding to Yin and Yang" in *Plain Questions* of Paul U. Unschuld's translation version.

16.3 Special Cases in Medicine: Three Yin and Three Yang

In ancient medicine, yin-yang theory has a special manifestation, with yin being further divided into greater yin, lesser yin and reverting yin, while yang is divided into greater yang, lesser yang and yang brightness. These conceptual terms hold significant importance in medicine, and the core of meridian theory is named accordingly, that is, each hand and foot has three yin and three yang meridians, totaling the "twelve regular meridians". In the chapter "Discourse on Heat" of *Plain Questions* (*Su Wen*, 《素问》), the process of exogenous febrile diseases is divided into six stages, corresponding to the six meridians: greater yang, lesser yang and yang brightness (exterior pattern); greater yin, lesser yin and reverting yin (interior pattern). The *Treatise on Cold Damage and Miscellaneous Diseases* (*Shang Han Za Bing Lun*, 《伤寒杂病论》), an important medical work written in the Eastern Han Dynasty (AD 25–200), is based on these six stages. Consequently, the diagnostic method derived from this work is commonly referred to as the "six-meridian pattern identification system" by later generations.

Compared with the binary classification that yin and yang are divided into greater yin, lesser yin, greater yang, and lesser yang, there are two new classifications in three yin and three yang, i.e., yang brightness and reverting yin. There are few such classification methods in other fields of traditional Chinese culture, and discussions on this topic can be traced back to the Han Dynasty. At the beginning of the chapter "Discourse on the Division and Unity of Yin and Yang" of *Plain Questions*, Huang Di asked: "Now, the three yin and the three yang of man do not correspond to the one yin and one yang of heaven. What is the reason for this?" Qi Bo answered, "As for the yin and yang correspondences in man, count their associations and you can reach ten; expand these associations further and you can reach one hundred. Count these associations and you can reach one thousand; expand them further and you can reach ten thousand. The association exceeding ten thousand are countless, and still their essential principle is one". It's not an explicit answer, which followed by an explanation of the twelve meridians. Zhao Hongjun, in his book *The Time of Huangdi's Internal Classic* (*Nei Jing Shi Dai*, 《内经时代》), interprets Qi Bo's reply as an evasion regarding the distribution of meridians, and the content about meridians is an irrelevant answer. Even Zhang Jingyue (AD 1563–1640), a scholar with profound medical knowledge, was not able to provide a direct response to this question. In the chapter "Comprehensive Categories" of *Classified Classic* (*Lei Jing*, 《类经》) and *Yin-Yang and Five Elements* (*Yin Yang Wu Xing*, 《阴阳五行》), he just quoted this paragraph without offering further explanation or personal insights, and then quoted a lot of texts from other classics. Similarly, in the chapter "Meaning of Medicine and Yi" of *Appendices to the Classified Classic* (*Lei Jing Fu Yi*, 《类经附翼》), there are many contents about the theory of Yi-ology, but it did not explicitly explain the concept of the three yin and three yang. Therefore, answering this question is indeed not easy". Zhao Hongjun believed that the most probable origin of the three yin and three yang is the representation of three sons and three daughters.[15]

In the chapter "On the Trigrams" of *Book of Changes* (*Yi*,《易》), there are records that:

> The Qian Trigram symbolizes heaven; hence it is referred to as the father; while the Kun Trigram symbolizes earth, hence it is referred to as the mother. The Zhen Trigram is a symbol of the first son obtained when the father seeks offspring, thus it is called the eldest son; and the Xun Trigram is a symbol of the first daughter obtained when the mother seeks offspring, thus it is called the eldest daughter. The Kan Trigram is a symbol of the second son obtained when the father seeks offspring again, thus it is called the middle son; and the Li Trigram is a symbol of the second daughter obtained when the mother seeks offspring again, thus it is called the middle daughter. The Gen Trigram is a symbol of the third son obtained when the father seeks offspring once more, thus it is called the youngest son; and the Dui Trigram is a symbol of the third daughter obtained when the mother seeks offspring once more, thus it is called the youngest daughter.

According to Zhao Hongjun, the concept of three yin and three yang originally emerged in the *Book of Changes* and was later absorbed into medical theory. However, throughout the extensive history of Yi-ology, the three yin and three yang concept is hardly mentioned except for explaining the formation of the Ba Gua (Eight Trigrams) as derived from Qian, Kun, and their six children. Tamba Motoyasu, a Japanese scholar renowned for his research in traditional Chinese medicine, briefly commented on the topic of three yin and three yang:

> Greater, Lesser, yin, and yang were originally known as the four seasons. ... However, subsequent scholars who studied the *Book of Changes* used these terms primarily to explain the practice of divination with yarrow stalks. It is in medical theory that greater yin, lesser yin, reverting yin, greater yang, lesser yang, and yang brightness are collectly referred to as the three yin and three yang.[16]

The concept of "three yin and three yang" is primarily associated with medical theory and has received limited attention from philosophers and historians studying the yin-yang theory. Even in traditional medicine, there is little mention of the origin or explanation of this concept; it is mainly applied without further elaboration.

In the *Pulse Classic* (*Mai Jing*,《脉经》) written by Wang Shuhe (AD 210–280) in the Western Jin Dynasty (AD 265–317), there are discourses on pulse by doctors of the pre-Qin period. Among these, two discourses mentioned three yin and three yang, making them the earliest known records on the topic.

> Bian Que said: When the pulse exits and enters at equal intervals, it is considered a normal pulse. If it exits once and enters twice, it is classified as lesser yin; if it exits once and enters three times, it is greater yin; and if it exits once and enters four times, it is reverting yin. Conversely, if it enters once and exits twice, it is lesser yang; if it enters once and exits three times, it is yang brightness; and if it enters once and exits four times, it is greater yang. The exiting phase of the pulse is associated with yang, while the entering phase is associated with yin.

---

[15] Zhao Hongjun. *The Time of Huangdi's Internal Classic* (*Nei Jing Shi Dai*,《内经时代》). Shijiazhuang Hongqi Printing House, 1985, pp. 133, 137.

[16] Tamba Motoyasu (Japan), the chapter "Three Yin and Three Yang" in the *Yi Sheng* (『醫賸』).

> The circulation of three yin and three yang meridians is as follows: Yinshi (03:00-04:59) to greater yang, Wushi (11:00-12:59) to yang brightness, Shenshi (15:00-16:59) to lesser yang, Xushi (19:00-20:59) to lesser yin, Zishi (23:00-00:59) to greater yin, and Choushi (01:00-02:59) to reverting yin. The pulse of lesser yang suddenly becomes small, large, short, or long, and vibrates within 60% of itself. It is exuberant at midnight on the first day of the 11th month as well as the first day of the first month or the second month... When the vibration reaches a range of 60% of the pulse or above, patients will have a symptom of headache and fullness in the rib side. If accompanied by vomiting, the patient can be cured. If accompanied by restlessness, the patient may die soon. Acupuncture on both sides of the end of ribs and points of gallbladder meridian at a depth of 70%.

The other five pulses also have similar descriptions as those of the Shaoyang pulse; however, for brevity, these discussions are omitted. The pulse theory attributed to Bian Que in the *Pulse Classic* can be found in three other Han Dynasty medical texts: the *Plain Questions*, *Miraculous Pivot* (*Ling Shu*, 《灵枢》) and *Classic of Difficulties* (*Nan Jing*, 《难经》). However, these texts do not explicitly mention Bian Que by name. As the chapter "Catalog of Literary Works" in *Book of the Former Han* include *Bianque's Internal Classic* (*Bian Que Nei Jing*, 《扁鹊内经》) and *Bianque's External Classic* (*Bian Que Nei Jing*, 《扁鹊外经》), the only explanation for this difference is that during the compilation of *Pulse Classic* by Wang Shuhe, the works attributed to Bian Que were available and the contents of his works were also included in the latest version of *Huangdi's Internal Classic*. Because if Wang Shuhe had extracted the content from *Huangdi's Internal Classic*, he would not have attributed it to Bian Que. These contents show that the pulse diagnosis and meridians were closely related and developed together in the pre-Qin period. Three yin and three yang can be used as the name of meridians, the quantity-oriented indicator of the exuberance and debilitation of yin and yang in pulse diagnosis, and the time standard that divides one day into six periods.

According to the chapter "The Ties between Yin and Yang Qi and Sun and Moon" in *Miraculous Pivot*, there are three yin and three yang meridians on the feet, totaling twelve meridians on both sides; while on the hands, there are two yin and three yang meridians totaling ten meridians on both sides, without meridians of hand reverting yin. These are consistent with the recorded meridians in medical books unearthed from the Mawangdui Han Dynasty Tombs, which indicates that they are the continuation of the earlier meridian theory. The explanation states: "The area above the waist is considered heaven, while the area below the waist is considered earth. Therefore, heaven is yang, and earth is yin. Thus, the 12 meridian pathways of the feet correspond to the 12 months of the year, while the ten fingers correspond to the ten days. The sun governs fire, and thus, what is above is yang". This explanation also provides insights into the naming of the meridians of the Hand and the Foot and the meanings of reverting yin and yang brightness:

> The earth branch Yin represents the yang qi generated in the first month. It controls the minor yang[17] conduit of the left foot. The earth branch Wei, the sixth month, controls the minor yang conduit of the right foot. The earth branch Mao, the second month, controls the major

[17] "Minor yang" refers to "lesser yang" in Paul U. Unschuld's translation version of *Plain Questions* and *Miraculous Pivot*.

yang[18] conduit of the left foot. The earth branch Wu, the fifth month, controls the major yang conduit of the right foot. The earth branch Chen, the third month, controls the yang brilliance conduit of the left foot. The earth branch Si, the fourth month, controls the yang brilliance conduit of the right foot. These two yang streams merge on the front side. Hence it is called "yang brilliance". The earth branch Shen represents the yin qi generated in the seventh month. It controls the minor yin[19] conduit of the right foot. The earth branch Chou, the twelfth month, controls the minor yin conduit of the left foot. The earth branch You, the eighth month, controls the major yin[20] conduit of the right foot. The earth branch Zi, the eleventh month, controls the major yin conduit of the left foot. The earth branch Xu, the ninth month, controls the ceasing yin[21] qi conduit of the right foot. The earth branch Hai, the tenth month, controls the ceasing yin qi conduit of the left foot. These two yin streams exchange all their contents. Hence it is called "receding yin qi".

Figure 16.3 illustrates the correspondence between the months and the three yin and three yang, clearly showing that the change law in the circulation of yin and yang, which progresses from weak to strong and then back to weak. It should be noted that this is only one method of matching three yin and three yang with time. In this matching scheme, only the two months associated with yang brightness and reverting yin are closely connected and can be considered as one "period". In contrast, the months related to lesser yang, lesser yin, greater yang and greater yin are each assigned to separate periods, not contiguous in time. Therefore, it is impossible to form the "Six Qi Circulation Circle" (Fig. 16.4). To solve this problem, ancient medical philosophers proposed conceptual framework. In this framework, reverting yin is understood as representing the imminent end of yin, and it is quantitatively weak, similar to lesser yin (Fig. 16.5).

Let's analyze Fig. 16.3 in more detail as it presents interesting observationst:

(1) The six concepts of "three yin and three yang" are depicted only once in Fig. 16.3, which is consistent with Figs. 16.4.
(2) In quantity, the sum of lesser yang and lesser yin or the sum of greater yang and reverting yin is equal to yang brightness or greater yin respectively.
(3) In the respective fields of yin and yang, there exists, or is realized, a smooth transition from weak to strong and then back to weak. By contrast, if the binary classification of "greater and lesser yin-yang" is displayed in the form of images (Fig. 16.6), it is obviously impossible to achieve such "perfection". As the quantitative change from weak to strong in yin and yang appears abrupt in the transition between them. In light of this, as the ancients said, the concepts of "greater, lesser, yin and yang" in the yin-yang school are merely synonymous with the seasons of spring, summer, autumn, and winter. Consequently, they

18 "Major yang" refers to "greater yang" in Paul U. Unschuld's translation version of *Plain Questions* and *Miraculous Pivot*.

19 "Minor yin" refers to "lesser yin" in Paul U. Unschuld's translation version of *Plain Questions* and *Miraculous Pivot*.

20 "Major yin" refers to "greater yin" in Paul U. Unschuld's translation version of *Plain Questions* and *Miraculous Pivot*.

21 "Ceasing yin" and "receding yin" refer to "reverting yin" in Paul U. Unschuld's translation version of *Plain Questions* and *Miraculous Pivot*.

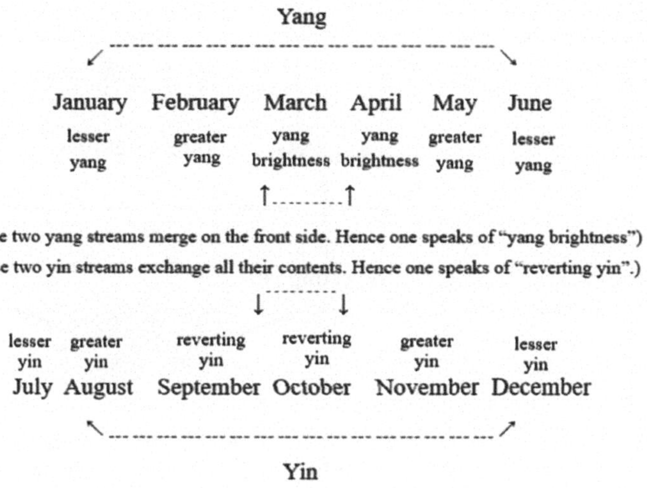

Fig. 16.3 Illustration of yin-yang and twelve months

Fig. 16.4 Six qi circle

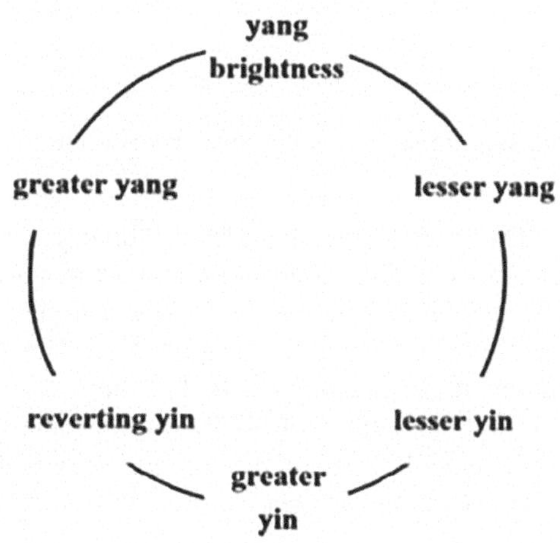

believed that there is no profound philosophical or theoretical significance to be appreciated in this framework.

The ternary classification of yin and yang is a more rational theory and philosophically grounded theory. However, it is intriguing to consider why this theory is only found in medicine. One explanation could be that "yin-yang theory has been extensively applied and developed in ancient medicine", as mentioned at the beginning of this chapter. Another possibility is that the culture of neighboring countries (see Chapter 6) influenced the division of one year into two parts of yin and yang, with each part further divided into three seasons.

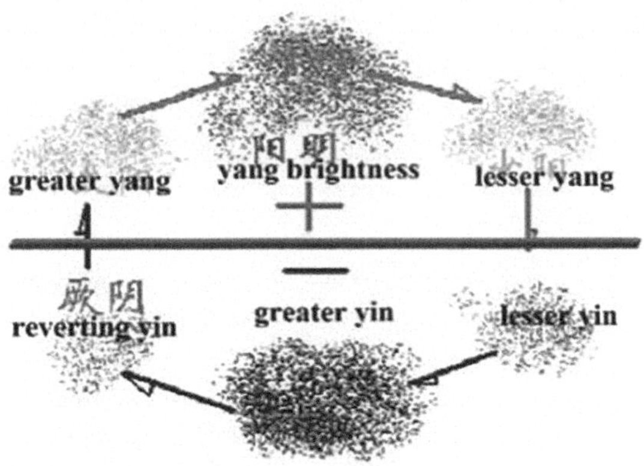

Fig. 16.5 Illustration of the quantitative change of "three yin and three yang"

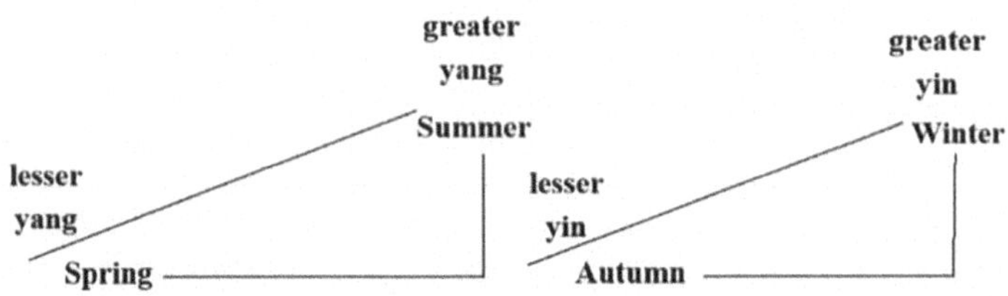

Fig. 16.6 Deficiency of the binary classification of yin and yang

There are many records of "three yin and three yang" in *Huangdi's Internal Classic*, but some questions and explanations in these texts remain unclear and unconvincing. This can be observed in the question and answer following the text mentioned earlier in the chapter "The Ties Between Yin and Yang Qi and Sun and Moon" in *Miraculous Pivot.* The question is that "Just now you mentioned that the celestial stem Jia controls the minor yang conduit in the left hand. That does not coincide with the regular numbers. Why is that so?"

However, the answer dwells on the abstract and avoids real issues.

> Here are meant the yin and yang qi of heaven and earth, not the sequence of the five elements in the course of the four seasons. Furthermore, yin and yang are designations, and have no physical appearance. The fact is: When counting them, one may reach the number ten, and leaving them one may reach 100; expanding them one may reach 1000, and further extending them one may reach 10,000. That is meant here.

It shows that the match of three yin and three yang with meridians is a unique mode, which cannot be consistent with the four seasons, yin-yang, and the Yi-ology system. As *Huangdi's Internal Classic*, being a compilation of medical treatises, naturally encompasses variations in the understanding and explanation of theories. Therefore, there are different understanding of three yin and three yang. For instance, the match of three yin and three yang with 12 months that recorded in the chapter "Explanations on the Vessels" in *Plain Question* is another mode that yin alternate with yang, rather than the first six months match with yin or yang and the second six months match with yang or yin. It can also be regarded as a manifestation of the transformation of yin-yang theory from circulation advocated by the yin-yang school to balance. Of course, circulation still existed in this mode.

However, from a different perspective, these explanations, often viewed as "prevarication" by later generations, can also be regarded as "correct understanding". For instance, some researchers regard the expressions like "qi-blood circulation", "circulating like loop", "circulating in the whole body" in ancient Chinese medical books as indications that ancient people had discovered blood circulation thousands of years ago, but in fact they are not. In ancient China, people did not conceive of the heart, an organ of monarch, beating day and night. They believed that the precordial beat that could be felt by hand is the beating of the great collateral of the stomach. Therefore, it is believed that lung is the origin of qi-blood circulation throughout the body. The circulation of 12 meridians starts with the lung meridian. During one exhalation, the contents of the meridians proceed three Cun (1 Cun = 1/3 dm); and the same occurs during one inhalation. Qi-blood circulates along the order of 12 meridians, which proceed six Cun during one breathing period and the total length is 16.2 Zhang (1 Zhang = 3 and 1/3 m). Guided by this thought, there arose beliefs in the "vessel qi" being located in certain meridian or certain place at certain time and the notion of correct and incorrect needling points. Whether dividing one day into six hours based on three yin and three yang, or dividing it into 12 h based on 12 months, both models are connected to the theory of meridians and the circulation of vessel qi. Therefore, as previously quoted, in the *Plain Questions*, Qi Bo answers the question concerning three yin and three yang with the distribution of meridians, which should not be regarded as the "irrelevant answer", but rather as the correct answer. Thus, the theory of three yin and three yang is indeed a special form of yin-yang theory in the field of medicine. In the Han Dynasty academia, the concept of three yin and three yang can only be traced back to the system of meridian theory.

In addition, as previously mentioned, there is the record of "six-meridian pattern identification system" for febrile disease in the chapter "Discourse on Heat" in *Plain Questions*. But the concept of yin and yang was not used to distinguish diseases. It is believed that the symptoms of each stage all belong to yang and are febrile disease. "Three yin and three yang" only relates to the "exterior and interior" of diseases. From the first day to the sixth day, the diseases gradually develop from the exterior (yang) to the interior (yin) in the order of greater yang, yang brightness, lesser yang, greater yin, lesser yin and reverting yin. And after another cycle from seventh day to the 12th day, the diseases gradually disappear. The *Treatise on Cold Damage and Miscellaneous Diseases* inherited the "six-meridian" system, but in which all the diseases of three yin belong to yin characterized by cold deficiency. Therefore, yin-yang pattern identification has been formed from this understanding. That is, "excess patterns belong to greater yang, while deficiency patterns belong to lesser yin; excess patterns belong to yang brightness, while deficiency patterns belong to greater yin". This is another typical manifestation of the transformation from the previous emphasis on circulation to a focus on opposition in yin-yang theory. It shows that yin-yang theory, the most important concept in the basic theoretical system of traditional Chinese medicine, was still in its infancy in the early classics. It seems that the theory of "three yin and three yang" also has the characteristics of starting from circulation and eventually focusing on opposition and balance.

16.4 Medicine and Yi-Ology

The saying that one who has no knowledge of *Book of Changes* cannot be called a great physician still be regarded as an old adage of medicine. It is believed to have originated from Sun Simiao (AD 541–682), a great doctor of the Tang Dynasty (AD 618–907). But it doesn't have such an extensive historical background, and actually it is from Zhang Jingyue, a doctor of the Ming Dynasty (AD 1368–1644). Sun Simiao's works do mention the importance of learning the *Book of Changes* for a great doctor:

> To be a great doctor, one must have a deep understanding of various medical classics such as the *Plain Questions*, *A-B Classic of Acupuncture and Moxibustion* (*Zhen Jiu Jia Yi Jing*, 《针灸甲乙经》), *Huangdi's Acupuncture Classic* (*Huang Di Zhen Jing*, 《黄帝针经》), *Ming Tang Liu Zhu* (《明堂流注》), as well as knowledge of the 12 meridians, three positions and nine indicators, five Zang-organs and six Fu-organs, exterior-interior correspondences and acupoints, materia medica and herbal combinations, and classical prescriptions from renowned doctors such as Zhang Zhongjing (around AD 150-219), Wang Shuhe, Ruan Henan, Fan Dongyang, Zhang Miao, Jin Shao. In addition, one should have a profound understanding of yin and yang, destiny through astrology, various physiognomies methods, and be skilled in divination methods such as burning tortoise-shell, *Book of Changes*, Wu Zhao and Liu ren (two ancient forms of Chinese divination).[22]

Therefore, Sun Simiao considered the *Book of Changes* as just one aspect of a doctor's extensive knowledge, which cannot be regarded as a medical work or a direct source of medical theory. Moreover, in Sun's works, there is no connection between medicine and the *Book of Changes*. Generally speaking, *Book of Changes* was rarely mentioned in medical works before the Song Dynasty (AD 960–1279). Even monumental medical works such as *Plain Questions*, *Miraculous Pivot*, and *Treatise on Cold Damage and Miscellaneous Diseases*, which revolutionized the field, there is no influence from the *Book of Changes* and no fundamental theory that is directly from it.

In order to truly understand the relationship between medicine and *Book of Changes*, it is necessary to examine its basic nature and its development throughout different times.

16.4.1 *The Basic Nature of the* Book of Changes

The *Book of Changes* is a book of divination. It employed various methods, including the use of tortoise shells and yarrow stalks, to predict auspicious or inauspicious

[22] Sun Simiao (the Tang Dynasty). *Essential Prescriptions Worth a Thousand Gold for Emergencies* (*Bei Ji Qian Jin Yao Fang*, 《备急千金要方》), Volume 1.

events through trigrams derived from numerical operations. The 64 hexagrams in the *Book of Changes* are composed of two symbols: "–" (yin) and "– –" (yang). Regarding the origin of these two symbols, Mr. Guo Moruo (AD 1892–1978) once interpreted them as symbols representing male and female genitals, symbolizing concepts such as male and female, heaven and earth, yin and yang.[23] This interpretation is undoubtedly influenced by the earlier scholars' use of the symbols Yang Yao (–) and Yin Yao (– –). Consequently, later scholars came to believe that, from their most basic symbols, both Ba Gua and the *Book of Changes*, embody philosophical concepts such as dichotomy, the unity and opposition of yin and yang, etc. However, this remains a subjective assumption and does not align with the objective laws of historical development.

Briefly speaking, in the development of human thinking, concreteness typically precedes abstraction. Abstract concepts summarizing the basic laws of unity and opposition and elaborating on the changes of things in accordance with those laws would not emerge at an early age or at the early stage of civilization. In fact, the male and female genitalia were hieroglyphics in the early stage (male: 且, female: 也). However, the symbols used in the divination with yarrow include not only "–" and "– –", but also "---". If the symbols of "–" and "– –" really represent male and female genitalia and possess the property of yin and yang, what kind of genitalia or other objective things does the symbol "---" represent? And what abstract conception does it reflect? Therefore, the key to understand the essential nature of the *Book of Changes* as a divination book lies in comprehending the origin and meaning of Yin Yao and Yang Yao.

In the chapter "The Survey Part A" in *Book of Changes*, there is the record of the formation of trigrams:

> The process involves the use of 50 yarrow stalks, but only 49 are actually used in operation. The diviner puts away one yarrow stalk at the beginning and randomly divides 49 yarrow stalks into two parts to symbolize two cardinal ways: heaven and earth. Then he puts one yarrow stalk between the ring and the small fingers of the left hand to symbolize man, the third cardinal way. The diviner utilizes a unit of subtracting four to subdivide both parts, symbolizing the four seasons. Then the diviner places the remaining stalks from the right part between the middle and ring fingers of the left hand, symbolizing a leap month. Another leap month occurs once every five years, thus for the left part, the surplus stalks are placed between the middle and index fingers of the left hand, initiating a new cycle of operations.

From this, we can learn the basic process of divination with yarrow in the *Book of Changes*. The diviner prepares 50 yarrow stalks, but actually, 49 are used in the operation. One yarrow stalk is taken out from the 49 stalks and put aside. Then the remaining 48 stalks are randomly divided into two parts. Four stalks are subtracted from each part until the surplus stalks are less than or equal to four. He puts the surplus stalks of each part together with the one initially taken out from 49 stalks. As the sum of the surplus stalks of the two parts would be eight or four, the stalks taken out would be 44 or 40 after the first operation. Then he repeats the process twice and the number of stalks taken out would be one of 36, 32, 28 or 24. Then

---

[23] Guo Moruo. *Research on Ancient Chinese Society* (*Zhong Guo Gu Dai She Hui Yan Jiu*, 《中国古代社会研究》). Beijing: People's Publishing House, 1954, p. 26.

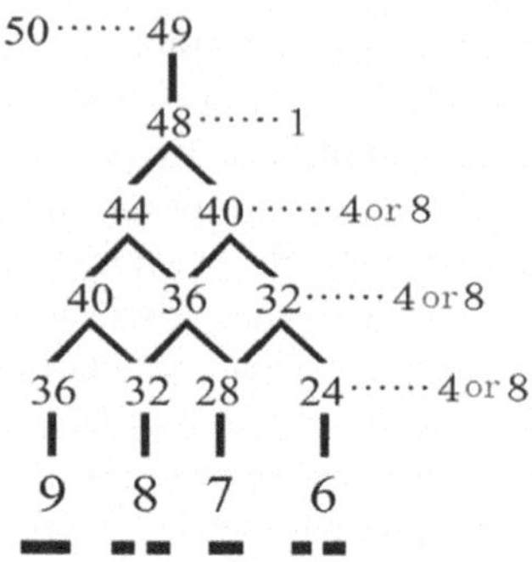

Fig. 16.7 Illustration of the formation of trigrams

the number is divided by four and the diviner gets a number of nine, eight, seven, or six. This number corresponds to a line in the trigram. A single trigram consists of three lines, while a double trigram consists of six lines. So, the process is repeated three or six times to compose a trigram. In the *Book of Changes*, "–" represents an odd number (9 or 7) and "– –" represents an even number (8 or 6), which formed the symbolic Ba Gua. Therefore, the symbols are actually derived from specific numbers and represent the abstraction of odd and even numbers, rather than the representation of yin and yang (Fig. 16.7).

Those who perceive the profound philosophy in the *Book of Changes* often point to the statement that "The Tai Ji (Supreme Ultimate) gives birth to the Liang Yi (Two Modes), the Liang Yi give birth to the Si Xiang (Four Constitution), and the Si Xiang to the Ba Gua", as a significant argument. They use this to support the idea that the philosophical concept of "one dividing into two" forms the foundation of the Ba Gua. However, "Si Xiang" are only four specific numbers "9, 8, 7, 6".[24] While these numbers are indeed used to compose the Ba Gua, they were not from the "Liang Yi" of yin and yang. Rather, they were categorized as the "Liang Yi" by "Si Xiang".

By examining the unearthed findings, we can gain a clearer understanding of how the divination with yarrow developed from specific numbers to symbols. In the Shang and Zhou Dynasties (1600–256 BC), there is evidence of the Ba Gua expressed as three or six numbers in oracle bone scripts, bronze inscriptions and pottery inscriptions. In addition, bronze wares from the same period exhibit patterns composed by four or five symbols of "–", "– –" or "---", like "≡".[25] For the trigram composed of these symbols, the number of yarrow stalks and the method used to form trigrams are different from those recorded in the *Book of Changes*, but their essence is completely identical. The source of the three symbols is that "When the remainder is below ten, seven is one line (–), eight is two lines (– –), and nine is

[24] Yang Xiong (53–18 BC, the Han Dynasty), the Volume 1 of *Tai Xuan Classics* (*Tai Xuan Jing*, 《太玄经》) says: "Xuan emphasizes the number of eight and nine just as the Yi highlights Si Xiang (Four Constitution)".

[25] Zhang Yachu, Liu Yu. "*Problems of Divination from the Eight Trigram Symbols of the Shang and Zhou Dynasties* (*Cong Shang Zhou Ba Gua Shu Zi Fu Hao Tan Shi Fa De Ji Ge Wen Ti*, 《从商周八卦数字符号谈筮法的几个问题》)", *Archaeology* (*Kao Gu*,《考古》), 1981, (2): 155.

three lines (---)".[26] Therefore, it is essential to recognize that the so-called trigram composed of three or six lines, and two or three symbols, should be regarded as a figure of the divination with yarrow, devoid of any inherent philosophy. Its authority lies in their association with ancestral worship, as the process must be carried out in the ancestral temple, so as to "draw wisdom from ancestors".[27] Therefore, it was believed that the emergence of a trigram is controlled by the spirit of the ancestors.

Moreover, besides using yin and yang to explain the two symbols of the Ba Gua, there are many scholars in both ancient and modern times who interpret the upper, middle and lower lines of the Ba Gua as heaven, earth, and human beings. In fact, they also made the same mistakes. Questions arise regarding their explanation of trigrams composed of four, five, or six lines. Are they truly attempting to correlate them with four seasons, five elements, and six qi? Therefore, when conducting historical research, it is crucial not to impose the ideology and culture of later generations onto those of the ancients, as this can transform simple matters into mysterious concepts.

16.4.2 The Development of Yi-Ology

After understanding the divination with yarrow, it can be known that *Book of Changes* is a book of divination in nature, which helped it to avoid being burned in the Qin Dynasty (221–207 BC). However, the basic nature of *Book of Changes* was changed by people in the Han Dynasty. Six Classics are all historical books. *Book of Changes* was also included as one of the Confucian classics and Boshi (Official Specialist) in *Book of Changes* were employed. Therefore, more and more texts expounding *Book of Changes* emerged. Because these classics are extensive, and often expand the very short and concise doctrines into thousands of words, which leads to the condition that "a person learns only one classic from a young age and can only expound it when he becomes an old man".[28]

Book of Changes is composed of two parts: text and annotation. The text contains 64 trigrams, the trigrams' names, line statements, and trigram statements. The earliest annotation of the *Book of Changes* were called "Shi Yi" ("Ten Wings"), including *Commentary on the Judgment (Part A and B)* (*Shang Xia Tuan Ci*, 《上下彖辞》), *Overall Image (Part A and B)* (*Shang Xia Xiang Ci*, 《上下象辞》), *Commentary on the Appended Phrases (Part A and B)* (*Shang Xia Xi Ci*, 《上下系辞》), *Commentary on the Words* (*Wen Yan*, 《文言》), *Sequence of the Trigrams* (*Xu Gua*, 《序卦》), the *Explanation of the Trigrams* (*Shuo Gua*, 《说卦》) and *Assorted or Miscellaneous Trigrams* (*Za Gua*, 《杂卦》). It was said that "Shi Yi" were written by Confucius, but subsequent generations proved that they were written in the Qin and Han Dynasties. In

[26] Yang Xiong (the Han Dynasty), the Volume 1 of *Tai Xuan Classics*.

[27] Ban Gu (AD 32–92, the Han Dynasty), "Yarrow Stalks and Tortoise Shells"of the chapter "Discourse on Virtues" in *Comprehensive Discussions in the White Tiger Hall* (*Bai Hu Tong Yi*, 《白虎通义》).

[28] The chapter "Category of Literary Works" in *Book of the Former Han*, Volume 30, No. 10.

the *Commentary on the Judgment*, there are no theories of yin-yang or five elements. Instead, it focuses on the concepts of firmness and yieldingness and conforms to the rules of heaven, earth, and four seasons. In the *Overall Image*, there are both yin-yang theory and concepts of firmness and yieldingness, but the content of yin-yang theory is less than that of the latter. And there is no five-phase theory. In the *Commentary on the Appended Phrases*, yin-yang theory takes center stage. In the *Explanation of the Trigrams*, there are matches between eight directions, four seasons, heaven, earth, thunder, wood, water, fire, mountain and lake. This is the concrete manifestation that *Book of Changes* was developed by Confucianism in the Han Dynasty. It can be said that the changes in *Book of Changes* and the process that medicine and other disciplines absorbed theories of yin-yang and five elements are synchronized. However, it should be noted that although there is yin-yang theory in *Book of Changes* and was regarded as the work of Confucius, it was not the main idea of Yi-ology in the Han Dynasty. Pi Xirui (AD 1850–1908), a scholar of the Qing Dynasty (AD 1636–1912), pointed out that "Although the books about Yi-ology in the Western Han Dynasty (202 BC–AD 8) have been lost, there are still verifiable statements". He consulted "The collation and study on Yi-ology by Liu Xiang (77-6 BC)", *Writings of Prince Huainan* (*Huai Nan Zi*, 《淮南子》) that "collected the various scholars' discourses on Yi-ology", as well as other comments on the *Book of Changes* by Jia Yi (200–168 BC), Dong Zhongshu (179–104 BC) and other prominent Confucians. Based on this research, Pi Xirui concluded that "*Book of Changes* mainly talks about philosophical connotations and human affairs, thus, yin-yang and divination are not the main topic of it".[29]

In the Han Dynasty, Yi-ology had two characteristics, or two branches. Firstly, as a Confucian classic, it was annotated and developed in the aspects of philosophical connotations and human affairs. Secondly, it was used for divination, which was its original role and was still applied. In the late Western Han Dynasty, Liu Xiang and his son Liu Xin (50 BC–AD 23) are masters of Confucian classics who also studied *Book of Changes*. They compiled *Seven Categories of Writings* (*Qi Lve*, 《七略》) and respectively categorized the two kinds of works of Yi-ology into *Bibliography of Confucian Classics* (*Liu Yi Lve*, 《六艺略》) and the "Divination" in *Bibliography of Calculation and Divination* (*Shu Shu Lve*,《数术略》), which are totally unrelated to the "Yin-Yang School" mentioned in *Bibliography of Masters and Philosophers* (*Zhu Zi Lve*, 《诸子略》) and the "Five Elements School" mentioned in *Bibliography of Calculation and Divination*. It shows that *Book of Changes* is not the origin of the yin-yang school. As Pi Xirui said, from Confucius to scholars in the early Han Dynasty, the *Book of Changes* was primarily used to explain human affairs rather than disasters caused by the imbalance of yin and yang. Meng Xi (about 90 BC–AD 40) and Jing Fang (77 BC–AD 37) first came up with using the *Book of Changes* to speculate on natural disasters. Their opinions have their own merits and benefits,

---

[29] Pi Xirui (the Qing Dynasty). *Comprehensive Discourses on Confucian Classics* (*Jing Xue Tong Lun*, 《经学通论》). Beijing: Zhonghua Book Company, 1954, pp. 17–18.

but in the end, their contributions were limited to the extra chapters of the *Book of Changes* and have little influence on later generations.[30]

However, in the Song Dynasty, the interpretation of the *Book of Changes* took a significant shift with the emergence of Neo-Confucianists as a creation mode for the creation of all things in the universe, marking a departure from its previous interpretations. Only by being clear on this stage in the development of Yi-ology can we grasp the essence of the *Book of Changes.*

In the pre-Qin period and the Han and Tang Dynasties, Confucianism only focused on ethics and lacked philosophical speculation. Therefore, Confucianism was unable to engage in debates with Buddhism and Taoism with its "theory of pre-existing destiny" in the Wei, Jin and the Southern and Northern Dynasties (AD 220–589), as well as the Sui and Tang Dynasties (AD 581–907). Therefore, since the rise of Neo-Confucianism in the Song Dynasty, Confucianism has experienced two self-revolutions: one is to deny the "theory of pre-existing destiny"; the other is to philosophize its ethical theory by absorbing the understanding on the world origin from Buddhism and Taoism and the logical structure of their philosophy. As a result, "Neo-Confucianism" came into being, which subsequently became the official philosophy. Over the following six or seven hundred years, Neo-Confucianism exerted a profound influence on the development of China's economy, politics, and cultural thoughts.[31]

During the Song and Ming Dynasties, Yi-ology was extensively studied by Neo-Confucianists. The pioneer Zhou Dunyi (AD 1017–1073) created the *Tai Ji Diagram* (*Tai Ji Tu*, 《太极图》) based on the *Wu Ji Diagram* (*Wu Ji Tu*, 《无极图》) by Chen Tuan (AD 871–989). *Wu Ji Diagram* illustrates the process of Taoist alchemy, with its order progressing from bottom to top, while *Tai Ji Diagram* depicts the generation and development of myriad things and the order is from top to bottom (Wu Ji—Tai Ji—yin and yang—five elements—male and female—myriad things). It seems that *Tai Ji Diagram* was to expound the thought that "in Yi-ology, there is the Tai Ji at first. The Tai Ji gives birth to the Liang Yi, the Liang Yi give birth to the Si Xiang, and the Si Xiang to the Ba Gua, which was recorded in *Commentary on the Appended Phrases* (*Part A*). However, from my perspective, there is no intrinsic relationship between them. The author of *Commentary on the Appended Phrases* is the first one who contradicted "the method of trigrams formation" and put forward the concept that "the Tai Ji gives birth to the Liang Yi, the Liang Yi give birth to the Si Xiang, and the Si Xiang to the Ba Gua". A more accurate comment on him is that he elucidated yin-yang theory using trigrams and lines rather than incorporating yin-yang theory into Yi-ology. The Neo-Confucianists of the Song and Ming Dynasties further reiterated this way. In nature, *Book of Changes* is the tool used by them to expound the theory of Neo-Confucianism, including image-numerology represented by Shao Yong (AD 1011–1077) and the integration of image-numerology and philosophical connotations by Zhu Xi (AD 1130–1200). As for the nature of the divination mentioned earlier in the

---

30 Pi Xirui (the Qing Dynasty). *Comprehensive Discourses on Confucian Classics*, Beijing: Zhonghua Book Company, 1954, p. 19.

31 Zhang Liwen. *Research on Neo-Confucianism* (*Song Ming Li Xue Yan Jiu*, 《宋明理学研究》). Beijing: China Renmin University Press. 1985, pp. 1–96.

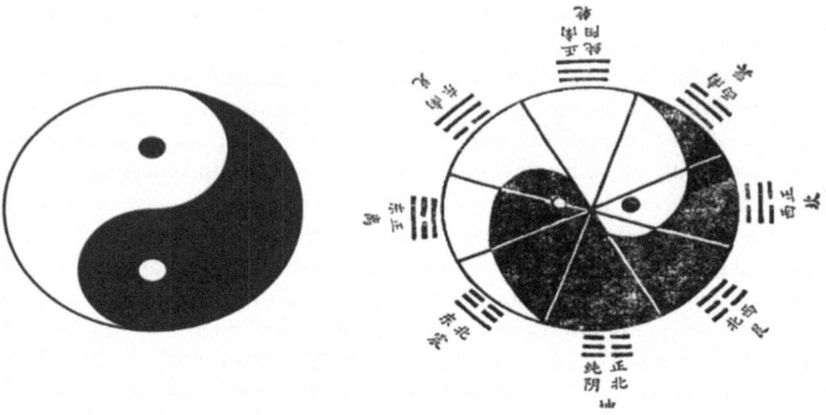

a b

Fig. 16.8 The symbol of Tai Ji. The history of the "Symbol of Tai Ji", commonly known as the "Yin-yang Fish", is relatively short. It was probably intended to show the interlocking of yin and yang, with the black and white "fish eyes" representing the presence of yang within yin and the yin within yang (**a**). The symbol drawn today quantitatively reflects the thought that "Qian Trigram and Kun Trigram give rise to the other six trigrams" (**b**). Among them, the pure white or black part symbolizes heaven (Qian Trigram, pure yang) or earth (Kun Trigram, pure yin); while the other six parts represent varying proportions of yin or yang, decreasing to two-thirds, one-half, and one-third, respectively. However, it is impossible to have the trigram of "half yin and half yang" in Ba Gua

Book of Changes, whether it has been negated or forgotten since the Song Dynasty can be answered by two examples. When Zhu Xi left office, he intended to write a remonstrance because Zhao Ruyu (AD 1140–1196) was dismissed from the prime minister, but due to the unfavorable divination result, he burned the remonstrance and named himself "Dun Weng"; Wang Fuzhi (AD 1619–1692) of the Qing Dynasty used divination twice to decide whether to participate in the anti-Qing movement.[32]

Due to the fact that the symbols in *Book of Changes* have the mathematical nature of even and odd, which can be connected with the philosophical concept of "one dividing into two", the book became a fertile ground for the development of yin-yang theory. Therefore, both the Confucianists who wrote *Commentary on Book of Changes* (*Yi Zhuan*, 《易传》) before the Song Dynasty and the Neo-Confucianists of the Song and Ming Dynasties chose it as the culture medium to introduce the Taoist philosophy and pictures into Confucianism. For instance, "Tao", "Taiji", the *Tai Ji Diagram* (Fig. 16.8) featuring in the interlocking yin-yang and the view that "one is the child of the divine law; after one comes two, after two comes three, and after three comes all things".

The evolution of Yi-ology throughout different time periods is a fascinating subject. In the pre-Qin period, its primary role was divination, which was closely related to the characteristics of that era. In the Han Dynasty, Confucian scholars

[32] Zhang Liwen. *Research on Neo-Confucianism*. Beijing: China Renmin University Press, 1985, pp. 390, 610.

advocated the philosophical connotations and human affairs in *Book of Changes*, elevating it to a position of great importance among the six classics. In the Song Dynasty, Neo-Confucianists expanded upon the *Book of Changes* to expound their philosophical thoughts and cosmology including "Li" (Tai Ji, Taoism, and heavenly principles), "Qi" (yin-yang and five elements are just two qi), and "universe" (all things follow the same principle). During the Han and Tang Dynasties, the theory of image and numerology also played a role, and were interweaved with *Book of Changes* in their interpretations. In modern times, with the widespread emphasis on science in society, Yi-ology has formed a deep bond in various scientific fields, shaping what is now known as "contemporary Yi-ology". This includes the correspondence between Ba Gua and the binary system, the correspondence between 64 trigrams and the periodic table of elements, as well as the new planets of the solar system, the particle flow and the biological field, etc. It can be predicted that as long as the social and cultural tendencies change, Yi-ology will also adapt and transform accordingly. Nevertheless, the study of Yi-ology remains a timeless classic that holds significance across all ages.

16.4.3 Medical Theory Has Nothing to Do with Book of Changes

The *Book of Changes* contains nearly 30-line statements concerning body and disease, but no medical knowledge. Instead, these statements predicting trauma, pregnancy, eye disease, and other diseases by divination. Although both medicine and Yi-ology have incorporated the yin-yang theory, there is no inheritance relationship between them. To illustrate this point, we can consult the different numerical systems associated with yin and yang in medicine and Yi-ology. In the *Book of Changes*, odd numbers refer to yang while even numbers refer to yin: "heavenly number one, earthly number two, heavenly number three, earthly number four, heavenly number five, earthly number six, heavenly number seven, earthly number eight, heavenly number nine, earthly number ten[33]". Especially number nine and six are the representatives of yin and yang, with all yang lines referred to as nine and all yin lines as six. In medicine, the emphasis on the yin-yang relationship of odd and even numbers is not consistent, and when these numbers are considered, they often contradict the Yi-ology system. In most cases, odd numbers refer to yin, and even numbers refer to yang. It can be seen in *Plain Questions* and other classics, such as "five Zang-organs and six Fu-organs", "seven impairments and eight benefits", "the heaven follows the six-six system and the earth follows the nine-nine system", etc. In the chapter "Catalog of Literary Works" in *Book of the Former Han*, the explanation of classical formulas mentions "discrimination of five bitter and six pungent flavors". It is evident that the "five" and "six" here are not modifier attributes used to

[33] Han Bo (the Jin Dynasty) annotator, Kong Yingda (the Tang Dynasty) commentator. *The Annotation and Interpretation of the Book of Changes* (*Zhou Yi Zhu Shu*, 《周易注疏》), Volume 11.

express numbers, because the bitter and pungent flavors cannot be divided into five or six categories. In terms of flavors, pungent and sweet flavors are yang because they disperse, and sour and bitter flavors are yin because they induce vomiting and purgation. Therefore, the numbers six and five act as digital symbols for yin and yang. Interestingly, in the silk medical texts unearthed from the Mawangdui Han Dynasty Tombs, two works on meridians record six yang meridians and five yin meridians, aligning with the rule. It would be incorrect to interpret the number "11" as indicating an incomplete meridian system, suggesting a missing meridian. Other works from that period can shed light on the importance of the numbers 5, 6 and 11. For example, the *Discourses of the State* states that "The heaven has six kinds of qi and the earth has five elements, and these numbers are the usual (the heaven indicates six and the earth indicates five)", but the "Catalog of Temperament and Calendar" provides further details:

> Generally, the heaven emphasizes the number six and the earth emphasizes the number five. There are six kinds of qi in heaven, and they generate five flavors as they descend. As heaven and earth intersect, the number six and five constitute –the cycle of Heavenly Stems and Earthly Branches, which the ancient Chinese used in daily life. Therefore, the cycle of Heavenly Stems includes six Jia days, and the cycle of Earthly Branches includes five Chen days. The cycle of Heavenly Stems and Earthly Branches completes after 11 steps, and then a new cycle begins.

At that time, there was no direct correlation between the meridians and the five Zang-organs and six Fu-organs. Therefore, we can only conclude that the correspondence of six with yang and five with yin is due to the numerological concept that odd numbers are associated with yin and even numbers with yang.

In ancient Chinese medicine, there was only one example that accepted the concept that odd numbers are yang and even numbers are yin. That is, women can conceive within six days after menstruation. It further specified that on the first, third, or fifth days, they are more likely to conceive boys, while on the second, fourth, or sixth days, they are more likely to conceive girls. The statement was initially recorded in the *Dong Xuan Zi* (《洞玄子》), which was quoted by *Formulary from the Heart of Medicine* (*Yi Xin Fang*, 《医心方》). When later generations quoted this statement, they often said that it is "from *The Daoist Canon* (*Dao Zang Jing Lun*, 《道藏经论》)",[34] and believed that "it is for this reason that in Taoism, there is the six-day abstinence of all sexual activities and after this period, women can't be pregnant".[35] Although it is absurd, this opinion was widely accepted and supported by most physicians in the Tang, Song, Yuan, Ming, and Qing Dynasties. It is possible that this misconception influenced the difficulties many emperors faced in fathering children.

Those who subscribed to this view were wrong for "learning without thinking". However, astute observers who paid attention to practice and didn't blindly believe the words of the ancient people would not be misled by this statement. For example,

[34] Zhu Su (the Ming Dynasty), *Prescription for Universal Relief* (*Pu Ji Fang*, 《普济方》), Volume 337.

[35] *Encyclopedia of Daily Lives* (*Mo E Xiao Lu*, 《墨娥小录》), Volume 10, Beijing: China Bookstore, the photocopy version of Ju Hao Tang, 1959.

Zhang Jingyue, a physician of the Ming Dynasty, believed that "Every word in the *Book of Changes* is a guide to medicine".[36] Nevertheless, he had different opinions on this statement: "The statement put forward by *The Daoist Canon* that women can conceive boys on the first, third, and fifth days after menstruation is a conjecture without real evidence, and I cannot agree with it". As he was aware that "There are women who conceive on the 10th, 15th, and even 20th day after menstruation".[37] This is the only idea in ancient medicine where the idea aligned with Yi-ology, yet it still misled people for thousands of years!

16.4.4 Ancient Chinese Medicine Using Trigrams as Metaphors

Ancient Chinese physicians, who were well-versed in classics, history, and Confucian texts, frequently employed trigrams as metaphors to expound physiology, pathology, medicine, and pharmacology. A notable example can be found in the chapter "The First-year Reign of Zhaogong" in the *Commentary of Zuo*, which records that the Duke Ping of the Jin States (?–532 BC) sent for doctor Yi He and was diagnosed with Gu disease. When asked what Gu disease was, he replied: "Gu disease is caused by excessive sexual activities. Literally, Gu means poisonous insects or snakes that are deliberately kept in utensils. Similarly, when rice is stored for a long time, the small insects that fly out of it are also a kind of Gu. According to the *Book of Changes*, if an older woman confuses a young man (an improper relationship between a man and a woman), the result is like a violent wind toppling the grass and trees on a mountain. This condition is also referred to as Gu". This illustrates that Duke Ping of the Jin State fell ill due to his excessive indulgence, mirroring the symbolism of Trigram Gu where the wind is beneath the mountain. As the proverb goes, "Excessive drinking and promiscuity will ultimately exhaust a person".

However, such metaphor in medical literature were not common until the Song Dynasty, after which their occurrence increased significantly. This trend aligns with the flourishing of Neo-Confucianism during the Song and Ming Dynasties. One example of trigrams being used as a metaphor for human organs can be found in recorded literature. "Trigram Xun and Zhen represent dynamic, while Trigram Kun and Gen represent static. The composition of Trigram Guan can be explained by the physiological function and structure of the eye. Eyes provide vision. Mushang-gang (目上纲, upper eyelid) represents blinking (dynamic), and Musxiagang (目下纲, lower eyelid) represents non-blinking (static). Therefore, the Trigram Guan has Trigram Xu above and Trigram Kun below. Similarly, Trigram Yi has Trigram Gen above and Trigram Zhen below, because the lower jaw controls chewing (dynamic),

---

[36] Zhang Jiebin (the Ming Dynasty), Volume 1 of *Appendices to the Classified Classic*.

[37] Zhang Jiebin (the Ming Dynasty), *Complete Works of Jingyue* (*Jing Yue Quan Shu*, 《景岳全书》), Volume 38, 39.

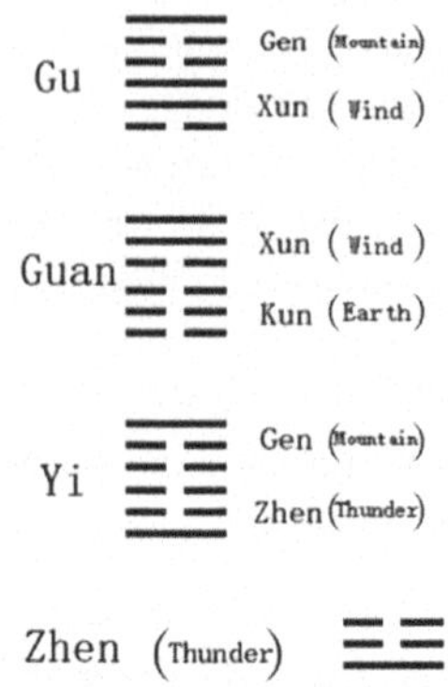

Fig. 16.9 Trigrams that can be used as metaphors for Chinese physicians

while the upper jaw controls non-chewing (static)".[38] Another example is the use of trigrams as a metaphor for pharmacology. It is described, "The hollow center of the lotus leaf is similar to the shape of Trigram Zhen. Trigram Zhen represents vibration, enabling people to perceive the power of growth through it" (Fig. 16.9).[39]

Historically, Zhang Jingyue is renowned as one of the foremost scholars who studied the relationship between medicine and the *Book of Changes*. His influential work, *Medicine and Book of Changes* (*Yi Yi*, 《医易》), also contains many such metaphors. For example, the Dui Trigram (☱) is shaped like a mouth, with its opening at the top; the Xun Trigram (☴) is shaped like a thigh, resembling two thighs hanging down. These metaphors are not essential, and in some cases, no metaphor is even better, as they have no practical significance. However, there are several trigrams that doctors use as metaphors, which can help later scholars understand the yin-yang theory.

(1) ䷊ and ䷋ (Tai Trigram and Pi Trigram), ䷾ and ䷿ (Jiji Trigram and Weiji Trigram)

The common feature of these two groups of trigrams is that if the yang is below and the yin is above, it is auspicious, and otherwise it is ominous. The Jiji Trigram (coordination between water and fire) and Weiji Trigram (incoordination of water and fire) composed of Kan Trigram (water) and Li Trigram (fire) are easier to understand, because they represent a natural phenomenon, that is, only water can be transpired by placing it above the fire. But the Tai Trigram and Pi Trigram composed of Qian Trigram (Heaven) and Kun Trigram (Earth) do not conform to natural phenomena. At the same time, they are also inconsistent with government affairs, because the emperor pertains to heaven and the people embody the earth. How can there be a principle that the emperor is below and the people are above? Therefore, later generations of Yi-ology scholars mostly interpret Tai Trigram and Pi Trigram as a kind

38 Zhang Congzheng (AD 1156–1228), *Confucians' Duties to Their Parents* (*Ru Men Shi Qin*, 《儒门事亲》), Volume 2.

39 Li Gao (AD 1180–1251, the Yuan Dynasty), the chapter "Spleen and Stomach Deficiency" in *Prescriptions with Collectible Value* (*Lan Shi Mi Cang*, 《兰室秘藏》), Volume 1.

of change or transposition. According to this, even the auspicious Tai Trigram, can reverse the position of the heaven (yang) and earth (yin) during transformation. This shows that the *Book of Changes* contains the philosophy of yin-yang transformation. However, such theory is unconvincing, as every trigram represents an independent and specific divination result. It can only be used for predicting weal and woe, embodying certain stability and practical meaning. If all can be transformed into each other, the diviners will become philosophers, and there is no weal or woe at all. Thus, all the results of divination can be interpreted as "misfortune might be a blessing in disguise". In fact, it is not difficult to explain why "the trigram that the trigrams with heaven below imply auspiciousness, and otherwise it is ominous". The reason lies in the fact that the lower trigrams are internal trigrams, superior to the external trigrams positioned above. These two trigrams are suitable for explaining some theory of TCM or the theory of health preservation because according to the TCM theory, the primordial yang of a healthy body should be at the lower part of it, or otherwise, the body will be in an abnormal state, such as "deficiency yang floating upward" or "upper heat and lower cold".

The famous formula "Jiaotai Pill" is often said to be derived from *Han's Clear View of Medicine* (*Han Shi Yi Tong*, 《韩氏医通》). However, Han only mentioned in his book that the formula can be used in combination with Huang Lian (Golden Thread, Rhizoma Coptidis), a bitter and cold medicine, without explicitly naming the formula. And here are the original words: "Huang Lian is the preferred choice for treating diseases caused by fire…Boil Huang Lian and Guan Gui (Chinese cinnamon, Cinnamomum cassia) with clear water, add honey, and then drink it on an empty stomach. This formula can quickly harmonize the coordination between the heart and kidneys".[40] This formula uses Huang Lian to clear the upper heart fire and uses cinnamon to warm up the lower yuan, reverse the imbalances of upper yang and lower yin, and upper fire and lower water, which is consistent with the auspicious and ominous changes observed in the two groups of trigrams mentioned earlier, hence name it as "Jiaotai Pill".

(2) ䷗, ䷒, ䷊, ䷡ and ䷪ (Fu Trigram, Lin Trigram, Tai Trigram, Dazhuang Trigram, Guai Trigram)

Yin-yang balance is an ideal physical condition, that is, the image of Tai Trigram. But in fact, absolute balance is rarely achieved. The physical condition tends to manifest as yin exuberance or yang exuberance in different degrees, and according to which can be divided into other four trigrams. The goal of a doctor is to, restore the patient's body to a relatively balanced state like the image of "Tai Trigram" as much as possible, through treatment.

This thinking is of great importance in TCM, that is, the general principle of "TCM practitioners should avoid the mistake of treating excess as excess and deficiency as deficiency when conducting treatment based on pattern differentiation".

[40] Han Mao (the Ming Dynasty): the chapter "Medicinal Property" in *Han's Clear View of Medicine* (*Han Shi Yi Tong*, 《韩氏医通》), Volume 2, Beijing: People's Medical Publishing House, 1989, p. 28.

Although pattern identification and treatment is widely recognized as the characteristic, strength and guiding principle of traditional Chinese medicine, but not all treatments consistently adhered to the principle in practice. For treating aphtha and toothache, the immediate response is to consider using Niu Huang (Calculus Bovis) for clearing the upper; for hemorrhoids, Di Yu (Radix Sanguisorbae) and Huai Jiao (Fructus Sophorae) are commonly used; for constipation, Da Huang (Radix et Rhizoma Rhei) is an essential medicine; and for fever, medicine with cold and cool nature is typically the first choice. In fact, all these diseases belong both to deficiency pattern and to excess pattern, requiring treatment through elimination or reinforcement. Understanding the transformation of the above five trigrams can greatly assist students in comprehending the waning and waxing of yin and yang. However, the fundamental problem of using these trigrams as metaphors is to explain the concept of yin and yang, rather than delve into the nature of divination as described in the *Book of Changes*. Therefore, after studying the relationship between medicine and *Book of Changes*, Zhang Jingyue came to a conclusion that the relationship between medicine and Yi-ology only lies in the yin-yang theory.[41]

The theory of Yi-ology is profound, and it is not possible to definitively state how a doctor can become a great practitioner simply by understanding *Book of Changes*. This enigmatic nature of Yi-ology has led to a certain level of blind admiration and worship. Therefore, the belief that "medicine originates from Yi-ology" is very popular today. Some people may have mistakenly identified certain similarities between isolated passages in the classics that interpret the *Book of Changes* and those found in medical classics, leading them to believe that ancient medical theory originated directly from Yi-ology.

The reason why later generations thought that *Book of Changes*, a book of divination, has an inseparable relationship with medical theory is that it was traditionally regarded as the origin of yin-yang theory. Among them, the earliest and most well-known reference to this connection can be found in the chapter "The World" in *Zhuangzi* (《庄子》), where it states that "The *Book of Changes* is for expounding yin and yang". Over time, people began to interpret the lines of "–" and "– –" as the representatives of yin and yang, constantly elucidating the changes of yin and yang implied in two primary forms (Yin Yao and Yang Yao), Ba Gua and 64 Trigrams. However, researchers who pay attention to the cultural development process have already pointed out that "there is no concept of yin-yang in the *Book of Changes*, and it emerges in the Warring States period (475–221 BC) when people expounded yin and yang with the *Book of Changes*".[42] Therefore, why did people's understanding of *Book of Changes* change so much? It is because that in the course of the development

[41] Zhang Jiebin (the Ming Dynasty), the chapter "Medicine and Yi" in *Appendices to the Classified Classic-Meaning*, Volume 1.

[42] Xie Songling, *Heaven, Human and Phenomenon: An Introduction to the History of Yin-Yang and Five Phases Theory* (*Tian Ren Xiang: Yin Yang Wu Xing Xue Shuo Shi Dao Lun*, 《天人象:阴阳五行学说史导论》), Jinan: Shangdong Art Publishing House, 1989, pp. 26–27. And Hu Weijia listed a large number of arguments and the same views of predecessors when discussing the origin of the concept of yin-yang in his article "The Formation and Significance of the Concept of Yin-Yang, Five Elements and Qi: Exploration on the Scientific Thought System in the Pre-Qin Period" in

of Yi-ology, the origin nature of *Book of Changes* was increasingly blurred, and the *Book of Changes* became a tool for ideologists to exert their own theories. While other Confucian classics have a unified content and form that restrict interpretation, the *Book of Changes* has the characteristics that its content and form can be separated and developed in many ways. The philosophical connotations are the content of *Book of Changes* and image-numerology is the form of the *Book of Changes*. So, different interpretations and statements arise from those who study the *Book of Changes*. In the chapter "Fangshu" in the *Book of the Later Han*, Confucius is mentioned as stating that there are four aspects of principles the sages expound in the *Book of Changes*, which was recorded in *Commentary on the Appended Phrases*: Those who rely on it to guide their speech-making admire its rhetoric; those who rely on it to guide their actions admire the philosophy of changes; those who rely on it to guide them in tool-making appreciate the trigram and linear images; those who rely on it to guide their divination with tortoise shells and yarrow stalks admire its principles of divination.

Therefore, it can be said that "The history of Yi-ology is closely tied to the *Book of Changes*, which has undergone a process of separating form and content as well as continuously moving toward formalization and abstraction"; "The trigram theory focusing on the contradictory relations played a vital role for the development of ancient Chinese dialectical thoughts".[43] It is widely believed that *Commentary on Book of Changes*, which was greatly influenced by yin-yang theory, was written in the mid-and-late period of the Warring States, or even in the last stage of the Warring States. Both medicine and Yi-ology adopted yin-yang theory during this period, but there was no direct inheritance relationship between them. The *Book of Changes* simply serves as a carrier of the yin-yang theory.

Throughout the development history of yin-yang school, yin-yang theory and its relationship with medicine, we can observe that the theory of yin-yang school is a kind of social philosophy that discussed political affairs with the order of yin-yang and four seasons, and did not completely involve the dialectical thinking of the pre-Qin period, which emerged and developed naturally along with the advancement of human thinking. And Yi-ology is an important carrier of this rational thinking.

As mentioned earlier, the five-element theory was promoted during the Han Dynasty, while the yin-yang theory gained prominence in the Song Dynasty. This is supported by Zhou Dunyi's Tai Ji Diagram, which illustrates that the two fundamental forces of yin and yang transcend the five elements and the myriad things. In the field of medicine, both the yin-yang theory and the five-element theory were emphasized during the Han Dynasty. Their strength lies in their ability to inherit the concept of the circulation of yin and yang, as well as their focus on the opposition and balance between these two forces. However, during the Jin and Tang Dynasties, the

---

Studies in the History of Natural Sciences (*Zi Ran Ke Xue Shi Yan Jiu*, 《自然科学史研究》), 1993, (1): pp. 16–28.

[43] Xu Zihong, translator and annotator, the foreword of *Translations and Annotation to Book of Changes* (*Zhou Yi Quan Yi*, 《周易全译》), Guiyang: Guizhou People's Publishing House, 1991, pp. 2, 6.

focus of medicine shifted toward practical application. In contrast, during the Song Dynasty, physicians placed greater emphasis on theory. As a result, the yin-yang theory gained widespread recognition in the medical classics of the Han Dynasty, leading to a surge in explanations and annotations. In this context, the yin-yang theory gradually became the foundation of theoretical medicine and spurred discussions on the relationship between medicine and the *Book of Changes*. Today, the notion that "medicine originated from the *Book of Changes*" is promoted by modern scholars, as evidenced by the numerous related works available in bookstores.

The current abundance of works on "medicine and Yi-ology" can be partly attributed to Zhang Qicheng's explanation in the foreword of a book collection studying Yi-ology. He states that "the Book of Changes is unique in that it is composed of both a symbol system and a writing system, is the only text revered by both Confucianism and Taoism, and has had a profound impact on humanity and science". As a result, "The Book of Changes is regarded as the primary source of Chinese civilization and the central focus of Chinese culture".[44]

Throughout history, various fields of ancient Chinese science and technology and culture have intersected. Therefore, the philosophical thought has interpenetrated with science and technology, ethics, and morality, instead of emerging separately from physics. However, if we study the essence of these relationships, it is not hard to find that some of them are mutually integrated, such as medicine and yin-yang theory, while others coexist without a direct connection, such as medicine and the *Book of Changes*.

[44] Zhang Qicheng, the foreword of *Yi-ology and Traditional Chinese Medicine* (*Yi Xue Yu Zhong Yi*, 《易学与中医》). Beijing: Cathay Bookshop, 1999.

Chapter 17
Development and Evolution of "Pulse Diagnosis"

The diagnostic method of traditional Chinese medicine (TCM) is to diagnose a disease or a pattern by a comprehensive analysis based on the information obtained from inspection, auscultation and olfaction, inquiry, and palpation. The four methods of obtaining information like medical history and signs are the "four diagnostic methods" of TCM.

The question of which diagnostic method holds the most value is a subject of debate. Generally, most people believe that doctors who "discern the cause of the disease without asking extensive questions" are highly skilled and respected by the patients, which seems to underscore the fascination and importance of pulse diagnosis. But in TCM classics, inspection is regarded as the foremost skill. There is a saying that "doctors who can diagnose patients by palpation are clever; while "those who can do it by inspection are considered masters in the field".

However, it is inquiry, somewhat underestimated, that can provide doctors with the most invaluable information for diagnosis, which cannot be obtained by other methods. Therefore, a comprehensive diagnosis should combine the four diagnostic methods, ensuring the best possible care for patients. It is crucial to emphasize that diagnosis should be approached with honesty and integrity, without any form of deceit or showmanship.

When discussing the development of four diagnostic methods, each possesses its own distinct characteristics. Palpation should not be solely attributed to pulse diagnosis, as all the diagnostic methods involving direct contact with patients fall under the category of palpation. Therefore, assessing the firmness of the abdomen, detecting abscess and identifying sore spots should be considered as forms of palpation. And pulse diagnosis does not entirely depend on the three fingers (the first finger, middle finger, and the third finger).

Pulse diagnosis plays an important role in TCM and contributes to the enigmatic nature of the practice. Pulse diagnosis is commonly depicted in various artforms as a symbol of the doctor's role. Then, what is the origin of this diagnostic method using three fingers?

Y. Liao, *Reconstructing the Image of Medicine in the Qin and Han Dynasties*, Understanding China, https://doi.org/10.1007/978-981-95-6801-7_17

17.1 Pulse Diagnosis

Pulse diagnosis in TCM refers to the identification of the meridian to which a disease corresponds according to the clinical symptoms. Obviously, this diagnostic method is not the earliest diagnostic method employed in TCM. The most basic and original way to treatment involves directly addressing the symptoms presented by the patient. Thus, only when the meridian theory comes into being can it be possible to attribute symptoms to a certain meridian and then reach therapeutic principles and methods.

This origin of diagnostic method can be traced back to the medical silk manuscripts excavated from Mawangdui Han Dynasty Tombs. These ancient texts provide evidence of the utilization of pulse diagnosis in TCM. For example, in the *Moxibustion Classic of the Eleven Foot-Arm Meridians* (*Zu Bi Shi Yi Mai Jiu Jing*, 《足臂十一脉灸经》), it states that "applying ignited moxa to … meridian to treat …"; in the *Moxibustion Classic of the Eleven Yin-Yang Meridians* (*Yin Yang Shi Yi Mai Jiu Jing*, 《阴阳十一脉灸经》), there are also expressions that "the diseases transmitted by X meridian include … that belong to Y meridian; the diseases produced by viscus include …". Although this method of pulse diagnosis is considered the most basic one, the connection between the meridian theory and clinical manifestations is the closest and most direct. This diagnostic approach is largely independent of influences from the five-phase theory and the Zang-Fu organ theory. However, the palpation recorded in the Mawangdui silk manuscripts has no inner connection with the meridian theory. For example, in *Moxibustion Classic of the Eleven Foot-Arm Meridians*, it is recorded that "if the patient's pulse beats like the pounding of rice by three hands, the patient will die in no more than three days"; "if there are dropped-beat pulse, the patient will die in no more than three days". This diagnostic method is consistent with the method and principle of palpation used in Western medicine.

According to the findings in Mawangdui silk manuscripts, symptoms play a crucial role not only in diagnosing the pathological changes of meridians, but also in determining the patient's survival. These can be seen in the *Moxibustion Classic of the Eleven Foot-Arm Meridians*, *Moxibustion Classic of the Eleven Yin-Yang Meridians* and *Death Signs of the Yin-Yang Pulses* (*Yin Yang Mai Si Hou*, 《阴阳脉死候》), where symptoms alone were relied upon to diagnose whether the patient can survive or not. From this point, it shows that in the early stage of the meridian theory, the pulse diagnosis played a leading role, which can be regarded as the initial manifestation of the connection to pulse, disease, and diagnosis. This diagnostic method can be also found in certain chapters of the *Plain Questions* (*Su Wen*, 《素问》) and the *Miraculous Pivot* (*Ling Shu*, 《灵枢》), but it is worth noting that only the chapter "Explanations on the Vessels" in the *Plain Questions* and the chapter "The Conduit Vessels" in the *Miraculous Pivot* inherited almost all the different pulse types and symptoms recorded in the *Moxibustion Classic of the Eleven Yin-Yang Meridians* (see Tables 17.1 and 17.2). For example, in Table 17.1, focusing on the foot lesser yang meridian, it can be seen that the distinct differences in the understanding of various pulse and symptoms in other chapters compared to the *Moxibustion Classic of the Eleven Yin-Yang Meridians*. Table 17.2 provides further examples, such as deafness, heart pain, and pharynx disease, which are attributed to corresponding meridians in different ways in the *Moxibustion Classic of the Eleven Yin-Yang Meridians*, *Moxibustion Classic of the Eleven Foot-Arm Meridians*, *Plain Questions* and *Miraculous Pivot*. From this analysis, we can conclude that the diseases transmitted by the meridians recorded in the chapter "The Conduit Vessels" align primarily with those recorded in the *Moxibustion Classic of the Eleven Yin-Yang Meridians*; while when there are overlaps between the disease transmitted by meridian and diseases produced by viscus in the *Moxibustion Classic of the Eleven Yin-Yang Meridians*, the contents were modified in the chapter "The Conduit Vessels". Comparing the pulse and symptoms in the *Moxibustion Classic of the Eleven Foot-Arm Meridians* with those in the *Moxibustion Classic of the Eleven Yin-Yang Meridians*, we can see that the former is closer to the diseases produced by the viscus of the latter, and is obviously different from the diseases transmitted by meridian. This further emphasizes the point discussed in Chapter 2 of this book, that the disease transmitted by meridian and the disease produced by viscus are not classifications of diseases, but rather a result of earlier works on sphygmology containing more comprehensive contents. Later generations gradually disregarded other theories of pulse and symptoms as their systems were deemed inferior to those recorded in the chapter "The Conduit Vessels". Whereas the twelve meridians in the chapter "The Conduit Vessels" gradually became the classic method that has been used up to the present day.

With the advent of the meridian theory, the body surface was divided into different regions. The change of superficial veins in a certain region can also be regarded as the manifestation of the pathological changes of the meridian, leading to the emergence of the "method of diagnosing collaterals". This kind of method is extensively discussed in the chapter "Discourse on Skin Sections" in the *Plain Questions*. The expression "Observing the collaterals" in the *Plain Questions* and *Miraculous Pivot* generally refers to this diagnosis.

17.1.1 Ancient Pulse-Taking Methods

The ancient pulse-taking methods refer to the methods that are no longer in use. For example, the chapter "The Sixteenth Difficulty" in the *Classic of Difficulties* (*Nan Jing*, 《难经》) mentioned various approaches to examining pulse movements, "The movement in the pulses can be examined at the nine indicators in three positions, as well as through its yin or yang nature. One may, further, examine it by exerting light or heavy pressure, or by taking into regard the sixty informants, or one single movement in the pulses as it varies by the four seasons. The distance from the sages is long and far. Today, everybody selects but one of these methods and considers his choice to be correct. How can one distinguish what are correct and incorrect diagnostic methods?" This shows that these methods belonged to an earlier era when the *Classic of Difficulties* was written. In *Classic of Difficulties*, the explanation for the methods of those pulse diagnosis and terminologies are "to apply those ancient methods into the Cun, Guan, and Chi methods to develop its own theory in the name

Table 17.1 The records of various symptoms relating to foot lesser yang meridian

	Disease transmitted by meridian						
	Bitter taste in the mouth	Frequent sighing	Difficulty to turn the body due to hypochondriac pain	Slightly dirty complexion	Lusterless skin	Feverish sensation over the lateral side of foot	Yang reversal
Moxibustion Classic of the Eleven Yin-Yang Meridians			▲		▲	▲	▲
Chapter "The Conduit Vessels" in *Miraculous Pivot*	▲	▲	▲	▲	▲	▲	▲
Moxibustion Classic of the Eleven Foot-Arm Meridians							
Chapter "The Generation and Completion of the Five Depots" in *Plain Questions*							
Chapter "Discourse on the Essentials of Diagnosis and On Exhaustion in the Conduits" in *Plain Questions*							
Chapter "Discourse on Heat" in *Plain Questions*							
Chapter "To Pierce Lower Back Pain" in *Plain Questions*							
Chapter "Discourse on Recession" in *Plain Questions*			▲				
Chapter "Discourse on Opposition and Compliance in Piercing in the Course of the Four Seasons" in *Plain Questions*							
Chapter "Discourse on Misleading Piercing" in *Plain Questions*			▲				

(continued)

Table 17.1 (continued)

	Disease transmitted by meridian						
	Bitter taste in the mouth	Frequent sighing	Difficulty to turn the body due to hypochondriac pain	Slightly dirty complexion	Lusterless skin	Feverish sensation over the lateral side of foot	Yang reversal
Chapter "The Conduits and Their Sinews" in *Miraculous Pivot*							
Chapter "Various Diseases" in *Miraculous Pivot*							
Chapter "Oral Inquiry" in *Miraculous Pivot*		▲					

Disease produced by viscus

Headache	Mandible pain	Pain of outer canthus	Swelling and pain of the supraclavicular fossa	Swollen armpit	Scrofula	Chills following perspiration	Malaria	Pain in the lateral side of chest and hypochondrium, rib-side, hip and knee, the anterior side of tibia, fibula, external malleolus and all the joints	Inability of the fourth toe to move		The third toe impediment
▲						▲	▲	▲		□ □	▲
▲	▲	▲	▲	▲	▲	▲	▲	▲	▲		
		▲	▲	▲	▲			▲	▲	□	
								▲			

(continued)

Table 17.1 (continued)

Disease produced by viscus											
Headache	Mandible pain	Pain of outer canthus	Swelling and pain of the supraclavicular fossa	Swollen armpit	Scrofula	Chills following perspiration	Malaria	Pain in the lateral side of chest and hypochondrium, rib-side, hip and knee, the anterior side of tibia, fibula, external malleolus and all the joints	Inability of the fourth toe to move		The third toe impediment
	▲							▲			
		▲									
								▲	▲		

Table 17.2 Records of different meridian tropism of deafness, heart pain and pharynx pain

	Moxibustion Classic of the Eleven Foot-Arm Meridians	The chapter "Discourse on Recession" in *Plain Questions*	The chapter "Discourse on Misleading Piercing" in *Plain Questions*	The chapter "Various Diseases" in *Miraculous Pivot*	The chapter "The Generation and Completion of the Five Depots" in *Plain Questions*	The chapter "Discourse on the Essentials of Diagnosis and on Exhaustion in the Conduits" in *Plain Questions*	The chapter "Discourse on Heat" in *Plain Questions*
Hand yang brightness meridian (teeth meridian)			▲				
Hand greater yang meridian (shoulder meridian)	▲	▲					
Foot greater yang meridian	▲						
Hand lesser yang meridian	▲			▲			
foot reverting yin meridian					▲		
Hand great yin meridian	▲			▲			
Foot greater yin meridian		▲		▲			
Hand lesser yin meridian		▲					

(continued)

Table 17.2 (continued)

	Moxibustion Classic of the Eleven Foot-Arm Meridians	The chapter "Discourse on Recession" in *Plain Questions*	The chapter "Discourse on Misleading Piercing" in *Plain Questions*	The chapter "Various Diseases" in *Miraculous Pivot*	The chapter "The Generation and Completion of the Five Depots" in *Plain Questions*	The chapter "Discourse on the Essentials of Diagnosis and on Exhaustion in the Conduits" in *Plain Questions*	The chapter "Discourse on Heat" in *Plain Questions*
Foot lesser yin meridian	▲	▲	▲	▲			
Hand reverting yin meridian		▲					
Hand lesser yang meridian				▲			
Foot reverting yin meridian				▲			
Hand lesser yin meridian							
Hand greater yang meridian (shoulder meridian)							
Foot lesser yin meridian			▲	▲			
Hand lesser yang meridian		▲					

(continued)

Table 17.2 (continued)

	Moxibustion Classic of the Eleven Foot-Arm Meridians	The chapter “Discourse on Recession” in *Plain Questions*	The chapter “Discourse on Misleading Piercing” in *Plain Questions*	The chapter “Various Diseases” in *Miraculous Pivot*	The chapter “The Generation and Completion of the Five Depots” in *Plain Questions*	The chapter “Discourse on the Essentials of Diagnosis and on Exhaustion in the Conduits” in *Plain Questions*	The chapter “Discourse on Heat” in *Plain Questions*
Foot reverting yin meridian						▲	
Foot greater yin meridian							▲

of interpreting the *Huangdi's Internal Classic* (*Huang Di Nei Jing*, 《黄帝内经》)".[1] Therefore, these interpretations cannot fully capture the precise details of the ancient pulse-taking methods, which require further exploration and discussion.

17.1.2 Three Positions and Nine Indicators

The basic diagnostic method known as the "Three Positions and Nine Indicators" is to take arterial pulses from in three areas of the human body, namely the upper (head), middle (hand), and lower (foot) regions to diagnose the location of disease by comparing the size and change of pulses. This method has the following characteristics:

(1) The method of "Three Positions and Nine Indicators" can only be documented in the *Plain Questions*, but not in the *Miraculous Pivot*. In the *Plain Questions*, apart from the chapter "Discourse on Three Sections and Nine Indicators[2]" that provides a detailed introduction to this method, other chapters such as "Discourse on the Eight Cardinal Turning Points and on Spirit Brilliance", "Discussion on the Division and Union of True Qi and Evil Qi" and "Discussion on Regulating the Conduits" also make references to "Three Positions and Nine Indicators".

The reason for the difference is that although "Three Positions and Nine Indicators" is a representative and quite a common term that highlights the meticulousness of pulse diagnosis in traditional medicine, it specifically belongs to a certain medical school in ancient times.

(2) The proponents of "Three Positions and Nine Indicators" held a critical stance toward the pulse diagnosis of Cunkou. It is the pulse diagnosis of Cunkou that only provides information regarding the presence or absence of illness. Just as it is stated in the *Huangdi's Internal Classic*: "When it arrives at the Cunkou and strikes the hand, at times it is big, at times it is small. When it is big, then evil arrives; when it is small, then it is balanced". It further emphasizes, "It is in the yin and in the yang sections; one cannot quantify its advancement" Therefore, "Hence, one searches for the evil at the nine indicators in the three sections"; "Those who do not know of the three sections, they do not distinguish between yin and yang and they do not separate heaven and earth". "If one pierces without knowing the nine indicators in the three sections and the locations of the vessels affected by a disease, even in the event of significant transgression, the practitioner is unable to stop it. If one punishes where there is no transgression,

[1] Shang Qidong, "Research on Pulse-taking"; *Chinese Journal of Medical History* (*Zhong Hua Yi Shi Za Zhi*, 《中华医史杂志》), 1995, (2): 95.

[2] "Three Sections and Nine Indicators" refers to "Three Positions and Nine Indicators" in Paul U. Unschuld's translation version of *Plain Questions*.

this is called a great error …He does not know the three sections and the nine indicators. Hence, he cannot continue for long”.[3]

The proponents of the “Three Positions and Nine Indicators” method considered it to be the best diagnostic method, asserting that disease can be treated preemptively according to this method. They even went so far as to claim that “The nine indicators in the three sections: treat them as the origin. The discourse on the nine needles: it is not necessary to keep it”.[4] Therefore, this viewpoint is quite different from the perspective of the author of the *Miraculous Pivot*, who deemed the “Nine Needles” as pivotal in diagnostic method. These varying opinions in the field of medical academics may be a contributing factor explaining why “Three Positions and Nine Indicators” is absent from the *Miraculous Pivot*.

(3) The proponents of “Three Positions and Nine Indicators” did not refer to the theories of twelve meridians and five Zang-organs and six Fu-organs in their development of this theory. Instead, they introduced an idea that “The perfect numbers of heaven and earth, they begin with one and they end with nine. One is heaven; two is the earth, three is man. Subsequently, they are multiplied by three - three times three is nine - to correspond to the nine fields. Hence, man has three positions. Each position has three indicators. They serve to decide about death or survival. They serve to manage the one hundred diseases. They serve to regulate states of depletion and repletion and to eliminate evil and disease”. Additionally, they proposed the concept of “Nine Zang-organs” as a novel contribution to the field of traditional medicine.

Later generations of scholars did not sufficiently appreciate the significance of numerology upheld by this school of thought, but mistakenly assigned the attributes “The heaven indicator of the upper position: the moving vessels on the two sides of the forehead. The earth indicator of the upper position: the moving vessels on the two sides of the cheeks. The main indicator of the upper position: the moving vessels in front of the ears”. To the foot lesser yang meridian, foot yang brightness meridian, and hand lesser yang meridian, respectively. Such an attribution is ill-advised and fails to align with the original teachings of the “Three Positions and Nine Indicators” methodology.

It is because the text only mentioned that “The heaven indicator serves to examine the qi at the corners of the head. The earth indicator serves to examine the qi of the mouth and teeth. The man indicator serves to examine the qi of ears and eyes”. Moreover, in order to align with the principle of “three times three makes nine”, the proponents of “Three Positions and Nine Indicators” had to add “the qi in the chest” to the middle part of the body where the heart and lung are located so as to correspond to the number of “three” (see Table 17.3).

This characteristic reinforces the notion that its theory is distinct from the theories of the twelve meridians and Zang-Fu organs. It is worth noting that the influence of

[3] The chapter “Discourse on the Division and Union of True Qi and Evil Qi” in *Plain Question*.

[4] The chapter “Discourse on the Eight Cardinal Turning Points and on Spirit Brilliance” in *Plain Question*.

Table 17.3 Pulse-taking method based on nine indicators in three positions

Three positions nine indicators		Pulse	Reflection
Upper position	Heavens indicator	The moving vessels on the two sides of the forehead	The qi at the corners of the head
	Man indicator	The moving vessels in front of the ears	The qi of ears and eyes
	Earth indicator	The moving vessels on the two sides of the cheeks	The qi of mouth and teeth
Middle position	Heavens indicator	Hand greater yin meridian	Lung
	Man indicator	Hand lesser yin meridian	Heart
	Earth indicator	Hand yang brightness meridian	The qi in the chest
Lower position	Heavens indicator	Foot reverting yin meridian	Liver
	Man indicator	Foot greater yin meridian	Spleen and stomach
	Earth indicator	Foot lesser yin meridian	Kidney

the worship of the number "nine" in the Western Han Dynasty (202 BC–AD 8) played a significant role in shaping this method. In particular, it had similarities with other practices of the time, such as the "Calendar Based on Three Governance" formulated during the periods of Wang Mang and Liu Xin, and the system that unifies weights, measures, and rhythms based on numerology. Based on these observations, it can be inferred that the method known as the "Three Positions and Nine Indicators" likely originated during the middle and late Han Dynasties.

The "Three Positions and Nine Indicators", based on numerology, has limited practical value due to its deviation from reality, which is also the primary reason why it did not gain widespread acceptance. Although "Three Positions and Nine Indicators" involves pulse diagnosis in different regions, its essence is different from the method of "taking pulse in different meridians", leading to potential confusion between the two approaches.

17.2 Pulse-Taking Method at Renying and Cunkou

This diagnostic method compares the size of the pulses at specific locations on the body, namely the "Renying" (pulse over the side of the neck) and "Cunkou" (pulse over the wrist), to determine the meridian to which the disease belongs. This method is mainly recorded in the *Miraculous Pivot*, including chapters such as "End and Beginning", "The Conduit Vessels", "The Measurements of the Vessels", "Heat Diseases", "Discussing Illness; Examining the Foot-long Section", "The Four Seasonal Qi", "Prohibition and Appropriation", "The Five Complexions", "The Yin and Yang Categorization and the 25 Human Types", etc. While in the *Plain Questions*, there are only two brief discussions on this method: "Discourse on the Six Terms of a Year and on Phenomena associated with the Condition of the Depots[5]" and "Discourse on Abdomen and Center". This method exhibits the following characteristics:

(1) "The movement in the vessels at the Cunkou is controlled by the center. The movement in the vessels at the Renying opening is controlled by the exterior": the internal body belongs to Yin and the external body belongs to Yang, so it is believed that "At the qi opening the condition of the yin qi shows. At the Renying opening that of the yang qi shows".[6]

At the same time, since the *Miraculous Pivot* has repeatedly emphasized that the pulses of Renying and Cunkou can reveal the relationship of exuberance and debilitation of yin and yang, it is plausible to speculate that the pulse-taking method described in the chapter "The Sixteenth Difficulty" of the *Classic of Difficulties* refers to the method based on yin and yang.

This provides a possible explanation for the absence of pulse-taking method at Renying and Cunkou which is very important in various ancient pulse-taking methods mentioned in the chapter "The Sixteenth Difficulty".

(2) Normal pulse: "Both correspond to each other; they leave together and they come together. This is as if one were to pull a rope that has the same strength all along. In spring and summer, the movement in the vessels at the *ren ying* opening is somewhat more massive than at the inch opening. In autumn and winter, it is somewhat more massive at the inch opening. Those who are like this, they are called "healthy persons".[7]

The "normal pulse" method of pulse-taking at Renying and Cunkou shares similarities with the pulse-taking method based on the "Three Positions and Nine Indicators". Both methods rely on the concept of "balance" to diagnose whether the body is healthy or not. However, the pulse-taking method at Renying and Cunkou is more abstract and general compared to the method based on "Three Positions and Nine Indicators". The reason is that the different foundations of the two methods. The "Three Positions and Nine Indicators" is rooted in the number "nine" of numerology, so "nine indicators should be like one"; while the pulse-taking method at Renying and Cunkou is based on the yin-yang theory, so it simply requires that "both correspond to each other".

(3) Twelve abnormal meridians:

> When the movement in the vessels at the *ren ying* opening is one time greater than that at the inch opening (Cunkou), the disease is in the foot minor yang[8] conduit. When it is one time greater and racing, the disease is in the hand minor yang conduit.

---

[5] "Depots" refer to "Zang-organs" in Paul U. Unschuld's translation version of *Plain Questions*. He also uses "palaces" to explain "Fu-organs".

[6] The chapter "The Four Seasonal Qi" in *Miraculous Pivot*.

[7] The chapter "Prohibition and Appropriation" in *Miraculous Pivot*.

[8] "Minor yang" refers to "lesser yang" in Paul U. Unschuld's translation version of *Plain Questions*. He also uses "major yang/yin" to explain "greater yang/yin", and uses "ceasing yin/yang" to explain "reverting yin/yang".

> When the movement in the vessels at the *ren ying* opening is two times greater than that at the inch opening, the disease is in the foot minor yang conduit. When it is two times greater and racing, the disease is in the hand minor yang conduit.
>
> When the movement in the vessels at the *ren ying* opening is three times greater than that at the inch opening, the disease is in the foot yang brilliance[9] conduit. When it is three times greater and racing, the disease is in the hand yang brilliance conduit.
>
> When the movement in the vessels at the inch opening is one time greater than that at the *ren ying* opening, the disease is in the foot ceasing yin conduit. When it is two times greater and racing, the disease is in the hand heart ruler conduit.
>
> When at the inch opening is two times greater than that at the *ren ying* opening, the disease is in the foot minor yin conduit. When it is one time greater and racing, the disease is in the hand minor yin conduit.
>
> When the movement in the vessels at the inch opening is three times greater than that at the *ren ying* opening, the disease is in the foot major yin conduit. When it is three times greater and racing, the disease is in the hand major yin conduit.[10]

The same text can be found in the chapter "End and Beginning" in *Miraculous Pivot*. The theoretical basis of this method is the yin-yang theory.

However, in practice, the highly generalized yin-yang theory alone may not suffice to distinguish the specific pathological changes of the twelve meridians. Therefore, when natural science intersects with philosophical thoughts and concepts, it may exhibit theoretical advancement but also become somewhat disconnected from reality.

(4) Changes in pulse manifestation: As mentioned earlier, the limitations of the pulse-taking method at Renying and Cunkou, which diagnoses by comparing the size of the two pulses, have been noted. To address these limitations, the concept of pulse manifestation has been introduced.

- Renying
  - If the pulse of Renying is strong, it indicates heat syndrome due to excess of yangqi;
  - If the pulse of Renying is weak, it indicates cold syndrome due to insufficiency of yangqi;
  - If the pulse of Renying is tense, it indicates pain syndrome due to obstruction of the channels by pathogenic factors;
  - If the pulse of Renying is slow, irregular and intermittent, it indicates that the disease is now alleviated and then aggravated.
- Cunkou
  - Qi abundance results in distension and a feeling of fullness. The center is cold, and food is no longer digested;
  - Qi depletion results in heat in the center. The patient emits phlegm, is short of breath qi, and the color of his urine changes;
  - Qi tightness results in pain and blockage-illness;
  - A movement with intermissions is associated with a disease that is at times severe, and at times less active.

---

[9] "Yang Brilliance" refers to "yang brightness" in Paul U. Unchuld's translation version of *Miraculous Pivot*.

[10] The chapter "Prohibition and Appropriation" in *Miraculous Pivot*.

(5) Guange (closure and obstruction): "When the movement in the vessels at the Renying is four times greater than that at the Cunkou, and it is both large and rapid, this is called 'overflowing yang'. Overflowing yang is an external barrier, and the patient will die; there is no cure"; "When the movement in the vessels at the Cunkou is four times greater than that at the Renying, it is called 'internal closure'. In the case of an internal closure, the movement in the vessels is both large and rapid. The patient will die; there is no cure".

The term "Guange (Closure and obstruction)" holds significance in the pulse-taking method at Renying and Cunkou. In fact, there are two words: "closure" and "obstruction", which correspond to the Cunkou pulse and the Renying pulse, respectively. Since the Cunkou reflects the internal changes of the Five Zang-organs and is also known as "Neiguan (internal closure)"; and the Renying reveals the external changes of the Six Fu-organs, it is referred to as "Waige (external obstruction)". In the *Classic of Difficulties*, "Guange" was interpreted as "if the movement of the Cunkou pulse extends upward to the Yuji, that constitutes an overflow. It signals external closure and internal resistance", indicating that the original meaning of "Guange" has been lost.

The chapter "The Conduit Vessels" in the *Miraculous Pivot* contains records of specific methods of the pulse-taking method at Renying and Cunkou following each meridian disease, but those records cannot be found in its original text, *Moxibustion Classic of the Eleven Yin-Yang Meridians*. Therefore, it can be inferred that this method likely developed after the *Moxibustion Classic of the Eleven Yin-Yang Meridians* and the *Pulse Classic* (*Mai Jing*, 《脉经》) were unearthed from Zhangjiashan, Hubei province.

Moreover, in the *Miraculous Pivot*, the chapters "The Conduit Vessels" and "Prohibition and Appropriation" are among the few medical treatises on Huang Di's teachings and doctrines. It is important to note that the current version of *Huangdi's Internal Classic* is not to the same as that recorded in the chapter "Catalog of Literary Works" in *Book of the Former Han* (*Han Shu*, 《汉书》). Instead, it possibly includes medical classics from various schools recorded in the "Catalog of Literary Works". Based on this understanding, we can infer a connection between the contents of chapters in the *Miraculous Pivot* that record Huang Di's teachings and the answers of his ministers and the records related to *Huangdi's Internal Classic* in the "Catalog of Literary Works" (see Chapter 3 of the book's Part II). Therefore, it is reasonable to consider that the doctors or the medical school that valued the book in the early period supported the pulse-taking method at Renying and Cunkou.

17.2.1 Pulse-Taking Method Based on Four Seasons

Pulse-taking method based on four seasons, according to pulse manifestation, matches each season with its corresponding five Zang-organs and holds that each season has a normal pulse (Spring corresponding to string pulse, summer to surging pulse, autumn to floating pulse, and winter to sunken pulse). Excess and insufficiency indicate disease.

The previous text did not include late summer as a single season, resulting in the absence of a corresponding season for the spleen in the relationship between the four seasons and the five elements. This omission indicates that this method likely came into being before the widespread respect for the five-phase theory in the middle period of the Western Han Dynasty.

The pulse-taking method based on the four seasons is elaborated in the chapter "Discourse on the Jade Mechanism and the True Qi of the Depots" in the *Plain Questions* (see Table 17.4). It is consistent with the records of the pulse-taking method based on four seasons recorded in the chapter "Difficulty Fifteen" in the *Classic of Difficulties*, and the *Classic of Four Seasons* (*Si Shi Jing*, 《四时经》) cited by Wang Shuhe of the Western Jin Dynasty (AD 265–317) in the *Pulse Classic*. Therefore, we can gain a comprehensive understanding of pulse-taking methods based on the four seasons.

The chapter "Wide Promulgation of the Five Qi" in the *Plain Questions* provides a record of the relationship between seasons and abnormal pulses and introduces the season "late summer". It shows that this relationship was developed after the prosperity of the five-phase theory. The introduction of late summer as a supplement to the pulse-taking method based on four seasons does not hold any diagnostic significance. Instead, its purpose is primarily to create a complete alignment between the four seasons, the five Zang-organs, and the five elements.

17.2.2 Pulse-Taking Method Based on the Pulse Location

This method recorded in the chapter "The Fifth Difficulty" in *Classic of Difficulties* involves gradually taking the pulse from the surface of the skin to the bone to diagnose the pathological changes in the lung, heart, spleen, liver, and kidney.

In the chapter "Discourse on the Essentials of Vessels and the Subtleties of the Essence" in *Plain Questions*, there is a passage that states: "In spring, it floats at the surface, like a fish swimming in a wave. In summer, it is in the skin, overflowing; the myriad beings are abundant. In autumn, it descends into the skin. The hibernating creatures will soon go into hiding. In winter, it is at the bones; the hibernating creatures are firmly sealed in. The gentleman resides in his chamber". Although the text mentions the four seasons, it is important to note that the method primarily belongs to the pulse-taking method based on the pulse location.

The aforementioned pulse-taking methods share a common characteristic: they have been separated from the meridian theory, and have been more closely related to the theories of yin-yang, five elements, and four seasons. These methods involve the use of fingers to feel the pulse, displaying a variety of changes in pulse manifestation to varying degrees, which is quite different from the later standardized "pulse manifestation" methods. These methods can be found scattered throughout certain chapters of the *Plain Questions* and *Miraculous Pivot*, which suggests that they were

Table 17.4 Pulse-taking method based on four seasons

Season	Normal pulse	Pulse manifestation	Abnormal pulse	Diseases
Spring	Xian (string)	The pulse in spring appears soft, weak, slippery, straight and long	Excess: the pulse appears full and strong	Exterior diseases. bad temper, absentmindedness, dizziness and epileptic diseases
			Insufficiency: the pulse appears not full and weak	Interior diseases. chest pain involving the back and hypochondriac distension and fullness
Summer	Gou (surging)	The pulse in summer comes forcefully and recedes mildly	Excess: the pulse comes forcefully and recedes forcefully	Exterior diseases. fever, pain of the skin and muscles and acute eczema
			Insufficiency: the pulse does not come forcefully but recedes forcefully	Interior diseases. dysphoria, cough, expectoration and passing flatus
Autumn	Fu (floating)	The pulse in autumn appears light and floating, rapid in coming and scattered in receding	Excess: the pulse appears floating at first, hard in the middle and deficient at both sides	Exterior diseases. hiccup, backache and discomfort
			Insufficiency: the pulse appears floating and weak	Interior diseases. shortness of breath, cough, hemoptysis due to adverse flow of qi and laryngeal stridor
Winter	Ying (deep)	The pulse in winter appears deep and throbbing	Excess: the pulse comes as hard as stone	Exterior diseases. lassitude, spinal pain, shortness of breath and no desire to speak
			Insufficiency: the pulse recedes rapidly	Interior diseases. solicitude like hunger, hypochondriac coldness, spinal pain, lower abdominal fullness and brownish urine

(continued)

Table 17.4 (continued)

Season	Normal pulse	Pulse manifestation	Abnormal pulse	Diseases
Spleen-pulse		The normal states of the pulse of the spleen cannot be felt, but the abnormal changes of the pulse of the spleen can be felt	Excess: the pulse comes like running of water	Exterior diseases. heaviness and lassitude of the four limbs
			Insufficiency: the pulse appears like the beak of a bird	Interior diseases. obstruction of the nine orifices known as Zhongjiang (heaviness and stiffness)

may not have been widely used, and eventually fell out of practice by the middle of the Eastern Han Dynasty (AD 25–220). As a result, the *Classic of Difficulties* mentions that due to the loss of the classics over time, the specifics of these methods cannot be fully known.

17.2.3 Pulse-Taking Method Based on Pulse Manifestation

It is a method to diagnose the nature, severity and prognosis of the disease based on the frequency, rhythm, and morphological changes of the pulse. There is no record of pulse manifestation in Mawangdui silk manuscripts, but the *Book on Meridians* (*Mai Shu*, 《脉书》) discovered later in Zhangjiashan, Hubei province, almost completely supplemented the missing contents of the *Rules of the Meridians* (*Mai Fa*, 《脉法》) from Mawangdui. The *Book on Meridians* includes descriptions of pulse manifestations such as "full, feeble, slippery, static, stirred and swift". Interestingly, this paragraph was not included in the chapter "The Conduit Vessels" of the *Miraculous Pivot* together with the course of the meridians and related diseases, but was collected in the chapter "Discourse on the Three Sections and Nine Indicators" in the *Plain Questions*. The texts are not identical and are listed below:

Book on Meridians: "The method of pulse diagnosis is to press …with the left hand and touch the inner ankle with the right hand. If other pulses are forceful, while the kidney pulse is feeble, the disease will occur. If other pulses are smooth, while kidney pulse is hesitant, the disease will occur. If other pulses are calm, while the kidney pulse is excited, the disease will occur. For some pulses, the excited pulse is the normal condition, such as the kidney pulse, lung pulse and heart pulse. However, if the pulse is swift, the disease will occur".

"Discourse on the Three Sections and Nine Indicators" in the *Plain Questions*: "With your left hand, place it on the foot, five inches above the ankle, and press the vessel there. With your right hand, place it on the ankle and flick the vessel. If the movement responding to one's finger extends over five inches or more, as if there were wriggling worms, this indicates no disease. If the movement responding to one's finger is quick and strikes the hand like a torrential stream, this indicates disease. If it strikes the hand slowly, this indicates disease. If the movement responding to one's finger upwards fails to extend over the entire distance of five inches, or if upon flicking the vessels it does not respond to one's finger at all, this indicates death". (*New Emendation of Huangdi's Internal Classic* (*Xin Jiao Zheng*, 《新校正》) mentioned that "Both *A-B Classic of Acupuncture and Moxibustion* (*Zhen Jiu Jia Yi Jing*, 《针灸甲乙经》) and the book annotated by Quan Yuanqi have another version that says to press the region five cun[11] above the internal ankle with the left hand and flick the ankle of the patient with the right hand".)

The sentence "There are moving pulses including…If they beat fast, the patient will fall ill" discusses the diagnosis of the moving pulse, and indicates that a fast pulse leads to disease. However, in the context of the whole paragraph, this sentence seems to be an interpretation of the aforementioned diagnostic method. For the palpation of a moving pulse, it is not necessary to press the region five cun above it with the left hand or flick it with the right hand.

Some argue that this method was used to observe the return of the great saphenous venous,[12] which sounds reasonable, but does not match the content such as "flicking the ankle" and "five cun above it". It is possible that the flicking method is meant to touch the nerve behind the ankle joint.

In ancient medicine, the concept of pulse has always been vague and lacked a precise definition. The pulse was believed to encompass various conduction and communication systems in human. Therefore, physicians can not only palpate the pulses at the ankle joint but also flick the nerves to observe the conduction. However, the two different methods are both called "pulse-taking", which may be the main reason why this text is difficult to understand. But in any case, it can be seen that pulse-taking existed at that time. Nevertheless, it is clear that pulse-taking was practiced at that time.

It is important to further clarify that the main distinction of Bian Que's sphygmology from other mentioned sphygmology is its focus on diagnosing the nature of the disease solely through changes of pulse manifestation. This pulse-taking method plays a more important role in the *Plain Questions* than other methods. It can be said that many of the pulse manifestations used in later generations can be traced back to this. Therefore, the statement "The trend to diagnose the disease with pulse-taking methods originates from Bian Que" in the chapter. "The Biographies of Bian Que and Cang Gong" in *Records of the Grand Historian* (*Shi Ji*, 《史记》) probably is to emphasize the widespread usage of pulse-taking method during that time and its close association with Bian Que.

By analyzing the pulse manifestation method described in the *Plain Questions* and *Miraculous Pivot*, significant differences between the two books become apparent. In the *Miraculous Pivot*, about 17 chapters focus on pulse-related content, but most of them are about pulse-taking methods at Renying and Cunkou. So, the most frequently

---

[11] Cun is a traditional Chinese measurement, and 1 cun is equal to the width of one's thumb.

[12] Jin Shiying, Research on Collateral Vessel Diagnosis; *Chinese Journal of Medical History*, 1987, (3): 160.

mentioned pulse manifestations include terms such as "big", "small" and "full", which are used for comparison between Renying Pulse and Cunkou Pulse.

Indeed, the reason why those pulse manifestations are frequently used is because they are the basic manifestations. For example, in the chapter "Transports" in the *Miraculous Pivot*, it simply states that "When someone has a yang disease and the movement in the yang vessels is minimal, that is a movement contrary to the norms. When in the course of a yin disease, the movement in the yin vessels is massive, that is a movement contrary to the norms". Moreover, in the chapter "Cold and Heat Sensations" in the *Plain Questions*, it only mentions that "If the evil qi abounds, they are to be drained. If the proper qi is depleted, they are to be supplemented".

Throughout the *Miraculous Pivot*, only one chapter titled "The Physical Appearances of Diseases Resulting from the Presence of Evil Qi in the Long-term Depots and Short-term Repositories" is dedicated to pulse-taking methods. The chapter, in conjunction with the five Zang-organs, fully demonstrates the main diseases of the six pulse manifestations (large, small, moderate, swift, slippery, and unsmooth) and the correlation between string, surging, regularly intermittent, floating, and sunken pulse with five colors to diagnose diseases of Zang-Fu organs. It is worth noting that the *Miraculous Pivot* does not include any other chapters or paragraphs about specifically discussing pulse manifestations.

However, *Plain Questions* offers more extensive coverage of pulse manifestation in disease diagnosis, such as chapters of "Discourse on the Essentials of Vessels and the Subtleties of the Essence", "Discourse on Phenomena Reflecting the Status of Qi in a Normal Person", "Discourse on the Jade Mechanism and the True Qi of the Depots" and "Discourse on Very Strange Diseases". For instance, the chapter "Discourse on Phenomena Reflecting the Status of Qi in a Normal Person" involves more than 30 pulse manifestations. While *Miraculous Pivot* has only one chapter dedicated to pulse manifestations with 11 pulse manifestations discussed, *Plain Questions* comprises nine chapters that cover over ten pulse manifestations.

The fundamental reason for the difference between the two books is their independent compilation and involvement of different medical schools. *Plain Question*s incorporates some of Bian Que's sphygmology, so there is more content about pulse-taking methods.

The inconsistency in pulse manifestation content, found in both *Plain Questions* or *Miraculous Pivot*, is due to the use of more than 50 nouns. This does not necessarily indicate complexity in pulse manifestation, but because of a lack of academic exchanges at that time, there was no standardized medical terminology. It was not until the publication of *Pulse Classic* by Wang Shuhe that more standardized names and definitions of pulse manifestations were developed. The development of pulse-taking method can be divided into three stages:

(1) Pulse-taking method at different meridians: examine different meridians by taking the pulse at different positions of the body and diagnose the nature of the disease based on the changes of pulse manifestation. For example, it is stated in the chapter "Discourse on Disease Manifestations" in *Plain Questions* that "The yang brilliance qi is in permanent movement. The great yang qi and the

minor yang qi do not move. While usually they do not move, in the present case they move with great speed. This is the manifestation of that disease".

Until the Eastern Han Dynasty, with the publication of *Treatise on Cold Damage and Miscellaneous Diseases* by Zhang Zhongjing, a specific method for diagnosing stomach qi by taking the anterior tibial pulse was introduced. Although the theoretical basis (numerology) of "Three Positions and Nine Indicators" is not considered, the three indicators of the lower position and the heaven indicator and man indicator of the middle position belong to the pulse-taking method at different meridians. This method maintains a close relationship with the meridian theory and is still used by a few doctors.

(2) Pulse-taking method at Cunkou sorely: only take the radial artery pulse on the wrist. At this stage of development, the pulse-taking method has completely separated from the meridian theory in nature and become an independent diagnostic method.
(3) Pulse-taking method at Cunkou: divide "Cunkou" pulse into three parts: Cun, Guan and Chi to diagnose the pathological changes of five Zang-organs and six Fu-organs. This method has gained widespread adoption after the publication of the *Classic of Difficulties.* Although expressions of "Chi" and "Cun" can be seen in *Plain Questions* and *Miraculous Pivot*, "Chi" actually refers to the palpation of forearm skin, rather than the Chi pulse. It can be considered that pulse-taking method only developed to the second stage in the times of *Plain Questions* and *Miraculous Pivot*, which is the pulse-taking method at Cunkou exclusively.

In summary, it is easy to conclude that the development of the pulse-taking method at Cunkou took a significant amount of time, with impractical methods being gradually abandoned. Practical experience gained from various changes of pulse condition and their correlation with diseases has been continuously incorporated into the pulse-taking methods, eventually forming an independent diagnosis method. The earliest notable book in this field was *Pulse Classic* by Wang Shuhe in the Western Jin Dynasty, which summarizes 24 different pulse manifestations and gives a concise explanation of them (see Table 17.5). Since then, subsequent books on sphygmology may have undergone changes, but they generally adhered to the basic pulse manifestations recorded in *Pulse Classic.*

It can be said that the development of pulse diagnosis originated from the meridian theory, but it eventually was independent of it. Throughout its development, the pulse diagnosis has transitioned from simplicity to complexity, and finally returned to a state of conciseness from being miscellaneous. The most refined pulse-taking method is the "pulse-taking method at Cunkou", which is a culmination of the harmonious integration between medical practice and the philosophical thought that the small universe (Cunkou pulse) contains the big universe (human body). However, the development of this method arose from many ancient pulse-taking methods that have been abandoned along with clinical experience.

Table 17.5 24 kinds of pulse manifestation in *Pulse Classic* by Wang Shuhe

Pulse	Pulse manifestation
Floating pulse	A superficially located pulse which can be felt by light touch and grows faint on hard pressure
Hollow pulse	A floating, soft and large pulse, but empty in the center
Surging pulse	A floating and large pulse, which comes on exuberant
Slippery pulse	A pulse coming and going smoothly like beads rolling on a plate, which is somewhat similar to rapid pulse
Rapid pulse	A pulse coming and going rapidly
Irregular-rapid pulse	A rapid pulse with irregular intermittence
String-like pulse	A pulse that cannot be felt with light or moderate pressure and felt like a tight string of a musical instrument when pressing hard
Tight pulse	A pulse feeling like a tightly stretched cord
Deep pulse	A deeply located pulse that cannot be felt with light or moderate pressure and can only be felt when pressing hard
hidden pulse	A pulse which can only be felt upon pressing to the bone
Drumskin pulse	A hard and large pulse that is somewhat similar to deep pulse and hidden pulse and is slightly string-like
Replete pulse	A large pulse felt forceful with pressure
Faint pulse	A thready and soft pulse, scarcely perceptible
Rough pulse	A thin, slow and scattered pulse coming and going unsmoothly with intermittence
thready pulse	A pulse as thin as a silk thread, feeble yet always perceptible upon hard pressure
Soft pulse	A floating and thin pulse felt softer than normal
Weak pulse	A deep and thin pulse felt softer than normal and forceless with pressure
Feeble pulse	A slow, large and soft pulse, felt empty with pressure
Scattered pulse	A large and scattered pulse felt floating with light pressure and impalpable with heavy pressure that is due to excessive qi and blood deficiency
Moderate pulse	A pulse coming and going slowly, but which is slightly faster than slow pulse
Slow pulse	A pulse with less than four beats to one cycle of the physician's respiration, coming and going very slowly
Irregularly intermittent pulse	A moderate weak pulse, pausing at irregular intervals
Regularly intermittent pulse	A rapid pulse pausing at regular and long intervals, which indicates that the patient will die. While the patient with an irregularly intermittent pulse that also pauses at intervals will not be die
Throbbing pulse	A forceful pulse at the Guan position, which feels like touching a bean and cannot be felt at the Cun and Chi position

Chapter 18
Establishment of the Theory of Acupuncture and Moxibustion Therapy in the Qin and Han Dynasties

Acupuncture and moxibustion therapy is a unique treatment in traditional Chinese medicine. The origin of these therapies is steeped in legends that have persisted since ancient times, such as the legend of "Shen Nong tasting hundreds of herbs and Fu Xi making nine needles". In addition, some attribute the invention of needling to Huang Di. These legends have led to the division of ancient Chinese medicine into two systems: one is the drug therapy system based on the legend that "Shen Nong tasting hundreds of herbs"; the other is the acupuncture and moxibustion therapy system based on the legend that "Fu Xi making nine needles and Huang Di inventing needling".[1] These legends serve to emphasize the significant role that acupuncture and moxibustion therapy hold in traditional Chinese medicine.

But when did the magical acupuncture and moxibustion therapy become so influential? Mounting evidence shows that the origin and historical development of acupuncture and moxibustion therapy may not be as mythical or ancient as they have been portrayed. It was when the medical theory, especially the meridian theory, was well developed that acupuncture and moxibustion therapy truly flourished and became recognized as effective treatment modalities.

[1] For example, Liu Fu mentioned in the preface of *Shennong's Classic of Materia Medica* (*Shen Nong Ben Cao Jing*, 《神农本草经》) that "I have studied the origin of medicine and found that there existed two viewpoints. One supported Shen Nong developed materia medica and the other supported Huang Di developed acupuncture and moxibustion. While it just because different theories separate different schools of thoughts". (Chinese Ancient Medicine Association, the edition published in the 31st year of the Republic of China < AD 1912–1949 >).

Y. Liao, *Reconstructing the Image of Medicine in the Qin and Han Dynasties*, Understanding China, https://doi.org/10.1007/978-981-95-6801-7_18

18.1 Acupuncture and Moxibustion Therapy in the Early Period

Acupuncture and moxibustion therapy is commonly believed to have originated from the primitive means of bloodletting and hot compressing, dating back to the time when ancient people started to use fire and ground stone tools. However, it is noteworthy that there is no record of treating diseases with acupuncture or moxibustion in the chapter "Government Officials" in *Rites of Zhou* (*Zhou Li*, 《周礼》), which provides detailed records of the medical system and treatment of the Western Zhou Dynasty (about 1046–771 BC). That shows, even if acupuncture and moxibustion therapy existed before the Western Zhou Dynasty, their application was likely limited at that time. On the other hand, one possible reason for the absence of acupuncture and moxibustion in the *Rites of Zhou* may be that the court doctors were either unable or unwilling to treat noblemen with these methods that could cause injury. Therefore, this therapy could not be accepted by orthodox medicine at that time. This phenomenon still existed in the later dynasties. For example, in the chapter "Longevity, Early Death, Hardness and Softness" of *Miraculous Pivot* (*Ling Shu*, 《灵枢》), it is said: "When piercing those wearing linen garments, one also applies fire to cauterize them. When piercing an eminent person, one also applies hot medical compresses". It indicates that the treatments varied depending on whether the patient was a dignitary or an ordinary individual.

There are generally two historical references cited for proving the use of acupuncture and moxibustion in the Spring and Autumn period, but unreliably. One is recorded in the chapter "Duke Cheng" of *The Commentary of Zuo* (*Zuo Zhuan*, 《左传》). A doctor called Huan was ordered to treat Duke Cheng of Jin. After diagnosing, he said, "The illness is above the diaphragm and below the heart, which cannot be treated by 'Gong' or 'Da'". Some believe that "Gong" refers to moxibustion and "Da" refers to acupuncture.[2] However, this interpretation can only be considered an assumption. The other is from the chapter "Li Lou" of *The Mencius* (*Meng Zi*, 《孟子》) that "to treat a serious disease lasting seven years, three-year Ai (艾) is not enough". Here, "Ai" is a polysemous word in Chinese. Some mistakenly supposed "Ai" to be moxa and interpreted the sentence as that three-year moxa is not effective enough for treating a seven-year serious disease. In fact, "Yi (艾)" is the right pronunciation; it should be taken to be a verb, meaning "to recuperate" or "to govern or to treat", rather than a noun "moxa".[3] That is to say, the seven-year chronic disease

---

[2] Li Ding, *Perplexities to Acupuncture and Moxibustion* (*Zhen Jiu Xue Shi Nan*, 《针灸学释难》), Shanghai TCM College Press, Shanghai, 1986, p. 1. See also *The Collection of Academic Papers by Tang Yongtong* (*Tang Yong Tong Xue Shu Lun Wen Ji*, 《汤用彤学术论文集》) Beijing: Zhonghua Book Company, 1993, p. 320.

[3] The chapter "Shi Gu" of *The Dictionary* (*Er Ya*, 《尔雅》) Volume 1: "Ai refers to nourishment". In the chapter "Minor Odes of the Kingdom-Xiao Min" of *The Book of Songs* (*Shi Jing*, 《诗经》), there is a sentence that is "Huo Su Huo Yi" in Chinese. Mao Heng of the Han Dynasty annotated that: "Yi refers to governance. This sentence means that some people are serious in nature, and some people have the governing capacity" in *Mao's Annotation to the Classic of Poetry* (*Mao Shi Zhu Shi*, 《毛诗注疏》), Volume 19. In the chapter "Wan Zhang" of The Mencius (Meng Zi), there is an idiom that

cannot be easily cured in a short period. Therefore, it is unconvincing to believe that acupuncture and moxibustion therapy has been continuously developing since the primitive society or that "acupuncture and moxibustion therapy has been commonly applied in clinical, including treatment for diseases, prevention and healthcare during the Spring and Autumn period".[4] The most reliable materials for studying the application of acupuncture and moxibustion in that period are the medical works unearthed from Mawangdui Han Dynasty Tombs, and *Plain Questions* (*Su Wen*, 《素问》) and *Miraculous Pivot* written in the Han Dynasty (202 BC–AD 220).

(1) Moxibustion: it is generally accepted that moxibustion appeared earlier than acupuncture because *Moxibustion Classic of the Eleven Foot-Arm Meridians* (*Zu Bi Shi Yi Mai Jiu Jing*, 《足臂十一脉灸经》) (hereinafter referred to as "*Foot-Arm*") and the *Moxibustion Classic of the Eleven Yin-Yang Meridians* (*Yin Yang Shi Yi Mai Jiu Jing*, 《阴阳十一脉灸经》) (hereinafter referred to as "*Yin-Yang*"), unearthed from Mawangdui, only recorded moxibustion and not acupuncture. However, there is no evidence to support that meridian theory predated the Warring States period (475–221 BC), so it cannot be concluded that moxibustion based on meridian theory originated before the Warring States period. Nevertheless, the moxibustion recorded in *Prescriptions for Fifty-two Diseases* (*Wu Shi Er Bing Fang*, 《五十二病方》), another medical silk book unearthed from Mawangdui Han Dynasty tombs, follows a different approach. It does not rely on meridian theory, and the locations and methods of moxibustion are relatively disordered. Some moxibustion methods are at the primitive level of heating large areas of the body surface.[5] Therefore, this type of moxibustion might have appeared earlier. To further illustrate this point, we can analyze the titles of the medical books unearthed from Mawangdui. Most of these books were found without titles, and titles used today were all given by archaeologists based on their contents. Only *Prescriptions for Fifty-two Diseases* retains its original catalogue. The book recorded as many as 52 kinds of diseases, several times the number of those recorded in works on ancient medical prescriptions in the Western Han Dynasty (202 BC-AD 8).[6] And the contents in it include medicines, decoctions, witch doctors treating diseases by prayer, moxibustion,

---

is "Zi Yuan Zi Yi (repenting and redressing one's errors)". Zhao Qi of the Han Dynasty annotated that "Yi refers to management", *Commentary and Sub-commentary on the Mencius* (*Meng Zi Zhu Shu*, 《孟子注疏》), Volume 9.

[4] Li Zheng'an, *Preliminary Study on the Moxibustion in the Qin and Han Dynasties*, Chinese Journal of Medical History, 1989, (1): 4.

[5] For example, in line 180 of *Prescriptions for Fifty-two Diseases* (*Wu Shi Er Bing Fang*, 《五十二病方》), there is the record that "burning dried firewood and straw to heat patient's back and letting two people massage the patient's buttocks". *Silk Manuscripts from Mawangdui Han Dynasty Tombs* (*Ma Wang Dui Han Mu Bo Shu*, 《马王堆汉墓帛书》),Volume 4, arranged by the working team for the silk manuscripts from Mawangdui Han Dynasty Tombs, Annotation and Note, p. 47).

[6] Among the books of ancient prescriptions recorded in the part of "Prescriptions and Techniques" of the chapter "Catalog of Literary Works" in *Book of the Former Han* (*Han Shu*, 《汉书》), there are "prescriptions for eleven diseases", "prescriptions for twelve diseases" (two), "prescriptions for sixteen diseases" (two).

and other therapies. It is obvious that *Prescriptions for Fifty-two Diseases* is a compilation of various ancient medical prescriptions at that time. Thus, it can be concluded that the advent and application of these therapies should be much earlier than the writing time of *Prescriptions for Fifty-two Diseases.* In addition, in comparison to acupuncture, moxibustion doesn't require techniques to make fine needles and is generally perceived as less fear and pain to patients. Therefore, it can be inferred that moxibustion, which is not based on meridian theory, appeared earlier than acupuncture. However, during the Spring and Autumn period (770–476 BC), the moxibustion was not as widely utilized as drugs, decoctions, or other therapies provided by witch doctors. Therefore, in *Prescriptions for Fifty-two Diseases*, moxibustion only occupied a small proportion of space.

(2) Acupuncture: there are records of the stone needle in medical books from Mawangdui. But they mention that the size of the stone needle must match the size and depth of the abscess.[7] It shows that at that time, acupuncture with stone needle was only used for treating the surgical disease of abscess, the nature of which is the same as that recorded in the chapter "Surgeon" of *Rites of Zhou.* After the metal needles of "nine needles" came into being, though in medical works "metal needle" and "stone needle" were often collectively referred to as "needles and stones",[8] their functions were still described separately.[9] "Nine needles" include "Pizhen"[10] with the same function as the stone needle. Based on the perspective that certain individuals were perplexed by the notion that abscesses could be treated using both stone and metal needles, it becomes apparent that these two medical instruments must coexist within distinct systems. Probably, the stone needle was first used in Shandong Province, while nine needles originated in the southern regions.[11] That is to say, according

---

[7] *Rules of the Meridians* (*Mai Fa*, 《脉法》): "Cutting the pus-containing tissue with... (the words were missed here). If the carbuncle contains pus, the size should be measured and... (the words were missed here)", which was followed by some cases of the non-consistent size of stone needle and abscess. See: *Silk Manuscripts from Mawangdui Han Dynasty Tombs* (Volume 4), arranged by the working team for the silk manuscripts from Mawangdui Han Dynasty Dynasty Tombs, Annotation and Note, p. 17.

[8] The chapter "Further Discourse on the Five Depots" of *Plain Questions*: "If someone has an aversion to needles and pointed stones, it is impossible to talk to him about perfect skill".

The chapter "Discourse on Moving the Essence and Changing the Qi" of *Plain Questions*: "toxic drugs to treat their interior and needles and pointed stones to treat their exterior".

[9] The chapter "Discourse on Different Therapeutic Patterns Suitable for Use in Different Cardinal Points" of *Plain Questions*: "their diseases are always yong-abscesses and ulcers. For their treatment, pointed stones are suitable"; "Their diseases are cramps and block. For their treatment, fine needles are appropriate".

[10] The chapter "The Nine Needles and the Twelve Origin Openings" of *Miraculous Pivot*: "The lancet needle: Its end is shaped like the pointed end of a sword. It serves to remove massive amounts of pus". The chapter "Official Needles" of *Miraculous Pivot*: "If a disease generates massive suppuration, one chooses for treatment a lancet needle".

[11] The chapter "Discourse on Different Therapeutic Patterns Suitable for Use in Different Cardinal Points" of *Plain Questions*: "it is for sure that the therapy with pointed stones originated in the East", "it is for sure that the nine needles have originated in the South".

to the information available at present, there is no evidence in Chinese medical history of the process that stone needles were first used to stimulate certain parts of the human body for treatment and then were replaced by metal needles. Although the bloodletting therapy of puncturing the surface of the human body with various sharp objects may be used in primitive society, it was not involved in the earliest works on pulse and in *Prescriptions for Fifty-two Diseases*, a compilation of various therapies. It suggested that the stone needle had no influence on medicine at that time and were mainly utilized in certain folk medicine cases. Therefore, it can be concluded that before the Spring and Autumn period, primitive moxibustion and acupuncture methods did exist, but they were not closely linked with acupuncture therapy that has gradually become a comprehensive system in traditional medicine.

Early acupuncture and moxibustion share similarities with later acupuncture in terms of administration methods, but lack the theoretical foundations applied in ancient traditional medicine. Therefore, it can be classified as empirical medicine. In terms of time, moxibustion, guided by theoretical principles, originated in the Warring States period, predating acupuncture, which was also born in the Warring States period. Empirical medicine, despite its value, does have inherent limitations as it often relies on repetitive patterns and lacks comprehensive analysis. Although this experience can be accumulated and enriched over time, it remains at an early stage of medical development. Various primitive methods such as bloodletting, hot compressing, and pressing were widely used in the traditional medicine across various regions and nationalities in the world. Then, why did they develop into systematic acupuncture and moxibustion therapy only in China? The fundamental reason for this lies in the integration of these primitive treatments with the ancient Chinese meridian theory and other medical theories. By transcending the rudimentary level of empirical medicine, so it was able to develop rapidly.

18.2 Theoretical Development and Evaluation

(1) Meridian theory: the first and crucial step of acupuncture therapy toward theoretical medicine is to combine itself with meridian theory, which took place from the Warring States period to the Han Dynasty. In *Moxibustion Classic of the Eleven Foot-Arm Meridians* and *Moxibustion Classic of the Eleven Yin-Yang Meridians*, the two earliest monographs on sphygmology available, information on various meridian diseases mainly focuses on the pain along the courses of meridians with only limited mention of visceral organs. So there are few descriptions of visceral diseases. But in the chapter "Conduit Vessels" of *Miraculous Pivot*, which is based on the *Moxibustion Classic of the Eleven Yin-Yang Meridians*, the relevant information has been supplemented. The meridians were matched with certain organs, and more than 70 kinds of symptoms were supplemented, significantly broadening the application of acupuncture and moxibustion therapy and covering almost all kinds of internal

diseases. This marked a significant turning point because it enabled acupuncture and moxibustion therapy to separate from surgery in nature and develop its own definition as it is understood today.

The meridian theory essentially serves as a means to describe a conduction phenomenon other than the human nervous system. It became a medical theory when doctors used it as a basis to understand the interconnection of various parts of the human body and develop corresponding therapeutic schedules. Acupuncture and moxibustion therapy is more closely related to meridian theory than drug therapy due to its reliance on identifying the specific meridian related to the disease for treatment.[12] Whereas the theoretical system of drug therapy at that time focused on connecting diseases of various parts of the body through meridian theory, or directly, with the five Zang-organs. This was further linked to the five qi and five flavors of drugs based on the five-phase theory, so as to determine the medication principles and prescriptions.

(2) Yin-yang theory: as mentioned earlier, the meridian theory describes the conduction system of the human body. Thus, it is important to note that this theory initially emerged without any philosophical thought. The names of meridians in the *Moxibustion Classic of the Eleven Yin-Yang Meridians* and *Moxibustion Classic of the Eleven Foot-Arm Meridians*, the earliest books on meridians, demonstrated the attributes of yin or yang. However, it is crucial to clarify that those attributes were likely added by people at a later stage after the development of meridian theory.[13] That transformation contributed to the further theorization of acupuncture and moxibustion therapy. Based on "The principles of all piercing have their beginning in the conduit vessels",[14] the guideline that "For the application of the needles it is essential to know how to balance the yin and the yang qi",[15] and the therapeutic principle that "For a swelling, one chooses for the treatment the three yang conduits. For an outflow of undigested food, one chooses the three yin conduits"[16] were put forward from the meridian theory.

As the rational core of yin-yang theory is the most basic but high-level generalization and explanation of the laws and movement of the objective world, it penetrates into all fields of ancient medicine and has no particularity in itself. The significant

---

[12] For example, in Yin-Yang, after listing various symptoms, it says that "XXX meridian rules their treatment"; and in Foot-Arm, there is the statement that "such diseases should be treated by applying an ignited moxa on XXX meridian".

[13] In *Moxibustion Classic of the Eleven Yin-Yang Meridians*, the hand greater yang meridian, hand lesser yang meridian, and hand yang brightness meridian are called as shoulder meridian, ear meridian, and tooth meridian respectively, which seem to be the remnants of the original names of meridians.

[14] The chapter "Conduit Vessels" of *Miraculous Pivot*.

[15] The chapter "Root and Connection" of *Miraculous Pivot*.

[16] The chapter "The Nine Needles and the Twelve Origin Openings" of *Miraculous Pivot*.

impact on acupuncture and moxibustion is that both acupuncture[17] and moxibustion[18] have developed the techniques of "reinforcing" and "reducing" respectively to adapt to the therapeutic principle of "Regulate the patient's yin and yang. If there is an insufficiency, supplement it; if there is a surplus, drain it".[19]

(3) "Four Seasons" theory: the integration of the four seasons theory with medicine before the Han Dynasty can be considered as a medical school, with its influence extending to various aspects such as diagnosis,[20] health preservation,[21] etiology, and[22] acupuncture and moxibustion, etc. The development of the four seasons theory can be divided into two parts, characterized by the introduction of "late summer". Generally speaking, before the widespread adoption of the five-phase theory in the Western Han Dynasty, the four seasons theory was widely adopted and was closely related to Yin and Yang theory. While with the flourishing of the five-phase theory in the later period, there was a desire to establish a closer correspondence with the five-phase theory, the "late summer" season was introduced to complement the existing four seasons to match the "earth (center)" of the five elements.

---

[17] There are many reinforcing and reducing techniques. For example, in *Miraculous Pivot*, there are records that deep needling is reducing and shallow needling is reinforcing. "Draining is to confront them. Supplementing is to pursue them". (The chapter "End and Beginning"); "the needle is to be inserted slowly and to be withdrawn quickly" means reducing, and "the needle is to be inserted quickly and to be withdrawn slowly" means reinforcing (The chapter "Explanatory Remarks on the Small Needles"). In *Plain Questions*, there are records that reducing refers to that "When the patient inhales, insert the needle" and "When the exhalation is completed, the needle is removed"; reinforcing means that "When an exhalation is completed, insert the needle" and "Wait for an inhalation and pull the needle". (The chapter "Discourse on the Division and Union of True Qi and Evil Qi"); reducing refers to that "the left hand opens the needle hole"; and reinforcing refers to that "the left hand closes the needle hole" (The chapter "Discourse on the Aims of Piercing").

[18] The chapter "The Transport Openings on the Back" of *Miraculous Pivot*: "When one resorts to fire for supplementation, he must not extinguish the fire by blowing at it. He must wait until it fades by itself. When one resorts to fire to conduct a drainage, he must quickly blow at the fire to let it blaze up".

[19] The chapter "Discourse on Bone Hollows" of *Plain Questions*.

[20] The chapter "Discourse on Disease Manifestations" of *Plain Questions*: "As for 'assess,' one finds the location of the disease and assesses it in on the basis of the four seasons". Note: "Duo" refers to the ancient medical book *To Estimate and Measure* (*Kui Duo*, 《揆度》). In addition, the *Pulse Classic* by Wang Shuhe in the Western Jin Dynasty (AD 265–317) also belongs to this kind of work, as it cited the content of the ancient medical work *Classic of Four Seasons* (*Si Shi Jing*, 《四时经》).

[21] The chapter "Comprehensive Discourse on Regulating the Spirit in Accordance with the Qi of the Four Seasons" of *Plain Questions*: "Now, the yin and yang qi of the four seasons, they constitute root and basis of the myriad beings. Hence, the sages in spring and summer nourish the yang and in autumn and winter nourish the yin".

[22] The chapter "Comprehensive Discourse on Phenomena Corresponding to Yin and Yang" of *Plain Questions*: "If a person is harmed in winter by cold, he will suffer from warmth disease in spring. If he is harmed in the spring by wind, he will develop outflow of undigested food in summer. If he is harmed in summer by summer heat, he will suffer from jie and malaria in autumn. If he is harmed in autumn by dampness, he will develop cough in winter".

In *Plain Questions* and *Miraculous Pivot*, there are 10 chapters that combine acupuncture and moxibustion with the four seasons theory,[23] including the early literature that only mentioned four seasons and the later works in which "late summer" was introduced. However, in these chapters, the match between the four seasons theory and the five-phase theory was still not entirely accurate.[24] It was not until the second century, with the production of the *Classic of Difficulties,* that a more precise match between the two theories was achieved.[25] However, at that time, the four seasons theory had completely evolved into the five-phase theory, losing its original meaning. The fundamental reason for the combination of the four seasons theory and acupuncture and moxibustion therapy is that the qi of the four seasons has its own definite places,[26] so "In all cases, one must look for their location to apply needles and pointed stones".[27] However, this theory proved ineffective due to the complexity of diseases. Whereas the application of the four seasons theory in the fields of health preservation and etiology was completely different, which sought to establish the objective relationship between natural climate change and human physiological changes in the human body. It can be said that the application of the four seasons theory in the theoretical development and clinical treatment of acupuncture and moxibustion had no significance. It just served as a general medical thought from the pre-Qin period to the Han dynasty before being included in theories of acupuncture and moxibustion.

(4) Five-phase theory: it is widely recognized as an essential component of traditional Chinese medicine. Its significance is well-established, and therefore it's unnecessary to elaborate. However, its influence on the theoretical system of acupuncture and moxibustion therapy is relatively limited.[28] The only one that needs discussion is

---

[23] The chapter "Discourse on The Essentials of Diagnosis and on Exhaustion in the Conduits", "Discourse Thoroughly Deliberating upon Depletion and Repletion", "Discourse on Opposition and Compliance in Piercing in the Course of the Four Seasons", "Discourse on Holes to Treat Water and Heat" of *Plain Questions*; " To Consider the Transportation Openings as the Foundation", "The Four Seasonal Qi", "The Ties between Yin and Yang Qi and Sun and Moon", "Cold and Heat Diseases", "Official Needles", "The Qi Moving in Accordance with the Norms Divide a Day into Four Time Periods" of *Miraculous Pivot.*

[24] For example, spring should be matched with "sinew", but it is matched with the meridian whose nature is "fire" in the chapter "Discourse on Opposition and Compliance in Piercing in the Course of the Four Seasons" of *Plain Questions* that involves the season of late summer.

[25] The "the Seventy-fourth Difficulty" of *Classic of Difficulties*. The chapter begins with "the classic statement". So it shows that the content was quoted from other medical works and indicates that the four seasons theory appeared earlier than the *Classic of Difficulties*.

[26] The chapter "Discourse on Opposition and Compliance in Piercing in the Course of the Four Seasons" of *Plain Questions*: "In spring the qi is in the conduit vessels. In summer the qi is in the tertiary network vessels. In late summer the qi is in the muscles and in the flesh. In autumn the qi is in the skin. In winter the qi is in the bones and in the marrow".

[27] The chapter "Discourse on the True Words in the Golden Chest" of *Plain Questions*.

[28] The content in medical works unearthed from Mawangdui has no connection with the five-phase theory; The content of acupuncture and moxibustion in *Plain Questions* only involves five Zang-organs. And in *Miraculous Pivot*, there is only a few words concerning five phases, like "The twelve conduit vessels: they are differentiated in accordance with the five phases" and "All piercing falls into one of five categories, corresponding to the five Zang-organs".

the relationship between "five transport points"[29] and five elements. In *Plain Questions* and *Miraculous Pivot*, six chapters mentioned the five acupoints. In the chapter "Discussion on Acupoints" of *Miraculous Pivot*, the names of "Five transport points" on meridians are introduced. It is said that in the meridians related to five Zang-organs (yin meridians), the first acupoint of the inner side of limbs is the Well acupoint and pertains to Wood in five elements, and that in the meridians related to six Fu-organs (yang meridians), the first acupoint of the lateral side of limbs is Well acupoint and pertains to Metal in five elements. Apart from these references, there is no evidence indicating a direct connection between the five acupoints and the five-phase theory. But they do have a close relationship with the "four seasons". In the *Classic of Difficulties* (*Nan Jing*, 《难经》), another important medical classic written in the Eastern Han Dynasty (AD 25–220), we find evidence of a strong correlation between the five transport points and the five-phase theory (see Table 18.1).

Firstly, it explicitly shows the match relationship between the five transport points and the five elements. Secondly, the match between the five acupoints and the four seasons, which was also based on the five-phase theory, is more structured compared to the accounts in *Plain Questions* and *Miraculous Pivot*. In addition, on the basis of the five-phase theory, the main diseases of each acupoint are outlined.[30] The acupuncture and moxibustion theory based on the five-phase theory did not originate solely from the *Classic of Difficulties*, but was developed from the *Ten Changes* (*Shi Bian*, 《十变》).[31] Further evidence can be found in *Essentials of Acupoint Location and Acupuncture-Moxibustion Therapy* (*Ming Tang Kong Xue Zhen Jiu Zhi Yao*, 《明堂孔穴针灸治要》) written in about the first century.[32] It shows that acupuncture and moxibustion therapy, which initially did not require the five-phase theory as a foundation, underwent a transformative development in the middle of the Eastern Han Dynasty by incorporating the five-phase theory. This shift was not a superficial adaptation but rather a genuine embrace of the five-phase theory as the theoretical basis. After *Essentials of Acupoint Location and Acupuncture-Moxibustion Therapy* containing this theory was included in the

[29] "Five transport points" refer to the key acupoints located below the knee and elbow joints, which start at the tip of limbs and continue all the way the elbows or knees, including well, brook, stream, river, and sea, respectively.

[30] For example, "the river points can cure heavy body and the pain in one's joints", as river points whose nature belongs to earth in yin meridians, matching with the spleen of five phases and could control muscle.

[31] "The Sixty-third Difficulty" of *Classic of Difficulties*: "The Ten Changes states: The sequence of holes located on the streams associated with the body's five Zang-organs and six Fu-organs, including 'brooks,' 'rapids,' 'streams,' and 'confluences,' is always preceded by a 'well'". "The Sixty-fourth Difficulty" of *Classic of Difficulties*: "*The Ten Changes* states further: The yin wells are wood; the yang wells are metal; the yin brooks are fire; the yang brooks are water...".

[32] As this book didn't taboo on using the name of Emperor Shang of the Han Dynasty, it is considered that the book was written before Yanping's reign in the Eastern Han Dynasty. But according to its content, the book was later than *Plain Questions* and *Miraculous Pivot*. See Huang Longxiang: *The Compilation and Textual Criticism on Huang Di's Illustration for Acupoints and Meridians* (*Huang Di Ming Tang Jing Ji Xiao*, 《黄帝明堂经辑校》), China Medical Science and Technology Press, 1987, p. 241.

Table 18.1 The main statements on "five transport points" in *Classic of Difficulties, Plain Questions,* and *Miraculous Pivot*

	Well acupoint	Brook acupoint	Stream acupoint	River acupoint	Sea acupoint
"The Sixty-Fourth Difficult Issue" of *Classic of Difficulties*	Yin: wood Yang: metal	fire water	earth wood	metal fire	water earth
"The Seventy-Fourth Difficult Issue" of *Classic of Difficulties*	In spring, prick the well acupoints	In summer, prick the brook acupoints	In late summer, prick the stream acupoints	In autumn, prick the river acupoints	In winter, prick the sea acupoints
"The Sixty-Eighth Difficulty" of *Classic of Difficulties*	Through the well acupoints, one can master fullness below the heart	Through the brook acupoints, one can master body heat	Through the stream acupoints, one can master a heavy body and pain in one's joints	Through the river acupoints, one can master panting and coughing as well as alternating spells of cold and heat	Through the sea acupoints, one can master influences proceeding contrary to their proper course, as well as diarrhea
The chapter "Discourse on Holes to Treat Water and Heat" of *Plain Questions*	In winter, take the wells to bring the yin countermovement down, take the brooks to replenish the yang qi		In Autumn, take the transporters to drain the yin evil, take the confluences to clear the body of yang evil		
The chapter "The Qi Moving in Accordance with the Norms Divide a Day into Four Time Periods" of *Miraculous Pivot*	The Zang-organs are controlled by winter. In winter one pierces the well openings	The colors are controlled by spring. In spring one pierces the creek openings	The time periods of severity and weakness of diseases are controlled by summer. In summer one pierces the transport openings	The musical tones/voices are controlled by late summer. In late summer one pierces the stream openings	The flavors are controlled by autumn. In autumn one pierces the confluence openings

(continued)

Table 18.1 (continued)

	Well acupoint	Brook acupoint	Stream acupoint	River acupoint	Sea acupoint
The chapter "To Consider the Transportation Openings as the Foundation" of *Miraculous Pivot*	In winter one selects for treatment the section covering all the well openings and all the transport openings	In spring one selects for treatment the network vessels, and all the creek openings	In summer one selects for treatment all the transport openings		In autumn one selects for treatment all the confluence openings
The chapter "Nine Needles and Twelve Yuan - Primary Acupoints" of *Miraculous Pivot*	Where they exit, these are the wells	Where they move in swift currents, these are the creeks	Where they flow, these are the transport openings	Where they permit passage, these are the streams	Where they enter, these are the confluences

A-B Classic of Acupuncture and Moxibustion (*Zhen Jiu Jia Yi Jing*, 《针灸甲乙经》) compiled by Huangfu Mi in the Western Jin Dynasty, the theory has become an important theoretical principle in acupuncture and moxibustion therapy until today.

18.3 Development of Acupoints

The works on meridians unearthed from Mawangdui have no record of acupoints, only classifying various diseases into different meridians, and taking it as a treatment principle. It may be attributed to the fact that those books merely talk about moxibustion. That is to say, moxibustion can be broadly applied to the corresponding meridians according to the diseases, without requiring strict and accurate acupoint locations. The earliest record of human acupoints can be traced back to Bian Que, who treated the "deathlike reversal" for the Prince of Guo State with acupuncture during the Spring and Autumn period and the Warring States period. While the medical records of Chunyu Yi, a famous doctor in the Western Han Dynasty, documented certain acupuncture and moxibustion methods, some mentioned only the names of the meridians and others indicated specific locations without mentioning acupoint names.[33] It indicates that acupoints started to emerge around the late Warring States period but weren't of significant importance in acupuncture and moxibustion therapy until the early period of the Western Han Dynasty. This feature can also be

[33] The chapter "The Biographies of Bian Que and Cang Gong" of *Records of the Grand Historian* (*Shi Ji*, 《史记》), Volume 150.

evident in *Plain Questions* and *Miraculous Pivot*, where only about half of the therapeutic principles of acupuncture and moxibustion therapy focus on meridians rather than acupoints. It was until the Eastern Han Dynasty that the situation underwent a fundamental change, and the number of acupoints increased rapidly,[34] with specific names and accurate locations.[35] Additionally, the meridian theory reached a more advanced level. Generally speaking, it only serves as a guiding theory to distinguish the attribution of diseases and determine the therapeutic principle.

Another law of acupoint development is that the number of acupoints increases from the extremities to the trunk. For example, in *Plain Questions* and *Miraculous Pivot*, acupoints are mainly located at the superficial parts of the limbs below the knees and elbow joints. In later acupuncture and moxibustion works, there were 45 recorded acupoints in the stomach meridian of foot yang brightness (42 acupoints of which were recorded in *Essentials of Acupoint Location and Acupuncture-Moxibustion Therapy*). In *Miraculous Pivot*, only 10 acupoints below the knee joint and 6 acupoints above the neck were recorded, with no record of the trunk and thigh acupoints. This may be due to the fact that the needles used at that time were relatively coarse and were not suitable for deep insertion in the trunk. Only when the needle-making technology advanced to a certain level could the depth of needling be increased, allowing for the expansion of the acupoints available for treatment toward the chest and abdomen. Considering the distribution of acupoints in the early period and the prevalent use of collateral pricking and bloodletting in acupuncture methods,[36] it can be inferred that the use of the Haozhen (filiform needle) described as "3.6 cun in length (1 cun is equal to the width of ones' thumb)" and "as thin as the mouth of a mosquito"[37] was not as early as previously believed in acupuncture and moxibustion therapy. It was likely widely used in the late Western Han Dynasty. Therefore, in *Essentials of Acupoint Location and Acupuncture-Moxibustion Therapy*, not only did the number of acupoints in the trunk

---

[34] When it comes to the number of acupoints, it is generally 365 in *Plain Questions* and *Miraculous Pivot* so as to match with 365 days a year. In fact, only 160 acupoints can be listed, while 349 acupoints were recorded in *Essentials of Acupoint Location and Acupuncture-Moxibustion Therapy*. (The number of acupoints is quoted from *The Lecture of Acupuncture and Moxibustion* (*Zhen Jiu Xue Jiang Yi*, 《针灸学讲义》), edited by Nanjing College of Chinese Medicine, Shanghai: Shanghai Scientific and Technical Publishers, 1964, p. 12.)

[35] In *Plain Questions* and *Miraculous Pivot*, many acupoints only have locations without names, such as the side of the third vertebra on the back, Houzhong openings, the lower back and the sacrum intersect, the vessels below the tongue, tips of eyebrows, etc. While in *Essentials of Acupoint Location and Acupuncture-Moxibustion Therapy*, all acupoints have specific names and locations. For example, the location of "the side of the third vertebra on the back" recorded in *Miraculous Pivot* has been specified as: "Feishu (BL13), 1.5 cun lateral to the lower border of the third vertebra on the back".

[36] The chapter "Blood and Qi, Physical Appearance and Mind" of *Plain Questions*: "whenever one treats the disease, one must first remove their blood". And in the chapter "To Pierce Malaria" of *Plain Questions*, there is the record that "Pierce the major yin transverse vessel and let blood. Pierce the moving vessel on the instep and let blood. Pierce between the ten fingers and let blood. Pierce the two vessels below the tongue and let blood. Pierce the two eyebrows and let blood. Pierce into the cleft and let blood. Pierce the severed Bone and let blood".

[37] The chapter "The Nine Needles and the Twelve Origin Openings" of *Miraculous Pivot*.

substantially increase but also the depth of needling was set. And in *Plain Questions* and *Miraculous Pivot*, the depth of the needling is rarely mentioned, and needling should be repeated in one place many times.[38]

The relationship between acupoints and the meridian theory is highly intricate. On one hand, apart from the collateral pricking resembling bloodletting, acupuncture applied along the course of meridians required specific locations. Therefore, after the emergence of the meridian theory, acupoints naturally came into existence and continued to increase. But on the other hand, the acupuncture and moxibustion therapy belonging to empirical medicine didn't diminish due to the emergence of meridian theory, but was constantly enriching and developing. For example, in the treatment of febrile diseases, there are methods that involve determining the therapeutic principle according to meridian theory,[39] as well as methods of "Altogether 29 locations must be cauterized" from head to foot"[40] or needling at 59 points to "reduce heat".[41]

And the therapeutic principle outlined in the chapter "The Conduits and their Sinews" of the *Miraculous Pivot* is "The transport openings to be pierced are at the location of the pain", which is simpler and allows for an infinite number of acupoints. At the same time, empirical medicine serves as an important basis for theoretical medicine. For example, even before understanding the relationship between the spinal nerve of the back and viscera, doctors found that when the viscera is ill, there will be a specific reaction in the back. "When the patient feels a reaction there, and if an existing pain is resolved, then that is the transport opening".[42] As doctors gained a deeper understanding of this objective phenomenon, they were able to identify a set of "Back Acupoints" that can be used for diagnosing and treating the diseases of organs including five Zang-organs, six Fu-organs and diaphragm. Meanwhile, in clinical practice, if doctors discovered many acupoints that were not listed in the meridian theory, it prompted them to discover and summarize new rules. This process

---

[38] The deepest needling depth mentioned in the chapter "The Stream Waters" of *Miraculous Pivot* is no more than 6 fen and "The depth of a piercing must not exceed 2 fen" on the meridians of the two arms. But there are fixed times of needling. For example, in the chapter "Discourse Thoroughly Deliberating upon Depletion and Repletion" of *Plain Questions*, it is stated that "pierce three wounds to the side of the hand major yin conduit and two on each of the capstring vessels" and "Move five inches upward from the ankle and pierce with three needles there". And there are even some statements to match with numerology, like "The number of wounds to be generated follows the waxing and waning of the moon. At the time of crescent moon, generate one wound on the first day, two wounds on the second day, and so forth until 15 wounds on the 15th day, then 14 wounds on the 16th day, and so forth". (the chapter "Discourse on Misleading Piercing"). All these sentences indicate that people did not pay attention to the depth of needling at that time. They mainly treated diseases through regulating the stimulation intensity at a certain meridian or region.

[39] The chapter "Cold and Heat Disease" of *Miraculous Pivot*, the chapter "To Pierce Heat" of *Plain Questions*.

[40] The chapter "Discourse on Bone Hollows" of *Plain Questions*.

[41] The chapter "Discourse on Holes to Treat Water and Heat" of *Plain Questions*; The chapter "Heat Diseases" of *Miraculous Pivot*. But the location of needling is different. The former needles the patient's head, chest and back. The latter needles the patient's head and hands.

[42] The chapter "The Transport Openings on the Back" of *Miraculous Pivot*.

led to the emergence of new theories, such as the local diagnosis of "ear acupuncture therapy" and acupoint selection methods.

In short, as people had a deeper understanding of acupoints, the connection between acupuncture and moxibustion therapy and medical theories like the meridian theory would become less rigid. Doctors began to place greater emphasis on the special therapeutic effects of acupoints on certain diseases. While this may be considered a regression toward empirical medicine in essence, it is actually highly necessary in practical application.

The above passage provides a summary of the development stages of acupuncture and moxibustion therapy before the Han Dynasty and highlights the complex relationship between empirical medicine and theoretical medicine. Moreover, the summary basically aligns with the modern definition of acupuncture and moxibustion.[43] In fact, in ancient China, especially before the Han Dynasty, acupuncture and moxibustion were not used solely to balance the body's functions (yin-yang balance) for treating various diseases, but also to directly treat the focal points of diseases, ranging from the surface of the body to the internal organs. For example, moxibustion could be used to burn warts[44] and carbuncle-abscess[45] on the body surface, while acupuncture, performed with a "needle" that served as a surgical instrument, was employed for treating diverse diseases such as fistulas, scrotal edema, joint effusion, intestinal Ascaris lumbricoides, abdominal abscess, and ascites caused by various carbuncles, sores and lymphatic tuberculosis on the body surface.[46] And some diseases in the abdominal cavity could not be treated by acupuncture and moxibustion or purging therapy. Attempting such treatments could lead to adverse consequences, even death,[47] which was also gleaned from failure experiences. Similarly, at that time, the meridians were not distinguished from the visible tissues of the human body; instead, they were regarded continuously as one part of human tissue structure. Thus, the concept of "the meridian disease" pertains to the intermediate level between the body surface and internal organs. Therefore, there existed a difference in treatment: "toxic drugs must be administered to attack the center, while chisel stones, needles, and moxa are used to treat the exterior, and acupuncture

[43] The modern definition of acupuncture and moxibustion is: "applying acupuncture and (or) moxibustion to meridians and (or) acupoints to regulate the function of Zang-Fu organs and blood and qi, so as to treat diseases". (Quoted from the preface of *The Lecture of Acupuncture and Moxibustion*, edited by Nanjing College of Chinese Medicine, Shanghai: Shanghai Scientific and Technical Publishers, 1964).

[44] *Prescriptions for Fifty-two Diseases*, line 102. See *Silk Manuscripts from Mawangdui Han Dynasty Tombs* (Volume 4), arranged by the working team for the silk manuscripts from Mawangdui Han Dynasty Tombs, Annotation and Note, p. 39.

[45] The chapter "Obstruction- and Impediment-Illnesses" of *Miraculous Pivot*.

[46] They can be found in the following chapters of *Miraculous Pivot*: "Obstruction- and Impediment-Illnesses", "The Four Seasonal Qi", "Cold and Heat Sensations", "Piercing to Regulate True and Evil Qi", "The Nine Needles and the Twelve Origin Openings", "The Receding Qi Diseases", "Upper Barrier", "The Four Seasonal Qi".

[47] The chapter "Discourse on Strange Diseases" of *Plain Questions*, and the chapter "Discourse on Abdomen and Center" of *Plain Questions*.

Fig. 18.1 Sun Simiao giving acupuncture treatment. Yang Liuqing woodblock painted by Qian Huian of the Qing Dynasty (AD 1636–1912). It is about the legend that Sun Simiao, sitting on the back of a tiger, gave acupuncture treatment to the Dragon King

and moxibustion to treat external diseases".[48] Also, the depth of acupuncture was correspondingly specified, encompassing the layers from skin, flesh, pulse, sinew to bone. It is evident that there exists a gap between pre-Han Dynasty acupuncture and moxibustion and the acupuncture and moxibustion in later generations, both in terms of the objective scope of treatment and the subjective understanding of doctors. The difference indicates that, in the process of developing acupuncture and moxibustion into a systematic therapeutic method, morphology played a significant and even dominant role in both the objectivity and the doctors' mindset. Thus, it is unfounded to solely attribute the basic characteristics of traditional Chinese medical system to abstract philosophical ideas while disregarding their morphological basis, as is commonly believed.

[48] The chapter "Discourse on Decoctions and Wines" of *Plain Questions*. It is "Biqi (必齐)" instead of "Huoqi (火齐)" in the original text. Many commentators have pointed out that "Huoqi" should be the right one and offered explanation, which is quite reasonable. Therefore, I chose "Huoqi" here.

Chapter 19
The Theory of Five Circuits and Six Qi

According to ancient beliefs, true effectiveness in medical treatment could only be achieved by a medical practitioner who possesses comprehensive knowledge of the line patterns in heaven above, the structures of the earth below, and the affairs of man in the center. For medical practitioners, understanding "the line patterns in heaven above" in the early days indicates to gasp the alternation of seasons, or the sequence of the four seasons in the yin-yang school (i.e., in spring there is the generation, in summer the growth, in autumn the harvesting, and in winter the storage), and its relationship with the physiology and pathology of the human body.[1] In addition, it means to have insight into something about the mandate of heaven. It's like what Sun Simiao, a great doctor of the Tang Dynasty (AD 618–907) said, "As a medical practitioner, it is essential to not only study various medical books, but also excel in fortune-telling, physiognomy, and divination. Then can one be considered a master of medicine. As a medical practitioner, it is imperative not only to thoroughly study diverse medical texts but also to excel in fortune-telling, physiognomy, and divination. Only then can one truly be considered qualified to be a master of medicine".[2]

As time progressed, the theory of five circuits and six qi gradually emerged, building upon the belief that the sequence of the four seasons is the result of the movement of yin and yang in heaven and earth. According to the regular movement of five circuits and six qi, disease property in a certain year or period of the sexagenary cycle (a 60-year cycle) can be inferred.

In the Song Dynasty (AD 960–1279), medical examinations included a section on "five circuits and six qi", requiring students to answer about the location, application,

[1] Details can be found in the theory of yin and yang in the chapter "Comprehensive Discourse on Regulating the Spirit in Accordance with the Qi of the Four Seasons" in *Plain Questions* (*Su Wen*, 《素问》).

[2] Sun Simiao (the Song Dynasty), *Important Formulas Worth a Thousand Gold Pieces for Emergency* (*Bei Ji Qian Jin Yao Fang*, 《备急千金药方》).

Y. Liao, *Reconstructing the Image of Medicine in the Qin and Han Dynasties*, Understanding China, https://doi.org/10.1007/978-981-95-6801-7_19

and prescription for the five circuits and six qi of the specific year.[3] This marked the inclusion of the theory of five circuits and six qi as part of the basic theory of traditional Chinese medicine. For medical practitioners at that time, the so-called "knowing the line patterns in heaven above" largely referred to understanding the five circuits and six qi of each year. Even today, people often quote a statement from the *Treatise on Cold Damage and Syndrome Analysis* (*Shang Han Jian Zheng Xi Yi*, 《伤寒兼证析义》) written by Zhang Zhuo, a physician in the Qing Dynasty (AD 1636–1912), which still holds true today, "If a person lacks understanding of the five circuits and six qi, regardless of how many remedy books he has read, he will end up with failure".[4] It also shows that the importance of medical practitioners comprehending the "five circuits and six qi" for each year.

However, after carefully reading the following analysis, you will be astonished to find that the theory known as "ancient meteorological medicine" and often regarded as a hallmark of a knowledgeable medical expert, is completely a fantasy based on the theories of yin and yang and the five elements.

19.1 Basic Contents of the Theory of Five Circuits and Six Qi

The basic idea of the five circuits and six qi is that the universe is governed by the cyclic movement of "five circuits" (wood, fire, earth, metal, and water) and "six qi" (wind, cold, dampness, dryness, sovereign fire, and ministerial fire), which determine the nature of diseases each year.

The concept of the five circuits is the combination of five elements with time and can be classified as follows:

(1) Major Circuit: The major circuit is the core and the most crucial part of the five circuits. It determines the main influence of the year and is based on the combination of the five elements and the heavenly stems. However, the pairing of the five elements and the heavenly stems in the theory of five circuits and six qi differs from the conventional norms. Instead of the usual pairings where wood is ruled by Jia and Yi; fire by Bing and Ding; earth by Wu and Ji; metal by Geng and Xin; water by Ren and Gui, the pairings are as follows: earth is ruled by Jia and Si; metal by Yi and Geng; water by Bing and Xin; wood by Ding and Ren; fire by Wu and Gui. When the heavenly stems are arranged in odd numbers in sequence, the matched circuit for that year is considered "greatly excessive" (surplus). On the contrary, even numbers are "inadequate" (not enough) (as shown in Table 19.1).

[3] He Daren (the Song Dynasty), *Test Collection from Various Medical Departments of the Imperial Medical Bureau* (*Tai Yi Ju Zhu Ke Cheng Wen Ge*, 《太医局诸科程文格》).

[4] Zhao Zhuo (the Qing Dynasty), the chapter "Five Circuits and Six Qi" in *Analysis of the Concurrent Patterns of Cold Damage* (*Shang Han Jian Zheng Xi Yi*, 《伤寒兼证析义》).

Table 19.1 Attributes of five elements and Yin and Yang in circuits of year

Heavenly Stems	Jia	Yi	Bing	Ding	Wu	Ji	Geng	Xin	Ren	Gui
Five circuits	Earth	Mental	Water	Wood	Fire	Earth	Mental	Water	Wood	Fire
Greatly excessive (+)/in adequate (−)	+	−	+	−	+	−	−	−	+	−

(2) Principal Circuits: Within a year, the five elements are divided into "five periods", known as the principal circuits. According to the sequence of wood, fire, earth, metal, and water, each phase governs 73 days and 5 marks,[5] a pattern that remains consistent year after year..

The corresponding relationship between the five elements and the four seasons originated in an earlier era. To make them match with each other, the last 18 days of each season were interpreted as "ruled by earth". Therefore, each of the five elements rules an average period of 72 days annually. Yet as the five elements cannot be perfectly matched with the four seasons, ancient thinkers introduced the concept of "long summer" to encompass the remaining days and made it associated with earth. The sentence "Spring dominates late summer; late summer dominates winter; winter dominates summer; summer dominates autumn; autumn dominates spring" can be seen in the chapter "Discourse on the True Words in the Golden Chest" in *Plain Questions* (*Su Wen*, 《素问》). "The five periods succeed each other; each of them governs one of these sections of the year. When the days of an annual cycle are completed, this constitutes one cycle, after which it commences anew" can also be found in the chapter "Discussion on Six-Plus-Six System and the Manifestations of the Viscera" in *Plain Questions.* All of these suggest that the coordination of "domination in circuit", namely, the coordination between the five elements and relevant period of the year, has existed before the formation of the theory of five circuits. Based on this, the five elements evolved from "season ruling" to "year ruling", thereby losing the direct relationship with the similarity between seasonal features and the attributes of the five elements.[6] However, this evolution led to the establishment of the inherent definition of the theory of five circuits and six qi—determining the disease nature of a specific period (such as a year, half year, 1/5-year, 1/6-year) solely based on the yin and yang and five elements properties of the heavenly stem and earthly branches.

On the other hand, although the sequence of the "five elements ruling seasons" remains consistent annually, the method of their dominance has changed. The "domination in circuit" within the theory of the five circuits and six qi incorporates an additional calculation rule known as "the alternation of excess and deficiency" (alternating yin and yang). According to this calculation rule, the yin and yang attributes of the "domination in circuit" are determined by the "great excess" or "inadequacy"

[5] A day is divided into 100 marks, and the dominating time of five phases makes a year.

[6] For example, "summer" in the four seasons and "fire" in the five phases both have the characteristic of "heat".

Table 19.2 Comparison of "Domination in Circuit" between Jia and Ji (earth circuit) years

Five seasons	Spring (73.05 days)	Summer (73.05 days)	Long summer (73.05 days)	Autumn (73.05 days)	Winter (73.05 days)
Heavenly stems: Ji year, the major circuit refers to "inadequacy of earth period"					
Domination in circuit	Inadequacy of wood period	Great excess of fire period	Inadequacy of earth period	Great excess of metal period	Inadequacy of water period
Heavenly stems: Jia year, the major circuit refers to "great excess of earth period"					
Domination in circuit	Great excess of wood period	Inadequacy of fire period	Great excess of earth period	Inadequacy of metal period	Great excess of water period

of the major circuit. For example, as Table 19.1 shows, the major circuit in the Year of Ji is "inadequacy of the earth circuit", thus the earth in the "domination in circuit" also belongs to yin. Following the principle of "alternated yin and yang", the five circuits of the year can be calculated as follows: inadequacy of the wood circuit, great excess of the fire circuit, inadequacy of the earth circuit, excess of the metal circuit, and inadequacy of the water circuit (as shown in Table 19.2). In summary, the "domination in circuit" not only maintains the traditional coordination of the five seasons and five elements but also incorporates the calculation rule of "the alternation of excess and deficiency" to connect with the major circuit and to integrate the five elements with yin-yang.

(3) Guest circuit: The only difference between guest circuit and the domination in circuit is that the first circuit of each year is uncertain and is determined by the major circuit. Then, the subsequent circuits are calculated according to the sequence of the generation of the five elements. The guest circuit is not extensively detailed in the discussion of the five circuits and six qi in the received version of *Huangdi's Internal Classic* (*Huang Di Nei Jing*, 《黄帝内经》). According to Zhang Jingyue, a physician in the Ming Dynasty (AD 1368–1644), there are relevant statements in the *Methods for the Jade Calendar according to the Celestial Origin* (*Tian Yuan Yu Ce Jie Fa*, 《天元玉册截法》): "The guest circuit operates differently from the domination in circuit, and follows the same principles as the 'dominant qi' and 'guest qi' in the six qi". Therefore, "Although there is no explicit elaboration in *Huangdi's Internal Classic*, the generation and transformation of qi and circuits have inherent laws and should follow the principles discussed in *Jade Tablets on the Celestial Origin* (*Tian Yuan Yu Ce*, 《天元玉册》).[7]".

[7] Zhang Jiebing (the Ming Dynasty), "Guest Circuit in the Five Circuits" of the chapter "Five Circuits and Six Qi" in *The Illustrated Wings of the Classified Canon (Lei Jing Tu Yi,* 《类经图翼》*)*, Volume 2.

Table 19.3 Corresponding relationships between six qi, three yin and three yang and the earthly branches

the earthly branches	Zi and Wu	Chou and Wei	Yin and Shen	Mao and You	Chen and Xu	Ji and Hai
Yin and yang	Lesser yin	Greater yin	Lesser yang	Yang brightness	Greater yang	Reverting yin
Six qi	Sovereign fire	Dampness	Ministerial fire	Dryness	Cold	Wind

The phrase "six qi" appeared in the pre-Qin period, but its meaning has been interpreted differently.[8] For example, in the chapter "Free and Easy Wandering" of *Zhuangzi*, it states: "Take advantage of the natural order between heaven and earth, and master the variations of the six qi". However, the interpretation of "six qi" in this context varies among commentators. As recorded in the chapter "First Year of Lord Zhao" in *The Commentary of Zuo* (*Zuo Zhuan*, 《左传》), when the physician He diagnosed a disease, he said: "There are six [i.e., environmental] heavenly qi", which refers to "yin, yang, wind, rain, clouds and light" and has no direct connection with the "six qi" as used by later medical experts. The "six qi" in the theory of five circuits and six qi is probably derived from the "five qi" (wind, cold, dryness, dampness, and summer heat) attributed to the five elements (wood, water, metal, earth, and fire). "The summer heat (heat)" was further categorized into "sovereign fire" and "ministerial fire" respectively, thus transforming the original "five qi" becomes the "six qi" containing two "fires". This modification was made to align with the twelve earthly branches and the three yin and three yang,[9] addressing the issue of the imbalance between the number of yin and yang elements (as shown in Table 19.3).

The most basic method of calculating "qi" that governs a year is similar to that of the qi associated with an annual period, that is, the qi corresponding to the earthly branch of a certain year. However, this kind of "qi" was later completely disregarded by people and was not even given a name. What people place importance on is the qi dominating each solar term which can be categorized into the "dominant qi" and "guest qi".

(1) Dominant qi: The "six qi", which are arranged in the order of wood, sovereign fire, ministerial fire, earth, metal, and water, each govern four solar terms starting from the Great Cold in December (the 24th solar term) and remain constant from year to year. They are commonly referred to as the "dominant qi". The corresponding relationship of "qi" and the 24 solar terms, described as "the structures of the earth corresponding to the positions of the qi during the six seasonal terms" in the chapter "Comprehensive Discourse on the Subtle Significance of the Six Qi" in *Plain Questions*, is characterized by the five

[8] Refer to *A Variorum of the Zhuangzi* (*Zhuang Zi Ji Shi*, 《庄子集释》) written by Guo Qingfan in the Qing Dynasty, published by Zhonghua Book Company in 1961. pp. 20–21.

[9] Ban Gu (the Han Dynasty), "Five Phases" of the chapter "Discourse on Virtues" in *Comprehensive Discussions in the White Tiger Hall (Bai Hu Tong Yi,《白虎通义》).*

Fig. 19.1 Images of "dominant qi" (the structures of the earth corresponding to the positions of the qi during the six seasonal terms)

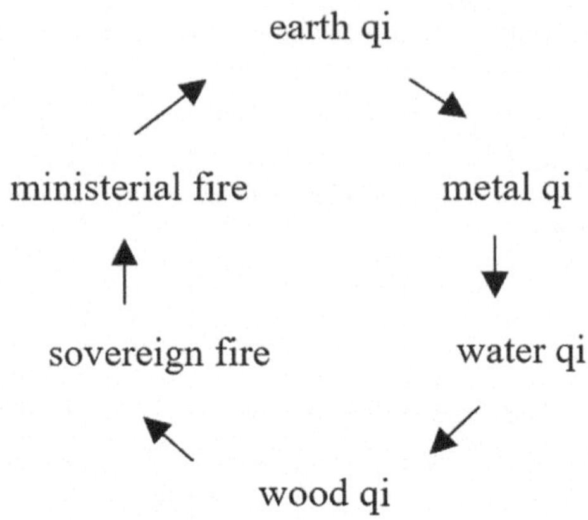

elements. Because if the sovereign fire and the ministerial fire are considered as a single entity, the sequence and mutual relationships follow the order of the mutual generation cycle of the five elements (Fig. 19.1). Moreover, the correspondence between each qi and the 24 solar terms is nearly the same as the untransformed one, i.e., it still conforms to the sequence of spring wood, summer fire, late summer earth, autumn metal, and winter water.

(2) Guest qi: Guest qi is determined by the relationship between the six qi and the earthly branches. It is called "Guest qi" because the correspondence between the "qi" and the 24 solar terms varies every year. The calculation method is as follows: a year is divided into "six steps" (six parts), each lasting 60.875 days; each step is governed by one of the six qi; the names and sequence of the six qi are: lesser yin sovereign fire → greater yin dampness-earth → lesser yang ministerial fire → yang brightness dryness-metal → greater yang cold water→ reverting yin wind-wood; every year, the earthly branch determines the "third qi" (i.e., the C, the A which is the initial qi and the F which is the final qi), the interrelationship of which is shown in Table 19.3. However, this relationship no longer follows the mutual generation cycle of the five elements. Instead, it is based on the theory of three yin and three yang (Figs. 19.2, 19.3).

However, it must be noted that as a medical theory, the theory of five circuits and six qi emphasizes the nature of the "six qi", rather than the position of the three yin and three yang, because only the nature of the six qi can provide a theoretical explanation of the interrelationship among diseases, pathogenesis and treatment methods, etc. Therefore, in the theory of five circuits and six qi, the six qi are called the foundation and six principals (qi), while the three yin and three yang are only the "tip" of the six qi[10] and are set up for "naming their positions".[11] Despite this, the position of "three

[10] The chapter "Comprehensive Discourse on the Subtle Significance of the Six Qi" in *Plain Questions.*

[11] The chapter "Comprehensive Discourse on the Progression of the Five Periods" in *Plain Questions.*

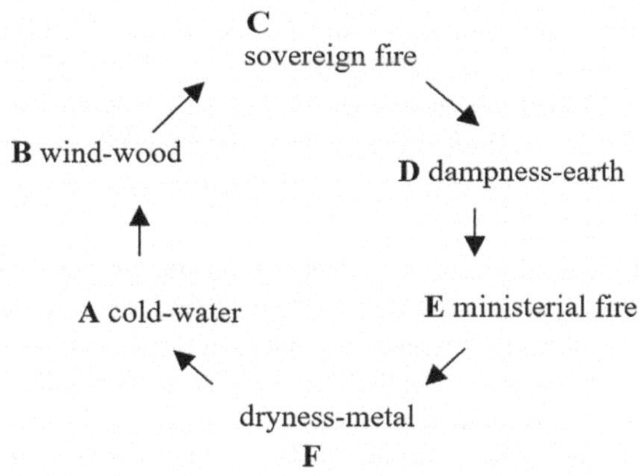

Fig. 19.2 Images of "guest qi" (the course of the six times six-ten days terms of the Way of Heaven)(Each qi above is of the first and seventh of the twelve Earthly)

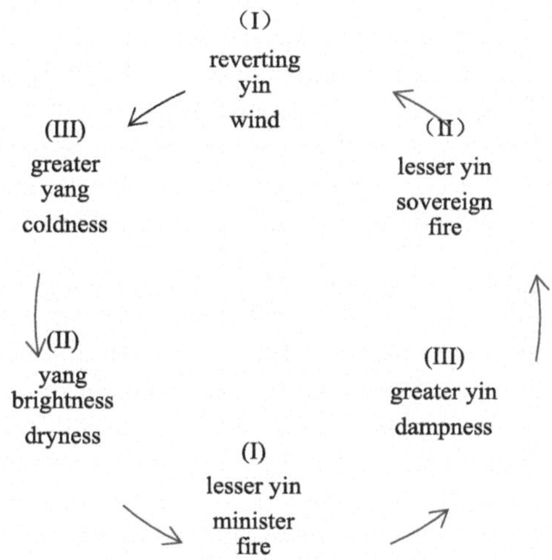

Fig. 19.3 The position of "three yin and three yang"

yin and three yang" still provides a reasonable basis for the deduction of the six qi in the theory of five circuits and six qi. First of all, the figure shows that yin and yang each occupy half of the complete circle and are in balanced harmony. The order of yin and yang follows the number one, two and three, respectively. Between yin and yang, the first yin is opposite to the first yang, the second yin is opposite to the second yang, and the third yin is opposite to the third yang. With the development of the theory of five circuits and six qi, the application of the cycle of six qi has become increasingly diverse. It is stipulated that the "above" is "heaven", so the "qi" corresponding to it is named "the qi controlling heaven". Correspondingly, the "below" is "terrene", so the qi corresponding to it is named "the qi controlling terrene[12]", while the positions of other qi can be inferred from the positioning of three yin and three yang.

Concerning the interaction between "five circuits" and "six qi" (which can also be seen as the interaction between the heavenly stems and earthly branches or between the five elements and yin and yang), the theory of five circuits and six qi also defines some concepts:

1. Celestial correspondence: When the five-phase attribute of the qi associated with a specific year (the heavenly stem) is consistent with the five-phase attribute of the earthly branch stipulated in the "six times six-ten days terms of the Way of heaven", it is called "celestial correspondence".
2. Convergent year: When the qi associated with a specific year coincides with the five-phase attribute of the earthly branch as stipulated in "the structures of the earth corresponding to the six seasonal positions" and is located at the center position of the "five orientations", it is called a convergent year.
3. Tai-yi celestial correspondence: This refers to a year that simultaneously meets both the conditions of celestial correspondence and convergent year.
4. The same celestial correspondence: When "great excess" is joined from below by the qi controlling terrene, that is referred to as the same celestial correspondence.
5. The same convergent year: When "inadequacy" is joined from below by the qi controlling terrene, that is referred to as the same convergent year.

These are some basic concepts and calculation methods in the theory of five circuits and six qi. On this basis, by calculating the attributes of the "circuits" and "qi", it becomes possible to predict the diseases people may suffer from and determine appropriate medications for treatment in a certain year or at a certain season. This practical application is one of the significant values of the theory of five circuits and six qi. However, the theory of five circuits and six qi, with its complex contents and systematized framework, is actually a consolidation of various calculation methods of "circuits" and "qi" employed by people after the Song Dynasty, based on earlier practices predating the Tang Dynasty. To solve the contradictions between different calculation methods, various terms such as dominant circuit, guest circuit, middle circuit, dominant qi and guest qi were introduced. When using the systematized theory to analyze the historical materials related to the theory of five circuits and six qi before the Tang Dynasty, it is evident that there is a wealth of related content. This might lead one to assume that the theory of five circuits and six qi has been a complex system since its inception and marvel at the wisdom of ancient sages. However, this perception does not reflect the true historical context. Therefore, to gain a genuine understanding of the formation and development of the theory of five circuits and six qi, it is necessary to approach it from a historical perspective.

[12] "Terrene" means "on the earth (or under the ground)". For example, in the chapter "Five Phases of the Comprehensive Discussions" in the *White Tiger Hall* (*Bai Hu Tong*, 《白虎通》), it is said that "yin qi locates under the terrene and nurtures everything".

19.2 Formation and Evolvement of the Theory of Five Circuits and Six Qi

The earliest existent documents that record the theory of "five circuits and six qi" are seven chapters entitled "Comprehensive Discourse on…" in *Plain Questions*, one of the two independent works that constitute the *Huangdi's Internal Classic*. Since all seven chapters are about the same theory, later generations call them "the seven chapters of five circuits and six qi" or "the seven comprehensive discourses" in *Plain Questions*. These are: Comprehensive Discourse on Arrangements of the Principal Qi of Heaven, Comprehensive Discourse on the Progression of the Five Periods, Comprehensive Discourse on the Subtle Significance of the Six Qi, Comprehensive Discourse on Changes Resulting from Qi Interaction, Comprehensive Discourse on the Five Regular Policies, Comprehensive Discourse on the Policies and Arrangements of the Six Principal Qi, and Comprehensive Discourse on the Essentials of the Most Reliable. The theory of five circuits and six qi can barely be found in other chapters of the book. The seven comprehensive discourses were revealed by Wang Bing, the first person who annotated *Plain Questions* in the Tang Dynasty. In his annotation, he noted that "The seventh volume of the book was kept by Master Zhang, thus the received version of the book consists of eight volumes.[13] Fortunately, I found the lost version of *Plain Questions* collected by Zhang at Guo's school". Issues about the origins of the seven comprehensive discourses and whether they are the missing parts of *Plain Questions* have been debated among scholars in the medical history community. Some hold that the theory has already been manifested in the book, and the seven comprehensive discourses should be the lost chapters of *Plain Questions*, as the concept and wording of the seven chapters are consistent with those of the book.[14] Others, however, insist that these discourses were written after the third century, or at the latest, the seventh century, based on astronomical studies.[15] There is also a perspective suggesting that neither the seven chapters nor the theory of five circuits and six qi was included in the original version of *Plain Questions*, and that they were written at a later age. This viewpoint is supported by the absence of any mention of the seven comprehensive discourses in the first annotated version of *Plain Questions* by Quan Yuanqi in the Qi and Liang periods, and the *Great simplicity of Huangdi's Internal Classic*[16] by Yang Shangshan in the Sui Dynasty (AD 581–618). Some even claim that the theory did not exist until the tenth century, hence remained unmentioned by the medical experts in the Sui and Tang Dynasties. Therefore, the

[13] Wang Bing, the preface of *Plain Questions*.

[14] Wang Shufen, Textual research of seven comprehensive discourses of five circuits and six qi, *Chinese Journal of Medical History* (*Zhong Hua Yi Shi Za Zhi*, 《中华医史杂志》), 1987, (4):229.

[15] Hao Baohua, Wang Jin, "New discussion about the completion time of the seven comprehensive discourses", *Chinese Journal of Medical History*, 1984, (1):46.

[16] *Great Simplicity of Huangdi's Internal Classic* (*Huang Di Nei Jing Tai Su*, 《黄帝内经太素》) is the Genus-book which changes *Plain Questions* and *Miraculous Pivot*, and its completion time is still unknown, but Yang Shangshan has annotated it in the Sui Dynasty (581–618).

seven comprehensive discourses and Wang Bing's annotations could have been later fabrications.[17]

Actually, the issue of when the seven comprehensive discourses were composed and whether they are part of the extant version of *Huangdi's Internal Classic* are separate matters. If the assumption that *Huangdi's Internal Classic*, consisting of *Plain Questions* and *Miraculous Pivot*, was completed before the first century AD was true, then the seven comprehensive discourses couldn't be included in *Plain Questions.* This is because the method to designate the year by the heavenly stems and earthly branches is derived from the Quarter-Remainder Calendar in the Eastern Han Dynasty (AD 85). There are several common features shared by the other chapters in *Plain Questions*: of the correspondence between the five elements and the heavenly stems is used only for designating days; the five elements correspond to the four seasons; only the concepts of "generation, growth, harvesting, and storage" are present; and the concept of "transformation" associated with earth and earth's dominance during the long summer is only found in a few chapters. This correspondence aligns with the explanation from the chapter "Five Elements" in *White Tiger Hall* written in the first century, which states "The earth, the lord in five elements, is too distinguished to take charge of a season like the other four phases. So, there are only four seasons".

In addition, the concept of how the three yin and three yang correspond with twelve months is only mentioned in the chapter "Explanations on the Vessels" in *Plain Questions*, which states:

January	March	May	July	September	November
Greater yang	Reverting yin	Yang brightness	Lesser yin	Lesser yang	Greater yin

* months in the table are on the lunar calendar.

If the table was exhibited as a cycle, greater yang corresponds with lesser yin, reverting yin with lesser yang, and yang brightness with greater yin. It is in line with the corresponding relationship between three yin and three yang in Zang-Fu Organs theory[18] and meridian and collateral theory, rather than the relationship between yin-yang and six qi in the theory of "five circuits and six qi". Meanwhile, the concept of sovereign fire and minister fire is not mentioned in *Plain Questions.* All of these indicate that the seven comprehensive discourses are different from other chapters

[17] Fan Xinghuai., "The source of the theory of five circuits and six qi", *Journal of Medical History* (*Yi Shi Za Zhi*, 《医史杂志》), 1951, (1), 3.

[18] The corresponding relationship between three yin and three yang in Zang-Fu Organs is as follows: greater yin corresponds with yang brightness, such as the hand greater yin (lung) and hand yang brightness being exterior and interior, and foot greater yin (spleen) and foot yang brightness being exterior and interior; lesser yang corresponds with greater yang, such as hand lesser yang (heart) and hand greater yang(small intestine) being exterior and interior, foot lesser yang (kidney) and foot greater yang(bladder) being exterior and interior; reverting yin corresponds with lesser yang, such as hand reverting yin (heart enclosure) and hand lesser yang (triple burner) being exterior and interior, foot reverting yin (liver) and hand lesser yang (gall bladder) being exterior and interior.

in theoretical construction. Therefore, we can infer that the seven comprehensive discourses and *Plain Questions* are not formed in the same era.

However, in *Plain Questions*, there is already a tendency to transform the four seasons into five seasons for perfectly matching the five elements. It is noteworthy that the addition of the concept of "late summer" not only resolves the imbalance between the four seasons and the five elements, as well as the five qi (i.e., wind, cold, dampness, heat, and dryness), but more importantly, lays the foundation for the thought of "five circuits cycle". This further develops into the new theory of "five circuits ruling years". Moreover, in the discussion of "deficiency and excess" in the chapter "The Dew of the Year" in *Miraculous Pivot*, the concept of "year" is also mentioned:

> When one happens to live in a year with a weakness of qi, faces new moon, and has lost the harmony with the seasons, that is called a "triple condition of depletion".
>
> When one encounters a year of plenty of qi, is confronted with full moon, and exists in harmony with the seasons, robber wind and evil qi may be present, but they are not at all dangerous.

Therefore, if the received version of *Huangdi's Internal Classic* was assumed to be completed at the turn of the Western Han (206 BC–AD 24) and Eastern Han Dynasties (AD 25–220), or slightly later, then the conditions for the emergence of the theory of five circuits and six qi was basically in place in the middle and late Eastern Han Dynasty. It is likely that the authors of the five circuits and six qi works were individuals who had expertise in both astronomy and medicine, as physicians were serving in the astronomical institutions at that time.[19] Besides, according to the studies on phonological changes, some people believe that the seven comprehensive discourses were finished in the Eastern Han Dynasty.[20] Obviously, whether these discourses were added to *Plain Questions* by Wang Bing or were already included in the treasured private copy of his Master is not of significance.

Although ancient and modern scholars have different views on the time of the completion of the seven comprehensive discourses and the establishment of the five circuits and six qi theory, they generally regard them as an integrated system, believing that "each of the seven discourses has its particular emphasis, and is linked

[19] As *recorded in the Book of the Later Han* (*Hou Han Shu*, 《后汉书》) "The salary of the imperial astronomer (an official who holds astronomy and calendar) is 600 bushels of grain. The note says: "he is in charge of the seasons/time of Heaven (tian shi) and the sequence of the stars (xing li)". In Liu Zhao's annotations quoted from *Ceremonial of Han Offices* (*Han Guan Yi*, 《汉官仪》), it is said that "There are thirty-seven Expectant Appointees (太史待诏) in the imperial astronomer's office, i.e., six Calendarists (治历), three Diviners by Tortoise Shell (龟卜), three Directors of Buildings (庐宅), four Experts in the Phases of the Sun (日时), three Directors of Sacrifices to Expel Evil Influences (《易》筮), two Directors of Sacrifices to Expel Evil Influences (典禳), three Experts in the Tradition of Master Ji (籍氏), three Experts in the Tradition of Master Xu (许氏), three Experts in the Tradition of Masters Dian and Chang (典昌氏), two Specialists of Methods (嘉法), two Supplicants for Rain (请雨), two Elucidators (解事) and one physician (医)". (Refer to "Officials Part Two" of the chapter "Officials" in the volume 35 of the *Book of the Later Han)*

[20] Qian Chaochen: *Research on the Language of Huangdi's Internal Classic* (*Nei Jing Yu Yan Yan Jiu*, 《内经语言研究》), Beijing: People's Medical Publishing House, 1900, p. 293.

with each other as an integral whole".[21] However, upon closer examination of the seven comprehensive discourses, it becomes evident that though they share the topic of five circuits and six qi, they are different in the calculation methods and the conception of the core theory. What's more, through the careful study of *Huangdi's Internal Classic*, scholars found that this version was not compiled by a single individual or at the same period, as were the seven comprehensive discourses. By comparing the corresponding relationship between the five elements and the heavenly stems, and between the six qi and the earthly branches, as well as their different applications in ruling seasons and years, Table 19.4 suggests that the calculation of five circuits and six qi at that time was carried out in different ways. The following are the main points:

(1) Calculate five circuits and six qi according to stems and branches: This plain and clear method is adopted in the chapters "Comprehensive Discourse on Arrangements of the Principal Qi of Heaven" and "Comprehensive Discourse on the Progression of the Five Periods". This method involves matching the five elements with the heavenly stems to determine the circuit, distinguishing between the great excess and inadequacy according to the even and uneven numbers of the heavenly stems, and determining the qi by matching the earthly branches with six qi. Then there is only a single circuit and a single qi every year. Hence, it's unnecessary to define the conceptual differences of domination in circuit, guest circuit, dominant qi, and guest qi. Both heavenly stems and earthly branches are used to arrange the years.

This method effectively incorporates the basic idea of combining the five elements with yin-yang, making it a simple yet powerful approach. The arrangement of years based on the combination of the five elements and the heavenly stems involves matching the five elements with the heavenly stems. This practice is derived from the five-phase theory. However, the integration of the five elements and yin-yang is accomplished by identifying the significant surplus or deficiency in the "circuit" of a particular attribute. This identification is done by considering the evenness or oddness of the heavenly stems' numbers. Based on the proposal in *Huangdi's Internal Classic,* which suggests linking the five circuits to the manifestations of the three yin and three yang, a strong correlation between stems and branches is established. This correlation fully realizes the integration of the five elements and yin-yang by transforming the "five qi" generated by the five elements into the "six qi", which include sovereign fire and ministerial fire. These correspond to the three yin and three yang and the earth branches (Fig. 19.4).

(2) Calculate six qi according to six steps: This method is adopted in the chapter "Comprehensive Discourse on the Subtle Significance of the Six Qi". Firstly, each of the six qi rules over 60.875 days in a year. The qi designated by the

---

[21] Wang Qi et al., *Study and Research of the Theory of Five Circuits and Six Qi* (*Yun Qi Xue Shuo De Yan Jiu Yu Kao Cha*, 《运气学说的研究与考察》), Beijing: Knowledge Publishing House, 1989, p. 206.

Table 19.4 Application methods of "five circuits" and "six qi" in the seven comprehensive discourses

Discourses Correspondence	The Principal Heaven Qi	Progression of the Five Periods	Subtle Significance of the Six Qi	Changes Resulting from Qi Interaction	Five Regular Policies	Policies and Arrangements of the Six Principal Qi	Essentials of the Most Reliable
Five elements -the heavenly stems	Ruling years	Ruling years				Ruling years	?
Five elements-five seasons				Only four seasons		For excess and deficiency	
Six qi-earthly branches(in the way of yin-yang)	Ruling years	Ruling years	Ruling seasons		Ruling years	Ruling years Ruling seasons	?
Six qi-earthly branches(in the way of five elements)			Ruling seasons				
Remarks			Observe the qi	Astrology	The great excess and inadequacy of the five elements creating terrestrial effect and qi controlling heaven	Transforms the circuit of year to the qi associated with an annual period to match the qi controlling heaven and terrestrial effect.	It mainly talks about the mechanism of diseases

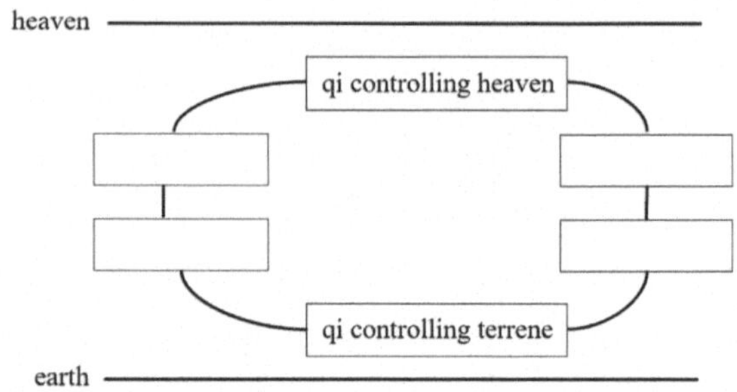

Fig. 19.4 The assumption of the circulation of six qi between heaven and earth

earthly branches method is the third qi of the year, and the sequence of qi in the rest step can be calculated by the cycle of six qi. However, when the manifestations are greatly excessive or inadequate, the method of "the luminaries of heaven moving the position of their light" is used. It determines the manifestations in three cases, which are "when its time has arrived and the qi arrives (balanced qi)", "when its time has not yet arrived and the qi has arrived (greatly excessive)", "when its time has arrived and the qi does not arrive (inadequate)", according to the discrepancy between the constant value of "balanced qi" and the measured shadow of the sun. This is referred to as "the six times six-ten days terms of the way of heaven". When combining this method with the theory of the five elements, the sequence of six qi is transformed into the sequence of generation of the five elements, namely wind, sovereign fire, minister fire, dampness, dryness, coldness, and each of them rules four seasonal terms, resulting in what is known as "the structures of the earth correspond to the positions of the qi during the six seasonal terms". It's important to note that there are significant differences in the connotations of the heaven and the earth between these two methods. In the method of "Calculating five circuits and six qi according to stems and branches", five elements are associated with heaven and the six qi is associated with earth. While in the method of "Calculating six qi according to six steps", six qi, following the sequence of the cycle of yin and yang, is associated with heaven, while the six qi, following the sequence of generation of five elements, is associated with the earth. The former method contains the circuits and qi, while in the latter one, the concept of five elements is just implicit, or in other words, there is only qi without circuits. That is to say, the method of "calculating six qi according to six steps" was merely established for experts of the heavens (the study of astronomy) to await qi and to calculate steps.

(3) Distinguish between heaven and earth according to six qi: It is a pattern of five circuits and six qi that can best reflect people's conception of the universe. The most significant change happened in the use of the cycle of six qi as follows: the cycle of six qi composed of three yin and three yang is just used to illustrate that the six qi complete a cycle every six years according to "Calculate Five circuits and six qi according to stems and branches". However, this method does not provide much insight into how people at that time perceived the existence of the six qi in the universe. Does qi in the universe change its attributes every

year? Or do the six qi with different attributes rule the year in turn? If the latter hypothesis is true, where are the other five qi that are not in the dominant position?

However, in the theory of "Distinguish between heaven and earth according to six qi", it is assumed that the six-qi cycle is located between heaven and earth. The top position is believed to control heaven, referred to as qi controlling heaven, while the bottom position that controls terrene, called qi controlling terrene. The remaining qi in the rest of the four positions are of less importance. The determination of the qi controlling heaven in a given year is based on the earthly branches. The calculation of five circuits and six qi according to stems and branches also believes that the earthly branch rules six qi, in which the co-domination of five circuits and six qi functions as the core of its theory. On the contrary, this perspective of the six qi governing the universe does not take into account the five elements and five circuits, rather it applies qi to complete its assumption for the existence of the universe and its movement pattern. As stated by Wang Chong, a philosopher in the Eastern Han Dynasty, in his book *On Balance* (*Lun Heng*, 《论衡》) "Discussive Weighing", "Qi is spread in the universe". The other intriguing fact is that in the qi cycle, although the six positions are opposite in terms of yin and yang, the six qi are not contradictory in nature. For example, the greater yang cold water opposites the greater yin wet earth, and the lesser yin monitorial fire opposites yang brightness dry metal. In terms of yin and yang, while the pairs of greater yang and greater yin, and lesser yin and yang brightness are opposite in nature, the pairs of cold and wet, and fire and dry have similar natures. The similarities between the qi controlling heaven and the qi controlling terrene provide a valuable explanation for the relationship between the attributes of circuits and features of diseases. In the years of Chou, and Wei, the circuits are characterized by dampness and coldness: "dampness qi comes down, mist and clouds form; the earth stores yin qi; massive cold arrives, hibernating creatures hide early, standing water increases, and the ground cracks with ice".

In the chapter "Comprehensive Discourse on the Policies and Arrangements of the Six Principal Qi", there is a further development on the theoretical framework of qi controlling heaven and qi controlling terrene ruling the year together. The writer put five circuits between heaven and earth and called it "middle circuit", thereby combining the yin-yang theory and five-phase theory. In order to coincide with the new theoretical concept, the writer modified the concept of "celestial correspondence", originally used to describe "the coincidence of heavenly stems and earthly branches", to signify the "the coincidence of middle circuit and qi controlling heaven". In addition, he supplemented new concepts like the iso-convergent year, and like "the same celestial correspondence" for describing "the coincidence of middle circuit and qi controlling terrene". Therefore, it's essential to recognize these concepts that emerged through the development of theory of five circuits and six qi, and to understand their definitions and application scopes. For example, if the concept of "middle circuit" was classified into "the method of circuit calculation by heavenly stems and earthly branches", it would be difficult to understand.

Because although there are six qi circulation circles in this method, the concepts of qi controlling heaven and qi controlling terrene are not incorporated.

(4) Determining circuit by astrology: The chapter "Comprehensive Discourse on Changes Resulting from Qi Interaction" is the embodiment of this kind of system. Although the chapter extensively discusses the question of "In the transformations caused by the five periods, if there is great excess or deficiency, how is that?", the thought of determining five circuits by the corresponding relationship between heavenly stems and five elements is absent. This system determines circuits by "the five corresponding stars above", that is,

> "When the qi associated with an annual period is greatly excessive, then the star associated with this period moves away to the North. When the period and the qi agree with each other, then each star moves along its path."
>
> "The fact is, when the qi associated with an annual period is greatly excessive, then the star that is in fear of the dominant agent loses its color and adopts that of its mother. When the qi associated with an annual period is inadequate, then it adopts the color of that agent which it cannot dominate."

It is very similar to the sayings in astrology "The five stars are different in color and size. They act in line with the time and solar terms according to their trajectories. If these stars don't lose their colors and match well with the four seasons, it's an auspicious omen. If not, an ill omen". "If the appearance or disappearance, haunting or running, obedience or disobedience, and slowness or quickness of these stars are in line with their times, they correspond to the laws and thus the country stays peaceful. If they violate their times, they will act disorderly and become evil stars. In this circumstance, the country will be in decline and vicissitude may occur as well as accompanied calamities like the chaos of war, famine, etc."[22]Therefore, this circuit calculation based on astrology is just an application in the medical domain. Thus, the concept of "the corresponding of human and nature" mainly serves as a description of human diseases.

The seven comprehensive discourses in *Plain Questions* are actually a collection of treatises dealing exclusively with the theory of five circuits and six qi. Each discourse is relatively independent, providing later generations with various calculation methods of five circuits and six qi in the early days. Meanwhile, the seven discourses share the same emphasis on the combination of the five elements and yin-yang, allowing us to grasp the distinctive characteristics of the times as described in the five circuits and six qi theory. To further elaborate:

The completion time of the two medical classics, *Plain Questions* and *Miraculous Pivot* of the received version of *Huangdi's Internal Classic,* is uncertain, but they were undoubtedly completed before the theory of five circuits and six qi. One of their remarkable characteristics is that the independent application of the yin-yang theory and the five-phase theory in different chapters. For example, in the chapter "The Yin and Yang Categorization and the 25 Human Types" of *Miraculous Pivot* written

[22] *The Book of Jin* (*Jin Shu*, 《晋书》), Volume 12.

in a question-and-answer format between Huang Di and Bo Gao, individuals were divided into five types: wood, fire, earth, gold and water, which is totally irrelevant to yin-yang. While in the chapter "To Penetrate Heaven" of *Miraculous Pivot* structured as a dialogue between Huang Di and Shao Shi, people were also divided into five types: greater yin person, lesser yin person, greater yang person, lesser yang person, and persons with a harmonious balance of yin and yang qi, which is only based on the yin-yang theory. In addition, in the received version of *Huangdi's Internal Classic*, the five-phase theory is limited in its application. For instance, it barely involves the five-phase theory when talking about acupuncture and moxibustion. Whereas in *Huangdi's Bright Hall Classic* (*Huang Di Ming Tang Jing*, 《黄帝明堂经》), the first monograph on acupuncture and moxibustion compiled in the Eastern Han Dynasty, explores the five-phase theory extensively in relation to the attributes of acupuncture points. And in *Classic of Difficult Issues* (*Nan Jing*, 《难经》), the indications of acupoints are almost discussed on the basis of the five-phase theory.[23] Besides, it elucidates that there are six Fu-Organs (Fu-Organ pertain to yin and yin numbers are even) and five Zang-Organs (Zang-Organ pertain to yang and yang numbers are odd) in the human body, and "one pulse showing ten changes" results from the five pathogens (five elements) of the soft and hard nature (yin and yang) meeting together ($5 \times 2 = 10$). It also divides accumulation-gathering into yin and yang, and further divides accumulation into five categories according to the orientations of the five elements—the accumulation of liver (wood) is located at the left, the accumulation

[23] Refer to the discussion on "the relationship between acupuncture and moxibustion and the five-phase theory" in chapter 10 of Part II of the book. In addition to the examples given, *Classic of Difficulties* also embodies the integration of yin-yang theory and five-phase theory. For example, it is said in The Seventy-third Difficult Issue, "All wells are located where the flesh is shallow and thin, and where the qi is too few to be employed for any treatment. How does one proceed if one has to prick them? It is like this. All wells are associated with wood. All brooks are associated with fire. Fire is the child of wood. Thus, when one has to prick the wells, one should drain them through the brooks". Based on the generation of five phases, if the qi of "child" reduces, the qi of "mother" is bound to supplement the "child". Therefore, the brooks can be pricked instead of the wells. The Seventy-ninth Difficult Issue states: "It is said in the *Huangdi's Internal Classic*, 'To confront those who come and attempt to remove them without sufficient strength, how could it be that this will not result in a depletion of proper qi? To pursue those who go away and invariably assist them, how could it be that this will not result in a repletion of evil qi? To generate a depletion or a repletion, this is as if one had made a gain, as if one had suffered a loss; What is called repletion and depletion, these are conditions as if something were present, and as if something were absent.' What is that to say? It is like this. 'To confront those who come and attempt to remove them' is a reference to draining the child; 'To pursue those who go away and invariably assist them' is a reference to supplement the mother. In the case of disease in the heart, draining at the rapids hole of the hand heart ruler conduit is considered as moving against and removing, while supplementing at the well hole of the hand heart ruler conduit is considered as following and assisting". For the original wording of the quotation of *Huangdi's Internal Classic*, see the chapter "The Nine Needles and the Twelve Origin Openings" of *Miraculous Pivot*. Annotators of previous generations called this a needling method named directional reinforcement and reduction. Namely, needling at the hole at which the qi arrives is called "to move against" and "to drain", while needling at the hole from which the qi departs is called "to follow" and "to supplement". And the author of the *Classic of Difficult Issues* interpreted it as "draining the child" and "supplement the mother" based on the five-phase theory.

of heart (fire) in the above, the accumulation of spleen (earth) in the center, the accumulation of lung (metal) in the right, and the accumulation of kidney (water) in the below. The pathogens transmit and transmute according to the law of the five elements. When the organ they restrain is hyperactive, they fail to transmit further, and they reside and accumulate, leading to the manifestation of accumulation. All the above highlight the significance of the five-phase theory and its combination with the yin-yang theory.[24] As mentioned before, the seven comprehensive discourses on the theory of five circuits and six qi were completed later than other chapters in the received version of *Huangdi's Internal Classic*, and they obviously indicate the tendency to integrate the five-phase theory with the yin-yang theory. From this point of view, it is consistent with the development characteristics of the medical theories in the Eastern Han Dynasty.

The theory of five circuits and six qi is undoubtedly supported by the five elements theory and the yin-yang theory. During the establishment of the theory, the five-phase theory held greater significance than the yin-yang theory, as the latter had not yet developed into a complete ideological system governing the five elements with yin and yang, like the Neo-Confucianism of the Song and Ming Dynasties. So, the five-phase theory was valued more than yin-yang theory by people. Therefore, the early works on five circuits and six qi were written in the name of "five elements" despite their specific contents. For example, the chapter "Comprehensive Discourse on Arrangements of the Principal Qi of Heaven" starts with "Heaven has the five agents and they control the five positions. Man has the Five-Zang Organs and they transform the five qi"; "The five periods succeed each other and each of them governs one of these segments of the year"; "The five periods complete their course in heaven. The Great Void spreads the true magical power of qi, and it exerts control over the principal qi of the earth" and so on. Although mainly focusing on the six qi, the chapter "Comprehensive Discourse on the Policies and Arrangements of the Six Principal Qi" is still full of expressions like "the transformations caused by the five periods" and "the five periods enacting their policies". Additionally, when considering the calculation methods, "five circuits ruling seasons" appeared first, then the theory of five circuits and six qi gradually developed after the transformation of the "five circuits ruling seasons" into "five circuits ruling years" by Wei Yuqu, based on the text of *Book on the Supreme Beginning and on the Principal Qi of Heaven* (*Tai Shi Tian Yuan Ce*, 《太始天元册》). What's more, though the six qi corresponding to the earth branches and the three yin and three yang is different from the five elements in cycle sequence and the "fire" has been transformed into the sovereign fire and ministerial fire, it is still the "five qi" (cold, summer heat, dryness, dampness, and wind) that closely related to the five elements in essence. Because both the nature of the disease and the characteristics of the natural climate need to be expressed through the framework of the five-phase theory.

[24] More details about how the yin-yang theory and five-phase theory are applied in the pulse diagnosis, viscera and bowels, acupuncture points, diseases and other areas can be found in Chapter 11 of Part II in *Classic of Difficulties*.

The theory of five circuits and six qi, although not widely popular before the Tang Dynasty, can be attributed to two points. First, if these works were written by the physicians serving in the royal astronomical institute, they were likely kept hidden away in inaccessible chambers, such as golden or stone chambers, making them unavailable to the ordinary people. Furthermore, following the prohibition of divination texts in the third year of Emperor Taishi's reign in the Jin Dynasty (AD 267), subsequent dynasties implemented similar bans. These successive prohibitions, which often resulted in severe punishments such as decapitation for unauthorized study of astronomy, were enforced to discourage the dissemination of forbidden knowledge. Secondly, although various medical classics were passed down through generations before the Song Dynasty, physicians paid more attention to practical skills such as prescription and pharmacy. For example, Huangfu Mi, a famous scholar and physician in the Jin Dynasty, believed that the received version of *Huangdi's Internal Classic* "was very profound but discussed more about theories than about practices". Therefore, he "cut out the meaningless and useless words and sentences" and selected the contents closely related to acupuncture and moxibustion therapy to compile the *A-B Classic of Acupuncture and Moxibustion* (*Zhen Jiu Jia Yi Jing*, 《针灸甲乙经》).[25] Another example is the encyclopedic medical masterpiece *Arcane Essentials from the Imperial Library* (*Wai Tai Mi Yao*, 《外台秘要》), compiled by Wang Tao of the Tang Dynasty. It adopted the formulas instead of the theories in the *Treatise on Cold Damage and Miscellaneous Diseases* (*Shang Han Za Bing Lun*, 《伤寒杂病论》) by Zhang Zhongjing, the medical sage of Chinese medicine. It was after the Song Dynasty that physicians began to pay attention to medical theories. The rise of Neo-Confucianism in this period led scholars to view medicine, previously considered a "humble occupation", as "one of the Confucians' ideal careers to make a thorough study on things". With the increasing number of intellectuals, a group of "Confucian doctors" with high cultural refinement appeared. Moreover, the rulers attached more importance to medicine. It was during this time that the Six-Meridian Syndrome Differentiation in the *Treatise on Cold Damage and Miscellaneous Diseases* became the guideline of clinical treatment in TCM. Zhang Zhongjing, once referred to as the "Second Sage", was now respected as the "Medical Sage". The *Treatise on Cold Damage and Miscellaneous Diseases* was no longer taken as a "formula book" but a comprehensive "medical classic" containing TCM principles, methods, formulas, and medications. The theory of the five circuits and six qi also began to receive more attention.

To sum up, there are three main schools about the theory of five circuits and six qi since the Song dynasty (AD 960). The first school regarded the seven comprehensive discourses as the words of ancient sages and extensively elaborated on and expanded upon them. The seven comprehensive discourses gradually form a complete system, so that they can correspond to each other and also can be elucidated into a new

[25] *A-B Classic of Acupuncture and Moxibustion* (*Zhen Jiu Jia Yi Jing*, 《针灸甲乙经》), compiled from *Plain Questions, Miraculous Pivot* and *Acupuncture-Moxibustion Treatment Essentials of the Bright Hall of Point* (*Ming Tang Kong Xue Zhen Jiu Zhi Yao*, 《明堂孔穴针灸治要》), is honored as the first existing treatise on acupuncture and moxibustion. Details of its compilation purport can be found in the "Preface" of the book.

meaning. It seems that people before the Tang Dynasty used to calculate the circuits and qi in a complex system. In the Song Dynasty, each volume of the medical examination had one question related with "five circuits and six qi", requiring students to answer the location, application and prescription of five circuits and six qi for that particular year, which contributes greatly to the spread of this theory. What's more, the formulas and medicine in the comprehensive remedy book *Prescriptions for Universal Relief* (*Pu Ji Fang*, 《普济方》) compiled in Ming Dynasty are listed according to the five circuits and six qi. Moreover, all the prescriptions arranged in the order of the sexagenary cycle (60 years) can be found in the book. During the Qing Dynasty (AD 1636–1912), there was a popular saying in the medical community, "One who has no understanding of the five circuits and six qi, regardless of how many remedy books he has read, will end up with failure. So, whoever engaged in the medical field always seeks to master the understanding of the circuits". That proves the high position of the circuits and qi in the medical theory.

The second school took an empirical approach to criticize the theory of circuits and qi. They argued, "The sages merely described the periodic progression and rotation of the qi in the universe. How could human diseases always correspond precisely? Is it possible that not a single person in a year would suffer from other disease other than those predicted?"[26] The *Complete Books of the Four Storehouses* (*Si Ku Quan Shu Zong Mu*, 《四库全书总目》) provided critical commentary on Wang Ji's *Readings of Five Circuits and Six Qi* (*Yun Qi Ding Lun*, 《运气易览》), stating, "The diagrams and explanations derived from the theory do have some insights. However, treating diseases should primarily focus on pulse diagnosis and symptom examination. Blindly adhering to the concepts of qi controlling heaven and terrene will ultimately miss the point of effective treatment". Similarly, the commentary on Dong Shuo's *Discourse on Five Circuits and Six Qi* (*Yun Qi Ding Lun*, 《运气定论》) remarks, "Diagnosing diseases through the four methods of examination (i.e., inspection, inquiry, auscultation and olfaction) and considering seasonal changes and earth qi is sufficient. There is no need to delve into unverifiable and intangible aspects of the theory of five circuits and six qi".

The third school brought about a transformation in the theory of the five circuits and six qi. Some medical practitioners, who neither admitted that the attributes of diseases could be deduced from the annual circuit of qi nor were willing to abandon their unwavering belief in the words of ancient sages. Therefore, to reconcile the contradictory ideas, they presented a new interpretation of the theory by focusing on the application of the six channels in their own ways. Representatives of this school include Liu Wansu and Zhang Congzheng, who were renowned medical experts of the Jin (AD 266–420) and Yuan (AD 1271–1368) Dynasties. They believed that "some diseases are just symptoms (signs), but some diseases are caused by qi", and "All diseases can be attributed to qi". As a result, the theory of the five circuits and six qi, transformed into a theory of pathogenesis. That means the relationship among the branches and circuits and qi was abandoned, and the attributes of the six qi (i.e.,

[26] Xu Dachun (the Qing Dynasty): Volume II "Discourse on Circuits and Qi Controlling Heaven" in *The Origins of Medical Science* (*Yi Xue Yuan Liu Lun*, 《医学源流论》).

wind, cold, summer heat, dryness, and fire) merely turned into the treatment principles without deduction. "Circuits and qi" mentioned in Liu Wansu's books *Explanation of Mysterious Pathogeneses and Etiologies Based on the "Basic Questions"* (*Su Wen Xuan Ji Yuan Bing Shi*, 《素问玄机原病式》) and *Collection of Writings on the Mechanism of Disease, Suitability of Qi, and the Safeguarding of Life as Discussed in the "Basic Questions"* (*Su Wen Bing Ji Qi Yi Bao Ming Ji*, 《素问病机气宜保命集》) are the perfect examples.

19.3 Science and Mystification—Two Incorrect Interpretations

Most TCM practitioners regard the theory of five circuits and six qi as a daunting mystery. It is because contemporary people tend to interpret the theory of yin-yang and the five elements from a philosophical perspective. From this point of view, yin and yang represent the dualistic nature of all things, with their unity, opposition, and mutual transformation. Five elements, sometimes referred to as five elements by extreme materialists, are used to classify the attributes of things, and describe the relationship of generation and restriction of things. Therefore, although TCM learners know yin, yang and five elements, few of them recognize that ancient scholars perceived yin and yang as the true essence of "qi" existing between heaven and earth, and the five elements as both basic attributes and noumenon (five virtues) of all things. In addition, the ancients believed that yin qi, yang qi and five virtues operate according to a certain law (also called natural law). Meanwhile, like most individuals, many TCM learners nowadays possess a scientific understanding of the universe, which does not include the traditional concepts of a round heaven and square earth or the idea of qi operating according to natural laws across the heavens and earth. Consequently, understanding the cosmological model underlying the theory of five circuits and six qi becomes challenging.[27] When encountering those who boast about their mastery of this mysterious theory, they can only feel humbly inadequate in their knowledge. However, since applying these theories directly in clinical practice is not necessary, they are often left on the shelf for admiration alone. Another common approach among practitioners is to acknowledge the objective connection between climate changes and pathogenesis, while recognizing that:

> "The theory of five circuits and six qi in ancient China is a theory on climate changes and their impacts on the human body. Based on the natural climate changes and the responses of organisms (especially human beings) to these changes, the theory integrates the natural climate phenomena with the biological phenomena and the natural climate changes with the

[27] The four directions above and below are called Yu (space), and the past and present are called Zhou (time). The theory of five circuits and six qi is indeed closely related to the conception of the Yu-Zhou (universe) in this sense.

> law of disease onset, which aims to explore the impact of climate changes on human health and diseases according to the laws of the universe.[28]"

Above is a typical explanation for the connotation of the theory of five circuits and six qi. However, it is important to note that this interpretation lacks scientific basis, because the theory of five circuits and six qi does not specifically deal with the seasonal and climatic changes that objectively exist in a year, but the imaginary climate changes in a cycle of six decades. In this case, human diseases and natural changes including climate changes are the manifestations of the cyclic movement of the circuit and qi between the heaven and earth. In other words, medical experts are not required to directly observe the natural climate changes or employ various diagnostic methods to understand the nature of a patient's disease. Instead, their focus lies in mastering the "circuits and qi" of a certain period of time, they can effectively treat diseases under the guidance of the theory. Although the theory of five circuits and six qi embodies the application of "the unity of the heaven and humanity" in Chinese medicine, it should be noted that the concept of "heaven" here does not refer to the natural world in general, but the "circuit" controlling all changes in nature, which is "the way of heaven" of a higher level. Even though the six qi may appear to be related to the changes of yin and yang, it must move steadily and regularly following the cycle of six qi in time and way. This is because its movement is determined by the "circuit". Therefore, the theory of five circuits and six qi, based on a completely transcendental and mechanically cyclic guideline, adopted the contents of the previous medicine that reflects the objective connection between climate changes and human diseases (the five elements matching the four seasons, i.e., domination in circuit). But its overall framework stands in contrast to the idea of "turning experience induction into theories" in the previous medicine. As a result, some skeptical medical experts who emphasize empiricism and positivism will inevitably deny the essence of the five circuits and six qi theory and view its concepts as abstract. They take the six qi as the abstract concept of the mechanism of disease and the mutual opposition and restriction between yin and yang in the six qi as the therapeutic principle. This kind of "five circuits and six qi theory" without calculation procedure reserves the terminologies in form, but its substance has been changed into a reasonable core of "diagnosing the nature of the diseases according to the empirical evidence obtained from the four diagnostic methods (observation, auscultation and olfaction, inquiry, palpation)".

To gain a comprehensive understanding the five circuits and six qi theory, it is important to explore its cultural and historical background, neither mystifying nor overly scientizing its principles. The theory is believed to have emerged during the Eastern Han Dynasty (AD 25–220), influenced by divination combined with mystical Confucianist belief prevalent during that era. For instance, the Japanese Tamba Motoyasu (Danbo Yuanjian) said that the five circuits and six qi theory was

[28] Edited by Cheng Shide, Appendix II "A Brief Introduction to the Theory of Five circuits and Six qi" in the *Annotated Collection of the Plain Questions* (*Su Wen Zhu Shi Hui Cui*, 《素问注释汇粹》), Beijing: People's Medical Publishing House, 1982, Volume II, p. 532.

"formed from the integration of Wei books[29] and medical books".[30] It is also pointed out that "five circuits" and "six qi" were put together firstly in the Yiwei.[31] And there are many similarities between Wei books and works discussing the five circuits and six qi theory.[32] However, if one asks about the source of the divination combined with mystical Confucian belief, it may be hard to explain. Or someone may ask whether the books of five circuits and six qi theory can be seen as one kind of the Wei books since there are so many similarities between them. The divination combined with mystical Confucianist belief was so prosperous after the Han dynasty (206 BC-AD 220) that the masters of Confucian classics annotated for it and interpreted the classics by extensively referring to Wei books. Considering these historical facts, we can understand the sociocultural context of that era. When studying the cultural context of the Eastern Han Dynasty from a scientific historical perspective, it is crucial to observe the notable advancements in academic trends and thinking methods compared to those of the preceding Western Han Dynasty (206 BC-AD 24). It integrated the Yiology with the theories of yin-yang and five elements and explained the formation of heaven and earth, politics, human relations (according to feudal ethics), the transformation and generation of all things in the universe, the human affections, and the rules of the world assisted by the numerical deduction. The theory of five circuits and six qi is the product as well as a component of the trend of the time.

Finally, let's summarize the theory of five circuits and six qi with three philosophical terms coined by Fang Yizhi:"交 (jiāo)", "轮 (lún)", "几 (jī)".[33]

交 (jiāo) means "interaction of opposites", such as the interaction of heaven and earth, yin and yang, movement and resting. Hence there are interactions of stems and branches, five circuits and six qi, dominant qi and guest qi, the qi controlling heaven and the qi controlling terrene.

轮 (lún) means "the cycle of motional transformation", such as the cycle of five circuits and six qi. Hence there are the generation and restriction of five elements and the alternative preponderance of the six qi, and the calculation of the circuit of the year, domination in circuit, guest circuit, dominant qi, and guest qi.

几 (jī) means "sign". The chapter "Comprehensive Discourse on Arrangements of the Principal Qi of Heaven" in *Plain Questions* states, "The mechanism of the perfect numbers is at hand and yet subtle. That which arrives, it can be seen. That which leaves, it can be pursued. Those who respect them, they prosper. Those who treat them with contempt, they perish". And the chapter "Comprehensive Discourse

---

[29] Books making a strained interpretation of Confucian classics and predicting the social stability and dislocation as well as the rise and fall with good or bad luck in the Han Dynasty also include many ancient myths and legends and cultural knowledge.

[30] Tamba Motoyasu (Japan), the chapter "Five Circuits and Six Qi" in the *Yi Sheng* (『醫賸』).

[31] One kind of Wei books, which is relative to the *Book of Changes*.

[32] Wang Shifu, "Discussion on the Origin of the Theory of Five Circuits and Six Qi", *Medical History and Health Organization*, 1958(2), p. 127.

[33] Fang Yizhi (the Ming Dynasty), the chapter "Three Signs" in *The Philosophy of East and West* (*Dong Xi Jun*, 《东西均》): "The intersection is the combination of two opposites; the cycle is the link of head and tail; where there is movement, intersection and cycle will appear, and signs could be made clear". Shanghai: China Book Bureau, 1962.

on the Essentials of the Most Reliable" in *Plain Questions* states, "carefully examine the requirements of the qi; do not neglect the trigger of the disease".

The core of the theory of five circuits and six qi lies in understanding the virtues (noumenon) and natural law (rule) of the movement of the universe, and the method of detecting signs (jī) in the movement and applying them to the medical practice. It is just because the "ultimate truth" of the theory of five circuits and six qi is entirely constructed by people's "intention". As a result, there are limitations to fully grasping this truth, and it may need to be set aside or approached with caution.

Chapter 20
Study on *Moxibustion Classic of the Eleven Yin-yang Meridians*

In recent years, researchers have diverted their focus to other objects, as it has been over a decade since the Mawangdui medical silk manuscripts were unearthed in 1973. Lately, however, as I was reading the *Moxibustion Classic of the Eleven Yin-yang Meridians* (*Yin Yang Shi Yi Mai Jiu Jing*, 《阴阳十一脉灸经》), I stumbled upon something intriguing that captured my attention.

20.1 Titles and Therapeutic Principles

Mawangdui medical silk manuscripts were unearthed without specific names, and the research team assigned names based on their contents. Two of the books on the circulation of meridians, indications and treatment are named as the *Moxibustion Classic of the Eleven Foot-arm Meridians* (*Zu Bi Shi Yi Mai Jiu Jing*, 《足臂十一脉灸经》) and *Moxibustion Classic of the Eleven Yin-yang Meridians* respectively. The difference in their names arises from the sequence of meridians described. In the *Moxibustion Classic of the Eleven Foot-arm Meridians*, the foot meridian is described first, and then the hand meridian. While in the *Moxibustion Classic of the Eleven Yin-yang Meridians*, the yang meridian is mentioned first, then followed by the yin meridian. It's appropriate to name the former as *Moxibustion Classic*, as there is a sentence that "A certain disease can be treated with the moxibustion on a certain pulse" following the description of the circulation and disease of meridians. The latter, however, does not extensively cover moxibustion therapy, with only one section briefly mentioning it, which proves to be supplemented by later generations. Therefore, further study is warranted to determine the suitability of the book's nomenclature of and to explore the viewpoint that "there is no such therapies as needles or other drugs in it".

The structures and formats as well as therapeutic principles applied in the *Moxibustion Classic of the Eleven Yin-yang Meridians* can be seen in Table 20.1. It shows

Y. Liao, *Reconstructing the Image of Medicine in the Qin and Han Dynasties*, Understanding China, https://doi.org/10.1007/978-981-95-6801-7_20

that the records of each meridian follow a consistent sequence—the name of meridians, the course of meridians, syndromes, treatment, and diseases produced by viscus (further details will be provided later). Given that there is only one paragraph specifically mentioned "the way of moxibustion" in the parts of kidney meridian, it is believed to be supplemented or annotated by later generations.

Upon closer examination, it becomes evident that the Chinese character "灸" (moxibustion) is only mentioned in one paragraph of the *Moxibustion Classic of the Eleven Yin-yang Meridians*. Hence, it may not be appropriate to classify it into "Moxibustion Classic" like *Moxibustion Classic of the Eleven Foot-arm Meridians*. The above analysis is not to rectify the name of the silk manuscripts, but to clarify that the whole passage just outlined the general treatment principles for a certain meridian, without providing specific guidance on how to treat it. Therefore, the aforementioned view that "there are no any other treatments like acupuncture and herbs except for moxibustion in the silk manuscripts" is obviously inaccurate. Further evidence can be found in the *Book on Meridians* (*Mai Shu*, 《脉书》) unearthed in Zhangjiashan, Hubei province, which indicates that *Moxibustion Classic of the Eleven Yin-yang Meridians*, *Rules of the Meridians* (*Mai Fa*, 《脉法》), and *Death Signs of the Yin-yang Pulses* (*Yin Yang Mai Si Hou*, 《阴阳脉死候》) are derived from a single classic book. This inspires us to speculate that the saying "teaching students the pulse law" in the prologue of *Rules of the Meridians* refers to the contents from *Moxibustion Classic of the Eleven Yin-yang Meridians*. The subsequent paragraphs in the *Rules of the Meridians* discussing the application principles of moxibustion and stone needle therapy, are also made for the meridians described in the *Moxibustion Classic of the Eleven Yin-yang Meridians*. It is generally acknowledged that the completion of the book *Moxibustion Classic of the Eleven Yin-yang Meridians* is later than that of the *Moxibustion Classic of the Eleven Foot-arm Meridians*. The names of the two classics reflect a shift from specific descriptions like hands and arms to more abstract descriptions like yin and yang. What's more, *Moxibustion Classic of the Eleven Yin-yang Meridians* merely outlines the general treatment principles without providing specific treatment methods, symbolizing a progression from practice to theory and gradually forming its overall comprehensive for physiology and treatment.

20.2 Meaning of "Disease Transmitted by Meridian" and "Disease Produced by Viscus"

Before the excavation of the silk manuscripts from Mawangdui, the concepts of "disease transmitted by meridian" and "disease produced by viscus" were firstly introduced in the chapter "The Conduit Vessels" in *Miraculous Pivot* (*Ling Shu*, 《灵枢》), and have traditionally been taken as two categories of diseases. That is, diseases of each meridian can be classified into meridian-induced disease (disease transmitted by meridian) and viscus-induced disease (disease produced by viscus) or disease caused by external pathogenic factors (disease transmitted by meridian) and disease

Table 20.1 Structure and format of *moxibustion classic of the eleven Yin-yang meridians*

Meridians	Course of the meridians	Diseases transmitted by meridian	Indications	Diseases produced by viscus	Supplementary explanations
Bladder meridian (BL)	……	……	Bladder meridian (BL)	……	
Gallbladder meridian (GB)	……	……	Gallbladder meridian (GB)	……	
Stomach meridian (ST)	……	……	Stomach meridian (ST)	……	
Small intestine meridian (SI)	……	……	Small intestine meridian (SI)	……	
Triple energizer meridian (TE)	……	……	Triple energizer meridian (TE)	……	
Large intestine meridian (LI)	……	……	Large intestine meridian (LI)	……	
Spleen meridian (SP)	……	……	Lung meridian (LU)	Accompanied by predictions of death	
Liver meridian (LR)	……	……	Liver meridian (LR)	Accompanied by predictions of death or cure	
Kidney meridian (KI)	……	……	Kidney meridian (KI)	……	When the kidney meridian (KI) disease is treated with moxibustion, the patient is advised to take nutritious food, wear loose and comfortable clothes with hair unbraided and walk with a stick and heavy shoes. In this way, the pathogenic factors can be dispelled and the diseases can be cured after obtaining qi.
Lung meridian (LU)	……	……	Lung meridian (LU)	……	
Heart meridian (HT)	……	……	Heart meridian (HT)	……	

caused by internal damages (disease produced by viscus). Actually, the statement here is doubtful. However, the accuracy of this classification remains uncertain, and there is no definitive explanation for the two terms. So temporary consensus has been reached on this matter. In fact, the essential differences between the symptoms of the "diseases transmitted by meridian" and "diseases produced by viscus" of each meridian are scarcely discernible from the chapter "The Conduit Vessels" in the *Miraculous Pivot*. Therefore, it is challenging to establish clear criteria for the classification of the two kinds of diseases. And the classification seems to hold little clinical meaning. Notwithstanding these challenges, ancient medical experts made efforts to interpret the concept of "disease transmitted by meridian" and "disease produced by viscus" from different perspectives. This is because the descriptions of each meridian in the chapter "The Conduit Vessels" in *Miraculous Pivot* have already been quite systematic, including the following six aspects:

> name of the meridian→ course of the meridian→ "disease transmitted by meridian"→ "disease produced by viscus"→ therapeutic principle→ pulse condition

Take "lung meridian of hand greater yin (lung meridian <LU>)" as an example. This phrase indicates that the meridian associates with the "lung", belonging to the meridians of the "hand" and attributing to the "major yin" (greater yin). Moreover, the description of the course of the meridian shows where the meridians start from and terminate at, how they associate with the corresponding organs, how they interact with the organs internally and externally through the collateral vessels, and what the directions and routes of the meridians are. Additionally, it covers the diseases related to the meridians, i.e., "disease transmitted by meridian" and "disease produced by viscus". Still take the example of "lung meridian of hand greater yin":

> When these vessels are excited, then this will result in a disease of lung swelling with a feeling of fullness. The swelling goes along with cough……These are diseases generated by the lung: Cough with rising qi, panting and thirst……[1]

Following the discovery of the medical silk manuscripts from Mawangdui, the concepts of "disease transmitted by meridian" and "disease produced by viscus" are also found in *Moxibustion Classic of the Eleven Yin-yang Meridians*. However, despite their presence in this text, pre-existing notions still strongly influence the perception of these concepts as a classification method of diseases. Whereas the classification criteria for them in this work is more ambiguous than that in the chapter "The Conduit Vessels" in *Miraculous Pivot*, so the study on the criteria is often overlooked. However, the key point of the research lies precisely in the classification criteria. Comparing the two ancient works, it can be seen that the records about "disease transmitted by meridian" and "disease produced by viscus" in the chapter "The Conduit Vessels" in *Miraculous Pivot* were modified:

1. Overlapping content of the two concepts was greatly reduced;
2. Abundant descriptions about "disease produced by viscus" were added;

[1] The chapter "The Conduit Vessels" in *Miraculous Pivot* of Paul U. Unschuld's translation version.

3. The content of "therapeutic principle" was placed after that of "disease produced by viscus";
4. Theoretical words like "These are diseases generated by…"[2] were added to the content of "disease produced by viscus".

Consequently, the later generations consider it as a classification method of diseases and make various speculations about the classification criteria. However, upon examining the silk manuscripts from Mawangdui, there are many instances of repeated symptoms of "disease transmitted by meridian" and "disease produced by viscus". The typical examples are shown in Table 20.2.

By examining these examples and exploring other meridians, relationships between *Moxibustion Classic of the Eleven Yin-yang Meridians* and the chapter "The Conduit Vessels" in *Miraculous Pivot*, as well as between "disease transmitted by meridian" and "disease produced by viscus" can be summarized.

1. The content of "Disease transmitted by meridian" in the chapter "The Conduit Vessels" in the *Miraculous Pivot* largely mirrors that of the *Moxibustion Classic of the Eleven Yin-yang Meridians.* The descriptions of the meridians in these two classics are nearly identical.
2. The concept of "Disease produced by viscus" in the chapter "The Conduit Vessels" in *Miraculous Pivot* was further developed based on the ideas presented in the *Moxibustion Classic of the Eleven Yin-yang Meridians.*
3. There are numerous symptoms of "disease produced by viscus" recorded in *Moxibustion Classic of the Eleven Yin-yang Meridians.* Some of them are exactly the same as "disease transmitted by meridian" (e.g., small intestine meridian <SI>); some of them are developed on the basis of "disease transmitted by meridian" (e.g., large intestine meridian <LI)>); and some of them are completely different from "disease transmitted by meridian" (e.g., heart meridian <HT>). These observations indicate that "disease transmitted by meridian" and "disease produced by viscus" are not two categories of diseases. Moreover, by examining the structure and format listed in Table 40, we can deduce that the content of "disease produced by viscus" is put after that of the indications. It may indicate that the content of "disease produced by viscus" was annotated by the people who had read and studied the original book. Therefore, he copied what he fully agreed

[2] Diseases dominated by meridians involve five Zang-organs. That is, the diseases dominated by the lung meridian (LU) are generated by the lung; the diseases dominated by the spleen meridian (SP) are generated by the spleen; the diseases dominated by the heart meridian (HT) are generated by the heart; the diseases dominated by the kidney meridian (KI) are generated by the kidney; the diseases dominated by the liver meridian (LR) are generated by the liver. And diseases dominated by other meridians do not involve Six Fu-Organs. The diseases dominated by the large intestine meridian (LI) are generated by the fluid and humor; the diseases dominated by the stomach meridian (ST) are generated by the blood; the diseases dominated by the small intestine meridian (SI) are generated by the humor; the diseases dominated by the bladder meridian (BL) are generated by the sinews; the diseases dominated by the pericardium meridian (PC) are generated by the vessels; the diseases dominated by the triple energizer meridian (TE) are generated by the qi; the diseases dominated by the gallbladder meridian (GB) are generated by the bones.

Table 20.2 Comparison of "disease transmitted by meridian" and "disease produced by viscus"

		Moxibustion Classic of the Eleven Yin-yang Meridians	The chapter "The Conduit Vessels" in *Miraculous Pivot*
Small intestine meridian (SI)	Diseases transmitted by meridian	"Pain in the throat and a swollen chin. Patients are unable to turn their view. The shoulders feel as if they were pulled. The soft parts of the upper arms feel as if broken"	"Pain in the throat and a swollen chin. Patients are unable to turn their view. The shoulders feel as if they were pulled. The soft parts of the upper arms feel as if broken"
	Diseases produced by viscus	"Aching chin, throat blockage-illness, pain in arms and elbows"	"The ears are deaf. The eyes are yellow. The cheeks are swollen. Neck, chin, shoulders, soft parts of the upper arms, elbows, as well as the posterior edge of the outer side of the arms ache"
Large intestine meridian (LI)	Diseases transmitted by meridian	"Toothache and swelling of the gluteus"	"Toothache and swelling of the neck".
	Diseases produced by viscus	"Toothache, swelling of the gluteus, yellow coloring of the eyes, a dry mouth, aching soft parts of the upper arms"	"Yellow coloring of the eyes, a dry mouth, nasal flow and nosebleed, throat blockage-illness. Pain in the soft parts of the shoulders. The index fingers ache and cannot be used. When the qi abound in exuberance, then heat swelling sets in at locations crossed by the vessels. In the case of depletion, the patients shiver from cold and fail to recover"
Heart meridian (HT)	Diseases transmitted by meridian	"Accompanied with panting, the patients wish to rise when they sit, and their eyes gaze into the void. The stomach feels like suspended, as if they were hungry. The breath qi is insufficient and causes anger. The heart is frightened, as if someone were to grab him. The patients suffer from hunger, and yet they do not wish to eat. Their face is as dark as burnt candlewick. They cough with blood"	"Dry throat, pain in the heart, thirst and a desire to drink"

(continued)

Table 20.2 (continued)

		Moxibustion Classic of the Eleven Yin-yang Meridians	The chapter "The Conduit Vessels" in *Miraculous Pivot*
	Diseases produced by viscus	"□□□□□□, fissured tongue, dry throat with rising qi, choke, pain in the throat, yellow dan-illness, desire to sleep, cough, loss of voice"	"The eyes are yellow. The flanks ache. The posterior edge at the inner side of the soft parts of the arms aches with receding qi. The center of the palm is hot and aches".[3]

with, supplemented what he thought was inadequate, and wrote down his own opinions when he disagreed with the views presented in the original book.

Although we cannot differentiate the handwriting between "disease transmitted by meridian" and "disease produced by viscus" from the Mawangdui silk manuscripts, it is believed that people at that time understood and valued the distinctions between them. For example, the author of the chapter "The Conduit Vessels" in *Miraculous Pivot* essentially replicated all the texts of "disease transmitted by meridian", while making revisions and supplements to the content of "disease produced by viscus". The supplements of diseases and syndromes in the chapter "The Conduit Vessels" in *Miraculous Pivot* were mainly reflected in the content of "disease produced by viscus" (see Table 20.3).

From this table, it becomes apparent that the chapter "The Conduit Vessels" in *Miraculous Pivot* introduces only six additional types of "disease transmitted by meridian" compared with the *Moxibustion Classic of the Eleven Yin-yang Meridians*. On average, each meridian has less than one additional type. But in terms of "disease produced by viscus", the number is almost doubled, with a maximum of 12 of each meridian. Such a significant difference cannot be ignored.

One of the key insights gained from analyzing the relationship between "disease transmitted by meridian" and "disease produced by viscus" is that the Mawangdui medical silk manuscripts, while appearing antiquated today and requiring the expertise in ancient characters for interpretation, are not a solitary undertaking. Compared with *Moxibustion Classic of the Eleven Foot-arm Meridians*, *Moxibustion Classic of the Eleven Yin-yang Meridians* has already made a progress. And compared with "disease produced by viscus", "disease transmitted by meridian" in the *Moxibustion Classic of the Eleven Yin-yang Meridians* signifies another development. If compared with the chapter "The Conduit Vessels" in *Miraculous Pivot*, we can see a greater leap forward.

Only by carefully analyzing these details can we gradually comprehend how the current version of *Huangdi's Internal Classic* (*Huang Di Nei Jing*, 《黄帝内经》) was formed. Otherwise, people might erroneously perceive it as "the creation of sages".

[3] The chapter "The Conduit Vessels" in *Miraculous Pivot* of Paul U. Unschuld's translation version.

Table 20.3 Changes in the number of "disease transmitted by meridian" and "disease produced by viscus"

	"Disease transmitted by meridian"		"Disease produced by viscus"	
	Moxibustion classic of the eleven yin-yang meridians	Chapter "the conduit vessels" in *miraculous pivot*	*Moxibustion classic of the eleven yin-yang meridians*	Chapter "the conduit vessels" in *miraculous pivot*
Lung meridian (LU)	4	5	5	13
Large intestine meridian (LI)	2	2	5	9
Stomach meridian (ST)	10	11	10	24
Spleen meridian (SP)	5	7	10	15
Heart meridian (HT)	4	4	1	5
Small intestine meridian (SI)	4	4	4	10
Bladder meridian (BL)	10	9	12	17
Kidney meridian (KI)	9	9	10	14
Triple energizer (TE)	2	3	3	10
Gallbladder meridian (GB)	5	7	12	17
Liver meridian (LR)	5	5	5	7
Total	60	66	77	141

20.3 Number of Meridians and Concept of Divination

Because there are five yin meridians and six yang meridians in both silk books, researchers infer that there were signs of connecting meridians with five Zang-organs and six Fu-organs.[4] Rather than making bold assumptions, it is beneficial to analyze this issue from the perspective of yin-yang and divination. As to this issue, it has already been discussed in the Chapter 8 of Part II. The words about the Zang-organs and Fu-organs are occasionally seen in the description of the course of meridians in the *Moxibustion Classic of the Eleven Yin-yang Meridians*, but it is completely

[4] Literature Research Centre of Academy of Chinese Medical Sciences: "Formation and Development of the Meridian and Collateral Theory from Three Ancient Meridian Literature". It is contained in *Prescriptions for Fifty-two Diseases* (*Wu Shi Er Bing Fang*, 《五十二病方》) compiled by the sorting group of the silk manuscripts from Mawangdui Han Dynasty Tomb, p. 175.

different from that of later classics represented by the chapter "The Conduit Vessels" in *Miraculous Pivot.* For example, there is mention of the "lung meridian enters into the heart".

20.4 Characteristics of the Meridian and Collateral Theory

The above analysis places an emphasis on the treatment of diseases as described in the original ancient books and their derivative texts, pointing out that "disease produced by viscus" and the following words of the lesser yin vessel were described with the characteristic of health preservation. ("When the kidney meridian <KI> diseases were treated with moxibustion, the patient is advised to take nutritious food, wear loose and comfortable clothes with hair unbraided".) These were formed after times of textual copying and transmission. In addition, the parenthesis for the greater yin vessel that "it's the stomach vessel, covering the stomach" is quite different with other sentences of the book in literary style. These observations suggest that these are annotations being applied in text. All of these demonstrate the notion that the Mawangdui medical silk books were compiled, copied, transmitted, and supplemented over time. However, they cannot validate the claims that "the two moxibustion classics were compiled in the Spring and Autumn and the Warring States periods (770–221 BC) or even earlier"[5] and "were formed and developed in the clinical practice of the people".[6]

The completion time is met with strong opposition from Mr. Luo Genze. He advocated that it is impossible for a single individual to write a book before the Warring States period (475–221 BC).[7] The view has also faced criticism from Japanese scholars who assert that "The inference is unfounded. Regardless of the conclusion will be made, it is inappropriate to deduce the completion time of *Huangdi's Internal Classic* based on the uncertain assumption that it was composed during the Warring States Periods. On the contrary, the writing process and the completion time of *Huangdi's Internal Classic* should be inferred from substantiated facts found in the medical silk books".[8]

[5] Literature Research Centre of Academy of Chinese Medical Sciences: "Introduction to Four Lost Ancient Medical Books in Mawangdui Silk Manuscripts". It is contained in *Prescriptions for Fifty-two Diseases* compiled by the sorting group of the silk manuscripts, p136.

[6] Literature Research Centre of Academy of Chinese Medical Sciences: "Formation and Development of Meridian and Collateral Theory from Three Ancient Meridian Literature". It is contained in *Prescriptions for Fifty-two Diseases* compiled by the sorting group of the silk manuscripts from Mawangdui Han Dynasty Tomb, p174.

[7] Luo Genze's article is contained in Volume 4 of *Debate on Ancient History* (*Gu Shi Bian*, 《古史辨》), Shanghai: Shanghai Chinese Classics Publishing House, 1982, p8.

[8] Keiji Yamada (Japan): "Writing of *Huangdi's Internal Classic*"; *Translations of the History of Science* (*Ke Xue Shi Yi Cong*, 《科学史译丛》), 1987, (3): 1.

Determining the exact completion date of the medical silk manuscripts is a challenging task, and it remains uncertain whether earlier meridian classics will be discovered. However, theoretical discussions regarding this topic are welcomed. Firstly, if advanced acupuncture and moxibustion skills were applied in diagnosis and treatment based on meridian and collateral theories in the Spring and Autumn period (770–476 BC) or earlier, we would expect to find related details recorded in descriptions of medical issues in some ancient books such as the *Rites of Zhou* (*Zhou Li*, 《周礼》). But there is no record about the meridian and collateral theory in these books. At that time, diseases were diagnosed by inspection and auscultation from nine orifices and were treated with medicines in five flavors. The original stone needle therapy and moxibustion, as evidenced by records like "moxibustion on the back by using burned wood" and "moxibustion on the middle finger of the foot" in *Prescriptions for Fifty-two Diseases* (*Wu Shi Er Bing Fang*, 《五十二病方》) did not rely on the guidance of meridian and collateral theory. These primitive medical methods were widely used in both eastern and western countries. If the meridian system had originated from these practices, it would have developed independently in different countries. On the contrary, though various conceptions of the courses of qi and blood can be found in different traditional medical systems, these views have long been abandoned. While the meridian and collateral theory in traditional Chinese medicine, despite lacking definitive proof, continues to be applied in clinical practice. Consequently, it seems that it is the emergence of meridian and collateral theory that leads to the development of acupuncture and moxibustion, and this original theory and technique can be developed uniquely in China before gradually gaining acceptance among foreign practitioners.

Secondly, the development course of the meridian theory follows a unique path, as it is different from other branches of medicine or the natural sciences in growth law, i.e., accumulation of experience to theory, and further to perfection by continuous refinement. Therefore, it can't be explained by general laws like clinical practice and accumulation of experience. To illustrate this, let's compare the development of the theory of meridian with that of materia medica theory. Generally, *Huangdi's Internal Classic* and *Shennong's Classic of Materia Medica* (*Shen Nong Ben Cao Jing*, 《神农本草经》) are believed to have been completed at the same period. The latter is regarded as a landmark achievement in pharmacology and has undergone continuous development and innovation by later generations. While the former, as the classic of the meridian theory, has remained primarily limited to teaching and application. Despite the immense medical experts and accumulated clinical experience over the course of two millennia—spanning from the Han Dynasty (202 BC-AD 220) to the present—it remains a mystery why there has been a lack of substantial innovation in this field.

Last but not least, the meridian theory is a complicated comprehensive system. And when it is combined with other aspects of medical knowledge, its essence becomes even more intricate. For instance, the whole system of meridian and collateral involves not only the meridians themselves but also the meridian sinews and cutaneous regions. The twelve meridian sinews divide human muscles based on the

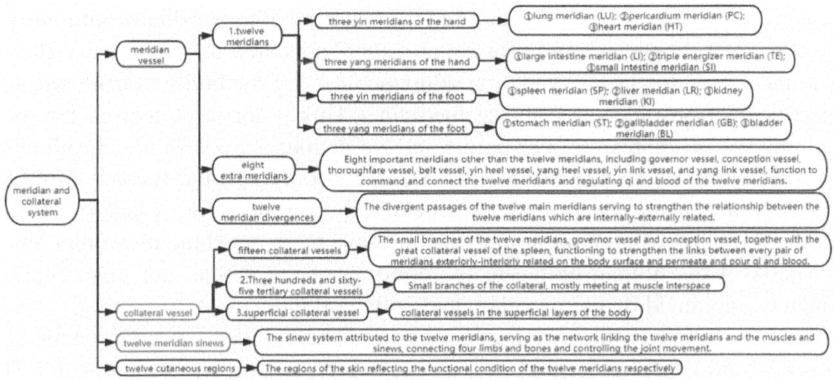

Fig. 20.1 Meridian and collateral system in contemporary traditional Chinese medicine

courses of the meridians. And according to the "originating" and "connecting" positions recorded in the chapter "The Conduits and Their Sinews" in *Miraculous Pivot*, the routes of meridian sinews are basically described at the anatomical level of that time. During a period of limited anatomical understanding, though the size of organs measured was nearly consistent with the modern measurement results, it is impossible to separate the organs from the vascular system and neural system. Due to the low level of morphology and limited cognition at that time, concrete things observed and phenomena perceived were expounded on together, resulting in some conceptual confusion in the meridian and collateral theory. As stated in the chapter "The Conduit Vessels" in *Miraculous Pivot*, "the twelve conduit vessels extend hidden in the partings of the flesh; they are in the depth and cannot be seen" and "all the vessels that can be seen are network vessels".[9] It is also pointed out that there are greenish and red vessels, which is obviously an objective description of the superficial vessels. Yet the "vessels" that cannot be seen are not great vessels in the depth. In other words, small vessels and capillaries of arteriovenous system correspond to certain collateral vessels and tertiary collateral vessels in the meridian and collateral theory, while great vessels are not in conformity with the twelve meridians (Fig. 20.1).

Moreover, as acupuncture and moxibustion became widely practiced, clinicians gradually discovered many acupuncture points that are effective for the treatment but are not located on the meridians, which are called "extra points". With the increasing discovery of these extra points, it was found that they could be able to form the epitome of the whole human body in some body parts, giving rise to the local acupuncture. Take the ear point as an example. In 1946, the American F.L. Potter reported a certain correspondence between the congenital development of kidney and the shape of auricle. In 1957, the French physician P. Nogier made a significant discovery when he unveiled a map of ear points distribution. This distribution map, resembling an

[9] The chapter "The Conduit Vessels" in *Miraculous Pivot* of Paul U. Unschuld's translation version.

inverted fetal projection, quickly captured the attention of the medical community.[10] Thereafter, "ear diagnosis" and "ear acupuncture" appeared in China. Nevertheless, "ear point", "ear diagnosis" and "ear acupuncture" are generally considered to be handed down from ancient Chinese medicine. This is inconsequential, but when explaining the mechanism of ear points and ear acupuncture with the meridian and collateral theory, many practitioners in TCM can only cite the records of certain meridians that reach to or in front of or behind the ear, revealing a lack of rigorous scientific thinking. Contemporary physicians have recognized the phenomenon that many parts of the human body are the epitome of the whole, but this cannot be adequately explained by the meridian and collateral theory.

Another example is the back transport point. There will be specific reactions on the back when the Zang-organs and Fu-organs are in disorder, as the chapter "The Transport Openings on the Back" in *Miraculous Pivot* states: "…one presses it with the finger. When the patient feels a reaction there, and if an existing pain is resolved, then that is the transport opening".[11] This objectively describes the corresponding relationship between spinal nerve roots and internal organs. When the organs undergo pathological changes, it is common that the spinal nerve roots at the corresponding horizontal positions become more sensitive, which can be treated with various physical therapies. This is drawn from the clinical experience rather than the meridian and collateral theory. These transport points are all taken as the points of the bladder meridian (BL) as they are just located on the course of the meridian. However, it is worth noting that this perspective has the potential to obscure simple concepts and create a sense of mystery, which could obstruct the understanding of the later generations regarding the pathological changes of organs. Some people believe that meridians and collaterals can be physically seen based on the visibility of back transport points and collaterals, whereas the above analysis has shown that their ideas are incomplete and wrong. Overall, as an important part of Chinese medical theory, the meridian and collateral theory remains shrouded in mystery. This is because the true nature of meridians and collaterals has not been fully understood. Additionally, this intricate system, which intricately connects the entire human body like a network, appears to sometimes refer to blood vessels, but not always. Besides, it is commonly acknowledged that meridian phenomena cannot be explained by neurology, but some indeed coincide with that of neurology. After experts study the science history to make clear the origin and development process of the meridian and collateral theory, these puzzles have been almost solved, while the essence of meridians and collaterals needs to be further explored by the future natural science.

---

[10] Editorial Committee of Ear Point Diagnosis: *Ear Point Diagnosis* (*Er Xue Zhen Duan Xue*, 《耳穴诊断学》), Beijing: People's Medical Publishing House,1990, pp. 1–2.

[11] The chapter "The Transport Openings on the Back" in *Miraculous Pivot* of Paul U. Unschuld's translation version.

Chapter 21
Study on the Newly Unearthed Document II: Exorcism Therapy

When the establishment of the modern natural science system occurred, it was primarily characterized by the use of "empirical evidence". As a result, all theories and doctrines explaining nature and its processes, as well as the methods of transforming and utilizing nature, were subjected to rigorous testing by the scientific community. This test not only evaluates the correctness, error, or inadequacy of some theoretical doctrines, methods, and techniques of the past but also establishes an insurmountable gulf between the "natural" and the "supernatural". As a result, witchcraft, a cognitive system that has been prevalent in various cultures for thousands of years and still survives among primitive tribes, has been labeled as "pseudo-science" due to their inability to differentiate between concepts of the "natural" and the "supernatural".

The pseudo-scientific nature of witchcraft stems from its distortion of natural laws and its flawed guidelines for action. In this system, various types of spells are created with the aim of achieving certain desires, but these spells are ultimately ineffective because they rely on illusory "supernatural forces" or the medium of force transmission that are considered "supernatural".

The system and structure of witchcraft can be summarized as "theoretical witchcraft" (pseudo-science) and "magic arts" (pseudo-technique). On a deeper level, it becomes evident that its mode of thinking—the theory of witchcraft—has a system similar to scientific understanding of things. However, this system and mode of thinking fail to correctly reflect the "truth" of the objective world, so it is called "pseudo-science". In addition, the specific spells that constitute witchcraft are called "pseudo-technique" because they were created by the inventors' intelligence and ingenuity. The "theoretical witchcraft" is not a guiding rule that the creators of the magic arts must follow, but rather an abstract principle synthesized by researchers who have explored the commonalities of various spells. Since these "theories" are essentially summaries of human thinking, they cannot solely serve as principles of magical (pseudo-technique) thinking but are also abstract principles on which certain ancient civilizations, now known as ancient sciences and technologies, were founded.

Y. Liao, *Reconstructing the Image of Medicine in the Qin and Han Dynasties*, Understanding China, https://doi.org/10.1007/978-981-95-6801-7_21

Therefore, the purpose of studying witchcraft is not to criticize its "untrue" nature but to further study its position and role in the history of human culture by analyzing the structure of various spells, clarifying the meanings behind behaviors and ways of thinking, and uncovering the respective characteristics and commonalities that exist across different cultural regions.

21.1 The Position of Exorcism Therapy in Medicine

Although exorcism therapy is no longer recognized or practiced in modern Western or Chinese medicine, it was a part of medicine in ancient times (both in China and other foreign countries). Despite the development of medical theories and other effective healing techniques, its position and application have not been diminished. There is no need to recount the records of the longstanding relationship between witchcraft and medicine in the pre-Qin period (before 221 BC).[1] Let's take what Sun Simiao, a TCM master of the Tang Dynasty (AD 618–907), said about the composition of the medicine as an example. He stated: "There are decoction, acupuncture and moxibustion, exorcism, magic figure and seals, and conduction exercise. Those five methods are all life-saving techniques".[2] Since the Sui Dynasty (AD 581–618), the Imperial Medical Academy established a position known as "Invoking and exorcising Boshi[3] (learned scholar)". In the Tang Dynasty, doctors were divided into four categories, one of which was the "exorcism masters", alongside physicians, acupuncturists, and masseurs.[4] The Song Dynasty (AD 960–1279) Imperial Medical Bureau divided medicine into nine categories, one of which was the "specialty of war-wound and exorcism".[5] The Yuan and Ming Dynasties (AD 1271–1644) of the Imperial Medical Academy also included the specialty of "Invoking".[6] For example, Sima Qian of the Han Dynasty (202 BC-AD 220) held that "believing in witchcraft but not medicine"

[1] Han Ying (the Han Dynasty), *The Unauthorized Biography of Han Poem* (*Han Shi Wai Zhuan*, 《韩诗外传》), Volume 10. For example, "When Yu Fu practices medicine, he uses the Nuo wood as the head and Angelica dahurica as the body, blows the orifice of the dead patient to stabilize the brain, and thus can bring the dead back to life". Another example, In the Han Dynasty (202 BC-AD 220), Liu Xiang's book, *History Novels in the Han Dynasty* (*Shuo Yuan*, 《说苑》), Volume 18, says: "When Miao Fu practiced medicine, he used the sugan as a mat and the grass as a dog, faced the north to invoke and spoke ten sentences, and those who were carried to ask for help restored to health".

[2] Sun Simiao (the Tang Dynasty), *Supplement to the Essential Prescriptions Worth A Thousand Gold Pieces* (*Qian Jin Yi Fang*, 《千金翼方》), Volume 29.

[3] *History of the Sui Dynasty* (*Sui Shu*, 《隋书》), Volume 28.

[4] *The Old Tang's Historical Book* (*Jiu Tang Shu*, 《旧唐书》), Volume 44.

[5] *The Collation and Research of the Compilation of Song's Regulations* (*Sui Hui Yao Ji Gao*, 《宋会要辑稿》), Beijing: Zhonghua Book Company, 1957, p. 2878.

[6] Tao Zongyi (the Yuan Dynasty), *History Notes of Song and Yuan Dynasties* (*Chuo Geng Lu*, 《辍耕录》), Volume 15; *History of the Ming Dynasty* (*Ming Shi*, 《明史》), Volume 74.

was "one of the six reasons that a doctor refused to treat a patient".[7] Even though Emperor Huizong of the Song Dynasty once "prohibited witchcraft",[8] such measures could not alter the prevalence of exorcism therapy or the practice of witchcraft. This is due to the fact that, for people lacking a view of natural science, there is no conceptual difference between the natural and the supernatural. People saw the failure of spells as a rejection of those specific spells, not of the entire belief system.[9] As Joseph Lee states: "The distinction between magic arts and science was only possible at a later stage in the history of human society … after the advent of modern science and technology in the early 17th century - something that Chinese culture had never achieved independently".[10] This understanding helps to illuminate how important the mindset of witchcraft and related spells were in ancient Chinese medicine.

It is believed that before the development of medical technologies, due to a lack of scientific understanding of nature and the limited means to fight against the disease, people had to beg for help from the deities for treating illnesses, so it was the period when witch doctors were dominant and witchcraft therapies were prevalent. This view, although commonly found in medical history books, lacks sufficient historical evidence. One of the reasons for the popularity of this view is that many people, consciously or unconsciously, accept the "animism" proposed by the British anthropologist E.B. Tylor (AD 1832–1917), who believed that primitive man, lacking a proper cognition of nature and incapable of fighting against it, would have generated the idea that "everything has a soul" before the formation of religion. But this view is not recognized by modern religious scholars. Because early human thinking had not yet developed to the level of being able to synthesize and abstract different phenomena, and no examples have been found to prove that the concept of "animism" existed in early civilizations.[11] In the case of exorcism therapy, for example, since the basic function of language is to express thoughts, and therefore, its intended recipients are supposed to be those who can "understand it", that is, individuals capable of grasping its meaning. However, as people's perceptions of this scope varied across different historical periods, the application of the invoking spells was also applied in diverse ways. From the medical classics unearthed in the Mawangdui Han Dynasty Tombs, it can be seen that during that period, exorcism therapy was applied to a narrower range of diseases compared to the later generations. This can be attributed to the limited availability of objects capable of receiving linguistic information in

---

[7] "The Biographies of Bian Que and Cang Gong" in *Records of the Grand Historian* (*Shi Ji*, 《史记》), Volume 150.

[8] *History of the Song Dynasty* (*Song Shi*, 《宋史》), Volume 21.

[9] *Old History of the Five Dynasties* (*Jiu Wu Dai Shi*, 《旧五代史》), Volume 29. For example, in the Five Dynasties, Emperor Zhuang Zong of the Later Tang "heavily praised and recruited people who could destroy the enemy ships. A dozen of crafty men came forward and all said they could spit fire to burn ships, or use spells to control blades. The emperor ordered them to have a try but none functioned".

[10] Joseph Lee, *Science and Civilisation in China* (*Zhong Guo Ke Xue Ji Shu Shi*, 《中国科学技术史》), Volume 2, p. 36.

[11] *Britannica Concise Encyclopedia* (*Jian Ming Bu Lie Dian Bai Ke Quan Shu*, 《简明不列颠百科全书》), Beijing: Encyclopedia Press of China, 1985, Volume 3, p. 15.

terms of both the "source" and the "object", which were not as abundant as in later times. For example, in the case of the "Xie (scorpion)" disease, which is thought to be stung by a scorpion, the people at that time would say: "fèng niǎo jiāng guàn ěr xīn.[12] (the "Phoenix bird" will be "peck through" your heart)", in order to dispel the evil. The "Phoenix bird" may mean "phoenix", but its thinking model is probably based on an analogy to the observation that chickens can peck at insects. Among the nearly 30 exorcism therapies in the *Prescriptions for Fifty-two Diseases* (*Wu Shi Er Bing Fang*,《五十二病方), the following are just a few deities that serve as sources of deterrence. ① Lightning, the King of the East, and □[13] the Lord of the West Hades (line 66); ② the Phoenix Bird (lines 82 and 84); ③ the God and Goddess of Heaven (line 204); ④ the Yellow God (lines 308 and 427); ⑤ the Five Soldiers (line 381); ⑥ the Witch (line 443). In addition to calling out these names, one can only rely on the power of the caster himself. Acts such as spraying, spitting, etc., which will be discussed later, are clearly important sources of deterrence, as the spell says: "pēn zhě jù pēn, shàng rú huì xīng, xià rú pēi xuè. (The practitioner spit furiously, above like a comet and below like blood.)" (line 51); spit and say: spit and say: "pēn qī (call out lacquer)," three times (line 380).

On the other hand, various punishments were used as deterrents in reality such as "beheading", "crushing the corpse", "stoning", "axe beheading", "slicing", "cutting off feet", "smearing pig shit", and "throwing into □ water".

The diseases that are treated with exorcism in the *Prescriptions for Fifty-two Diseases* can be categorized into the following causes:

1. Animal and insects: nests,[14] (scorpions, 2 articles), (snakes, 3 articles), carbuncles (2 articles), sore on the body;
2. Fox: Includes various pathologies of the vulvar area, such as urine retention, decadence, etc. (7 articles).
3. Demons: Qí (children's ghosts), lacquer (sore caused by lacquer king, 3 articles).
4. Unknown: Infants' blisters,[15] blood from wounds, (warts, 6 articles).

By analyzing these materials, it becomes evident that in ancient cognition, animals were the first objects capable of receiving linguistic information, followed by deities.

---

[12] Editing Group of Silk Book in Mawangdui tombs, Prescriptions for Fifty-two Diseases (*Wu Shi Er Bing Fang*,《五十二病方》), Line 84, *Silk Manuscripts from Mawangdui Han Dynasty Tombs [Part Four]* (*Ma Wang Dui Han Mu Bo Shu [Si]*,《马王堆汉墓帛书》〔四), "Notes on Interpretation", p. 36.

[13] □ refers to a missing word from the original text.

[14] For "Nests", See Editing Group of Silk Book in Mawangdui tombs, *Silk Manuscripts from Mawangdui Han Dynasty Tombs [Part Four]*, "Notes on Interpretation", p. 34; "Body Odor" seems inexact. In line 261 of the book, "If there is no nest, … the worms come out", which means "worm's nest", so we know that the disease is caused by worms.

[15] There is no mention of the cause of the disease in the text, but Mr. Yamada's detailed examination of this article suggests that the disease is caused by the blood of birds falling on children's clothing. For details, see Yamada Keiji: *The Bird Sings in the Night* (『夜鳴き鳥』), published in *Selected Translations of Works on Chinese History by Japanese Scholars* (*Ri Ben Xue Zhe Yan Jiu Zhong Guo Shi Lun Zhu Xuan Yi*,《日本学者研究中国史论著选择》), Volume 10, Beijing: Zhonghua Book Company, 1992, pp. 231–269.

And these deities were not entirely abstract. For example, Lightning, Phoenix Bird, the Witch, and Qí are all specific. Thus, the common belief in medical history that supernatural explanations and treatments primarily appeared in early human societies and were gradually replaced by scientific (physical and chemical) treatments, and that their relationship is antagonistic, should be properly revised. By comparing the application and targets of exorcism therapy during the Han and Tang Dynasties, it is clear that with the progress of culture and the complexity of human thought, the supernatural methods of explaining and treating diseases—exorcism—also underwent a process of development. In the case of China, at least from the Han to the Tang Dynasty, the scientific content of the medical system (physical and chemical therapeutic methods and related theories) developed in parallel with exorcism therapy. Although some spells can be seen to renew and update[16] in subsequent applications, they remained rooted in the same guidelines.

When discussing the role of witchcraft in the development of medicine, it should also be noted that, in addition to the exorcism therapy with explicit expressions, the basic ways of witchcraft thinking—the "Law of Similarity" and the "Law of Contact"—were, in fact, the foundation for much of the "scientific content" in the ancient Chinese medical system. This not only forms a certain way of thinking (e.g., association and analogy) that can generate both witchcraft and scientific outcomes but also leads to the uncertainty in the differentiation between witchcraft and science. The following is an example of each:

(1) Law of Similarity: For example, destroying a puppet that resembles one's enemy, or stabbing a paper figure with someone's name and birth date to bring about their failure or misfortune are typical manifestations of this witchcraft principle. Among the malaria treatment with exorcism therapies, the following spells can be seen:

When afflicted with malaria, one would stand by the water with a stone, and say the mantra: "yuān yuān yuán yuán, háng lù fēi nán, zhuō qǔ nuè guǐ, sòng yǔ hé guān, jí jí rú lǜ lìng.[17] (The road is not difficult to walk. Catch the malaria ghost and send it to the water god. Obey orders promptly.)" Then, throw the stone into the water and leave immediately without looking back.[18]

The basic thinking behind this witchcraft principle is that the same cause (or form) will inevitably result in the same effect, and its applications in medicine are numerous. For example, "harelip" is "caused by a pregnant woman seeing a rabbit and eating its flesh".[19] It is easy to find the fallacy of this statement, but not so easy to judge certain

---

[16] For example, *General Medical Collection of Royal Benevolence* (*Sheng Ji Zong Lu*, 《圣济总录》) in the Song Dynasty there are more than 60,000 words on the part of "forbidden technique of magic figures" and more than 300 magic figures.

[17] This is the pronunciation of pinyin, which mostly has no meaning. Its sounds may threat the evil or disease when use the incantation.

[18] Ge Hong (the Jin Dynasty), *Handbook of Prescriptions for Emergencies* (*Zhou Hou Bei Ji Fang*, 《肘后备急方》), Volume 3, People's Health Publishing House, 1963.p. 60,

[19] Chao Yuanfang (the Sui Dynasty), *Treatise on the Pathogenesis and Manifestations of All Diseases* (*Zhu Bing Yuan Hou Lun*, 《诸病源候论》), Volume 30.

medicinal theories, such as the effect of stalactite as "promote lactation"[20] according to their morphology and formation process; that eating an organ from an animal can nourish the corresponding organ in the human body (supported by the similarity in their chemical composition as revealed by modern biochemistry); that substances with "slippery" properties (e.g., rice soup and talc) can facilitate the opening of bodily passages (the former has fallen out of use, and the latter is a diuretic according to the principle of solute molecular infiltration, etc.). Li Shizhen's belief in the curative properties of a wedge for difficult childbirth was obviously incorrect, but the use of a comb to "comb the breast for more than a hundred times" to "promote lactation"[21] proved to be effective. Whether right or wrong, all these medical applications were based on the Law of Similarity.

(2) Law of Contact: For example, "In many parts of the world, we can see that the umbilical cord, especially the placenta, is regarded as a living object … where the child's guardian spirit or part of his soul resides. It is also generally accepted that the manner to handle a child's umbilical cord or placenta will also affect his or her future identity or career".[22] In the Silk Books unearthed from Mawangdui Han Dynasty tombs, there is the *Book on Obstetrics* (*Tai Chan Shu*, 《胎产书》), which draws a diagram of how to bury the uterus; for the explanation of the text, refer to the chapter "Yu's way of burying the placenta" in the *Recipes for Various Cures* (*Za Liao Fang*, 《杂疗方》).[23] Law of Contact means that once objects come into contact, they will retain a certain connection with each other, even if they are physically apart. In the chapter of "Peak-illness and Madness" of *Miraculous Pivot* (*Ling Shu*, 《灵枢》), there is an example of using a patient's blood to predict a seizure: "Place the blood in a gourd container. When the seizure occurs, the blood alone will move".[24] One of the most noteworthy "successful cases" in medicine is the invention of smallpox inoculation, where a child who had not contracted smallpox was made to wear the clothes of an infected child or was inoculated with scabs taken from a recovered patient, with the aim of "drawing out the fetal toxin". It is clear that the motivation for this behavior cannot be explained by the "transfer of disease" as a method of treatment, and that those who had not contracted smallpox-free were generally willing to receive such "transplants". The practitioner's intention was simply to use items from the patient's body (clothing, scabs, or pus) to guide the "fetal toxin" hidden in the uninfected person out. This explanation of "fetal

---

[20] The chapter "Stalactite" of *Shennong's Classic of Materia Medica* (*Shen Nong Ben Cao Jing*, 《神农本草经》) quoted from *Classified Materia Medica* (*Zheng Lei Ben Cao*, 《证类本草》), Volume 3 by Tang Shenwei (the Song Dynasty).

[21] Li Shizhen, Compendium of Materia Medica (Ben Cao Gang Mu, 《本草纲目》), Volume 38.

[22] James George Frazer (Britain), *Golden Bough* (*Jin Zhi*, 《金枝》), Chinese Version, Beijing: China Folk Literature and Art Publishing House, 1987, p. 32.

[23] Editing Group of Silk Book in Mawangdui tombs, *Silk Manuscripts from Mawangdui Han Tombs [Part Four]*, "Notes on Interpretation", p. 126, pp. 133–135.

[24] The chapter "Peak-illness and Madness" in *Miraculous Pivot* of Paul U. Unschuld's translation version.

toxin" persisted even after the introduction of cowpox inoculation in China, with the most important book on the subject being titled *The Introduction of Cowpox Inoculation* (*Yin Dou Lve*, 《引痘略》). The reason why smallpox vaccination become a great invention without any virological or immunological theories to support it is due to the simple "Law of Contact" mindset.

There are many similar treatments, but due to limited knowledge about ancient ways of thinking, researchers often analyze them using the theories of modern medicine. For example, the *Handbook of Prescriptions for Emergencies* is considered the first recorded text to mention the use of rabid dog brain tissue as a treatment for rabid dog bites. The record "kill the rabid dog, take the brain and apply it to the wound, and the disease will not recur" can be seen as at least in accordance with the principles of modern medicine on immunity, regardless of its therapeutic effect.[25] But if we knew that "there was a general belief in a connection between the injured person and the causative agent, so that any action taken on the causative agent would result in a good or bad outcome for the injured person accordingly".[26] It is not difficult to gain a basic understanding of the role of "witchcraft" in the development of ancient civilizations. As Dampier said, that "the relation of witchcraft to religion and to science is still under debate", but "whatever the actual relation of the three, witchcraft seems to be, after all, the cradle of religion and science".[27]

21.2 The Function and Energy Transferring Rite of Exorcism Therapy

(1) Spray and spit—the role of "qi": When performing a spell, the practitioner often exhales in a manner of spraying or spitting at the subject. For example, the following exorcism therapy can be found in the Silk Book in Mawangdui Han Dynasty tombs: "Spit and say: 'xiōng fù chǎn dà shān, ér jū shān gǔ xià.' (My brother and father live in the mountain, and you dwell below the valley…)". "Exhale and say: 'jiē nián, shì shā rén, jīn zī yòu fù zhī.' (Poisonous insects start to kill man again.)" This action is mostly performed before the chanting of a spell, though it is rarely used as a standalone practice, it holds its own independent meaning.

In Chinese tradition, spit has always been associated with despise or insult. For example, in the chapter of "Xi Gong" of *The Commentary of Zuo* (*Zuo Zhuan*, 《左传》), it is mentioned, "Spit without regard". Similarly, in the chapter "Zhao State" in *Strategies of the Warring States* (*Zhan Guo Ce*, 《战国策》), it is stated,

[25] Zhu Renkang, *Chinese External Medicine* (*Zhong Yi Wai Ke Xue*, 《中医外科学》), Beijing: People's Medical Publishing House, 1987, p. 5.

[26] James George Frazer (Britain), *Golden Bough*, Chinese Version, Beijing: China Folk Literature and Art Publishing House, Chinese version in 1987, p. 63.

[27] W. C. Dampier (Britain), *The History of Science* (*Ke Xue Shi*, 《科学史》), p. 479.

"The Empress Dowager directly told the people around her that if anyone suggested that the emperor's son should go to Zhao state as a hostage, she would spit on his face". However, in the context of exorcism, the "spit" seems to have more meanings beyond contempt and insult. In the legal literature of the Qin Dynasty (221–207 BC), the "*Models for Sealing and Physical Examinations* (*Feng Zhen Shi*, 《封诊式》)," there were cases of accusations of toxins in the mouth, called "speech toxins".[28] Furthermore, in discussing how to impart the appropriate medical skills according to the individual's characteristics, the chapter "Function and Competence" of the *Miraculous Pivot* says: "Those whose speech is swift and toxic and who look down on others can be made to spit on abscesses and exorcise diseases".[29] In addition, Wang Chong (about AD 27–97) of the Han Dynasty also mentioned in his book *On Balance* (*Lun Heng*, 《论衡》): "In the land of the great yang (longtime exposure to sunlight), the people are quick-tempered; their mouths and tongues are poisonous. Therefore, the people of Chu and Yue states are quick-tempered and quick to act. If they spit at someone, it will cause swelling of the veins and injury to that person; in the extremely hot land of the southern counties, people can make trees wither by spitting at them". Thus, the use of "spit" in exorcism therapy is most likely the result of seeing "spittle" as a carrier of the toxin from the mouth and tongue. Although the concepts of "speech toxins" and "poisonous mouths and tongues" are no longer popular, the practice of "spit" has persisted in exorcism.

The actions of spitting and spraying, although different, share the commonality of transmitting some kind of energy to the subject. Wang Chong explained: "The Wizard could invoke to cure people's illnesses and heal their wound, that because he (the Wizard) was born in the south of the Yangtze River and contained fierce qi[30] in his mouth". Later, Ge Hong of the Jin Dynasty (AD 266–420) wrote a similar view in his book *The Inner Analects of Bao Puzi* (*Bao Pu Zi Nei Pian*, 《抱朴子·内篇》): "A person who is good at using qi can regulate the qi movement inside to nourish the body, while the healthy qi outside the body can drive away evil and disease. However, common folks use it every day without fully comprehending its nature. Wu state and Yue state have forbidden exorcism spells, which are very effective due to the master use of qi".

In comparing the arguments presented by Wang and Ge, it is evident that Wang explains that the geographical region south of the Yangtze River, where the sunshine is more abundant, contains more fierce qi while Ge shifts his focus to the inherent qi of the human body which "people use but know little about" even though he cites Wu and Yue as an example. Although Ge Hong's explanation of what can be "controlled by qi" is quite broad does not involve religious or immortal elements. Instead, it is based entirely on using the energy of qi to constrain other things. For example, the

[28] Shuihudi bamboo slips editing group of Qin tombs, *Shuihudi Bambaoo Slips of Qin Tombs* (*Shui Hu Di Qin Mu Zhu Jian*, 《睡虎地秦墓竹简》), Beijing: Cultural Relics Publishing House, 1987, p. 276.

[29] The chapter "Function and Competence" in *Miraculous Pivot* of Paul U. Unschuld's translation version.

[30] Wang Chong (the Han Dynasty), *On Balance*, Volume 23.

sentence "resist the blade with your qi and you will be able to walk without being hurt or stabbed without being penetrated" bears resemblance to modern practices known as "Hard Qigong".[31]

From the perspective of intellectual history, Ge Hong's theory represents a significant development that took place after Confucianism replaced Taoism in the Han Dynasty, "a group of intellectuals who studied Taoist doctrines shifted their focus to scholarship and witchcraft respectively",[32] thus making "witchcraft" gradually theoretical and systematic, and displaying a tendency of philosophical thought with a theoretical explanation. In relevant literature of the Tang Dynasty, it can be seen that the application of qi in medical contexts became even more prominent, forming qi exorcism one of the "six methods" of incantation—pronouncing six characters "Chuī (吹), Hū (呼), Hē (呵), Xū (嘘), Xī (嘻), Sī (呬)" based on pinyin. The application of these is specific to certain conditions, for example, "If you spit at a heat disease, use Chuī to blow on it with cold air, then exorcise it; if you spit at a cold disease, use Chuī to blow on it with warm air, then exorcise it".[33] It is self-evident that the actions of "Chuī" and "Hē" elicit distinct sensations of heat and coolness on the human skin, and regardless of the therapeutic value of this approach, its medical significance is apparent. Spitting, on the other hand, is still an important part of the exorcism therapy, but with the addition of a "receiving" procedure in which the spell-caster first begs for the help from the deities and then conveys this "divine power" to the target person by spitting.

(2) Role of words: Reciting the spell words is an important part of exorcism therapy. However, the most common and misguided explanation is to mistakenly label exorcism therapy as a form of psychotherapy.[34] What makes this explanation untenable is that it merely observes the common utilization of language for therapeutic purposes in both exorcism therapy and psychotherapy, without delving into the fundamental distinctions between the two approaches.

In modern psychiatry, various methods, including "words", are used because "In many diseases, such as neurological disorders, and psychosocial factors, are causative factors that play an important role in the development of the disease. These

[31] Ge Hong (the Jin Dynasty), *The Inner Analects of Bao Puzi*, Volume 1.

[32] Zhu Yueli, *General Theory of the Tao Te Ching* (*Dao Jing Zong Lun*, 《道经总论》), Shenyang: Liaoning Education Press, 1991, p. 9.

[33] Sun Simiao (the Tang Dynasty), *Supplement to the Essential Prescriptions Worth A Thousand Gold for Emergencies*, Volume 29.

[34] Hebei Medical College, *Miraculous Pivot: Collated and Annotated* (*Ling Shu Jing Jiao Shi*, 《灵枢经校释》), Beijing: People's Health Press, 1982, Volume 2, p. 313. Another example is Liu Boji's *History of Chinese Medicine* (*Zhong Guo Yi Xue Shi*, 《中国医学史》), Taipei: Hwa Kang Publishing Center, 1974, p. 1. It states: "Therefore, their witchcrafts, spells, songs or dances, which are said to exorcise evil spirits for the sick, are nothing more than a kind of hypnotism and psychotherapy according to modern concepts". Recently, Yuan Wei's *Chinese Ancient Zhu You (Invoke Therapy) Exploration* (*Zhong Guo Gu Dai Zhu You Liao Fa Chu Tan*, 《中国古代祝由疗法初探》), *Studies in the History of Natural Science* (*Zi Ran Ke Xue Shi Yan Jiu*, 《自然科学史研究》), 1992, (1): 48) also affirmed the scientific value of Zhu You from the perspective of "psychotherapy".

psychological factors cannot be effectively addressed through medications or physical therapy and must be addressed through psychotherapy. Psychotherapy involves using therapeutic dialogue to influence a patient's mental activity. Its primary goal is to enhance patients' awareness of the underlying causes, clinical manifestations, and progression of their condition, with the assistance of skilled healthcare professionals. It guides patients to take the initiative to relieve anxiety, build confidence in overcoming the disease, and promote their physical and mental health, ultimately achieving treatment goals".[35] Although exorcism spells involve a "verbal" component, they are different from psychotherapy in that the following ways: ① the object of the spell is not the patient themselves, but another being capable of receiving the verbal message, such as animals, ghosts, spirits, etc. ② The spell does not aim to influence the patient's mental activity or consider "psychosocial factors". Its purpose is to request divine power to deter the targeted being and achieve the "deterrence" objective (treating illness is the only one of its purposes). ③ As a therapeutic method, its application is not limited to mental disorders but mainly targets physical disorders. Therefore, it can be asserted that exorcism spells, and especially the role of "words" in them, lack any elements associated with "psychotherapy" in modern medicine.

After correcting the "scientific evaluation" of the psychotherapeutic value of exorcism spells, it is necessary to correct another view that simply evaluates exorcism spells as a means of religious superstitious deception: "In witchcrafts, it is believed that divinely inspired spells, which have magical powers, can only be understood and employed by witches".[36] In fact, the spells are all composed of common language with clear meanings that anyone can understand and practice. In societies where the existence of spirits and deities is acknowledged by both the practitioners and the patients, and even the wider community, the use of words for communication is seen as natural, without any notion of "cheating" or "being cheated". Therefore, the point of analyzing the spell words is to understand how people of that time viewed the world and understood things, rather than judging what was right or wrong. As mentioned earlier, words themselves are not a force but rather a message that must be understood by those who receive them. Apart from the literal deterrent effect of the spells, what is worth studying and noting is the "pivot (deep meaning)" contained in these words. In an essay on the concept of "pivot" and its significance in the history of ancient Chinese thought, Mr. Li Zhichao[37] argues that the development of "pivot" as an abstract concept "contributed to the ancients' understanding of information, and first elucidated the role of consciousness and language as pivots". In the context of exorcism therapy, this control is exerted through the use of "name". Dong Zhongshu (179–104 BC) stated in his book, in the chapter "An In-Depth Examination of Names and Designations" of *The Luxuriant Dew of the Spring and Autumn Annals* (*Chun*

---

[35] Shen Yuqian, *Psychiatry* (*Jing Shen Bing Xue*, 《精神病学》), People's Health Press, second edition, 1990, p. 195.

[36] Zhu Yueli, *General Theory of the Tao Te Ching*, Shenyang: Liaoning Education Press, 1991, p. 21.

[37] Li Zhichao, *Triggerism - A Promising Outlook on Science* (*Ji Fa Lun-You Wei De Ke Xue Guan*, 《机发论——有为的科学观》), Studies in the History of Natural Science, 1990, (1):1.

Qiu Fan Lu, 《春秋繁露》): "The name is the first indication of the Big Tao … Its (pivot) is connected to heaven and earth". The application of this concept can be seen in the words of exorcism spells, as exemplified by the *Miscellaneous Prescriptions* (*Za Liao Fang*, 《杂疗方》) in the Silk Book in Mawangdui Han Dynasty tombs:

In cases where someone is unfortunate enough to be attacked by a snake or a wasp, use an exorcism therapy, spit on him three times, call him by the name of the insect/animal that attacked him, and say: "mǒu, rǔ dì xiōng wǔ rén, mǒu suǒ zhī míng, ér chù shuǐ zhě, ér chù tǔ zhě wéi qí, qī mù zhě wéi fēng sī, fēi ér zhī jīng nán zhě wéi yù. ér jìn ☑wèi □□, ěr jiāo wéi zōng sūn. Mǒu zéi, ěr bú shǐ mǒu zhī bìng yǐ, qiě fù □.[38] (You have five brothers, and those who live in water are □, those who live in the earth are Qi, those who live in wood are Wasp, and those who live in Jingnan are Wormy. But Jin □ has not □, and you are the descendant. Thief, if you do not make□'s illness go away, you will □.)

This spell poses some difficulties in making pauses due to missing words, making it challenging to understand its exact meaning. But one clear aspect conveyed by the spell is the importance of knowing the name of the attacker. In some cases, the spellcaster cannot name the attacker, and can only find a pretext that the spirits know: "Goddesses six Ding and gods six Jia know the name of the attacker".[39] It was Ge Hong of the Jin Dynasty who explicitly explained the effect of these spell words: "Knowing the name of a thing prevents it from causing harm".[40] This belief is reflected in the spell, when one encounters evil things in the mountains or forests or hears loud voices at night, "if one knows and calls them by name, these things dare not commit any crime."; "if one sees them, and calls them by name, they dare not do any harm". Although the method of "deterrence" by calling a thing by its name without using threatening words has a long history, Ge Hong, as a representative of Taoism and an outstanding ancient thinker, apparently regarded this method as a principle because he used the exorcism to explain the "metaphysics" of "words" and "name" as "pivot". As explained by Li Zhichao in his essay on "Pivot Theory", the scientific view of Taoism: "This is very important and should not be ignored in the study of traditional Chinese culture, the view of nature, and the history of scientific thought".

(3) Force Transfer: In the field of medical treatment, various kinds of internal and external medicines, massage, acupuncture and moxibustion, etc., are used to apply specific forces to the patient, aiming to achieve the desired treatment outcomes. As part of ancient medicine, exorcism therapy is also a way to use all the "available forces" to achieve the purpose of treatment. The difference between the two lies in the selection of the "external force" that is an objectively existing "force of nature" that can exert an influence on the object, and it becomes an integral part of the medical practice; if the chosen "external force" or the

---

[38] Editing Group of Silk Book in Mawangdui tombs, *Silk Manuscripts from Mawangdui Han Tombs [Part Four]*, "Notes on Interpretation", p. 128.

[39] Sun Simiao (the Tang Dynasty), *Supplement to the Essential Prescriptions Worth A Thousand Gold for Emergencies*, Volume 29.

[40] Ge Hong (the Jin Dynasty), chapter "Climbing" in *The Inner Analects of Bao Pu Zi*.

means of transfer is illusory, then it falls within the realm of witchcraft treatment methods.

For example, "rub moxa leaves with your hands so that the moxa sap is in your hands. Do not wash your hands for seven days and keep a vegetarian diet, then it becomes exorcism .... and say: 'yuàn wǒ cǐ shǒu yī qiē yōng zhǒng, yī qiē zhū dú, nǎi zhì yī qiē bìng, shǒu zhe jí chà.[41] (I wish a full recovery from the furuncles, carbuncle or any other poison of my hand.)' The idea behind this practice is to imagine this hand as a burning iron fork when treating a disease, and the patient's disease as snow, with the belief that touching the patient will remove the disease". Although "moxa" is a genuine herb and widely used in external treatment, the "transfer pathway" described above cannot transfer this external force to the subject, so the art of medicine cannot be established. In exorcism therapy, the force transfer sometimes requires nature things, such as "When the lightning comes, raise your eyes to the lightning, hold a knife in your right hand and rub it with your left hand, and say: 'zhù wǒ xíng jìn, zhèn shēng rú léi hǒu, wàn dú fú bì qì. (Help me perform the exorcism, the sound of thunder roar, let all poisons dissipate.)' Repeating till the sound of thunder ceases".[42] Another example: "When the swallows come in spring, take out a piece of paper, look at the swallows and dot ink on the paper without stopping until the swallows fly out of sight".[43] Since then, the knife and paper have become healing tools with divine power.

Due to the development of culture, a plethora of illusory deities and powerful spirits emerged, so the spellcaster gradually became the medium between "divine power" and "disease" (but their role is not "to convey the will of heaven"). Their spells primarily targeted deities capable of understanding it, rather than the cause of disease. "dì zǐ mǒu yǐ yuàn chí jìn fǎ, jìn duàn xié è guǐ dú zhī qì, jiù lǐ rén mín, fú yuàn jiàng zhēn qì liú bù chén shēn, lìng chén suǒ háng fú jìn yīng shēng chú chà, yīng shǒu chú yù.[44] (I would like to perform the imprison spells, to expel the evil ghosts as well as poisonous qi, and save the people. Please send down the true qi spreading my body, so that the magic figures draw by my hand can cure all diseases.)" The spellcaster "borrows" the divine power via various means, including words, and then transfers it to the patient by means of a knife, water, or saliva to cure diseases. Therefore, in this context, the exorcism spell can treat everything, far beyond the range of "being able to receive verbal information".

---

[41] Sun Simiao (the Tang Dynasty), *Supplement to the Essential Prescriptions Worth A Thousand Gold for Emergencies*, Volume 29.

[42] Sun Simiao (the Tang Dynasty), *Supplement to the Essential Prescriptions Worth A Thousand Gold for Emergencies*, Volume 29.

[43] ibid.

[44] ibid.

21.3 Exorcism Therapy and Religion

It is generally accepted that the division between medicine and witchcraft during the Western Zhou period (about 1046–771 BC) marked an important turning point in the development of TCM.[45] From the division of TCM recorded in the *Rituals of Zhou* (*Zhou Li*, 《周礼》), Sima Qian's criticism of those who "believe in witchcraft but not in medicine" and the evaluation of "invoking" therapy in the medical classic "Plain Questions" (*Si Wen*, 《素问》), it appears that since then, TCM has indeed embarked on a path of natural science, while gradually distancing itself from the art of exorcism and forbidden spells. However, as mentioned earlier, exorcism therapy has reappeared in the medical system since the Sui Dynasty (AD 581–618). This resurgence continued until the Ming Dynasty (AD 1368–1644), during which exorcism therapy still occupied a distinct position in the medical subspecialty. It is important to understand that this phenomenon does not signify a regression in medicine. Rather, its occurrence is closely tied to the introduction of Buddhism and the growing prominence of Taoism in medicine following the Eastern Han Dynasty (AD 25–220).

In discussing the relationship between exorcism therapy and religions, it is essential to focus on the mutual influence and interplay among these systems, rather than delving into the broader debates concerning the introduction of Buddhism, the rise of Taoism, and how Confucianism, Buddhism, and Taoism learn from each other, absorb each other and reject each other. It should be noted that since any religion is shaped by absorbing cultural elements inherent to the region and the nation, the examination of the relationship between exorcism therapy and religions is merely a study of the mutual influence and penetration among these systems, not a discussion of "witchcraft-religion" from the standpoint of philosophy and cultural history.

Let's start with Taoism: as a religion native to China, it has absorbed a lot from the divine and folk spells throughout its development. For example, the widespread use of magic figures by Taoists can be traced as "appearing no later than the first year of the Huan Di Yuan Jia (AD 151)",[46] as evidenced by the Silk Books unearthed from the Mawangdui tombs. "Let the patient drink the magic figure ash with mutton soup and then go to take a bath".[47] Magic figures are abstract figures used to deter ghosts and monsters. For example, it is said that there is a large peach tree in the East China Sea, under which there are two gods, known as "Shen Tu" and "Yu Lei", who can devour all kinds of ghosts.[48] Therefore, peach-related items such as peach sedges, peach stalks, and peach soup, etc., are popular among common people, gradually evolving into "peach talisman" paintings of two gods on the peach wood boards. Even painting

[45] Du Shiran et al. *Science and Civilization in China* (*Zhong Guo Ke Xue Ji Shu Shi Gao*, 《中国科学技术史稿》) Beijing: Science Press, 1982, p. 78.

[46] Wu Rongzeng, "Tao-Witchcraft Relations in the Eastern Han Dynasty as Seen in the Tombs Text", *Cultural Relics* (*Wen Wu*, 《文物》), 1981, (3): 56.

[47] Editing Group of Silk Book in Mawangdui tombs, *Silk Manuscripts from Mawangdui Han Dynasty Tombs [Part Four]* "Notes on Interpretation", p. 73.

[48] Wang Chong: *On Balance*, Volume 23. It is said that this is from the *Classic of Mountains and Seas* (*Shan Hai Jing*, 《山海经》), but there is no such text in the present version.

their figures on ordinary paper as "New Year paintings", they still have their symbolic content and meaning. But it is true that the practice of drawing "magic figures" has been gained significant prominence in Taoism.[49] By the Six Dynasties period (AD 222–589), "Magical figure" therapy could be found in medical writings, such as the treatment of children's cries of fear. "Write the character '贵 (Gui)' under the umbilicus for great fortune".[50] Another example is the treatment of women breast swelling caused by heat pathogen: "Use vermilion to write the character '鱼 (Yu)' on the patient's breasts"[51] etc. Subsequently, the "exorcism spells" have transformed into "book forbidden section" in Imperial Medical Bureau of the Song Dynasty, which is closely related to the social background of "The Maoshan sect during the Song Dynasty transitioned predominantly toward a talismanic focus, with rulers assigning significance to Taoist practices involving talismans, in contrast to the emphasis on such practices during the Tang Dynasty".[52]

Second, Taoism inherited the pre-Qin Taoist ideas of emptiness and non-action in its theoretical system, which Joseph Li summarized that Confucianism values yang and Taoism values yin. [53]While not required for the exorcism itself, this philosophy is also reflected in the exorcism theory of the Tang Dynasty: "*Scripture of Immortal* (*Shen Xian Jing*, 《神仙经》) said: yang is strong but impermanent, while yin is soft but enduring. When the sage closes his mouth, all things can be concealed … When the crowd engages in play, I stay alone; when the crowd wrangles, I keep quiet; when the crowd argue, I keep silent. This is the way of cultivating forbidden exorcism skills".[54] Therefore, exorcism in the Tang Dynasty is more complex than the records from Mawangdui. To acquire it, practitioners need to adhere to numerous precepts simultaneously. For example, they are required to refrain from speaking with others for seven days in order to cultivate the necessary abilities. Additionally, "recite the forbidden text silently, with only their own ears can hear the spell, otherwise their spells will fail".[55]

In terms of the source of power, the exorcism spell in the Tang Dynasty does not rely on the inner power of spitting, spraying, and the intimidation of words as Mawangdui's medical books recorded and is also different from Ge Hong's philosophical theory of being "forbidden by qi" and "if one knows its name, it cannot do

49 For example, Zhang Jiao was good at "magic figures and water spell to cure diseases", *Book of the Later Han* (*Hou Han Shu*, 《后汉书》), Volume 110, "The Sixty-first Biography of Huangfu Song"; and Ge Hong's *The Inner Analects of Bao Puzi*, also recorded many methods of drawing magic figures.

50 Ma Jixing, *Textual Research and Interpretation of Dunhuang Ancient Medical Books* (*Dun Huang Gu Yi Ji Kao Shi*, 《敦煌古医籍考释》), p. 193.

51 ibid. p. 272.

52 Zhu Yueli, *General Theory of the Tao Te Ching*, Shenyang: Liaoning Education Press, 1991, p. 95.

53 Joseph Lee, *Science and Civilisation in China*, Chinese version, Volume 2, p. 33.

54 Sun Simiao (the Tang Dynasty), *Supplement to the Essential Prescriptions Worth A Thousand Gold for Emergencies*, Volume 29.

55 ibid.

harm". Rather it begs all the deities to give the spellcaster power. A few examples of the names of deities contained in the *Incantation Classic* (*Jin Jing*, 《禁经》):

"The First Law of Upholding Fast Commandment": "All the deities help, and the forbidden law is practiced";

"The Second Law of Receiving Imprison Incantation": "The Three Divine Masters, the Divine Child, the Jade Maiden, the Heavenly Doctor, the Lu Doctor, all the Masters, the Supreme Elder, the Divine Kings of the Immortals, the Five Stars of the Sun, the Moon, the Twenty-eight Stars, the Big Dipper, the Divine Officers of the Immortals belonging to the Great Divine Kings";

"All Reception of Imprison Incantation": "Eastern wood Jin, Southern fire Jin, Western gold Jin, Northern water Jin, Central earth Jin, Heavenly Master, Eastern King, and the Queen Mother of the West";

"Reception Incantation on Lunar March 3": "All the water officials of the rivers, and the great dragon of the four seas";

"Reception Incantation of Immortal Taibai": "The Five Emperors of the Five Directions, the Five Forbidden Masters, and the Five Essence Swallowing and Poison Biting Night Deities Kings";

"Receiving Imprison Incantation of Heavenly Emperor Taiyi" : "Green Dragon, White Tiger, Vermilion Bird, Black Tortoise, the Seven Stars of the Big Dipper, the Great Forbidden Master of the Eastern Rule, the Hussar General of the Marquis of Wan Shi Zhao, the Duke of Su Pingnan, the General of the Eight Departments, the Seventy-two Forbidden Masters, the Master of Chen, the Master of Zhao, the Small Official of the Straight Symbol, the Child of the Straight Sun, all being the Protector on Today".

The influence and infiltration of Taoism on exorcism therapy is evident, as seen in the names of these deities involved. However, regardless of its religious influence, its essence remains unchanged, and it is still a system that falls under the category of "pseudo-arts" guided by "pseudo-science". While exorcism therapy, during a specific phase of its evolution, placed a greater emphasis on acknowledging the existence of deities compared to its primitive stage, its essence remains centered on subjugating these deities to fulfill the intentions of the spellcaster. Instead of aligning with the will of the deities. This essential difference sets witchcraft apart from religion.

Let's return to Buddhism: The circulation of Buddhist exorcism techniques can be seen in the remnants of medical books unearthed from Dunhuang. For example, when a pregnant woman is in difficult labor, the spellcaster can recite the following words: "There was a formula, there was a spell: 'nán wú gàn shī pó, tiān shǐ wǒ guǎng shuō cǐ zhòu jì[56] ... ... (Namo Ganesh Prabhupada, the spirit I say this mantra widely...)'". However, the introduction of this foreign exorcism method was not important to the traditional Chinese witchcraft system, just like adding a few more grains to the bucket of rice. The use of the number "7" in Buddhist medicine is noteworthy: "Recite the spell seven times in the morning", "pound and take seven times", and "recite the spell seven times a day for seven days".[57] Similarly, the "7" is

[56] Ma Jixing, *Textual Research and Interpretation of Dunhuang Ancient Medical Books*, p. 288.
[57] ibid. p. 491.

also widely used in traditional Chinese exorcism therapy, such as "keep your mouth shut for seven days and do not speak with anyone". "Do not wash your hands for seven days", "fast for seven days", "rinse your mouth with saltwater seven times and repeat three times"[58] and so on, to name but a few.

Comparing the use of the number "7" in Chinese and Indian medical systems, there are two differences: Firstly, Indian medicine uses the number "7" in a wider range than Chinese medicine, like, the dose of medicine. For example, "Take seven pills … … gradually increase to fourteen pills … can be up to twenty-one pills, which is originally from *Prescription Book of Monk Shen* (*Seng Shen Fang*, 《僧深方》).[59] Indian medicine also suggests treatments like when there is an eye disease, "washing the eyes for a duration of seven days" to treat eye diseases etc. In contrast, Chinese medicine emphasizes the number "7″ only in the context of exorcism therapy. Secondly, although the Chinese exorcism therapy (including medical and other purposes) is widely used with the emphasis on the number "7", there is no cultural background attributed to it. Neither in the realms of astronomy nor mathematic, nor in the Chinese mystical number beliefs (or numerology) system, does the number "7" hold any particular meaning. The numbers 1, 2, 3, 4, 5, 6, 8, and 9, each have their specific meanings, except for "7".[60] Thus, it is difficult to comprehend the frequent use of "7" in exorcism therapy without any source or foundation, in isolation from the ancient Chinese culture. Unlike in China, in places such as India, Persia, Babylon, Assyria, Egypt, the number "7" carries great mystical meaning and is used as a symbol of God.[61] Scholars of medicine have noted that the mysterious significance of the number "7" can be used as a clue to shed light on the interplay between various ancient medical theories. For example: "The great influence of ancient Eastern cultures on Hippocratic medicine can be seen in *The Book on the Number Seven*"[62]; "The influence of Eastern Mesopotamia is again demonstrated by the theory that 'talent matures through two seven-year periods' (referring to Aristotle's *Physiologia*)".[63] The belief in the magic of the number "7" can indeed be traced back thousands of years to the Mesopotamian civilization, and is very closely related to both medicine and witchcraft. They called a sacred disease magician, the water deity "Ea" as 7, who was worshipped as an ancestor by the

[58] Sun Simiao (the Tang Dynasty), *Supplement to the Essential Prescriptions Worth A Thousand Gold for Emergencies*, Volume 29.

[59] *Textual Research and Interpretation of Dunhuang Ancient Medical Books*, p. 152.

[60] Feng Chengjun, *Survey of the Popularity of Manichaeism in China* (*Mo Ni Jiao Liu Xing Zhong Guo Kao*, 《摩尼教流行中国考》), Shanghai: Commercial Press, 1937, p. 1. "From the examination of the use of numbers in our country, the numbers that are usually used are three and five, such as the three principles and five rules, the three lights and five elements and so on. Seven is used less, while only the people of the West commonly use, such as seven deaths, seven births, seven difficulties, seven treasures, and seven sounds.

[61] Yang Ximei, "On the Mysterious Number Seventy-two", *Journal of Archaeological Anthropology* (*Kao Gu Ren Lei Xue Kan*, 《考古人类学刊》), National Taiwan University, 1974, (35–36): 12–45.

[62] Gerhard Venzmer (Germany), *5000 Years of Medicine* (*Shi Jie Yi Xue Wu Qian Nian*, 《世界医学五千年》), p. 35.

[63] ibid, p. 41.

doctors.[64] If there is indeed a connection between "7" in Chinese exorcism therapy and foreign cultures, it must be noted that this influence was not introduced together with Buddhism after the Eastern Han Dynasty, because the comprehensive use of "7" can already be seen in the Mawangdui medical books. There are two possible routes of Babylonian influence on Chinese culture, namely, through Persia-India, or directly from Persia. However, exploring this influence extends beyond the realm of exorcism alone and touches upon broader issues of cultural exchange between Babylon and ancient China. Just as many other disciplines inevitably encounter this problem, but it is not advisable to draw definitive conclusions lightly, the mysterious number "7" in exorcism as a subject for future research. It should be noted that even if the mysterious number "7" in the exorcism is considered to have been influenced by foreign cultures, it does not carry any "monism" of cultural origin. That is to say, culture needs to "communicate", without which no culture can be completely "original".

[64] ibid, p. 7.

Chapter 22
Prescriptions for Fifty-Two Diseases and Early Internal Medicine Dosage Forms

During the Han Dynasty (202 BC-AD 220), there was a significant transition in Chinese internal medicine dosage forms, from grinding and swallowing the medicine powder to drinking the herbal soup after decoction. Before this shift, the decoction-drinking method was mainly used for food-based medicines. At that time, these two methods did not signify a difference in dosage forms. The decoction-drinking method had its unique guiding theory and formed a special category of medicine, hence the term "decoction" (汤液) originated. Since this type of medicine has been abandoned since the Han Dynasty, the meaning of the term "decoction" is no longer widely known, and it is generally interpreted as "preparation" (汤剂).[1] Jia Dedao said: "It is a liquid made by cooking grains, which was used as medicinal in ancient times, and can be considered the early form of the decoction as known today".[2] This description provides a fairly accurate understanding. Another explanation is from Mr. Zhao Pushan, who said: "In particular, the interrelationship between heat and taste. The correct combination of these two elements will achieve the best effect of the 'taste' produced by the decoction".[3]

The above two explanations focus on food and taste, indicating an awareness of the special nature of "decoction", but due to a lack of historical data, further exploration becomes challenging. In fact, the unearthed medical books in Mawangdui have provided valuable clues to trace the origins of the decoction-drinking method and its guiding theory. By examining these sources, we can gain a clearer understanding of the true meaning and historical evolution of decoction in the Han Dynasty.

---

[1] *The History of Chinese Medicine* (*Zhong Guo Yi Xue Shi*, 《中国医学史》) edited by Beijing College of Traditional Chinese Medicine, Shanghai: Shanghai Science and Technology Press, 1978, p. 8. *Historical Materials of Chinese Pharmacy* (*Zhong Guo Yao Xue Shi Liao*, 《中国药学史料》), edited by Xue Yu, Beijing: People's Medical Publishing House, 1984, p. 20. Fan Xingzhun: *A Brief History of Traditional Chinese Medicine* (*Zhong Guo Yi Xue Shi Lve*, 《中国医学史略》), p. 11.

[2] Jia Dedao: *A Brief History Traditonal of Chinese Medicine* (Zhong Guo Yi Xue Shi Lve, 《中国医学史略》), p. 10.

[3] Zhao Pushan: *Ancient Chinese Medicine* (*Zhong Guo Gu Dai Yi Xue*, 《中国古代医学》), p. 13.

Y. Liao, *Reconstructing the Image of Medicine in the Qin and Han Dynasties*, Understanding China, https://doi.org/10.1007/978-981-95-6801-7_22

Table 22.1 Medical books in the chapter "Catalog of Literary Works" in *Book of the Former Han* (Category of Classical Formula)

No.	Book Name	Vol Number
1	*Prescriptions for Twelve Diseases on Impediment* (*Wu Zang Liu Fu bi Shi Er Bing fang*, 《五藏六府痹十二病方》)	30
2	*Prescriptions for Sixteen Diseases on Hernia* (*Wu Zang Liu Fu Shan Shi Liu Bing fang*, 《五藏六府疝十六病方》)	40
3	*Prescriptions for Twelve Diseases on Pure Heat* (*Wu Zang Liu Fu Dan Shi Er Bing fang*, 《五藏六府瘅十二病方》)	40
4	*Prescriptions for Sixteen Diseases on Wind, Cold, and Heat Diseases* (*Feng Han re Shi Liu Bing fang*, 《风寒热十六病方》)	26
5	*Prescriptions of Bian Que. and Yu Fu in the Period of Huangdi* (*tai Shi Huang Di Bian Que. Yu Fu fang*, 《泰始黄帝扁鹊俞拊方》)	23
6	*Formula for Eleven Diseases Caused by Pathogen Invaded in the Middle* (*Wu Zang Shang Zhong Shi Yi Bing fang*, 《五藏伤中十一病方》)	31
7	*Formula for Manic-Depressive Psychosis Caused by Pathogen Invaded in the Five Viscera* (*Ke Ji Wu Zang Kuang Dian Bing fang*, 《客疾五藏狂颠病方》)	17
8	*Formula for Trauma* (*Jin Chuang Zong chi fang*,《金疮疭瘛方》)	30
9	*Formula for Women and Infants* (*Fu Ren Ying Er fang*, 《妇人婴儿方》)	19
10	*Models of the Classic of Decoction*	32
11	*Dietary Prohibitions of Shennong and Huangdi*	7

22.1 What Is "Decoction"

The term "decoction" is first found in the chapter "Catalog of Literary Works" in the *Book of the Former Han* (*Han Shu*, 《汉书》), where it is mentioned in the category of "Medical Texts—Classical Formula": "There were 32 volumes of *Models of the Classic of Decoction* (*Tang Ye Jing Fa*, 《汤液经法》)". Unfortunately, this book has been lost over time, and its contents remain unknown. However, according to the order of the 11 ancient remedy books contained in the "Classical Formula" category (see Table 22.1), the book *Models of the Classic of Decoction* is arranged after the nine remedy books of the four main clinical disciplines (internal, external, gynecological, and pediatric) and is next to the *Dietary Prohibitions of Shennong and Huangdi* (*Shen Nong Huang Di Shi Jin*, 《神农黄帝食禁》). This suggests that its content may be closely related with food-based medicines.

This speculation can be substantiated by the chapter "Discourse on Decoctions and Wines" in the *Plain Questions* (*Su Wen*, 《素问》), where it discusses the preparation of "decoctions" and "wines". It states, "To make decoctions and wines from the five grains, one must use rice and cook it with rice straw".[4] This speculation can

[4] The chapter "Discourse on Decoctions and Wines" in *Plain Questions* of Paul U. Unschuld's translation version.

be proved. The following are some of the references to "decoction" in the *Plain Questions:*

> Now, in high antiquity, when they made decoctions, they produced them as a rule, but did not consume them. In the middle antiquity, the adherence to the Way and virtue had somewhat declined, and evil qi occasionally invaded the body. People consumed decoctions and achieved success in myriad cases.... In the present day, toxic drugs must be administered to attack the interior, and chisel stones, needles, and moxa must be used to treat the exterior.[5]

When the people in middle antiquity treated a disease, once it had arrived, they treated it by means of decoctions. It took ten days to remove the diseases caused by the eight winds and five blocks. If the symptoms had not resolved after ten days, they treated them with *cao-su* and the branches of *cao-gai*, and supported this with roots and ends. When tip and root match, the evil qi will yield.[6]

If a faint complexion appears, decoctions will be used to treat it, and it will be cured within ten days. If a deep complexion appears, medicated wines must be used to treat it, and it will be cured within one hundred days.[7]

From this, we can discern that the "decoction" is different from general medicine such as *cao-su* and *cao-gai*, or the combination of various Chinese medicines and formulas. "Decoction" is specifically applicable to the initial stage of disease and has already lost its therapeutic effect, as mentioned in the chapter "People of Today" in *Plain Questions*. Thus it can be concluded that "decoction" is not equal to preparation. In addition, although both "decoction" and "wines" involve the use of cereal as material, the two are not the same thing. The difference lies in the fact that "wines" refers to the brewing process for alcoholic medicine, but "decoction" doesn't require this step. Accordingly, it may be inferred that "decoction" refers to grain soup that has been subjected to the decoction process without undergoing brewing.

22.2 The Medicinal Principle of Decoction

The principles and commonalities of the various "decoction" can be summarized as follows: first, due to their food-based ingredients, the decoction has the function of nourishing and strengthening the body. According to the *Plain Questions*, "decoction" is considered suitable for the initial stage of the disease, that is "it took ten days to remove the diseases of the eight winds and five blocks", and to use various medicines only if this method does not work, perhaps with this in mind. This method is extensively used in the *Health-preservation Prescriptions* (*Yang Sheng*

[5] The chapter "Discourse on Decoctions and Wines" in *Plain Questions* of Paul U. Unschuld's translation version.

[6] The chapter "Discourse on Moving the Essence and Changing the Qi" in *Plain Questions* of Paul U. Unschuld's translation version.

[7] The chapter "Discourse on Decoctions and Wines" in *Plain Questions* of Paul U. Unschuld's translation version.

Fang, 《养生方》) of Mawangdui medical book to treat impotence, sexual dysfunction and other diseases, and as a tonic to strengthen the body. Its theoretical origin can be traced back to *The Rites of Zhou* (*Zhou Li*, 《周礼》), which states that practitioners use "five flavors, five grains, and five cures to treat disease".[8] However, the nourishing and strengthening effects of food-based medicines are not limited to the decoction-drinking method, but can also work if taken in other ways. Therefore, the real peculiarity of the decoction made from food lies in the second point: the slimy, liquid-like soup.

The main theoretical basis for the decoction treating method is "nourishing the orifices with lubricating medicines". In *The Rites of Zhou* on medicine, alongside the principle of "The sour nourishes bones, the pungent nourishes tendons, the salty nourishes veins, the bitter nourishes qi, and the sweet nourishes flesh", there is another principle of "using lubricating medicines to nourish orifices". It was according to this principle that what today seems to be only ordinary food soups could be used to treat diseases in ancient times. Similarly, if certain medicines can be made into a viscous liquid, they are also used as decoction. Ban Xia Decoction (Pinellia Decoction) in the chapter "Evil Visitors" in *Miraculous Pivot* (*Ling Shu*, 《灵枢》) is an example:

> The recipe for the decoction is as follows: Take water that has flown for a distance of at least 1000 li: 8 sheng. Ladle it ten thousand times. Let the turbid contents sink down and take from the clear parts above 5 sheng. Boil it with fire burning with reed as fuel. When it bubbles add 1 sheng of shu mi and 5 ge of processed Ban Xia. Slowly add heat and diminish the liquid to 1.5 sheng. Remove the dregs and have the patient drink the juice from a small cup.[9]

According to medical theory during that time, insomnia was believed to be caused by the blockage of yin and yang and pulse channels, so after a long time of decoction, the smoothy of it could "make the paths passable". "As soon as the yin and yang conduits are passable again, he (the patient) will immediately fall asleep".

Without the theory of "nourishing the orifice with lubricating medicines", it is impossible to understand the medical cases and prescriptions of cereal decoction applied in the Han Dynasty and earlier periods. For example, rice (or japonica rice, old rice, etc.), in later materia medica, was mainly used to "stop dysentery and tonify the middle and benefit the qi",[10] but it was used to treat the protuberance illness (retention of urine) in the *Prescriptions for Fifty-two Diseases* (*Wu Shi Er Bing Fang*, 《五十二病方》) unearthed at Mawangdui.

> I. With a bucket of water, boil one-third of a liter of Ejiao (donkey-hide gelatin) or deer-hide gelatin and a liter of rice, drink it slowly after cooking it. Do not have dinner. (line 181).
>
> I. Boil millet with vinegar and wine. Boil three times and then let it cool for the patient to take. (line 189).

[8] Zheng's commentary (the Han Dynasty); Jia Gongyan's note (the Tang Dynasty): *An Annoation on "The Rites of Zhou"* (*Zhou Li Zhu Shu*, 《周礼注疏》), Volume 5.

[9] The chapter "Evil Visitor" in *Miraculous Pivot* of Paul U. Unschuld's translation version.

[10] Meng Shen (the Tang Dynasty): *Dietetic Materia Medica* (*Shi Liao Ben Cao*, 《食疗本草》), from *Compendium of Materia Medica* (*Ben Cao Gang Mu*, 《本草纲目》) by Li Shizhen (the Ming Dynasty), Volume 25.

Another example is in the case of anal pruritus:

> I. Take a fist-sized stone and slowly heat it with fire. Boil rice with water, then put the stone in it, □□ (these two words are missing), and slowly drink the rice porridge. (line 270).

With the exception of those examples, similar formulas can be found in the treatment of conditions such as "gangrene" and "paint dermatitis" (lines 286, 292, 388), where the diseases involve hairy orifices (pore). The use of cereal soup in these diseases is obviously not focused on "tonifying the middle and strengthening the spleen and stomach", so it should be analyzed from the perspective of "nourishing the orifices with lubricating medicines".

Among the 25 medical cases of Chunyu Yi (also called Cang Gong, about 215–140 BC), a famous physician in the Western Han Dynasty, listed in the chapter "The Biographies of Bian Que and Cang Gong" in *Records of the Grand Historian* (*Shi Ji*, 《史记》), there were six cases in which "Huo Qi Decoction" (or "Huo Qi Porridge" or "Huo Qi Rice Juice") was used. That is, Example 3: "The patient had a hernia and was unable to urinate or defecate. After drinking Huo Qi Decoction once, he could urinate, and after drinking it again he could defecate, and after drinking three times he was cured". Example 4: "A patient got this disease because he took a cold shower in cold weather and got cold, that's why he got the fever, and he was cured after drinking Huo Qi Decoction three times". Example 5: "When wind evil invades, the patient has difficulty in urinating and defecating, with red urine. As soon as he drank Huo Qi Decoction once, the urine and stool resumed. He was cured after drinking it again, and the color of urine became normal". Example 10: "Qi hernia causes difficulty in urination and defecation, and the patient's urine is red. Moxibustion on the meridian of the foot reverting yin, also cooked Huo Qi Decoction for the patient to drink, and he was cured after three days". Example 20: "The patient suffered from severe wind and would spit out food after swallowing it. Let the patient take the Huo Qi Decoction with rice juice, and he recovered in seven or eight days". Example 23: "The patient cannot eat with reverting qi. Let the patient drink Huo Qi Decoction. In six days, the qi will be down". Among the above cases, the third, fifth and tenth cases were bladder disorders, with difficulties in urination and defecation; the fourth case was an obstruction of the hairy orifice (pore); while the 23rd case was an obstruction of the stomach orifice (pylorus); and the 20th case was associated with the treatment method of "relieving urination to stop diarrhea". All of these cases demonstrate the use of cereal soup to open the orifices as a therapeutic approach.

22.3 How to Make Decoction

The treatment principle of "nourishing the orifices with lubricating medicines" mentioned in *The Rites of Zhou* raises the question of whether corresponding medicines can be identified. In *The Rites of Zhou* we can see: "There are four things to drink, the first is clear wine, the second is cloudy wine, the third is vinegar flavored wine, and the fourth is yi (酏)". "Jiangren (a kind of position) is responsible for serving six kinds of drinks to the emperor: water, vinegar flavored wine, sweet wine, clear wine, cloudy wine, and yi".[11] The nature and the production method of "yi" are specified in the chapter "Inner Rules" in *The Book of Rites.*

Ground the rice into powder, and then mix it with water. The final production is called yi.

Take the grains of rice and steep them in prepared rice-water. The fat from the wolf's breast was then cut into small pieces and mixed with steeped-rice to make yi.

The former scholars interpreted yi as congee, but there were other names for congee at that time, such as porridge, mi, etc.[12] Therefore, according to the two specific texts describing yi in the *The Book of Rites*, it can be inferred that the former is a rice flour paste; while the latter, when used as a drink, should refer to rice soup, which is closer to the meaning. In Zheng Xuan's commentary on the "Four Drinks", "Yi is referred to as the congee of today", but he also explains, "yi is the clear part of the porridge"; The word "yi" is also mentioned in *Origin of Chinese Characters* (*Shuo Wen Jie Zi*, 《说文解字》): "Jia Kui said that yi is clear porridge", referring to rice soup. *The Book of Rites* also mentions "millet (yellow rice) yi" in the section of "Drinking", so it is clear that the ingredient of cereal soup is not limited to rice alone. It can also be seen that the meaning of "yi" has gradually evolved from a specific reference to a particular kind of grain soup to an abstract one. When the two characters "millet" and "yi" are used together, "millet" is the definite article, while "yi" becomes a generic term for cereal soup. The same applies to the word "beverage", which was originally used to refer to plum beverage in particular but has later become a general term. *Plain Questions* says "the production of five grains decoction must use rice", so we can only assume that this is also an example.

Why cereal soup is called "yi" in *The Rites of Zhou* and *The Book of Rites*, but in *Plain Questions*, it is changed to "five grains decoction"? This is probably due to the fact that as Chinese characters have developed, the use of radicals and the connection between character form and meaning have become closer and stricter. "You (酉)" is basically no longer used as the radical for any liquid substance that has not been brewed. "Yi (酏)" and "liquid (液)" are morphemes, and they likely had the same pronunciation in ancient times. For example, the pronunciation of the word "liquid" in the chapter "Diverse Crafts" in *The Rites of Zhou* is "yi". Since the pronunciation of yi and ye is so similar that it is almost indistinguishable, there are many words with phonetic transcriptions between them. For example, 亦 (yi)—也 (ye), 诣 (yi)—谒 (ye), 曀 (yi)—夜 (ye),[13] etc. As we all know, the chapter "Diverse Crafts" was added by later generations, and it is precisely in this chapter that we see the use of "liquid", so we may consider the word "yi" and "liquid" as having the same meaning but used in different times.

---

[11] Zheng' commentary (the Han Dynasty); Jia Gongyan's note (the Tang Dynasty): *An Annoation on "The Rites of Zhou"*, Volume 5.

[12] Zheng' commentary (the Han Dynasty); Kong Yinda's note (the Tang Dynasty): *An Annotation on "The Book of Rites"* (*Li Ji Zhu Shu*, 《礼记注疏》), Volume 6.

[13] Dai Weiqing, *Elaboration on Phonetic Transcendentalism* (*Yin Zhuan Xue Fa Fan*, 《音转学发凡》), Hong Kong: Sing Chau World Book, 1962, p. 43.

In addition to using food as the main raw material, the most important feature of the preparation of decoction is to pay attention to the "balance of water and fire" in the process. In short, there are special requirements for water, fire and wood, as well as the decoction method itself. For example, in the Pinellia Decoction in the chapter "Evil Visitors" in *Miraculous Pivot*, it is required to use "water flowing a thousand miles away" and "reed fire", because the designer of this formula believed that flowing water and reed, which are hollow, had more "flow" and "pass" properties, which could correct insomnia caused by the blockage of yin and yang and the pulse channels. Although the properties of these objective materials cannot be directly translated into therapeutic effects on the human body, these specific requirements, after all, stem from a certain way of thinking that does not seem correct today.

Likewise, we can observe these specific requirements in the *Prescriptions for Fifty-two Diseases* and the *Health-preservation Prescriptions*. Among the special requirements for "wood" are: "pure wood", "mulberry", "wattle"[14] and the method involves heating 14 fist-sized stones and placing them in rice-soaked water.[15] The requirements for "fire" are: "Cease when boiling vigorously, wait for the water to calm down and then add heat to boil"; "Stop heating when the water boils, then repeat the above steps again",[16]etc. The more detailed formulas even specify the water-to-ingredient ratio and the specific steps of the preparation process. For example, "Use fresh millet to cook porridge with a water-to-rice ratio of 15:1, which means using 15 parts water to 1 part rice to yield 5 parts porridge. Then, after cooling the porridge until no steam is visible, it is transferred to a new crockpot and sealed with a three-layered cloth strip and a 2 cun (about 2.62 inches) thickness of blended mud. The crockpot is then placed in a circular pit and slowly heated until the sealed mud is dry and the fire goes out. Then take out the porridge and let the patient drink it several times. After drinking the porridge, the wound will heal".[17] Regardless of whether these concocting processes have scientific basis or not, it seemed that they must be followed at that time. That is because practitioners at that time believed that the therapeutic effect did not depend solely on the material itself, but on all external factors related to pharmaceuticals. These special preparation processes that must be followed, together with the predominantly food-based ingredients, constitute the conditions for a separate method or book. It is highly likely that the 32 volumes of the *Models of the Classic of Decoction* contained in the chapter "Catalog of Literary Works" in *Book of the Former Han* were a monograph of this kind, thus earning its placement after the clinical monographs and next to the *Dietary Prohibitions of Shennong and Huangdi*.

---

[14] *Prescriptions for Fifty-two Diseases*, lines 388, 402, 435. Editing Group of Silk Book in Mawangdui tomb: *Silk Manuscripts from Mawangdui Han Tomb [Part Four]* (*Ma Wang Dui Han Mu Bo Shu [Si]*,《马王堆汉墓帛书》〔肆〕), "Notes on Interpretation," pp. 69, 70, 73.

[15] *Prescriptions for Fifty-two Diseases*, lines 402. Editing Group of Silk Book in Mawangdui tomb: *Silk Manuscripts from Mawangdui Han Tomb [Part Four]*, "Notes on Interpretation," p. 70.

[16] *Health-preservation Prescriptions*, lines 4, 108. Editing Group of Silk Book in Mawangdui tomb: *Silk Manuscripts from Mawangdui Han Tomb [Part Four]*, "Notes on Interpretation," pp. 99, 109.

[17] *Prescriptions for Fifty-two Diseases*, line 92–93. Editing Group of Silk Book in Mawangdui tomb: *Silk Manuscripts from Mawangdui Han Tomb [Part Four]*, "Notes on Interpretation," p. 38.

The nature of "decoction", characterized by its use of food as the main ingredient and special processing methods, has sometimes led to confusion with food cooking,[18] giving rise to the legend that "Yi Yin created decoction". For example, some writings claim, "'Catalog of Literary Works' says that the *Models of the Classic of Decoction* is also known as *Yi Yin Decoction* (*Yi Yin Tang Ye*, 《伊尹汤液》), so subsequent generations inherited the legend that 'decoction' was created and invented by Yi Yin".[19] However, there is no such statement in the "Catalog of Literary Works". In the various written records about Yi Yin's decoction, he only talks about the taste but not about the therapeutic effects of decoction, so the works attributed to Yi Yin are not included in the category of "Medical Texts" but in the categories of "Daoist" and "Novels". The first person to associate Yi Yin with the book titled *Decoction* (*Tang Ye*, 《汤液》) was Huangfu Mi (AD 215–282), a literary scholar, historian and medical scientist of the Western Jin Dynasty (AD 265–317). As he said: "Yi Yin, with the talent of a sage, wrote and used the *Shennong's Classic of Materia Medica* (*Shen Nong Ben Cao Jing*, 《神农本草经》) as 'decoction' ...... Zhang Zhongjing (around AD 150-219) said that *Yi Yin Decoction* has dozens of volumes".[20] That's why various legends emerged claiming that Yi Yin created "decoction" or wrote the book *Decoction*.

In the history books, Yi Yin is mentioned as "carrying a rice pot and anvil to talk about the taste of decoction"[21]; "(Yi Yin) said soup has a nice taste"; "the foundation of all flavors lies in water. Five flavors and three materials, nine boilings and nine changes, fire for the discipline, sometimes fast, and sometimes slow",[22] etc., that is, a detailed discussion of the selection of cooking materials and processing steps. The "three materials" are "water, fire and wood",[23] which coincide with the requirements for water, fire and wood in the processing method of medicine decoction. It seems that during the period from the Three Kingdoms (AD 220–280) to the Western Jin Dynasty, when Huangfu Mi lived, there were still writings related to "decoction", and he associated "decoction" with the name of Yi Yin mainly on the basis of its actual content, which he considered to be the same as Yi Yin's cooking method. As for why Huangfu Mi mentioned that "Zhang Zhongjing said *Yi Yin Decoction* has dozens of volumes", it is mainly because until the end of the Eastern Han Dynasty (AD 25–220) when Zhang Zhongjing's *Treatise on Cold Damage and Miscellaneous Diseases* (*Shang Han Za Bing Lun*, 《伤寒杂病论》) was written, the traditional method of

[18] The *History of Chinese Medicine*, edited by Beijing College of Traditional Chinese Medicine: "Legend has it that Yi Yin, who invented the decoction, was a master of cooking techniques, which indicates that the invention of the decoction was closely related to food processing". Shanghai: Shanghai Science and Technology Press, 1978, p. 8.

[19] *Historical Materials of Chinese Pharmacy*, edited by Xue Yu, Beijing: People's Medical Publishing House, 1984, p. 20.

[20] Huangfu Mi (the Jin Dynasty, AD 265–420), Preface of *A-B Classic of Acupuncture and Moxibustion* (*Zhen Jiu Jia Yi Jing*, 《针灸甲乙经》).

[21] Records of the Grand Historian, Volume 3.

[22] *Master Lyu's Spring and Autumn Annals* (*Lyu Shi Chun Qiu*, 《吕氏春秋》), Volume 14.

[23] Chen Qiyou: *Commentary on 'Master Lyu's Spring and Autumn Annals'* (*Lyu Shi Chun Qiu Jiao Shi*, 《吕氏春秋校释》), Shanghai: Shanghai Xuelin Publishing House, 1984, pp. 739–767.

taking medicine had transitioned from the "powder and swallowing" method to the "boiling and decoction" method, which resembles the "decoction-drinking" method.

22.4 Evolution of Internal Medicine Dosage Forms

Once we grasp the unique characteristics of medicine decoction, it is easy to consider that the processing and administration methods of these decoction are not representative of ordinary medicines. From this point of view, if we examine the existing medical formulas before the Han Dynasty, we can find that most of them used the method of grinding and swallowing. First of all, if we analyze the medical books of Mawangdui, which can be identified as medical works from before the Western Han Dynasty, and categorize the internal medicine formulas into two categories based on "medicine" and "food", and then further distinguish them according to the dosage form, the differences are obvious (see Tables 22.2 and 22.3).

Table 22.2 Comparison of the two dosage forms of decoction and powder with the two ingredients of medicine and food in the *Prescriptions for Fifty-two Diseases*

Usage		Medicine	Food
Grinding, and drinking with wine and vinegar		(1), (5), (8), (23), (25), (41), (60), (72), (87), (90), (100), (114), (115), (172), (203), (215), (223), (236), (237), (248), (271), (275), (283), (441)	(438) chicken, snake
Drinking	Cooked in water		(3) Red bean, (19) Plum, (92) Foxtail millet seed, (94) Chicken, grains, rabbit meat, (99) Venison or wild boar meat, (100) Mutton, (168) (170) (171) (192) Violets, (173) Jujube, violets, honey, (181) Rice, (182) Snail, scallion, (187) Bean leaf, (201) Meat, (270) Rice, (286) Bean, (295) □□half pipeful
	Cooked in wine	(176) Sedum, (185) Tongue Fern	(43) Scallion, (410) Water caltrop, (189) Millet
	Cooked in vinegar		(161) Black Bean, (202) Egg
Unknown	(28), (36), (74), (149), (156),(158), (186), (225), (248), (459)		

Note The numbers in parentheses are the palace line items. Only the main ingredients of "food" are indicated in the table, and "medicine" is a complex formula, so please refer to the original text

Table 22.3 Classification of medicine and dosage forms in the *Health-Preservation Prescriptions*

Usage	Medicine	Food
Grinding and swallowing	(18), (33), (34), (37), (70), (74), (103), (105), (107), (112), (121), (124), (126), (131), (148), (152), (172)	(39) Bird egg, (51) Pork, (65) Radix Asparagi, Chicken, (108) Beef, (169) Dried meat slice
Drinking		(3) Radix Asparagi, husked sorghum, (11) Millet, Rice, (13) porridge,(28) Wheat, (32) Hornet pupae, (35) Eggs, (162) Millet, Rice, Ju, and Ginger, etc.
Unknown	(1), (9), (21), (24), (26), (30), (40), (42), (57), (96), (98), (102),(111), (113), (115), (116), (117), (120), (133–147), (154), (159),(175), (178–188)	

Note The numbers in parentheses are the palace line items. Only the main ingredients of "food" are indicated in the table, and "medicine" is a complex formula, so please refer to the original text

Furthermore, the *Wuwei Han Dynasty Medical Slips* (*Wu Wei Han Dai Yi Jian*, 《武威汉代医简》),[24] tentatively inferred to date back to the early Eastern Han Dynasty, also exhibit this feature. Except for the missing ones, there are 28 formulas on the slips. Of these, 27 are made into powder, pills, or medicated paste and then swallowed, while only one formula (No. 80 in the brief list) is a decoction. In addition, the *Juyan Han Dynasty Medical Slips* (*Ju Yan Han Jian*, 《居延汉简》),[25] also dating from the Han Dynasty, exhibit this feature. Despite the severe damage to the bamboo slips, the medicinal dosage form is not the decoction-drinking method, as shown by the text such as "If your abdomen does not regulate, take one pill a year". More complete formulas, such as: "Cold Damage. Four ingredients: Wu Hui (Coracoid) ten fens (1 fen=0.3 g), Xi Xin (Herba Asari) six fens, Zhu (Atractylodes) ten fens, and Gui (Cassia Twig) four fens. Take the powder with warm water", clearly demonstrate the grinding and swallowing method.

The *Plain Questions* and the *Miraculous Pivot*, written in the Han Dynasty, contain 12 formulas. In addition to the formula for boiling forsythia and herb root drink in the chapter "Obstruction–and Impediment-Illnesses" in *Miraculous Pivot*, other medicinal formulas are also taken by swallowing. For instance, Chapter 40 "Discussion on Abdomen and Center" in *Plain Questions*: "Take four black cuttlefish bones and one measure/root of madder, combine these two items and form pills the size of small beans using sparrow eggs"; Chapter 46 "Discourse on Disease Manifestations" in *Plain Questions*: "Take ten parts each of ze-xie and zhu and five parts of mi-xian. Mix and take before meals as much as can be taken up with three fingers"; Chapter 63 "Discourse on Misleading Piercing" in *Plain Questions*: "Shave a square inch of hair from the left corner of the patient's forehead, process the hair by burning it, and let the patient drink one cup of the processed hair with good wine". Medical cases recorded

[24] *Wuwei Han Dynasty Medical Slips* (*Wu Wei Han Dai Yi Jian*, 《武威汉代医简》), edited by Gansu Provincial Museum, Beijing: Cultural Relics Press, 1975.

[25] *Juyan Han Dynasty Medical Slips* (*Ju Yan Han Jian Jia Yi Bian*, 《居延汉简甲乙编》), The Chinese Academy of Social Sciences, Institute of Archaeology, China Books, 1980.

by Chunyu Yi also employ the method of grinding and swallowing, such as "taken a pinch of belladonna" and "taking a pinch of Flos Genkwa",[26] etc. This phenomenon persisted until the late Eastern Han Dynasty with the writing of Zhang Zhongjing's *Treatise on Cold Damage and Miscellaneous Diseases*, where the majority of the medicines were taken by decoction. Since then, people's understanding of the effects of medicine has been mainly focused on medicines themselves. The dosage form is only used as a reference factor for better efficacy, with decoctions generally used for urgent and severe diseases, and pills used for prolonged diseases.

After the introduction of the five-phase theory in the Han Dynasty, medical theory only retained the five tastes of "pungent, sweet, sour, bitter and salty", while "lubricant" was no longer considered a "taste". Consequently, the theory of "nourishing the orifices with lubricating medicines" gradually disappeared. The decoction is also not used by practitioners. However, it is worth noting that in ancient China, where there was no modern scientific or technological theory to guide medical practices, traditional medicine relied on considering the natural properties of objective things as therapeutic effects that can be applied on the human body. For example, the use of talcum as a diuretic is still focused on its natural property of "slippery". Since talcum can increase the osmotic pressure of urine (principle of osmotic diuresis of solute molecules) and inhibit the secondary reabsorption of renal tubules, resulting in a potent diuretic effect. The correct application of medicine under the guidance of an incorrect theory can only be described as a coincidence.

One final question worth considering is what is the reason for the shift in the dosage form of internal medicine? Although decocting medicinal plants can easily produce a liquid with medicinal flavor, why was this method rarely used in the Han Dynasty or before that time? Instead, the *Treatise on Cold Damage and Miscellaneous Diseases* suddenly extensively used this dosage formy? There are various speculations regarding this matter: It is possible to consider that during the Eastern Han Dynasty, there was a gradual change in the medicinal dosage forms, but the medical writings from this period were not transmitted. For example, the *Treatise on Cold Damage and Miscellaneous Diseases* contains two formulas with exactly the same efficacy and grouping, but one is a decoction and the other is a pill.[27] Most later commentators thought that pills were suitable for mild cases, reflecting a tendency to repeat the old pattern of looking at the ancients through the lens of thinking. From the basic point of view that no representative medical work can be written in isolation, the epoch-making medical work *Treatise on Cold Damage and Miscellaneous Diseases* was also built upon the inheritance and absorption of previous experience. Therefore,

---

[26] The chapter "The Biographies of Bian Que and Cang Gong" in *Records of the Grand Historian*, Volume 150.

[27] Didang Decoction (抵当汤, Resistant and Withstanding Decoction) and Didang Pill (抵当丸, Resistant and Withstanding Pill). Both formulas consist of Shui Zhi (Hirudo), Meng Chong (Tabanus), Tao He (Walnut and Radix) et Da Huang (Rhizoma Rhei) and are suitable for blood amassment syndrome with "fullness in the lower abdomen and spontaneous urination".

it is highly likely that these two formulas, with the same name, efficacy, and combination, but different dosage forms, are derived from different medical works. Besides, alchemy was already quite popular during the Eastern Han Dynasty, and "elixir" should be distinguished from general therapeutic drugs, regardless of whether they were praised or criticized. Therefore, the change of dosage form of general medicine may also be a passive response to differentiate it from elixirs. Another bolder scenario is that this shift was caused by the influence of foreign medical cultures during the Eastern Han Dynasty.

Chapter 23
The Relationship Between Taoism and Life Nurturing

To different extent, ancient Chinese philosophies, especially Taoism, influenced the formation of theoretical system of traditional Chinese medicine (TCM). However, nowadays there are some misunderstandings on the relationship between TCM and Taoism. For example, some scholars believe that "the TCM theory of yin-yang and five elements is originated from Taoism"; and that TCM, with its medical theoretical system containing the thoughts of ancient Chinese philosophies, is "an innovative medical theory created by Taoism". Others even elaborate TCM through the "medical thought of *Tao Te Ching* (*Laozi*, 《老子》) and *Zhuangzi* (《庄子》)".[1] These opinions, however, go against the thought of Lao Zi and Zhuang Zi, and are not commonly accepted by historians. However, it is widely acknowledged that the pursuit of immortality, "a tenet of Taoism", initiated traditional Chinese life nurturing. This is actually a prejudice originated from the misunderstanding of the essence of Taoism. It fails to fully cover the relationship between Taoism and traditional Chinese life nurturing, as well as the influence of Taoism on the development of ancient Chinese medical theoretical system. In the following sections, all these issues are to be further elaborated and differentiated.

23.1 Taoist View of Life

The pursuit of immortality through various life nurturing practices is a distinctive aspect of Taoist philosophy. This is how the traditional Chinese life nurturing was initiated.[2] From the perspective of the origin of traditional Chinese life nurturing, it is believed that "the concept of achieving everlasting life is derived from philosophy,

[1] Li Liangsong, *Chinese Traditional Culture and Medicine* (*Zhong Guo Chuan Tong Wen Hua Yu Yi Xue*, 《中国传统文化与医学》), Xiamen, Xiamen University Press, 1990, pp. 87–94.

[2] Editorial Board of China Encyclopedia of Medicine: the chapter of "Medical History" in *China Encyclopedia of Medicine* (*Zhong Guo Yi Xue Bai Ke Quan Shu*, 《中国医学百科全书》), p. 33.

Y. Liao, *Reconstructing the Image of Medicine in the Qin and Han Dynasties*, Understanding China, https://doi.org/10.1007/978-981-95-6801-7_23

especially the theory of Lao Zi and Zhuang Zi".[3] Joseph Needham, a famous historian on history of Chinese science, emphasized in his book *Science and Civilization in China* (*Zhong Guo Ke Xue Ji Shu Shi*,《中国科学技术史》) that "Taoism is the foundation of China's science and technology", and "studying the Taoist philosophy is vital for the understanding all aspects of Chinese science and technology". He also believed that:

> From its inception, ancient followers of Taoism were obsessed with the pursuit of immortality. They yearned to retain eternal youth, with robust bodies and radiant skins; and they believed that it was possible to discover ways to stop or even reverse the process of aging.[4]

It is evident that Joseph Needham and many historians of science naturally attributed the origin of the concept of pursuing immortality and various specific methods of life nurturing, including taking medicine for longevity, dao yin, Taoist sexual practices, and moving qi, to the Taoism and Taoist philosophy.

Upon closer examination of pre-Qin (before 221 BC) Taoists' views on life and death, we will find that these views are not in accordance with the doctrine of Taoism. The *Tao Te Ching*, while not directly discussing the issue of life and death, offers insight into this subject. One passage states, "High wind does not last long, nor does heavy rain. If nature's words do not last, why should those of man?"[5] This suggests that followers of Lao Zi would not believe in immortality. In the later-compiled *Zhuangzi*, the views on life and death were more explicit: "One who knows what is of Man may still use his knowledge of the known to develop his knowledge of the unknown, living till the end of his days and not perishing young. This is the fullness of knowledge; "life comes and goes, following its own laws"[6]; "life and death are inevitable and far beyond our control".[7] There are some fables and stories about immortality in *Zhuangzi*. For example, when it mentioned Peng Zu (Taoist immortal who is known for his longevity), it is stated in the book that "Peng Zu obtained it, and lived from the time of Shun until the time of the Five Princes". Furthermore, in a dialogue between Nanpo Tsek'uei and Nv Yu (or Female Yu), "You are of a high age, and yet you have a child's complexion. How is this?"[8] In the third story, Chien Wu said that on the Miao-ku-yi mountain there lives a divine one, whose skin is white like ice or snow and whose grace and elegance are like those of a virgin.[9] However, these fables and stories cannot serve as the examples of Taoists' pursuit for immortality. They are quoted to stress a Taoist philosophy: nothing (including the immortality)

---

[3] Editorial Board of China Encyclopedia of Medicine: History of Chinese Traditional Medicine Volume in *China Great Encyclopedia* (*Zhong Guo Da Bai Ke Quan Shu*,《中国大百科全书》), the editorial committee, p. 558.

[4] Joseph Needham, *Science and Civilization in China* (*Zhong Guo Ke Xue Ji Shu Shi*,《中国科学技术史》), Volume II, Chinese edition, pp. 154–155.

[5] *Laozi* (《老子》) annotated by Wang Bi (the Wei Dynasty), Chapter 23.

[6] Guo Xiang (the Jin Dynasty), *Commentaries on "Zhuang Zi"* (*Zhuang Zi Zhu*,《庄子注》), Volume 3.

[7] Guo Xiang (the Jin Dynasty), *Commentaries on "Zhuang Zi"*, Volume 7.

[8] Guo Xiang (the Jin Dynasty), *Commentaries on "Zhuang Zi"*, Volume 3.

[9] Guo Xiang (the Jin Dynasty), *Commentaries on "Zhuang Zi"*, Volume 1.

shall be deliberately pursued: "the men of the world think that to nourish the body is sufficient to preserve life; and when such nourishment is not sufficient to preserve the life, what can be done in the world that will be sufficient?".[10] In addition, the dialogue between Fengren and Yao (a legendary monarch in ancient China in *Zhuangzi*) has fully expounded the Taoist belief that "the true man of old did not know what it was to love life or to hate death. Such a one may be called a true man".

> The warden said, "Long life, wealth, and numerous offspring are what men wish for; then how is that you alone do not wish for them?" Yao replied, "Having many offspring brings many fears; wealth brings many troubles; and long life gives rise to many obloquies. These three things do not contribute to the cultivation of virtue; and therefore, I wish to decline them."[11]

Numerous similar examples further reinforce the notion that the pre-Qin Taoism, a profound philosophical system, was not supportive of the pursuit of immortality but rather opposed it. Taoists believe that to nourish the body is inadequate for preserving the life, and they have elaborations on cyclical nature of life, stating, "where there is birth, there must be death; where there is death, there must be birth".[12] By all these, the Taoists made it clear that the pursuit of immortality is absurd. Moreover, the Taoist pursuit of aligning with the law of nature was exemplified by their understanding of life and death which best shows the irresistibility of the law of nature.

"Preserving the body" is in accordance with the values of Taoism. However, "to preserve the body" does not mean to pursue longevity. This idea is exemplified by several stories from *Zhuangzi*. King Shun offered his throne to Zizhou Zhibo, who declined, stating, "I'm suffering from anxiety, and I'm about to give it a serious treatment. That leaves me no time to rule over the kingdom". Besides, the state of Han and the state of Wei contended for the land on their border. Zihua Zi persuaded Marquis Zhaoxi not to attack the state of Wei for the contentious land. By these stories, values of Taoism were presented, such as "Individuals who value life, regardless of their wealth, will not seek possessions at the cost of life; even if they are poor, they will not damage their health for the sake of personal gains". And that men shall not "lightly sacrifice their lives for material gains".[13]

If we refer to the discussion on "body" and "longevity" in the chapter "Noted School of Thought" in *Hanfeizi* (《韩非子》), we can see the difference between the two concepts more clearly.

> Now supposing someone addressed a person, saying, "I will make you wise and live long". People would certainly view this as deceptive. Indeed, wisdom is inherent to one's nature, while longevity is determined by fate. As nature and fate cannot be taught or acquired from others, asserting the ability to bestow these qualities on someone is considered deception by society.
>
> Now suppose there is a man, who, holding fast to his self-righteous principle, would not enter any endangered city, would not stay in military camps, and would not exchange a hair from

[10] Guo Xiang (the Jin Dynasty), *Commentaries on "Zhuang Zi"*, Volume 7.

[11] Guo Xiang (the Jin Dynasty), *Commentaries on "Zhuang Zi"*, Volume 5.

[12] Guo Xiang (the Jin Dynasty), *Commentaries on "Zhuang Zi"*, Volume 1.

[13] Guo Xiang (the Jin Dynasty), *Commentaries on "Zhuang Zi"*, Volume 9.

> his shin for any great profit in all under Heaven. Then be sure the sovereign of this age will respect him, regarding him as a gentleman who values life over material things……Yet as long as the superior honors the gentlemen who despise material trifles and esteem meaningful life, it is impossible to expect the people to sacrifice their lives for his royal cause.

So firstly, it is important to clarify that the discussions mentioned earlier revolve around life values, including doctrines of Lao Zi and Zhuang Zi, as well as "valuing one's body" (i.e., never serve the interests of all the people in the world even at the cost of a hair on his shin, a view proposed by Yang Zhu) mentioned in Han Feizi. Later, we will delve into how these life values mentioned above evolved into the theory of life nurturing (which focuses on a pure heart and few desires, a long-lasting life) as embraced by later followers of Yang Zhu and Taoism.

23.2 The Essence of Taoism and Life Nurturing

The term "life nurturing" was first found in Zhuang*zi*, the Taoist classic. This significantly explains why traditional Chinese life nurturing is thought to originate from Taoism. This term appeared three times in *Zhuangzi*, but it had little to do with later generations' understanding of "life nurturing" (i.e., to maintain the body and to promote health and longevity).

Firstly, the term "life nurturing" was mentioned in the chapter "The Preservation of Life" in *Zhuangzi*. The saying goes, "Human life is limited, but knowledge is limitless. If one aspires to pursue all the knowledge within his life, it is impossible". It was followed by a story called "Pao Ding Jie Niu" (a butcher skillfully dissecting an ox). The cook said, "A good cook changes his chopper once a year—because he cuts. An ordinary cook, once a month—because he hacks. But I have had this chopper for nineteen years, and although I have cut up many thousand bullocks, its edge remains as sharp as if it were freshly sharpened from the whetstone". Upon hearing the butcher's words, Prince Hui expressed admiration and stated, "From the words of this cook I've learned how to take care of my life". The story means that by the butcher's words, Prince Hui understood that it would be much easier and more efficient for him to rule his country in accordance with the law of nature; and that he did not need to follow the way of Yao and Shun, who "wore the hair off their legs and the flesh off their arms in endeavoring to feed their people. They tortured themselves mentally in order to conform to charity and duty. They exhausted their own energy to set up laws and regulations. Even then they did not succeed".[14]

Secondly, the term "life nurturing" appeared again in the chapter "The Full Understanding of Life" in *Zhuangzi*, "They who skillfully nourish their life are like shepherds, who whip up the sheep that they see lagging behind". Here "life nurturing" is almost equal to the meaning of the way of governing the people. Then, the stories of Shan Bao and Zhang Yi are used to illustrate the principle that those governing the world should not be one-sided. (In the State of Lu, there was a Shan Pao who lived

[14] Guo Xiang (the Jin Dynasty), *Commentaries on "Zhuang Zi"*, Volume 4.

among the rocks, and drank only water.) Shan Bao "only focused on his internal cultivation, and his physical form was eaten by a tiger". Zhang Yi "only focused on his external behavior, but a disease struck him from the inside". They died because neither of them attended to their weak links.

Lastly, the term "life nurturing" was mentioned in chapter "Kings Who Have Wished to Resign the Throne" in *Zhuangzi*.

> Hence it is said, "The true object of Tao is the regulation of the person. Quite subordinate to this is its application in the governance of the state and the clan; while the government of the kingdom is merely a byproduct. From this we can see that the services of the Kings are a surplusage of the work of the sages, and do not contribute to complete the person or nourish the life. Yet the superior men of the present age will, most of them, sacrifice their lives in pursuit of material possessions. Is it not cause for grief?

There are some reasons why I arranged the above three examples in the order of "inner chapters → outer chapters → miscellaneous chapters". For one thing, historians generally believe the inner chapters to be the most essential, followed by the outer chapters and the miscellaneous chapters. For another, the interpretation of "life nurturing" is different in these three chapters. In the inner chapter "The Preservation of Life", "life nurturing" means to follow the law of nature, which serves as the core of Taoist political theory. In the outer chapter "The Full Understanding of Life", it refers to the specific methods of governing people ("keep whipping the stragglers up…"). In the miscellaneous chapter "Kings Who Have Wished to Resign the Throne", it refers to values, such as "cherishing life and preserving one's body" and disregarding fame and wealth. Among them, the so-called "completing the person or nourishing the life" attaches more importance to life, but it still has nothing to do with later generations' understanding of "life nurturing". (Fig. 23.1).

In a word, since the "Tao" emphasized in pre-Qin Taoist thought contains an understanding of the laws of nature, the original meaning of "inaction" is for people to conform to nature and refrain from doing things that go against it. Human beings, along with all other living organisms, like inorganic substances, are all components of nature. Therefore, both the human body (life) and the social order of humanity should follow nature, which is the worldly principle of "governing by non-interference". The term "nurturing life" here summarizes the specific methods between "inaction" and "achieving all through inaction", so it naturally appears in discussions about governing the country, ruling the people, and preserving one's own well-being. In *Zhuangzi*, matters related to the physical lifespan are often referred to as "cultivating the physical form", it's not hard to understand that the "life" in "nurturing life" mainly refers to the populace and social order, rather than life (physical body) itself.

To truly understand the essence of the Taoist ideological system from the pre-Qin period to the early Western Han Dynasty (202 BC–AD 8), one really shouldn't forget the comment in the chapter "Catalog of Literary Works" in the *Book of the Former Han* (*Han Shu*, 《汉书》).

> The Taoist school is generally believed to have originated from court historians. They recorded the principles of success and failure, survival and extinction, blessings and misfortunes throughout history. Then they learned to grasp the key points and fundamentals, keep

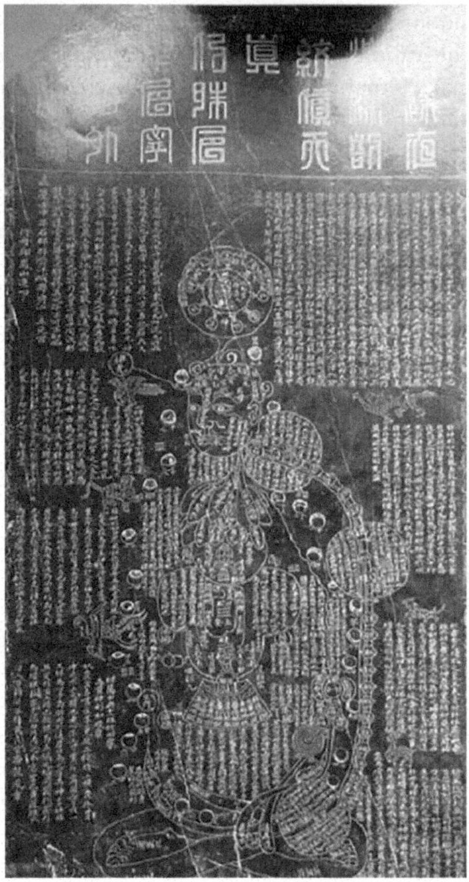

Fig. 23.1 Monument for exercising in Yuexiu Mountain, Guangzhou province. It was finished in 1811 during the reign of Emperor Jiaqing in the Qing Dynasty (AD 1636–1912). Researchers believe that "the content on the monument shares the same source with illustration for Self-cultivation in Chengdu and Beijing. It has elaborated on the affected body parts and functions of five Zang-organs, and emphasized crucial body parts and key points for practicing Neidan (internal alchemy) in particular. And this is of great significance to qigong practice". From this we can see that the understanding on Taoist philosophy of later generations is totally different from its original meanings

> themselves pure and calm, and maintain a humble and yielding attitude. This is the art of ruling for the sovereign.

Although the theories of governing the world in Taoism and Confucianism conflict with each other, in essence, it is just a matter of "people with different ways cannot work together".[15] One must not simply understand Taoist thought as a purely negative and "world-renouncing" learning.

---

[15] *Records of the Grand Historian* (*Shi Ji*, 《史记》), Volume 63.

23.3 Techniques of "Cultivating the Physical Appearance"

"Long life, wealth, and many offspring are what men wish for". This is a previously quoted saying from the chapter "The Heaven and the Earth" in *Zhuangzi*. The fear of death and the desire for longevity are deeply rooted in people's mind. As individuals strive to attain this desire, various methods of "life nurturing" were created. Like any other manifestations of culture and civilization, methods of "life nurturing" are also a product of human intelligence and an integral part of culture and civilization. In this sense, it is neither necessary nor possible to trace the exact origin of the methods of "life nurturing". The thought and behavior of "life nurturing" are born spontaneously, and they do not rely on the inspiration of any ideological system or wise philosophers.

Indeed, that is exactly the case. As previously explained, the ideas of pursuing long-life and eternal vision are incompatible with the Taoist ideological system. Now it is necessary to illustrate how these methods objectively existed in the society at that time, independent of the Taoist ideological system.

In the chapter "Increase Resilience and Develop Willpower" of *Zhuangzi*, it is said, "Panting and breathing, expelling the stale and taking in the fresh, imitating the postures of a bear hanging from a branch and a bird stretching its wings are merely for the sake of longevity. These are what the practitioners of qigong, those who focus on physical nourishment, and those who aspire to live as long as Peng Zu like". This passage is sufficient to illustrate two points. Firstly, Zhuang Zi put "the practitioners of qigong and those who focus on physical nourishment" on a par with four other types of people: "those who live in the mountains and disdain the world", "those who aim for a peaceful world and engage in teaching", "those who serve in the court and seek to honor the ruler and strengthen the state", and "those who live by the rivers and seas and shun the world". He considered them all as people who "deliberately pursue high-minded actions" but are without the Way, and they are not advocated by Taoism. Only when one reaches the state of "being noble without deliberate effort" and "living long without practicing qigong" can one embody "the way of heaven and earth and the virtue of sages". This clearly shows the attitude of pre-Qin Taoism toward the art of health preservation.

Secondly, this passage also shows that during that time, "those who practice physical exercises for health preservation" were objectively existing in the society, and their actions were to pursue "longevity". As for the specific techniques of health preservation, "panting and breathing, expelling the old breath and taking in the new" are techniques to move qi. Additionally, practices such as "hanging like a bear and stretching like a bird" are techniques of daoyin. Besides, according to the chapter "Free and Easy Wandering" in *Zhuangzi*, "there is a divine being who sustains themselves without consuming grains, but instead lives on wind (air) and dew". This suggests the popularity of bi gu (inedia) at that time. Further insight can be gained by examining silk books and bamboo slips unearthed in Mawangdui Tombs. While pre-Qin Taoists viewed these techniques as only used "for cultivating the physical appearance", later generations regarded them as "techniques for health preservation".

(1) Bi Gu (inedia): Bi Gu is a health-preservative method which involves abstaining from grains and living on certain plants.[16] In *Inedia and Breatharianism* (*Que Gu Shi Qi Pian*, 《却谷食气篇》) unearthed in Mawangdui Tombs of the Western Han Dynasty, there are descriptions on specific techniques and remedies for dealing with discomfort caused by "rejection of grains". According to the description, the practice of Bi Gu involves gradually replacing grains with Pyrrosia petiolosa from the first day of the month and increasing the portion till the full moon day, after which it is gradually decreased. If the practitioner experiences weakness or symptoms such as a heavy head and light feet, he shall inhale and exhale in a special way for a quick recovery. During the practice Bi Gu (inedia), people shall still eat certain plants and employ a special way of breathing. Thus, this way has something in common with "absorption of qi" and "taking medicine (for longevity)". According to the chapter "Zhang Liang, Marquis of Liu" in *Records of the Grand Historian* (*Shi Ji*, 《史记》), Zhang Liang, who was previously weak and susceptible to illness, learned and practiced Bi Gu (inedia) and daoyin, hoping to keep fit.

(2) Absorption of qi: literally it means to breathe in a certain way for "life nurturing". Theoretically, it refers to the "qi" of the heaven, earth, and nature as "the vital essence of the heaven and earth. This essence emerges, develops and matures in an intangible way. Those who keep the essence could expect a long life; those who lose it will face premature death". The key point is to take a deep breath: "the way is to inhale to the maximum extent…and make it deep and long"; and the breathing way in the morning shall be different from that in the evening, "the turbid qi dissipates during the night, and the fresh qi gathers in the morning. (To inhale the fresh qi) so as to dredge the nine orifices and nourish the viscera and bowels". Besides, in the process of absorption of qi, "there are taboos to avoid respectively in four seasons: to avoid turbid-yang in spring, hot wind in summer, frost smoke in autumn, and yin (cold) in winter".[17] With these words in mind, refer to the further elaborations on "breathing" in chapter 7—the qi inhaled through the mouth and nose will be circulated in the whole body; hence, "the way is to inhale to the maximum extent".

(3) Taking medicine (for longevity): This method has the most content. In the Mawangdui Silk Texts, there are records like "Eat pine and cypress, and drink the essence of wild animals. A bronze seal for health - preserving through taking medicine in the Western Han Dynasty. The inscription on the seal reads 'Seal of Zhao Xu Zichan. May fortune and blessings come daily. Ride on floating clouds, ascend Huashan Mountain, eat jade-like herbs, drink sweet spring water, take famous medicines, and become an immortal.' ([47] 415) By doing so, one can reverse aging, regain vitality, and have smooth and radiant skin"; "In the three

[16] *Inedia and Breatharianism*: people who practice inedia take Shi Wei (Folium Pyrrosiae). Editing Group of Silk Book in Mawangdui tomb, *Silk Manuscripts from Mawangdui Han Tomb[Part Four]* (*Ma Wang Dui Han Mu Bo Shu [Si]*, 《马王堆汉墓帛书》〔四〕), "Notes on Interpretation", p. 85.

[17] The chapter of "Huangdi Asking Rongcheng" in *Ten Questions* (*Shi Wen*, 《十问》). Editing Group of Silk Book in Mawangdui tomb: *Silk Manuscripts from Mawangdui Han Tomb [Part Four]*, "Notes on Interpretation", pp. 146–147.

months of summer, cook without fire and use the sun to cook. Then one will be wise and intelligent".[18] "Eat Xie (Allium chinense G.Don) in the three months of Spring, serious diseases will not occur frequently, and the tendons and bones will become stronger. Thus, Xie (Allium chinense G.Don) is called the "king of herb plants". Drink pure wine when lying in bed, as it is "the essence of the five grains"; Xie (Allium chinense G.Don) shall be eaten with eggs, because "the cock pertains to yang; its hair is bright and voice is loud and clear; when stretching its head, its feather stands…thus Taoist people eat eggs".[19]

(4) Daoyin (conduction exercise): it means to prevent diseases and promote health through body movements, so it is also called medico-athletics. In the Mawangdui Silk Texts, there is a colorful picture of men and women in various positions, with a total of 44 styles arranged in four layers. There are texts beside each posture, such as "Use a stick to regulate yin and yang", "Relieve deafness", "Relieve knee pain", "Bear – hanging", "Falcon - like posture", "Sitting to stretch in all directions", etc. From the postures of these figures and the annotations, it can be seen that many of them imitate the movement postures of animals, which is consistent with the record in chapter "Increase Resilience and Develop Willpower" in *Zhuangzi*: "Those who practice Daoyin and those who nourish their bodies" like "hanging like a bear and stretching like a bird".

(5) Receiving yin: It refers to the idea that men can achieve longevity through certain specific sexual behaviors. Receiving yin held an extremely important position in life nurturing at that time: In the *Ten Questions* (*Shi Wen*, 《十问》), a book recording various schools' strategies for longevity, six schools mentioned this matter ((Table 23.1)). Among these six schools, only Peng Zu (a figure in Chinese mythology renowned for his longevity) advocated complete abstinence from sexual activities to avoid diseases caused by "leakage of yin essence". The other five schools all focused on the relationship between sexual behavior and longevity. The key points of sexual behaviors aimed at longevity are as follows: ① No ejaculation. For example, "enjoy the pleasure (of sex) without ejaculation".[20] "Ensure that the vital essence does not spill, then various diseases will not afflict you, and thus you can achieve longevity"[21] ② During sexual intercourse, nourish the body by "ingesting" (such as "drinking the spring essence"

[18] The chapter of "Wang Qi Seeing King Zhao of Qin" in *Ten Questions*, Editing Group of Silk Book in Mawangdui tomb: *Silk Manuscripts from Mawangdui Han Tomb [Part Four]*,"Notes on Interpretation", pp. 151–152.

[19] The chapter of "Wen Zhi Seeing King Qi of Qi" in *Ten Questions*, Editing Group of Silk Book in Mawangdui tomb: *Silk Manuscripts from Mawangdui Han Tomb [Part Four]*), "Notes on Interpretation", p. 150.

[20] The chapter of "Yao Asking Shun" in *Ten Questions*, Editing Group of Silk Book in Mawangdui tomb: *Silk Manuscripts from Mawangdui Han Tomb [Part Four]*, "Notes on Interpretation", p. 148.

[21] The chapter of "Huangdi Asking Cao Ao" in *Ten Questions*, Editing Group of Silk Book in Mawangdui tomb: *Silk Manuscripts from Mawangdui Han Tomb [Part Four]*, "Notes on Interpretation", p. 146.

and "drinking the celestial nectar", which presumably refer to saliva).[22] Besides, in *Health-preservation Prescriptions* (*Yang Sheng Fang*, 《养生方》) and *Miscellaneous Prescriptions* (*Za Liao Fang*, 《杂疗方》) in the Mawangdui Silk Texts, there are many methods using drugs to treat "erectile dysfunction in the elderly", which shows the close relationship between the concept of "aging" and sexual function.

Is the concept of "receiving yin" the ancient saying of "fang zhong shu (Taoist sexual practices)"? This depends on the definition of "fang zhong shu (Taoist sexual practices)" given. If "fang zhong zhi shi (affairs in the bedroom)" is attributed to fang zhong shu (Taoist sexual practices) as it was in ancient times, then the art of "receiving yin" can be considered as part of fang zhong shu (Taoist sexual practices). However, upon deeper exploration, we can find that there are two totally different discussions on "fang zhong zhi shi (affairs in the bedroom)": one is about sexual knowledge and art; the other is about achieving longevity via sexual behaviors. Though they may appear similar in form, they are totally different in purpose and essence. To distinguish them, we temporarily attribute the discussion on sexual knowledge and art as "fang zhong shu (Taoist sexual practices)" in a narrower sense, while the art of "receiving yin" to the methods of life nurturing. As a natural manifestation of cultural progress, fang zhong shu (Taoist sexual practices) emerged for the harmony of sexual behavior between the sexes, so it is called "xi dao (art of sex)". In this art, the female climax is attached great importance, and genders are seen equal. The art of "receiving yin" was totally different from fang zhong shu (Taoist sexual practices); wherein male was the core, while female was instrumentalized. Men were seeking tonic benefits by the act of "inhaling" during sexual behaviors; from this point of view, women were actually regarded as objective materials for tonifying, like "medicine" in the tonifying method of "taking medicine (for longevity)", and "qi" in "absorption of qi". Scientifically, some ideas in fang zhong shu (Taoist sexual practices) are in line with modern physiology and promote health, such as "seven impairments and eight benefits" and "surplus essence shall be discharged". However, the idea "be sure not to ejaculate" in "receiving yin" is detrimental to people's health, which is harmful and useless.

Strictly speaking, for this batch of unearthed materials, we should consider their close inheritance relationship with pre-Qin culture. Besides, we shall take into account the fact that their burial date is 168 B.C., which shows that these materials also demonstrate the culture of the early Western Han Dynasty. Even so, there is no evidence of Taoism's influence on spiritual recuperation among the various methods of "life nurturing". Taoists accurately grasped the essence of these "life nurturing" methods. The therapeutical core of this life-nurturing science is to strengthen the body with materials from the outside world. Hence it can be concluded that at this time, the science of "life nurturing" is still at the primary stage with "life" being the

[22] The chapter of "King Pan Geng of Shang Dynasty Asking Gou Lao" in *Ten Questions*, and "Wang Qi Seeing King Zhao of Qin" in *Ten Questions*, Editing Group of Silk Book in Mawangdui tomb: *Silk Manuscripts from Mawangdui Han Tomb [Part Four]*,"Notes on Interpretation", p. 149 and p. 151.

Table 23.1 Life nurturing methods in chapters of ***Ten Questions***

No	Questioner	Respondent	Content						
			Receiving Yin	Taking pills (for longevity)	Absorption of qi	Abstinence	Conduction exercise	Lying techniques	Self-massage
1	Huang Di	Heavenly Master	▲						
2	Huang Di	Da Cheng		▲					
3	Huang Di	Cao Ao	▲						
4	Huang Di	Rong Cheng			▲				
5	Yao	Shun	▲						
6	Wang Zi Qiao Fu	Peng Zu				▲			▲
7	Pan Geng	Gou Lao	▲						
8	Yu	Gui					▲		
9	King Wei of the State Qi	Wen Zhi		▲				▲	
10	King Zhao of the Qin Dynasty	Wang Qi	▲	▲	▲				

subject. Then, how did a more advanced science of "life nurturing" come into being, which includes both xing (essence and spirit) and ming (body)?

23.4 The Evolution of Taoist Academic Thought

The "School of Song Xing and Yin Wen" is very important in the evolution of Taoist academic thought during the pre-Qin period. It is a branch of the Taoism school in the Warring States period (475–221 BC), represented by two scholars of Jixia Academy in the Qi State, namely, Song Xing and Yin Wen. Jixia Academy served as a place for scholars from different schools to "gather and discuss". Over time, ideas of different schools influenced each other. The "School of Song Xing and Yin Wen" focused primarily on "Tao", but its academic thought was influenced by other schools such as Confucianism, Mohism, Logicians, and Legalists. One characteristic of this school is to explain myriad things and spiritual activities with the ontological concepts of Tao, jing (essence), qi and shen (spirit). For example, "Tao is what fills the form"[23]; "Qi can be regarded as the stuffing of the body".[24] They also said: "The shapeless void is called Tao".[25] It should be noted that "the shapeless void" means "there is no fixed shape" for Tao. As for the complex forms of life phenomena such as psychology and thinking, they tried to explain with the theory of "vital energy and qi": "In the birth of human beings, Heaven produces the vital energy. Earth produces the form. These combine to form a human being. When they are in harmony, there is life. Without harmony, there is no life". Ontological "heart" is the residence for the "essence" bestowed by heaven—"the abode of the essence".[26] Therefore, the importance of "heart" is self-evident: "In the body, the mind holds the position of the king. The functions of the nine apertures resemble the separate responsibilities of officials".[27] Thus, it is vital to "cultivate the mind".

> With a stable mind in one's heart, one's ears are sharp and eyes are clear, and one's four limbs are strong and firm. Thus, one can serve as a dwelling place for the vital essence. Only after cultivating one's mind and quieting one's thoughts can one comprehend the Way. When our minds are well regulated, our sense organs are also well regulated. When our minds are at ease, our sense organs are also at ease. When the four limbs are in in proper position and the blood and qi are calm, with concentrated attention and undistracted ears and eyes, even distant things seem close. Thought gives rise to knowledge. Laziness brings sorrow. Fury breeds resentment. Melancholy causes illness. Severe illness leads to death. If one keeps thinking without a break, one will be troubled internally and weakened externally. If one doesn't make early plans, life will leave its abode.[28]

---

[23] The Tang Dynasty, *Guan Zi* (《管子》),Volume 16, annotated by Fang Xuanling.

[24] The Tang Dynasty, *Guan Zi*, Volume 13, annotated by Fang Xuanling.

[25] The Tang Dynasty, *Guan Zi*, Volume 13, annotated by Fang Xuanling.

[26] The Tang Dynasty, *Guan Zi*, Volume 16, annotated by Fang Xuanling.

[27] The Tang Dynasty, *Guan Zi*, Volume 13, annotated by Fang Xuanling.

[28] The Tang Dynasty, *Guan Zi*, Volume 16, annotated by Fang Xuanling.

The "School of Song Xing and Yin Wen" advocated the practice of staying in a state of "emptiness" and "tranquility" as a means of "cultivating the mind". Since this school was based at Jixia Academy, there are questions about whether it was influenced by the alchemists from the Yan and Qi regions (Yan and Qi refer to regions of Hebei and Shandong provinces). However, whether there was such an influence has been lost to history, and it's not very important. Strictly speaking, the theory of the "School of Song Xing and Yin Wen" was based on the noumenon of Tao, essence qi and the heart (mind). In other words, only "emptiness" can lead to achieving the "Tao". To achieve "emptiness", one must distance themselves from "desire" and "cognition". They believed that "when one acts rashly, one loses the right state; when one stays calm, one attains self-fulfillment".[29] By these means, the noumenon of Tao can be self-preserved and cultivated. Plenty of later-generation scholars over-interpret the word xu (emptiness), which is unnecessary. Xu (emptiness), in a simple sense, is comparable to not filling a bag or cupboard to its full capacity. This is a quite primitive analogical thinking, just like the "deficiency" of healthy qi and the "excess" of pathogen in etiology, "reinforcing the healthy qi" and "eliminating the pathogenic factors" in therapeutic strategies, or the "Tao" that can be accommodated only by eliminating the "desire" and "cognition" in the mind. However, later followers of the "School of Song Xing and Yin Wen" interpreted "attaining Tao" as "cherishing the body for longevity" and called this "life nurturing".

> Therefore, I say: If I wish to care for my body, I must first come to understand my inner reality. For this, I must observe everything within the six directions as a basis for examining what lies within the self. In such a way, after recognizing personal qualities, I hope that they come to understand how to manage the inner reality. When one understands how to manage one's inner reality, one may then come to understand how to nurture one's life. Do not change or add to it, and your life then will be preserved for a long time. Keep yourself in harmony in order to return to the mean, and your form and basic nature will then nurture each other. Concentrate your mind and never divide your attention, such is called coming to know the Way.[30]

In *On Posthumous Work of Song Xing and Yin Wen* (*Song Xing Yin Wen Yi Zuo Kao*, 《宋钘尹文遗著考》), Guo Moruo believed that 4 chapters of *Guanzi* were posthumous works of "School of Song Xing and Yin Wen", including chapter 36, 37 ("Art of the Mind I, II"), chapter 38 ("Purifying the Mind") (which was compiled relatively later), and chapter 49 ("Inner Workings"). He also believed that "Art of the Mind I, II", and "Inner Workings" were compiled or left by Song Xing; and "Purifying the Mind" was later compiled by Yin Wen. The importance of focusing on the later followers of "School of Song Xing and Yin Wen" lies in understanding the subtle internal change of Taoist academic thought. To be noted, "life nurturing" mentioned here is greatly divergent from that depicted in the story of butcher Pao Ding, and that in the saying, "life nurturing is like keeping a flock of sheep—keep whipping the stragglers up".

---

[29] The Tang Dynasty, *Guan Zi*, Volume 13, annotated by Fang Xuanling.

[30] The Tang Dynasty, *Guan Zi*, Volume 13, annotated by Fang Xuanling.

Thereafter, with critical acceptance of some ideas of "School of Song Xing and Yin Wen", Xunzi puts forward his "concept of physical form and spirit": "The physical form becomes whole and the spirit is born. Love and hate, delight and anger, sorrow and joy, are stored within—these are described as 'the emotions given us by nature.'"[31] This understanding contributes to the emergence of the life nurturing theory devoted to cultivating xing (essence and spirit) and ming (body), as well as gaining a right understanding of the meaning and purpose of "nourishing the physical appearance". Academic thoughts of Song Xing and Yin Wen were recorded in works of later followers of Zhuangzi; stories on cultivating the body by keeping a tranquil mind are widely quoted; and this way is honored the "Perfect Way" (to cultivate the body).

> See nothing; hear nothing; guard your spirit in quietude and your body will go right of its own accord. Be quiet, be pure; toil not your body, perturb not your vital essence, and you will live forever. If the eye sees nothing, and the ear hears nothing, and the mind thinks nothing, your spirit will stay in your body, and the body will thereby live forever…Therefore I have been able to live for twelve hundred years and my body has not grown old.[32]

Thus, in the late period of its development, the pre-Qin Taoism initiated another theory, which stressed the relationship between physical form and spirit, and the pursuit of longevity. Based on this theory, the art of "cultivating the physical appearance", which was originally independent of the Taoist thought system, developed into the art of "cultivating xing (essence and spirit) and ming (body)".

The Taoist concept of "preserving the body" shares similarities with the idea of "cherishing life" advocated by Yang Zhu School, as mentioned in the first section of this chapter. When elaborating the saying, "the two schools influenced each other in theories, leading to the unification of their later followers", Hou Wailu commented that "to a larger extent, the later followers of Zhuangzi were largely following the doctrine of Yang Zhu School". He also pointed out that words and deeds of followers of Yang Zhu School were recorded with admiration in outer and miscellaneous chapters of *Zhuangzi* (especially in chapter "Kings Who Have Wished to Resign the Throne").[33] Thus, the Taoist doctrine of "preserving the body" is rooted on the thought of Yang Zhu School. Compared with the doctrine of "abiding the law of nature" in *Tao Te Ching*, and "following the law of nature and taking no action" in *Zhuangzi*, "preserving the body" is a manifestation of the development and evolution of Taoism's academic thought. Moreover, unlike Mohist utilitarianism and Confucian ethics, "preserving the body" further evolved into the pursuit of "immortality" by eliminating desires.

> They are indulged with fat, mellow wines and cannot have enough of them, so these things should be addressed as "food and drinks leading to ulceration of the gut". They are also addicted to love affairs with pretty women and decadent tunes popular in Zheng and Wei, so these things should be addressed as "life-hacking axes".

---

[31] The Tang Dynasty, *Xun Zi* (《荀子》), Volume 11, annotated by Yang Liang.

[32] Guo Xiang (the Jin Dynasty), *Commentaries on "Zhuang Zi"*, Volume 4.

[33] Hou Wailu et al., *General History of Chinese Thought* (*Zhong Guo Si Xiang Tong Shi*, 《中国思想通史》), Volume 1, pp. 347–348.

> When clothes are too heavy and warm, the veins in the skin will be blocked. Thus, the vital energy cannot circulate very well. When too much nutritious food has been eaten, the stomach will distend, leading to gastrectasia. As a result, vital energy cannot circulate well. How can the lifespan be prolonged if this is the case?

> Ordinary sovereigns do not constrain their lusts, so it is quite normal for them to suffer from failure or perdition. Their ears cannot have enough to hear, their eyes cannot have enough to see, and their mouths cannot have enough to enjoy. As a result, all parts of their bodies will be swollen, their veins will be blocked, their blood vessels will be choked, all the nine apertures will be empty, and the function of all their organs will be twisted. If this is the case, even people like Peng Zu could not enjoy longevity.[34]

These excerpts are quoted from 3 chapters of *Master Lyu's Spring and Autumn Annals* (*Lyu Shi Chun Qiu*, 《吕氏春秋》), which are believed to have been compiled by later followers of the Yang Zhu School. They demonstrate the emphasis on "cherishing life over materials, later followers of Yang Zhu School pursued the art of life nurturing; in this sense, they were more like followers of *Zhuangzi*".[35] However, according to our previous discussion, "life nurturing" mentioned in *Zhuangzi* is not equal to immortality. Hence, let's consider it to be the combination of the value ("attaching greater importance to life over materials") and the Taoist pursuit of eliminating desires. Later followers, whether from the Yang Zhu School or Taoism school who accepted the idea of "attaching greater importance to life over materials", were criticized for their egoistic tendencies. This criticism can be seen in figures like Mencius, as exemplified in the chapter "Fan Lun Xun" in *Writings of Prince Huainan* (*Huai Nan Zi*, 《淮南子》). Therefore, it was quite natural (perhaps in the academic debate during the Warring States period) for academic debates to revolve around the way of "managing the body", which stressed the elimination of lust. Hence, the idea of "cherishing life" transformed into a pursuit of "longevity"; accordingly, human life was no longer weighed against the "thing" or the "world"; instead, people's life span was compared with that of Peng Zu (Taoist immortal known for his longevity). On the other hand, the changed social status of Taoism further promoted the shift of roles of "governing a country" and "managing the body" in its ideological system.

23.5 The Influence of Change of Social Status on Taoism

During the Warring States period, the school of Huang Lao (Huang Di and Lao Zi), which upheld Tao and rule of law, was spread to the Qi State and other regions. In the chapter "Mencius and Xun Qing" in *Records of the Grand Historian*, it is recorded that "Shen Dao, Huan Yuan, Jie Zi and Tian Pian, etc., studied the doctrines of the Huang Di and Lao Zi and propounded their own ideas according to these…The king of the Qi State favored these scholars". In the chapter "Le Yi Lie Zhuan" in *Records*

---

[34] The Han Dynasty, *Master Lyu's Spring and Autumn Annals*, Volume 1 and 2, annotated by Gao You.

[35] Hou Wailu et al., *General History of Chinese Thought*, Volume 1, p. 347.

of the Grand Historian, it is recorded that "Le Chen Gong studied the doctrines of the Huang Di and Lao Zi and had a good reputation in the Qi State, and was honored the virtuous and sagacious teacher". It is also recorded in *Records of the Grand Historian* that Cao Shen was honored the wise prime minister who assisted the ruling of the Qi State with the doctrines of this school. For example, it is recorded that during the administration of the Han emperors Wen and Jing, "Wang Sheng lived in retirement and was well versed in the teachings of the Huang Di and Lao Zi. Once he was summoned to court…"[36] "Empress Dowager Dou also favored the doctrines of the Huang Di and Lao Zi; hence, the emperor, the prince and all other members of Dou family had to read and follow the doctrines of Huang Di and Lao Zi"[37]; during the period of Jian Yuan (14–135 BC), "Deng Zhang was known among ministers".[38] Undoubtedly, Laozi's theory gained power at that time as a way of governing the country and the people. The popularity of Laozi's teachings played a role in the rise of Taoism, with Laozi being regarded as the founder of the later-emerged Taoism. However, at that time, it was the followers of Huang Di who were more focused on the pursuit of longevity. For example, Li Shaojun had told Emperor Wu of Han, "Apply the special gold to make food containers and use this container in your daily life. Then you can prolong your life. Only people with prolonged life like Huang Di can meet immortals in Penglai Mountain".[39] Besides, the chapter "Catalog of Literary Works" in *Book of the Former Han* recorded 10 chapters of immortals' stories, with 4 of them including the name of Huang Di in their chapter titles, along with others like Mixi, Shangsheng, and Shen Nong. In *Ten Questions* unearthed in Mawangdui Tombs of the Western Han Dynasty, Huang Di raised four questions, while Yao, Wang Zi Qiao Fu, Pan Geng, Yu, King Wei of state Qi, and King Zhao of the Qin Dynasty each raised one question; the respondents were Heavenly master, Da Cheng, Cao Ao, Rong Cheng, Shun, Peng Zu, Gou Lao, Gui, Wen Zhi, and Wang Qi successively.

Later, Emperor Wu of the Western Han Dynasty implemented a policy of rejecting the other schools of thought and respecting Confucianism exclusively. By this time, the Huang Lao School had already lost its power, though it continued to exist in a transformed and compatible form known as "Confucian-exterior-and-Taoism-interior". Hence, a number of intellectuals studying theory of Huang Lao school turned to academic research and Fangshu (Divination) respectively. Researchers of Taoism paid close attention to the influence of change (of social status) on Taoism. They thought that Huang Lao Dao, the rudiment of Taoism, came into being with the combination of Huang Lao scholars and Fang Xian Taoists. However, according to the chapter "On Taoist Scriptures" in *History of the Sui Dynasty* (*Sui Shu*, 《隋书》), there were 37 schools of Taoism in the Han Dynasty (202 BC–AD 220), with many emphasizing the importance of getting rid of greed and self-cultivation. Obviously, due to the decline of its political status, Taoist thinkers shifted their focus

[36] *Records of the Grand Historian*, Volume 120.

[37] *Records of the Grand Historian*, Volume 49.

[38] *Records of the Grand Historian*, Volume 110.

[39] *Records of the Grand Historian*, Volume 12.

toward "managing the body" instead of its previous ones, like "doing nothing and everything is done", and "following the law of nature". The *Annotation of Laozi by Heshanggong* (*Lao Zi He Shang Gong Zhu*, 《老子河上公注》), written in the Eastern Han Dynasty (AD 25–220) exemplifies this shift in emphasis. For example, it is recorded in Chapter Three in *Tao Te Ching* (words in brackets are notes written by He Shanggong) that: to empty their minds (less desire brings less disturbance), to fill their bellies (to concentrate to guard the five shen), to weaken their ambitions (being gentle and modest, not dealing with power), and to strengthen their bones (to cherish one's essence, to manage one's behavior, to supplement the marrow, so as to make the bone strong). Also, it can be seen in Chapter Six in *Dao Te Ching* that: gu shen enjoys immortality due to the cultivation of gu. Likewise, if the (five) shen is cultivated, a long life could be expected. The (five) shen here refers to hun (ethereal soul), po (corporeal soul), spirit, yi (ideation) and zhi (will). The liver stores hun, the lung stores po, the heart stores the spirit, the spleen stores yi, and the kidney stores zhi and essence. When all these five Zang-organs are exhausted, the (five) shen will be gone. In this way, the academic thought of *Tao Te Ching* has been transformed into the theory of life nurturing. Of course, Laozi himself and his thoughts were gradually regarded as the origin of this theory.[40] To be noted, this change triggered the combination of the spiritual realm of "emptiness and tranquility" of Taoism and the technique of "nourishing the physical appearance", forming the life nurturing theory which focuses on "cultivating xing (essence and spirit) and ming (body)".

In the Three Kingdoms period (AD 220–280), Ji Kang, a famous literati from the State of Wei, proposed in his book *Discourse on Health Preservation* (*Yang Sheng Lun*, 《养生论》) that:

> The gentleman knows that the physical appearance and spirit are inter-dependent. Therefore, he purifies his mind to secure the spirit, and stays tranquil to cultivate his body; besides, he breathes in a certain way, and takes Taoist pills to nourish his body; in this way, he keeps the balance between the physical appearance and spirit, and between the interior and exterior.

Ji Kang's ideas provide an insightful example of the principles underlying the concept of "life nurturing".

23.6 Life Nurturing and Medical Science

To gain health and longevity has long been a direct driving force behind the practice of life nurturing. In other words, because of people's desire for health and longevity, methods of life nurturing have been popular for thousands of years until today. Of course, from the ancient time till now, specific methods and means have been evolving. Some "unhealthy" methods have been abandoned. However, with continuous development through generations, some beneficial methods have stood the test of time and have become the mainstream methods of life nurturing and health care

[40] The chapter "Dao Xu" in *On Balance* (*Lun Heng*, 《论衡》), wrote by Wang Chong in the Han Dynasty, has already included various life-nurturing practices into that of Taoism and Lao Zi.

in later generations, such as daoyin, taking medicine (for longevity), absorption of qi (moving qi), etc. The beauty of life nurturing lies in the simultaneous cultivation of *xing* (essence and spirit) and *ming* (body). Health status shall also be measured from these two aspects. As an English saying goes, "Sound mind in a sound body"; *xing* (essence and spirit) and *ming* (body) are closely interrelated. However, early records on life nurturing, such as the medical books from Mawangdui, primarily focused on the cultivation of *ming* (body); this embodied time characteristic of the early methods of life nurturing, as well as the origin of life nurturing. The core of life nurturing theory at that time was to abide the law of heaven, earth and nature. "According to the law of heaven, yin and yang are vital; losing them, things vanish; securing them, things flourish"[41]; "Bluish food leads to bluish complexion; and yellow food leads to yellow complexion"[42]; and *daoyin* (conduction exercise) originated from imitation of animal movements, etc. Guided by these methods of life nurturing, people mainly focused on getting nourishment from the external world.

In terms of its relationship with medical science, Mawangdui medical works are seen as the overall manifestation of the medical culture of a certain era, although the exact compilation time and order of these medical works remain uncertain due to a lack of sufficient evidence. Surprisingly, ahead of the time of medical works like Cauterization Canon of the *Moxibustion Classic of the Eleven Foot-Arm Meridians* (*Zu Bi Shi Yi Mai Jiu Jing*, 《足臂十一脉灸经》), *Moxibustion Classic of the Eleven Yin-Yang Meridians* (*Yin Yang Shi Yi Mai Jiu Jing*, 《阴阳十一脉灸经》), and *Prescriptions for Fifty-two Diseases* (*Wu Shi Er Bing Fang*, 《五十二病方》), some terms were already seen in the discussion of the art of life nurturing,[43] as well as their initial attribution of yin and yang nature, such as five Zang-organs, six Fu-organs, and four seasons. This could be a clue to the origin of the theory of Zang-Fu organs, of four seasons, and of yin and yang, the important components of the basic theoretical system of traditional Chinese medicine, which were so commonly seen in medical works compiled after the Han Dynasty.

On the other hand, the idea of abiding the law of heaven, earth, and nature has become one of the most influential guiding principles in traditional Chinese medicine. For example, there are two sayings in *Plain Questions* (*Su Wen*, 《素问》): "Now, when the sages of high antiquity taught those below, they always spoke to them about the following. The depletion evil and the robber wind, there are specific times when to avoid them".[44] "Now, the yin and yang qi of the four seasons, they constitute root and basis of the myriad beings. Hence, the sages in spring and summer nourish

---

[41] Huang Di Asking Tianshi in *Ten Questions, Silk Manuscripts from Mawangdui Han Tomb [Part Four]*, commentary part, p. 145.

[42] Huang Di Asking Dacheng in *Ten Questions, Silk Manuscripts from Mawangdui Han Tomb [Part Four]*, commentary part, p. 145.

[43] Huang Di Asking Tianshi, Huang Di Asking Rongcheng, and Wang Qi Visiting King Zhao of Qin in *Ten Questions, Silk Manuscripts from Mawangdui Han Tomb [Part Four]*, commentary part, pp. 145,146 and 152.

[44] *Plain Question*, chapter 1 Shanggu Tianzhen Lunpian: Ancient Ideas on How to Preserve Natural Healthy Energy.

the yang and in autumn and winter nourish the yin"[45]; and these two sayings were developed from the saying "to avoid turbid-yang in spring, hot wind in summer, frost smoke in autumn, and yin(cold) in winter". And yin and yang had become the supreme guiding principle: "As for yin and yang, they are the Way of heaven and earth, the fundamental principles governing the myriad beings, father and mother to all changes and transformations, the basis and beginning of generating life and killing, the palace of spirit brilliance".[46]

23.7 Fang Zhong Zhi Shu (Taoist Sexual Practices)

Lastly, I will talk about this "taboo" topic theoretically. The term "fang zhong (Taoist sexual practices)" was first mentioned in the chapter "Catalog of Literary Works" in the *Book of the Former Han*: "fang zhong (Taoist sexual practices) is the peak of human nature and the Perfect Way". There are 186 volumes of works on fang zhong (Taoist sexual practices) by eight schools, forming a category in "Category of Prescriptions and Technique" in *Seven Categories of Writings* (*Qi Lve*, 《七略》). Although these works have been lost for a long time; thus, *Formulary from the Heart of Medicine* (*Yi Xin Fang*, 《医心方》) is the only available source for us to gain a general idea of ancient Chinese works on fang zhong (Taoist sexual practices). *Yi Xin Fang* was compiled at the end of the tenth century A.D, by Danbo Kanglai, an ancient Japanese medical expert. It includes references to various ancient Chinese works on fang Zhong, such as *Secret of Jade Chamber* (*Yu Fang Mi Jue*, 《玉房秘诀》), *The Maiden Scriptures* (*Su Nv Jing*, 《素女经》), *Jiutian Goddess Scriptures* (*Xuan Nv Jing*, 《玄女经》) etc. Records in bamboo slips unearthed in Mawangdui Tombs of the Western Han Dynasty such as *Methods of Intercourse between Yin and Yang* (*He Yin Yang*, 《合阴阳》), and *Discussion of the Perfect Way of Sexual Healthcare* (*Tian Xia Zhi Dao Tan*, 《天下至道谈》), share similarities with those in the chapter "On Arts of the Bedchamber" in *Formulary from the Heart of Medicine*; hence these records can be attributed to the category of fang zhong (Taoist sexual practices). Till now, these records from Mawangdui texts are probably the only historical sources available for us to study the fang zhong works during the Warring States period, Qin and Han Dynasties.

Methods of Intercourse between Yin and Yang and Discussion of the Perfect Way of Sexual Healthcare are works on human sexual behaviors, which can be summarized as follows:

---

[45] *Plain Question*, chapter 2 Siqi Tiaoshen Dalunpian: Major Discussion of Regulation of Spirit According to the Changes of the Four Seasons.

[46] *Plain Question*, chapter 5 Yinyang Yingxiang Dalunpian: Major Discussion on the Theory of Yin and Yang and the Corresponding Relationships Among All the Things in Nature.

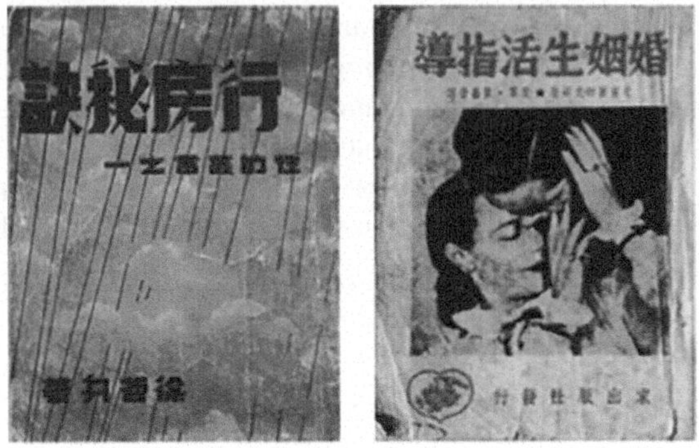

Fig. 23.2 Sex education books in the period of Republic of China

(1) To turn the female sex partner on with sufficient foreplay, including caressing and touching[47]; and the time shall be long enough.
(2) To strengthen the stimulation by various body movements and changes, so as to facilitate the orgasm of the female[48];
(3) To achieve sex harmony by paying attention to the sex partner's body responses.[49]

The emergence of Monographs of fang zhong (Taoist sexual practices) marked the progress of human civilization: it indicated that sex was no longer for procreation purposes only; it became an integral aspect of cultural and spiritual life of human being, as well as a vital component of interpersonal relationship. Particularly, the dominating feudal ethic at that time—men were superior to women—could not be seen in these works on fang zhong (Taoist sexual practices). Instead, women were vital partners: firstly, great emphasis was placed on women's will and happiness, acknowledging their pivotal role in fostering the harmony between sexes and among family members. Sexual behavior was viewed as a kind of spiritual and cultural life[50]; secondly, achieving female climax was prioritized through attention to foreplay and the whole process of the sexual experience, and the benefits of semen were attributed to women.[51] Sexual behavior was seen as a kind of physiological activity (Figs. 23.2 and 23.3).

In works on fang zhong (Taoist sexual practices), correct and incorrect sexual behaviors were summarized as "eight benefits" and "seven damages" respectively, representing the most important guidelines in this realm.

---

47 Methods of Intercourse between Yin and Yang in *Silk Manuscripts from Mawangdui Han Tomb [Part Four]*, commentary part, p. 155.

48 Methods of Intercourse between Yin and Yang in *Silk Manuscripts from Mawangdui Han Tomb [Part Four]*, commentary part, p. 156.

49 Discussion of the Perfect Way of Sexual Healthcare in *Silk Manuscripts from Mawangdui Han Tomb [Part Four]*, commentary part, p. 166.

50 Discussion of the Perfect Way of Sexual Healthcare in *Silk Manuscripts from Mawangdui Han Tomb [Part Four]*, commentary part, pp. 164 and 167.

51 Methods of Intercourse between Yin and Yang in *Silk Manuscripts from Mawangdui Han Tomb [Part Four]*, commentary part, p. 156.

Fig. 23.3 The first sex education magazine after the founding of the People's Republic of China. Cover of the first issue in 1988

> The eight benefits are: (1) Regulating qi; (2) Generating essence; (3) Knowing the right time; (4) Storing qi; (5) Harmonizing essence; (6) Accumulating qi; (7) Maintaining fullness; (8) Stabilizing the situation.
>
> The seven damages are: When sexual intercourse causes pain, it is called internal blockage; when it leads to sweating, it is called external leakage; when one keeps going without stop, it is called exhaustion; when one desires it but fails to perform, it is called impotence; when one pants and gets disorganized during the act, it is called annoyance; when one forces the partner who has no desire, it is called compulsion; when the act is done too hastily, it is called waste.
>
> Therefore, if one can make good use of the eight benefits and avoid the seven damages, they will have keen ears and bright eyes, a light and agile body, stronger yin energy, a longer lifespan, and a happy life.

In these works, the principle of "surplus essence shall be discharged"[52] was upheld. These works had nothing to do with "collecting yin to nourish yang" (the art of "receiving yin", which was followed and practiced by practitioners of life nurturing at that time). That is to say, a harmonious relationship between sexes, as well as the physical and mental health of both sides, could be achieved by normal sexual behavior.

Fang zhong shu (Taoist sexual practices) is different from the art of "receiving yin" mentioned above, and it cannot be classified within the scope of "sex medicine", either. It is a set of "sexual techniques" guiding people with normal sexual functions. Regarding sexual dysfunctions, a large number of internal and external use prescriptions are recorded in *Health-preservation Prescriptions* (*Yang Sheng Fang*, 《养生方》). Various eggs are used as the main raw materials in the internal use prescriptions, which obviously stems from the analogical thinking method. The external use medicines are mainly pungent stimulants, along with physical stimulation methods.

[52] Discussion of the Perfect Way of Sexual Healthcare in *Silk Manuscripts from Mawangdui Han Tomb [Part Four]*, commentary part, p. 163.

In the feudal society of ancient China, especially after the rise of Neo Confucianism of the Song (AD 960–1279) and the Ming (AD 1368–1644) Dynasties, significant constraints were imposed on people's normal sexual consciousness. As a consequence, open discussions about sex became a shame and were to be avoided, and sex knowledge was all attributed to fang zhong shu (Taoist sexual practices). Tao Zongyi of the Yuan Dynasty (AD 1271–1368) remarked that "Unorthodox practices such as nourishing essence and essential qi by plundering others' vital energy were called fang Zhong Shu (Taoist sexual practices)".[53] In this sense, fang zhong shu (Taoist sexual practices) contains a lot of sexual and obscene contents, which has been severely criticized by society and especially doctors.[54] On the other hand, due to the doctors' correct understanding of human sexual physiology and pathology, plenty of diseases related to sex have been fully understood and properly treated.[55] For example, there were special discussions and treatment methods recorded in ancient TCM medical works for physical and mental diseases of special groups, including the widower, the widow, the orphan and the childless, Buddhist monks and nuns, and maidens.[56] These efforts represent a triumph of humanity over the feudal ethic—"to preserve Heaven's laws and eliminate human desires".

---

[53] Tao Zongyi (the Yuan Dynasty), *A Record of Yuan Dynasty* (*Chuo Geng Lu*, 《辍耕录》), Volume 14, Sexual Practices.

[54] Han Mao (the Ming Dynasty), *Han's Clear View of Medicine* (*Han Shi Yi Tong*, 《韩氏医通》), Beijing: People's Medical Publishing House, 1989, p. 43.

[55] Chu Cheng (the Southern Qi Dynasty), *Chu's suicide note* (*Chu Shi Yi Shu*, 《褚氏遗书》) mentioned that for young girls with weak constitution, they need to nourish the blood and strengthen the spleen, and they are encouraged to get married soon; for young men with weak constitution, they need to strengthen the spleen, practice abstinence, and have a strong body before the marriage. They need to follow this essential principle.

[56] *Comprehensive Medicine According to Master Zhang* (*Zhang Shi Yi Tong*, 《张氏医通》), Shanghai: Shanghai Scientific and Technical Publishers, 1963, p. 524.

Epilogue: Repaying "Good Teachers" and "Opportunities"

Throughout the process of conceiving and writing this book, two words have constantly resonated in my mind: "Good Teachers" and "Opportunities". In my academic career, there were two teachers who have inspired me. It is essential for me to share these profound experiences, as they have shaped my perspective and contributed to the creation of this work.

I have never invited any celebrities to write the preface for my books. This decision comes from my observation of certain doctors who consistently emphasize their connections to famous practitioners, presenting themselves as "the son" or "the student" of these renowned figures. Whenever I see this, I can't help but wonder, "If this medical practitioner is truly skilled, or even more capable than their father or mentor, why would they feel the need to use such tactics to attract patients?" Furthermore, most prefaces and book reviews solicited from others are written out of a sense of obligation, driven by considerations of "face", making them difficult to decline. As a result, it's difficult to find any brilliant insights, let alone academic criticisms that hit the nail on the head. Actually, everyone knows this well but is just reluctant to say it directly. However, to create a good academic atmosphere, we all need to take practical actions. So, since we are aware of this, it's better to do it ourselves. In fact, every time I write a book, I often rewrite the preface several times. The reason is that publishing a book involves a great deal of effort from editors and financial investment from publishers. Then, readers also spend money and time on it. Therefore, I always want readers to gain something from the very first page.

A significant portion of this book is the result of my monographic study published in journals in the form of papers over the years after I began to study the history of medicine. Based on these, I added some information and published this book, which aims to embody the evolution from the "detailed case studies" to the "construction of interpretation system". Herein I shall take the "Epilogue" to introduce two key factors which play an important role in the process from the "origin" to the "forming system": Good Teachers and Opportunities.

Y. Liao, *Reconstructing the Image of Medicine in the Qin and Han Dynasties*, Understanding China, https://doi.org/10.1007/978-981-95-6801-7

Mr. Du's Edification Engraved in My Mind

At the end of 1982, I found myself in an unexpected situation. As one of the first batch of medical students after the "cultural revolution", unexpectedly, I was distributed to the Institute for History of Natural Sciences at Chinese Academy of Sciences. It was an extraordinary turn of events for me. I never imagined that, as a "barefoot doctor" raised in an academically inclined family—born into a household steeped in knowledge and having just completed a rigorous five-year formal medical education—I would ever consider venturing into the field of history, a subject I had never been interested in. The prospect was beyond anything I had ever imagined.

At the beginning, Mr. Chen Meidong from the "General History Group" of Research Office for Ancient urged me to read the *Draft for the History of Science and Technology in Ancient China* (*Zhong Guo Gu Dai Ke Xue Ji Shu Shi Gao*, 《中国古代科学技术史稿》) compiled by them and the *General History of China* (*Zhong Guo Tong Shi*, 《中国通史》) written by Fan Wenlan. To be frank, my initial experience with these books was somewhat dull and uninspiring! However, it wasn't long before Mr. Du Shiran, the vice director of the Research Office, returned from his visiting to Japan and offered me some "instructions" during our conversations:

(1) It's perfectly fine to "explore" in the first few years and observe how others conduct their research.
(2) Translate one article each year to improve your English skills while you're still young. Once middle age arrives, you'll likely be overwhelmed with various responsibilities and have little time for such pursuits.
(3) Thoroughly review the contents of the two journals, *Archaeology* (*Kao Gu*, 《考古》) and *Cultural Relics* (*Wen Wu*, 《文物》).
(4) When it comes to the subject of your own research, no matter how many times others have studied it before, you should study it yourself.
(5) After years of research, you'll accumulate a wealth of historical "topics." Then, you can integrate them and develop your own interpretive framework for history—that is the essence of historical research.

The first point surprises me to some extent: How can "a strict teacher" refrain from urging his students to study hard while providing funds for them to "travel around"? However, many years later, I came to realize that embarking on study tours can significantly broaden one's exposure to diverse experiences and knowledge. Mr. Du often says that one may learn as if "crying a book", meaning learning driven by external pressures such as occupation, tasks, or promotions, rather than genuine passion. On the other hand, one may learn as if "singing a book", where learning is fueled by personal interest. For him, study tends to be a form of amusement.

Regarding the second point, I followed Mr. Du's advice to continue translating articles. Through this process, I not only improved my foreign language skills but also learned something far more valuable. The key takeaways can be summarized as follows. Firstly, the concept of a "good teacher" goes beyond someone who simply offers earnest exhortations. Reading a well-written article is akin to receiving guidance from a teacher. This is particularly true for foreign scholars, whose different backgrounds and geographical locations lead them to have unique perspectives and approaches to observing and interpreting Chinese culture. Secondly, "translating" is not the same as "reading". Translation demands a meticulous analysis of each word, as well as a deep understanding of the author's main ideas and arguments in their entirety. Through this process, one gains more than just the experience of reading. This is why only "translation" can truly enhance one's academic literacy and knowledge. Thirdly, learning a foreign language is not simply about reading materials in another language, communicating with foreigners, or gaining the opportunity to "study abroad". By reflecting on the words of a cultural anthropologist, we can better understand the true purpose of learning foreign languages.

> Being confined to a particular society and speaking a particular language will make it extremely difficult for a person to understand and evaluate other ways of life. People tend to think their own way of life as superior and evaluate other ways of life based on their own standards. This bias is called ethnic egocentrism. To a certain extent, this attitude may be normal, because if people lose faith in their own society, they will get sick and even die. But too much ethnocentrism can get in the way of understanding the good, beautiful, meaningful things in other societies. The most important thing is to keep an open mind if you want to get the most out of your journey.[1]

If you appreciate the uniqueness of ancient Chinese medical theories, such as "cold damage" and "warm disease", along with their etiological explanations—treating heat with cold and vice versa, draining excess and tonifying deficiency—and their therapeutic formulas, consider the English proverb: "Feed a cold and starve a fever". You may notice striking similarities and overlaps between these concepts.

Likewise, when exploring the close relationship between "form" and "spirit" in traditional Chinese health preservation, reflect on the Western saying: "A sound mind in a sound body". This parallel reveals that Western thought also embraces a similar concept of holistic well-being. Here's another example: In China, there is a popular saying: "Eating white radish and drinking hot tea make the doctor crawl all over the floor with anger".[2] In Japan, a similar expression goes: "The orange becomes yellow, the doctor gets blue". This suggests that consuming oranges helps reduce illness, leaving doctors with less work, which, in turn, makes them feel "blue" (a metaphor for being idle or less busy). To explain further: "Modern science reveals that oranges are rich in vitamin C, which plays a crucial role in preventing colds and various other illnesses".[3]

---

[1] Writer:P. K. Bock (USA); Translator: Yu Xing'an et al. *Multiculturalism and Social Progress* (*Duo Yuan Wen Hua Yu She Hui Jin Bu*, 《多元文化与社会进步》), Shenyang: Liaoning People's Publishing House, Chinese version of 1988, p. 1.

[2] The saying about this proverb is very much the same, such as: "If one eats white radish under the snowstorm or has tea with white radish, then the doctor will be so angry and crawl all over the ground" and so on.

[3] Yoichiro Nakajima (Japan), *A History of Disease in Japan* (『病気の歴史』), Tokyo: Tomoyama, 2nd Ed., 1988, p. 341.

The third point is to review the "Archaeological Report", which I was previously exposed to and which played a significant role in deepening my studies and broadening my horizons. A prime example is an article published in *Archaeology* that discusses the numerous diagrams of hexagrams, composed of three symbols ("-", "—", "---"), found on bronze artifacts from the Shang and Zhou Dynasties (1600–256 BC).[4] Based on archaeological discoveries, this paper offers a fresh interpretation of the relationship between the *Book of Changes* (*Zhou Yi*, 《周易》), numbers, and the concepts of yin and yang. It provides a new perspective on understanding yin-yang theory, which is crucial for the study of medical history. Such interdisciplinary insights are difficult to obtain through papers focused solely on the history of medicine. However, the research could not continue at that time. Additionally, due to the limitations of the period, only cultural relics directly related to "medicine" were available for study. Until now, I have read with increasing interest and awareness, as I gradually discover the connections between medical science and non-medical fields. For the first time, I included numerous pictures of cultural relics in the newly compiled book. In a recently completed manuscript, I also incorporated a large number of cultural relic images. This decision is not driven by the market demand for "image-based reading"; rather, it is intended to broaden readers' perspectives and make the theory more engaging and vivid.

The last two points represent the fundamental paths that have guided us from "thematic research" to "systematic construction" over the past few decades. The importance and necessity of these points are woven throughout the entire book. Their significance becomes clear after reading it.

In short, I continue to keep Mr. Du's guidance in mind and consider it my personal motto. I frequently share these lessons with my colleagues and students. I am reiterating them here in the hope that more people, like myself, will benefit from them. Another key mentor in my academic journey is the renowned Japanese historian of science, Mr. Yamada Keiji.

What I've Known About Mr. Yamada

In 1983, Mr. Du asked Mr. Yamada to "take care of" me. Thereafter, under the guidance of Mr. Yamada, I began to translate papers by some famous Japanese medical historians like Saburou Miyashita, Okanisi Tameto, and even Minamoto Ryoen, a philosopher and thinker. And Mr. Yamada recommended my translation work to the journal *Collection of Translations of the History of Science* (*Ke Xue Shi Yi Cong*, 《科学史译丛》) sponsored by the institute at that time. Among these enlightening works, I found Yamada Keiji's article on the *Establishment* of *Huangdi's Internal Classic* particularly influential. At the end of the twentieth century, with the help of Mr. Li Jianmin of the Institute of History and Language of the "Academia

[4] Zhang Yachu, Liu Yu, "Some Issues of Divination from the Numerical Symbols of the Eight Diagrams of the Shang and Zhou Dynasties", *Archaeology*, 1981, (2): 155.

Sinica" in Taipei, we published Mr. Yamada's second collection of papers in Chinese in the *Series of Life Nurturing and Remedy* (*Yang Sheng Fang Ji Cong Shu*, 《养生·方技丛书》) of Dongda Book Co., Ltd. The preface, titled "What I've Learned from Mr. Yamada Keiji", offers a detailed account of his influence on me in various aspects. It remains "quite touching" to this day,[5] which is why I have transcribed it as follows:

In the autumn of 1983, Mr. Yamada Keiji visited China, coinciding with the Chinese Society for the History of Science and Technology's meeting in Xi'an. Less than a year after my graduation from medical school, I had limited knowledge of the history of science and was simply assigned to the airport to pick up Mr. Yamada as part of the conference staff. At that time, I never imagined that I would meet a mentor who would have such a profound influence on my academic career and benefit me so greatly.

During his stay, Mr. Yamada resided in the presidential suite, while I, as a member of the conference staff, stayed in a room with my colleagues. I had heard that he was very fond of Maotai, and I only visited his suite once during the symposium. I remember that he shared his educational background with us—he graduated from the Department of Cosmic Physics at the School of Science, Kyoto University, in 1955. He then went on to study the history of modern science in graduate school. After completing his master's program in 1959, he was appointed as a lecturer at the Institute of Human Sciences, Kyoto University. Mr. Yamada also explained the institute's professor-selection system: when positions for professors and associate professors become vacant, the institute issues a nationwide recruitment notice, and a temporary committee is set up to discuss the candidates. This system aims to select outstanding talent and minimize nepotism. After nearly seven years as a lecturer, Mr. Yamada left Kyoto University and became an associate professor at Doshisha University. He returned to Kyoto University in 1970 as an associate professor and was promoted to professor in 1978. Regarding his research, Mr. Yamada discussed their investigation into newly discovered historical materials on ancient Chinese science and technology, such as *the Great Simplicity of the Huangdi's Internal Classic* (*Huang Di Nei Jing Tai Su*, 《黄帝内经太素》) and the medical texts unearthed from the Mawangdui tombs, which he planned to publish as research results. In retrospect, it can be said that Mr. Yamada's research on the history of medicine is largely grounded in this work, with his series of research findings closely linked to *Huangdi's Internal Classic* (*Huang Di Nei Jing*, 《黄帝内经》) and Mawangdui medical books. His research achievements, *Study of the Historical Data of Newly Discovered Science in China Annotated Version of the Translation* (新發現中國科學史資料の研究·譯註篇) and *Discussed Version* (論考篇) (1985) are well-known among scholars of the history of ancient Chinese medicine. His works have inspired numerous scholars in the field.

---

[5] The manuscript was handed over to Mr. Li Jianmin of the Institute of History and Language. Everyone said it was "pretty touching", so it was published in the previous issue of the journal *On Balance in Ancient and Modern Times* (*Gu Jin Lun Heng*, 《古今论衡》) (No. 2 issue, 1999) of the institute.

During this period, Mr. Yamada also delivered a plenary lecture. His closing words remain vivid in my memory: "Of course, students should learn from their teachers, but students who cannot surpass their teachers are not truly excellent students!" Mr. Yamada meant that, as far as he was concerned, although he had learned a great deal from his predecessors in China, he also possessed the determination and confidence to surpass them. I heard these words several times in our subsequent conversations, especially when I wrote an article that presented views clearly differing from his. The genuine joy I felt when he praised my work is indescribable, and it remains a memory I will cherish forever. Two days later, Mr. Yamada left Xi'an for Henan province to visit the Dengfeng Observatory. After all, astronomy had been his original field of study.

Over the next few years, I tried to do some medical history research, but without any achievement, which left me feeling perplexed. Because so many scholars have been studying medical history for a long time, and they have repeatedly covered all the historical figures, important works, theories, and institutional events. I found myself wondering, where could I find untapped areas to explore for my research? During this search, I began to focus on translation as a way to learn a foreign language. I started reading and translating Mr. Yamada's articles. To my surprise, this translation process not only helped me acquire language skills but also provided me with new insights into a different research approach and mindset, leading to a sudden moment of enlightenment. So, what was this "research method and mindset" that I discovered? That is the approach of analysis. Instead of merely providing detailed descriptions of historical phenomena or elaborating on "achievements" and "scientific nature", I completely break away from this common "routine" and immerse myself in the rational analysis of historical events. Through analytical research, we uncover the ideological relationships underlying these phenomena, the connections and developments between different phenomena across ages, and the interactions between seemingly unrelated events. When a link is missing, hindering the discovery or establishment of a relationship, the interpretation system should be built by making bold assumptions (a working hypothesis), which is then carefully tested through extensive citation and verification. This type of research was invigorating, so there was no longer a need for me to search for an untapped area to focus on. It is difficult to quantify the potential hidden within these commonly cited historical materials, which have been repeatedly copied and quoted, yet now possess a newfound freshness and vitality. In fact, for Mr. Yamada, it is not easy to enter "the realm of freedom" in this domain of knowledge. As he himself put it,

> Given my educational background in natural science and my inclination toward a more Westernized way of thinking, delving into the realms of Chinese philosophy and science proved to be a challenging endeavor. To bridge this gap, I employed methods such as translating concepts, inductively formulating thinking approaches, constructing models, and reconstructing theories. Through these methods, I sought to make Chinese philosophy and science more accessible and comprehensible to myself. The ideological struggle between the East

> and the West, as well as between tradition and modernity, has given rise to the discoveries made along the way.[6]

Mr. Yamada was an avid smoker—five packages (100 cigarettes) per day. It is likely that he faced his own challenges in the realm of academic research.

In 1988, Mr. Yamada got a financial support from a consortium and invited me to visit and study. On December 6, 1998, I arrived in Kyoto, embarking on a year-long journey of living and studying in Japan.

At the Institute of Human Sciences in Kyoto University, where Mr. Yamada worked, there was a program of the history of Chinese science. Every Tuesday afternoon, we gathered to read the literature or report our own research findings. This practice had been in place for many years, becoming both a tradition and a habit. At that time, I thought that we should adopt a similar approach when we came back to China and carried out study and research in our institute. However, I soon realized that putting this into practice was not an easy task. This led me to reflect on the reasons behind this difficulty. There are some differences among them, and these differences all involve a core problem, that is, the difference in value orientation. First, every researcher at my home institute was busy with his own job. It doesn't matter for you to occasionally take half a day to listen to a presentation or read a book unrelated to one's specialization. But as the saying goes, there is no light package in the long way. Therefore, it is by no means easy to insist on it. Indeed, it can be seen as burdensome. It is the burden that gives people corresponding rewards—enriching their knowledge, broadening their horizons, learning other people's research methods and ideas, hearing other people's criticisms of their work and discovering new research topics... I think even Mr. Yamada, as the leader of the institute, has benefited greatly from this approach. Otherwise, he could not have been able to swiftly shift his research focus from astronomy to medical history and even technical history. I have observed that regardless of the subject he was reading, Mr. Yamada would diligently prepare in advance—looking up information on difficult points, terms, names of people and places for others to refer to. Similarly, other famous scholars who participated in the seminar, such as Tamato Tanaka (history of architecture), Kazuhiko Miyashima (history of astronomy), Hideki Kawahara (history of mathematics), Keizo Hashimoto (history of western science), also displayed an active interest in learning the knowledge beyond their major and actively engaged in the discussion. Perhaps it is only through this interdisciplinary approach that we can truly comprehend the ancient science and technology, and gain a comprehensive understanding of history.

Moreover, it is precisely because of the “burden” of the realities mentioned above that the success of such a workshop heavily depends on a convenor with appeal, cohesion, and, perhaps most importantly, personal charisma. I believe Mr. Yamada's personal charm largely stems from his academic achievements. When these achievements translate into social prestige, they naturally enhance his personal appeal. When I first arrived in Kyoto, I had the privilege of witnessing Mr. Yamada receiving the

---

[6] Yamada Keiji (Japan), *The Philosophical and Scientific Culture of the Ancient East Asia: Collected Papers of Yamada Keiji* ·Preface (*Gu Dai Dong Ya Zhe Xue Yu Ke Ji Wen Hua*, 《古代东亚哲学与科技文化》), Shenyang: Liaoning Education Press, 1996.

Osaragi Jiro Prize for his outstanding research on Miura Baien, an ancient Japanese philosopher and thinker. This accolade was widely celebrated. I once asked a scholar who had arrived by car, "Why are you here?" He replied that being associated with Mr. Yamada would offer certain benefits. For example, if Mr. Yamada were to recommend him, the press would be more likely to accept his manuscripts. Whether we admit it or not, fame and social influence are attractive to scholars. This is not inherently negative, as fame and influence can, in a sense, reflect one's accomplishments; in other words, people derive a sense of achievement from them. However, as scholars, it is not essential to expand one's social influence by engaging in extensive social activities. Mr. Yamada does not hold any honorary titles, such as director or member, yet his academic stature alone is sufficient to command admiration.

In March, I presented my research at the seminar for the first time. Although my research was largely inspired by Mr. Yamada's article, the *Establishment of Huangdi's Internal Classic*, I held a different view on the time when the *Huangdi's Internal Classic* was written. To tell the truth, I concerned that my viewpoint might upset Mr. Yamada, but how was the truth? At the end of the discussion, Mr. Yamada said that Mr. Liao should be entrusted with the important task of medical history research, and he should return to the field of astronomy as soon as possible. And he reiterated the "maxim" that students should excel their teachers. Then he took me out for a drink—obviously, I sensed his excitement, and felt a quiet glow of satisfaction; meanwhile, I had an opportunity to glimpse into Mr. Yamada's inner world for the first time—peered into his mind.

Mr. Yamada liked drinking and eating Chinese food. In the fore and aft of year, our seminar would gather at the Donghua restaurant in downtown Kyoto for a meal of Beijing cuisine. Afterward, we would continue our conversations over drinks and friendly chats at a traditional Japanese bistro in a nearby alley. It is common for the Japanese to drink. Friends often get together for drinking in several pubs until late hours. But Mr. Yamada was a moderate drinker, he would bid farewell to his friends and head home at about 11 o'clock.

In summer, in the letter from Chen Meidong, director of the Institute of the History of Natural Science, Chinese Academy of Sciences, Mr. Yamada was invited to be the honorary professor at the institute. Initially, I had thought that as a professor at the famous Kyoto University, he might not have much interest in this, but the fact was quite the opposite: Mr. Yamada was very happy and his reply letter in Chinese expressed his joy by saying, "Your letter makes my dream come true…At that time, I thought it was a foreigner's misnomer and I amended it. But five years later, when I received his "self-preface" for the Chinese version of the essays, I once again glimpsed into his inner world again—to understand why he was so moved. The "self-preface" begins with these words:

With profound respect and empathy for Chinese thought and culture, I have been devoted to the study of the history of Chinese science, technology and medicine for

30 years. It is really an honor and a pleasure for me to directly address Chinese readers.[7]

Through these words, one can clearly see Mr. Yamada's feelings toward China and Chinese culture.

In September, Mr. Yamada took me to Tokyo, where I visited the Tokyo Museum and some experts in medical history such as Otsuka Yasuo, Hiroshi Kosoto, and Mayanagi Makoto from the Toyo institute of medicine, Kitasato University. Over lunch, I asked Mr. Yamada what he thought of the value of studying the history of science, because in the Chinese mainland, there is often an emphasis on the practical applications of the history of science for society and the economy. It seemed that I was the first one to ask him this question, and after contemplating for a moment, he replied, "learning is learning. The value of learning is latent and far-reaching". I have come to know that apart from a few drinks in the evening, Mr. Yamada doesn't have any hobbies. Knowledge means almost everything to Mr. Yamada, but he never thought about the reward from knowledge. This is probably what we say, "Ancient scholars learn for improving themselves".

In 1989, Mr. Yamada played a key role in the establishment of the "International Center for Japanese Cultural Studies" and was appointed as a professor at the center (while still holding his position at Kyoto University). He often mentioned his desire to invite me to work there for another year, but at the time, I didn't take it seriously. To be honest, during my 365 days in Japan, I experienced something of a "culture shock", as it was my first time traveling abroad. Mr. Yamada suggested that I also conduct research on Japanese medical history in addition to Chinese medical history, but I dismissed it, thinking it was a matter for the Japanese. In fact, I didn't even write to Mr. Yamada for six months after my return. I feel deeply ashamed when I reflect upon this. But Mr. Yamada never minded for this and continued to provide me with learning opportunities. In 1991, 1993, and 1995, thanks to his recommendation, I had the honor to participate in the "International Conference on the History of Comparative Medicine between the East and the West" many times and made great progress in my knowledge. In 1995, I was invited to work for one year at the International Research Center for Japanese Studies.

After his transfer to the International Center for Japanese Cultural Studies, Mr. Yamada expanded his research focus from the history of Chinese medicine to the history of Japanese medicine. He organized a research institute on "Ancient Japanese Diseases and Medical History", which became a hub for both new and established scholars in Japanese medical history, as well as many international researchers. At this time, I had moved beyond my previous capricious and naive mindset. Not only did I no longer believe that the study of Japanese medical history was solely the concern of the Japanese, but I also developed a deep interest in the field and became passionate about competing with Japanese scholars. Whenever I completed a paper or presentation, Mr. Yamada would carefully help me revise it, which always made me feel his genuine concern for his students.

[7] Yamada Keiji (Japan), *the Philosophical and Scientific Culture of the Ancient East Asia: Collected Papers of Yamada Keiji*·Preface, Shenyang: Liaoning Education Press, 1996.

Mr. Ishida, an expert in the history of Chinese medicine, was also a member of the workshop and had been appointed as a visiting professor at the center. However, one day, as I was reading Mr. Ishida's book, I came across a chapter specifically dedicated to criticizing Mr. Yamada's ideas by name. I was truly shocked—but not by the criticism itself. What surprised me was Mr. Yamada's remarkable open-mindedness!

I knew Mr. Yamada's fondness for Chinese food, and since my dormitory was located at the research center, I often prepared food and drinks and invited him and other friends to my place for a get-together. However, at that time, Mr. Yamada had to refrain from drinking due to a heart attack, which saddened us all—his one great pleasure had to be given up. What a pity! But fortunately for Mr. Yamada, learning remained the greatest joy of his life!

Sometimes, when I would discuss Mr. Yamada with some middle-aged Japanese scholars of my own generation, they would perceive him as overly strict. I was puzzled because I had never felt this way before. So, they would say, "No, Mr. Yamada is special to you". There was a hint of jealousy in the words. This, of course, made me feel a sense of pride. However, upon reflection, I understood that his kindness for me and the unique look in his eyes was entirely due to a teacher's appreciation for his favorite pupil. I remembered that when someone was analyzing why students of Nobel Prize laureates often win Nobel Prizes too, in addition to the causal relationship, the mutual pursuit between teachers and students is cited as one of the important reasons. That is, when selecting students, teachers should purposefully pursue and nurture the ones worthy of cultivation. I thought that Mr. Yamada's continuous creation of opportunities for my studies stemmed from this perspective. When I was promoted as a researcher, he wrote the following recommendation: "From my perspective, Liao Yuqun is currently the most prominent scholar in the field of research on the history of Chinese medicine. I am sure that sooner or later the whole picture of Chinese medicine will be rewritten by him". Such compliments have put me under heavy pressure with no complacency. All of Mr. Yamada's efforts, no doubt, are for this knowledge and for the cultivation of talents. If we can have more teachers like Mr. Yamada in our real life, and if every one of us who has received the kindness from the past can make every effort to cultivate talents like Mr. Yamada, it will be a blessing for the academic community and the society. Personally, how can I live up to Mr. Yamada's expectations? I have never been able to surpass the teacher in learning as he expected, so I am not yet a good student in this sense. The publication of Mr. Yamada's collection of papers by the Taiwan Dongda Book Co., Ltd. brought me a different kind of hope—maybe there will be some promising people who are inspired by Mr. Yamada's study, and thus find the teacher or learn methods to become better students than teachers. If that happens, Mr. Yamada will be satisfied with that, we translators will feel relieved, and it is also a good thing for Dongda Book Co., Ltd.

Mr. Yamada retired in April 1997. Unlike most Japanese professors, who often spend a few years at a private university after retirement, he chose a different path, despite having the opportunity to earn a substantial income. However, by all accounts, a renowned scholar like Mr. Yamada remains in high demand. I once asked him about his plans for retirement, and he replied that he was free to do whatever he wanted. I could never quite guess what he intended to do—perhaps write some poetic literature?

In April of this year, I received a letter from Mr. Yamada, which read:

Long-term projects can generally be completed, and I now intend to focus on a new subject. My first step will be to write a book on ancient Chinese medicine, followed by research on the modern history of Japan, the history of China's Ming (1368–1644 AD) and Qing (1636–1912 AD) Dynasties, and the history of Europe after the Renaissance. Above all, I wish to break free from the heavy burden and allow the wings of my dreams to spread freely.

I knew that Mr. Yamada often saw things in his dreams that others did not. It made me contemplate my own future, and I had contemplated giving up translation. However, if Mr. Yamada were to document the exquisite scenery of his dreams, akin to entering a "peach garden", how could I bear to forgo the opportunity of sharing the joy and showcasing its beauty to everyone? It felt like a calling that had to be pursued. Meanwhile, I expressed my earnest desire to continue to cooperate with Dongda Book Co., Ltd, to share the responsibility of flourishing the academic community and serving the readers. (Written on December 31, 1998).

The Best Gift for Our Teacher Is Our Academic Achievements

The above may seem to focus solely on the topic of "good teachers", but it inherently encompasses the concept of "opportunities". Personally, I felt somewhat adrift when I was first introduced to the Institute for the History of Natural Science. However, this pivotal moment in my life became a transformative turning point. After that, I received the guidance of Mr. Du and the support of other mentors. Additionally, Mr. Yamada played a significant role by providing me with numerous international exchange opportunities, and one of his articles was a source of great inspiration.

I must also acknowledge the invaluable role of a more conducive social environment, marked by a freer academic atmosphere and a thriving publishing industry. Together, these factors created vital opportunities for my personal growth and achievements. Therefore, in my view, "good teachers" and "opportunities" are, in fact, inextricably linked.

At the time of completing the *Reconstructing the Medical Images in the Qin and Han Dynasties of China* (*Chong Gou Qin Han Yi Xue Tu Xiang*,《重构秦汉医学图像》), I am reminded of two gentlemen who have been my mentors in my academic career. Without their guidance and the opportunities they provided, I would never have reached this point. To Mr. Du, I would like to say that: after 30 years, I finally could pool historical events together and study them holistically and develop my own interpretive framework. To Mr. Yamada, I would like to say: your great expectations have urged me to move on. Today, the book can finally serve as a reward for you, my mentor.

Moreover, since teachers and students are not bound by blood ties, the long-existing schools and interests between them in the past have faded away and become a social relationship. As a result, good teachers are also members of society, striving for nothing more than to see their students surpass them, with their actions rooted in the value concept of "social utility". Both the teacher's guidance and the student's growth should undoubtedly be considered forms of social behavior.

The Institute for the History of Natural Science, which has supported my life and research for decades, is undoubtedly a social entity as well. Therefore, to "repay society for the opportunities you have received" carries a deeper meaning—namely, to love your society and embrace a sense of social responsibility.

The completion of the *Reconstructing the Medical Images in the Qin and Han Dynasties of China* certainly reflects personal effort and dedication, but more importantly, it is a result of the support of society, particularly the profound influence of good teachers and the opportunities they have provided.

The manufacturer's authorised representative in the EU is Springer
Nature Customer Service Centre GmbH, Europaplatz 3, 69115 Heidelberg,
Germany. If you have any concerns regarding our products, please
contact ProductSafety@springernature.com

Printed and bound by CPI Group (UK) Ltd, Croydon, CR0 4YY
21/09/2026
02220015-0016